Asymptotic Properties of Solutions of Nonautonomous Ordinary Differential Equations

Mathematics and Its Applications (*Soviet Series*)

Volume 89

Asymptotic Properties of Solutions of Nonautonomous Ordinary Differential Equations

by

I. T. Kiguradze

Institute of Applied Mathematics,
Tbilisi, Georgia

and

T. A. Chanturia †

SPRINGER-SCIENCE+BUSINESS MEDIA, B.V.

Library of Congress Cataloging-in-Publication Data

Kiguradze, I. T.
    [Asimptoticheskie svoĭstva resheniĭ neavtonomnykh obyknovennykh
differenfsial'nykh uravneniĭ. English]
    Asymptotic properties of solutions of nonautonomous ordinary
differential equations / by I.T. Kiguradze and T.A. Chanturia.
        p.   cm. -- (Mathematics and its applications. Soviet series :
89)
    Translation of: Asimptoticheskie svoĭstva resheniĭ neavtonomnykh
obyknovennykh differenfsial'nykh uravneniĭ.
    Includes index.
    ISBN 978-94-010-4797-5   ISBN 978-94-011-1808-8 (eBook)
    DOI 10.1007/978-94-011-1808-8
    1. Differential equations--Asymptotic theory.   I. Chanturiīa, T.
A. (Teĭmuraz Ambros'evich) II. Title. III. Series: Mathematics and
its applications (Kluwer Academic Publishers). Soviet series ; 89.
QA372.K5413 1992
515'.35--dc20   92-37359

ISBN 978-94-010-4797-5

This is the translation of the original work
*Asymptotic Properties of Solutions of
Nonautonomous Ordinary Differential Equations*
Published by Nauka, Moscow, © 1985

Printed on acid-free paper

SERIES EDITOR'S PREFACE

'Et moi, ..., si j'avait su comment en revenir,
je n'y serais point allé.'

Jules Verne

The series is divergent; therefore we may be
able to do something with it.

O. Heaviside

One service mathematics has rendered the
human race. It has put common sense back
where it belongs, on the topmost shelf next
to the dusty canister labelled 'discarded non-
sense'.

Eric T. Bell

Mathematics is a tool for thought. A highly necessary tool in a world where both feedback and non-linearities abound. Similarly, all kinds of parts of mathematics serve as tools for other parts and for other sciences.

Applying a simple rewriting rule to the quote on the right above one finds such statements as: 'One service topology has rendered mathematical physics ...'; 'One service logic has rendered computer science ...'; 'One service category theory has rendered mathematics ...'. All arguably true. And all statements obtainable this way form part of the raison d'être of this series.

This series, *Mathematics and Its Applications*, started in 1977. Now that over one hundred volumes have appeared it seems opportune to reexamine its scope. At the time I wrote

"Growing specialization and diversification have brought a host of monographs and textbooks on increasingly specialized topics. However, the 'tree' of knowledge of mathematics and related fields does not grow only by putting forth new branches. It also happens, quite often in fact, that branches which were thought to be completely disparate are suddenly seen to be related. Further, the kind and level of sophistication of mathematics applied in various sciences has changed drastically in recent years: measure theory is used (non-trivially) in regional and theoretical economics; algebraic geometry interacts with physics; the Minkowsky lemma, coding theory and the structure of water meet one another in packing and covering theory; quantum fields, crystal defects and mathematical programming profit from homotopy theory; Lie algebras are relevant to filtering; and prediction and electrical engineering can use Stein spaces. And in addition to this there are such new emerging subdisciplines as 'experimental mathematics', 'CFD', 'completely integrable systems', 'chaos, synergetics and large-scale order', which are almost impossible to fit into the existing classification schemes. They draw upon widely different sections of mathematics."

By and large, all this still applies today. It is still true that at first sight mathematics seems rather fragmented and that to find, see, and exploit the deeper underlying interrelations more effort is needed and so are books that can help mathematicians and scientists do so. Accordingly MIA will continue to try to make such books available.

If anything, the description I gave in 1977 is now an understatement. To the examples of interaction areas one should add string theory where Riemann surfaces, algebraic geometry, modular functions, knots, quantum field theory, Kac-Moody algebras, monstrous moonshine (and more) all come together. And to the examples of things which can be usefully applied let me add the topic 'finite geometry'; a combination of words which sounds like it might not even exist, let alone be applicable. And yet it is being applied: to statistics via designs, to radar/sonar detection arrays (via finite projective planes), and to bus connections of VLSI chips (via difference sets). There seems to be no part of (so-called pure) mathematics that is not in immediate danger of being applied. And, accordingly, the applied mathematician needs to be aware of much more. Besides analysis and numerics, the traditional workhorses, he may need all kinds of combinatorics, algebra, probability, and so on.

In addition, the applied scientist needs to cope increasingly with the nonlinear world and the extra mathematical sophistication that this requires. For that is where the rewards are. Linear models are honest and a bit sad and depressing: proportional efforts and results. It is in the nonlinear world that infinitesimal inputs may result in macroscopic outputs (or vice versa). To appreciate what I am hinting at: if electronics were linear we would have no fun with transistors and computers; we would have no TV; in fact you would not be reading these lines.

There is also no safety in ignoring such outlandish things as nonstandard analysis, superspace and anticommuting integration, p-adic and ultrametric space. All three have applications in both electrical engineering and physics. Once, complex numbers were equally outlandish, but they frequently proved the shortest path between 'real' results. Similarly, the first two topics named have already provided a number of 'wormhole' paths. There is no telling where all this is leading - fortunately.

Thus the original scope of the series, which for various (sound) reasons now comprises five subseries: white (Japan), yellow (China), red (USSR), blue (Eastern Europe), and green (everything else), still applies. It has been enlarged a bit to include books treating of the tools from one subdiscipline which are used in others. Thus the series still aims at books dealing with:

-   a central concept which plays an important role in several different mathematical and/or scientific specialization areas;
-   new applications of the results and ideas from one area of scientific endeavour into another;
-   influences which the results, problems and concepts of one field of enquiry have, and have had, on the development of another.

The shortest path between two truths in the real domain passes through the complex domain.

J. Hadamard

La physique ne nous donne pas seulement l'occasion de résoudre des problèmes ... elle nous fait pressentir la solution.

H. Poincaré

Never lend books, for no one ever returns them; the only books I have in my library are books that other folk have lent me.

Anatole France

The function of an expert is not to be more right than other people, but to be wrong for more sophisticated reasons.

David Butler

Michiel Hazewinkel

Contents

Preface

Over the last thirty years, intensive work has been carried out in the field of the theory of nonautonomous ordinary differential equations. During these years, new investigation methods were developed and results of principal importance were obtained. In particular, some subtle oscillation criteria for solutions of linear differential equations and equations of Emden–Fowler type were established, general theorems on the classification of equations with respect to oscillation properties of their solutions were proved, existence and nonexistence conditions for singular, proper, oscillatory and monotone solutions of various types were found, estimates of proper solutions were given in a neighborhood of infinity, asymptotic formulas for solutions of a sufficiently wide class of linear and nonlinear equations were derived, etc.

In this book an attempt is made to sum up these studies. The necessity of such an attempt is felt especially as the well-known monographs of R. Bellman [27] and G. Sansone [316, 317], devoted to related topics, reflect results which are about fourty years old.

The book consists of five chapters. In the first one we consider the linear differential equation

$$u^{(n)} = \sum_{k=0}^{n-1} p_k(t) u^{(k)} \tag{1}$$

with locally integrable coefficients p_k: $[a, +\infty[\to \mathbf{R}$ $(k = 0, \ldots, n-1)$. Here an apparatus of comparison theorems is developed, which allows us to establish optimal (in some sense) criteria for the equation (1) to have the Kneser properties A and B. Conditions for the existence of at least one oscillatory solution are found. These conditions cover the case when the coefficients p_k change sign. The Biernacki problem on the dimension of the subspace of solutions of (1) vanishing at infinity is solved. Existence criteria for bounded and unbounded solutions are given. Asymptotic formulas for solutions of certain classes of equations (1) are derived.[1]

Chapter II deals with quasilinear equations

$$u^{(n)} = \sum_{k=0}^{n-1} p_k(t) u^{(k)} + q(t, u, \ldots, u^{(n-1)}), \tag{2}$$

---

[1]Note that by 'solutions of the linear and nonlinear equations in question' we mean solutions that are maximally continued to the right.

where the nonlinear term $q\colon [a, +\infty] \times \mathbf{R}^n \to \mathbf{R}$ is small in a certain sense. Here we establish criteria for (2) to have a family of solutions asymptotically equivalent to solutions of the equation (1), and study the properties of this family.

In chapter III general nonlinear equations

$$u^{(n)} = f(t, u, \dots, u^{(n-1)}) \tag{3}$$

with right-hand sides $f\colon [a, +\infty[\times \mathbf{R}^n \to \mathbf{R}$ from the Carathéodory class are considered. In this chapter we solve the Kneser problem on monotone solutions and prove that under certain restrictions on f, the equation (3) possesses proper solutions with power asymptotics and fast growing solutions as well as so-called first and second kind singular solutions. For solutions of all these types, two-sided asymptotic estimates are derived. Besides, we develop a method allowing us to reduce the question of the oscillation of proper solutions of (3) to the question of the nonexistence of a positive solution of a certain specially constructed first order differential equation. For a wide class of equations the sufficient conditions obtained by this method turn out to be necessary as well. Finally, existence criteria for solutions vanishing at infinity are found.

In chapter IV we apply the results of the prvious chapters to the equation of Emden–Fowler type

$$u^{(n)} = p(t)|u|^\lambda \operatorname{sign} u \tag{4}$$

with a locally integrable coefficient $p\colon [a, +\infty[\to \mathbf{R}$ and an exponent $\lambda \in]0, +\infty[$. It is well known that equations of this type appear in many physical problems.

In the last, fifth, chapter we study in detail the equation (4) for $n = 2$. Here criteria for boundedness and unboundedness of all nonzero solutions are given, the Armellini–Tonelli–Sansone type theorem on the vanishing of all solutions at infinity is proved and asymptotic formulas for proper solutions are derived.

Some results are given in the form of problems to be solved by the reader. We also state problems whose solutions we do not know. These are marked with an asterisk.

Basic Notation

N is the set of positive integers;

$\mathbf{R} =]-\infty, +\infty[$, $\mathbf{R}_+ = [0, +\infty[$, $\mathbf{R}_- =]-\infty, 0]$;

if $I \subset \mathbf{R}$, then $I^k = I \times \ldots \times I$ (k times);

$C([a,b])$ and $\tilde{C}([a,b])$ are the sets of continuous and absolutely continuous functions $u: [a,b] \to \mathbf{R}$, respectively;

$C^k([a,b])$ is the set of k times differentiable functions $u: [a,b] \to \mathbf{R}$;

$\tilde{C}^k([a,b])$ is the set of functions $u: [a,b] \to \mathbf{R}$ which are absolutely continuous together with their derivatives up to the order k inclusive;

$L(I)$, for an interval $I \subset \mathbf{R}$, is the set of Lebesgue integrable functions $u: I \to \mathbf{R}$;

$\tilde{C}^k_{\mathrm{loc}}([a, +\infty[)$ and $L_{\mathrm{loc}}([a, +\infty[)$ are the sets of functions $u: [a, +\infty[\to \mathbf{R}$ whose restrictions to any interval $[a,b] \subset [a, +\infty[$ belong to $\tilde{C}^k([a,b])$ and $L([a,b])$, respectively;

$K([a,b] \times D)$, with $D \subset \mathbf{R}^n$, is the *Carathéodory class*, i.e. the set of functions $f: [a,b] \times D \to \mathbf{R}$ such that $f(\cdot, x_1, \ldots, x_n): [a,b] \to \mathbf{R}$ is measurable for all $(x_1, \ldots, x_n) \in D$, $f(t, \cdot, \ldots, \cdot): D \to \mathbf{R}$ is continuous for almost all $t \in [a,b]$ and

$$\sup\{|f(\cdot, x_1, \ldots, x_n)| : (x_1, \ldots, x_n) \in D_0\} \in L([a,b])$$

for any compact set $D_0 \subset D$;

$K_{\mathrm{loc}}([a, +\infty[\times D)$, with $D \subset \mathbf{R}^n$, is the set of functions $f: [a, +\infty[\times D \to \mathbf{R}$ whose restriction to $[a,b] \times D$ belongs to $K([a,b] \times D)$ for any interval $[a,b] \subset [a, +\infty[$;

$V([a,b])$ is the set of functions $u: [a,b] \to \mathbf{R}$ of bounded variation;

if $u \in V([a,b])$, then $\int_a^b |du(t)|$ is the total variation of the function u on the interval $[a,b]$;

$V_{\mathrm{loc}}([a, +\infty[)$ is the set of functions $u: [a, +\infty[\to \mathbf{R}$ whose restriction to any interval $[a,b] \subset [a, +\infty[$ belongs to $V([a,b])$;

$V([a, +\infty[)$ is the set of functions $u \in V_{\mathrm{loc}}([a, +\infty[)$ for which $\int_a^{+\infty} |du(t)| = \lim_{b \to +\infty} \int_a^b |du(t)| < +\infty$;

if $\phi: [a, +\infty[\to \mathbf{R}$ and $\psi: [a, +\infty[\to]0, +\infty[$, then the equalities $\phi(t) = o(\psi(t))$ and $\phi(t) = O(\psi(t))$ are equivalent to the conditions

$$\lim_{t \to +\infty} \frac{\phi(t)}{\psi(t)} = 0 \qquad \text{and} \qquad \lim_{t \to +\infty} \sup \frac{|\phi(t)|}{\psi(t)} < +\infty,$$

respectively;

   $[r]$ is the integral part of the constant r;

      $\square$ denotes the end of a proof.

CHAPTER I

LINEAR DIFFERENTIAL EQUATIONS

§1. Equations Having Properties A and B

Consider the linear differential equation

$$u^{(n)} = p(t)u \tag{1.1}$$

where $p \in L_{\text{loc}}(\mathbf{R}_+)$. As a rule we assume that either

$$p(t) \leq 0 \qquad \text{for } t \in \mathbf{R}_+ \tag{1.2}$$

or

$$p(t) \geq 0 \qquad \text{for } t \in \mathbf{R}_+. \tag{1.3}$$

A solution of (1.1) is said to be *oscillatory* if it has infinitely many zeros. Otherwise this solution is said to be *nonoscillatory*.

DEFINITION 1.1. The equation (1.1) has *property A* if every solution of (1.1) for n even is oscillatory and for n odd either is oscillatory or satisfies the condition[1]

$$|u^{(i)}(t)| \downarrow 0 \qquad \text{as } t \uparrow +\infty \qquad (i = 0, \ldots, n-1). \tag{1.4}$$

DEFINITION 1.2. The equation (1.1) has *property B* if every solution of (1.1) for n even is either oscillatory, satisfies (1.4) or satisfies the condition

$$|u^{(i)}(t)| \uparrow +\infty \qquad \text{as } t \uparrow +\infty \qquad (i = 0, \ldots, n-1), \tag{1.5}$$

and for n odd either is oscillatory or satisfies (1.5).

It is easy to verify that the equation $u^{(n)} = cu$ with $c < 0$ has property A and with $c > 0$ has property B.

---

[1] By writing '$u \downarrow 0$ as $t \uparrow +\infty$', we mean that u monotonically decreases in a certain neighborhood of $+\infty$ and tends to zero as $t \to +\infty$. The notation $u \uparrow +\infty$ as $t \uparrow +\infty$ should be similarly understood.

1.1. Classification of nonoscillatory solutions.

LEMMA 1.1. Let the inequality (1.2) (the inequality (1.3)) be fulfilled and let p differ from zero on a set of positive measure in any neighborhood of $+\infty$, i.e.

$$\text{meas}\{s \geq t : p(s) \neq 0\} > 0 \qquad \text{for } t \geq 0. \tag{1.6}$$

Suppose that u is a solution of (1.1) satisfying the condition

$$u(t) > 0 \qquad \text{for } t \geq t_0 \geq 0.$$

Then there exist numbers $t_1 \in [t_0, +\infty[$ and $l \in \{0, \dots, n\}$ such that $l + n$ is odd (even) and[2]

$$u^{(i)}(t) > 0 \qquad \text{for } t \geq t_1 \qquad (i = 0, \dots, l-1),$$
$$(-1)^{i+l} u^{(i)}(t) > 0 \qquad \text{for } t \geq t_1 \qquad (i = l, \dots, n-1), \tag{1.7$_l$}$$
$$(-1)^{n+l} u^{(n)}(t) \geq 0 \qquad \text{for } t \geq t_1.$$

PROOF. This lemma is a consequence of the following fact: if $v'(t) > 0$ and $v''(t) \geq 0$ for $t \geq t_0$, then we can find $t_1 \geq t_0$ such that $v(t) > 0$ for $t \geq t_1$. $\square$

REMARK 1.1. Corollary 6.4, stated below, implies that if the inequality (1.2) (the inequality (1.3)) holds and $k \in \{0, \dots, n-1\}$, then the condition

$$\int_0^{+\infty} t^{n-1} |p(t)| \, dt < +\infty$$

is necessary and sufficient for the existence of a solution u of (1.1) such that

$$\lim_{t \to +\infty} u^{(k)}(t) = 1.$$

Furthermore, if (1.6) holds, then by Lemma 1.1 this solution satisfies either (1.7$_k$) or (1.7$_{k+1}$), depending on whether the number $k + n$ is odd or even (even or odd).

REMARK 1.2. It follows from Remark 1.1 that if the inequality (1.2) (the inequality (1.3)) is fulfilled, then the condition

$$\int_0^{+\infty} t^{n-1} |p(t)| \, dt = +\infty \tag{1.8}$$

is necessary for (1.1) to have property A (property B).

Note that (1.8) is not sufficient for (1.1) to have property A (property B). Indeed, let M_n^* and M_{*n} be the largest local maxima of the polynomials

$$P_n^*(x) = -x(x-1)\dots(x-n+1) \tag{1.9}$$

and

$$P_{*n}(x) = x(x-1)\dots(x-n+1), \tag{1.10}$$

respectively, $c \in [-M_n^*, 0[\; (c \in]0, M_{*n}])$, and let α be a root of the equation $P_n^*(x) = c$ ($P_{*n}(x) = c$). Then $\alpha \in]0, n-1[$, $u(t) = t^\alpha$ satisfies the equation

$$u^{(n)} = ct^{-n}u, \qquad t \geq 1, \tag{1.11}$$

---

[2] For $l = 0$ ($l = n$) the inequality in the first (second) line should be omitted.

and, thus, this equation does not have property A (property B), although (1.8) holds in this case.

PROBLEM 1.1. If $c < -M_n^*$ ($c > M_{*n}$), then the equation (1.11) has property A (property B).

REMARK 1.3. Besides, Remark 1.1 implies that in Definitions 1.1 and 1.2 the conditions (1.4) and (1.5) can be replaced by the inequalities

$$(-1)^i u^{(i)}(t)u(t) > 0 \qquad \text{for } t \geq 0 \qquad (i = 0,\ldots,n-1)$$

and

$$u^{(i)}(t)u(t) > 0 \qquad \text{for } t \geq t_0 \qquad (i = 0,\ldots,n-1),$$

respectively, where t_0 is a nonnegative constant, which, in general, depends on the solution.

PROBLEM 1.2. If the inequality (1.3) holds, $t_0 \geq 0$, and u is a solution of the equation (1.1) under the initial condition $u^{(i)}(t_0) \geq 0$ ($i = 0,\ldots,n-1$), then

$$u^{(i)}(t) \geq u^{(i)}(t_0) \qquad \text{for } t \geq t_0 \qquad (i = 0,\ldots,n-1).$$

Furthermore, if $(-1)^n p(t) \geq 0$ for $t \geq 0$, $t_0 > 0$, and u is a solution of (1.1) under the initial condition $(-1)^i u^{(i)}(t_0) \geq 0$ ($i = 0,\ldots,n-1$), then

$$(-1)^i u^{(i)}(t) \geq (-1)^i u^{(i)}(t_0) \qquad \text{for } t \in [0,t_0] \qquad (i = 0,\ldots,n-1).$$

PROBLEM 1.3. If $(-1)^n p(t) \geq 0$ for $t \geq 0$, then the equation (1.1) has a solution satisfying the conditions

$$u(0) = 1, \qquad (-1)^i u^{(i)}(t) \geq 0 \qquad \text{for } t \geq 0 \qquad (i = 0,\ldots,n-1).$$

PROBLEM 1.4*. Is the solution mentioned in Problem 1.3 unique?

PROBLEM 1.5. If either the inequality (1.2) is fulfilled and the equation (1.1) has property A, or $n \geq 3$, the inequality (1.3) holds and the equation (1.1) has property B, then the equation (1.1) has an oscillatory solution.

LEMMA 1.2. Let the function u satisfy the condition (1.7_0). Then

$$\int_{t_1}^{+\infty} t^{n-1} |u^{(n)}(t)|\, dt < +\infty \tag{1.12}$$

and

$$u(t) \geq \frac{1}{(n-1)!} \int_t^{+\infty} (\tau - t)^{n-1} |u^{(n)}(\tau)|\, d\tau \qquad \text{for } t \geq t_1. \tag{1.13}$$

PROOF. By (1.7_0) the validity of the lemma immediately follows from the equality

$$u^{(i)}(t) = \sum_{j=i}^{k-1} \frac{(t-s)^{j-i}}{(j-i)!} u^{(j)}(s) + \frac{1}{(k-i-1)!} \int_s^t (t-\tau)^{k-i-1} u^{(k)}(\tau)\, d\tau, \tag{1.14_{ik}}$$

where $i = 0$, $k = n$ and $s > t$. $\square$

LEMMA 1.3. Let, for a certain $l \in \{1, \ldots, n-1\}$, the inequality (1.7_l) hold. Then

$$\int_{t_1}^{+\infty} \tau^{n-l-1} |u^{(n)}(\tau)| \, d\tau < +\infty, \tag{1.15}$$

$$u^{(i)}(t) \geq u^{(i)}(t_1) + \frac{1}{(l-i-1)!} \int_{t_1}^{t} (t-\tau)^{l-i-1} u^{(l)}(\tau) \, d\tau$$
$$\text{for } t \geq t_1 \qquad (i = 0, \ldots, l-1) \tag{1.16}$$

and

$$u^{(l)}(t) \geq \frac{1}{(n-l-1)!} \int_{t}^{+\infty} (\tau-t)^{n-l-1} |u^{(n)}(\tau)| \, d\tau \qquad \text{for } t \geq t_1. \tag{1.17}$$

If, in addition,

$$\int_{t_1}^{+\infty} \tau^{n-l} |u^{(n)}(\tau)| \, d\tau = +\infty, \tag{1.18}$$

then there exists $t_2 \geq t_1$ such that

$$u^{(l-1)}(t) \geq \frac{t}{(n-l)!} \int_{t}^{+\infty} \tau^{n-l-1} |u^{(n)}(\tau)| \, d\tau \qquad \text{for } t \geq t_2 \tag{1.19}$$

and

$$iu^{(l-i)}(t) \geq tu^{(l-i+1)}(t) \geq (i-1)u^{(l-i)}(t) \qquad \text{for } t \geq t_2 \qquad (i = 1, \ldots, l). \tag{1.20}$$

PROOF. By (1.7_l), the equality

$$\sum_{j=i}^{k-1} \frac{(-1)^j}{(j-i)!} t^{j-i} u^{(j)}(t) = \sum_{j=i}^{k-1} \frac{(-1)^j}{(j-i)!} s^{j-i} u^{(j)}(s)$$
$$+ \frac{(-1)^{k-1}}{(k-i-1)!} \int_{s}^{t} \tau^{k-i-1} u^{(k)}(\tau) \, d\tau \tag{1.21_{ik}}$$

with $i = l$, $k = n$, $s > t$ yields (1.15) and

$$\sum_{j=l}^{n-1} \frac{(-1)^{j-l}}{(j-l)!} t^{j-l} u^{(j)}(t) \geq \frac{(-1)^{n-l}}{(n-l-1)!} \int_{t}^{+\infty} \tau^{n-l-1} u^{(n)}(\tau) \, d\tau \qquad \text{for } t \geq t_1. \tag{1.22}$$

Applying (1.7_l) and (1.15), from (1.14_{il}) and (1.14_{ln}) we obtain (1.16) and (1.17), respectively.

Now suppose that (1.18) holds. Then in view of (1.7_l) and (1.21_{l-1n}),

$$\lim_{t \to +\infty} (u^{(l-1)}(t) - tu^{(l)}(t)) = +\infty \tag{1.23}$$

and

$$u^{(l-1)}(t) \geq \sum_{j=l}^{n-1} \frac{(-1)^{j-l}}{(j-l+1)!} t^{j-l+1} u^{(j)}(t) \qquad \text{for } t \geq t_2, \tag{1.24}$$

where t_2 is a sufficiently large number. The inequalities (1.22) and (1.24) immediately imply (1.19).

For $l = 1$ (1.20) is obvious. If $l \in \{2, \ldots, n-1\}$, we set

$$\rho_i(t) = iu^{(l-i)}(t) - tu^{(l-i+1)}(t)$$

and
$$r_i(t) = tu^{(l-i+1)}(t) - (i-1)u^{(l-i)}(t).$$
Since $\rho'_{i+1}(t) = \rho_i(t)$ $(i = 1, \dots, l-1)$, by (1.23) we have
$$\lim_{t \to +\infty} \rho_i(t) = +\infty \qquad (i = 1, \dots, l).$$
On the other hand, by (1.23),
$$\lim_{t \to +\infty} t^{1-i} u^{(l-i)}(t) = +\infty \qquad (i = 1, \dots, l).$$
So there exist $t_l \geq \dots \geq t_1$ such that $r_i(t_i) > 0$ $(i = 2, \dots, l)$. Since, in addition,
$$r'_i(t) = r_{i-1}(t) \qquad (i = 2, \dots, l), \qquad r_1(t) = tu^{(l)}(t) > 0 \qquad \text{for } t \geq t_1,$$
we easily obtain
$$r_i(t) > 0 \qquad \text{for } t \geq t_i \qquad (i = 1, \dots, l).$$
Therefore (1.20) is true. $\square$

Below we shall need two auxiliary statements concerning the unique solvability of certain multi-point boundary value problems.

Consider the equation (1.1) under the conditions
$$u^{(j-1)}(t_i) = 0 \qquad (j = 1, \dots, n_i; \ i = 1, \dots, k),$$

$$u^{(i_j - 1)}(t_0) = 0 \qquad (j = 1, \dots, n_0), \tag{1.25}$$

where $0 \leq t_0 < t_1 < \dots < t_k < +\infty$, $n_i \in \{1, \dots, n-1\}$ $(i = 0, \dots, k)$, $n_0 + n_1 + \dots + n_k = n$, $1 \leq i_1 < \dots < i_{n_0} \leq n$.

LEMMA 1.4. If the inequality (1.2) holds and n_i $(i = 1, \dots, k)$ are even numbers, then the problem (1.1), (1.25) has only the trivial solution.

PROOF. Suppose the problem (1.1),(1.25) has a nontrivial solution u. We assume that $p(t) \equiv -1$ in $]t_k, +\infty[$. Then, clearly, u is oscillatory. Let $t^0 > t_k$ be a zero of the function u and let u change sign exactly m times in the interval $[t_0, t^0]$, i.e. there exist $m+1$ intervals I_j $(j = 0, \dots, m)$, $\cup_{j=0}^m I_j = [t_0, t^0]$, such that $u^{(n)}$ does not change sign in each of them.

In the sequel we use the following notation.

If s_0 is an n_0-fold zero of a function v, then
$$\lambda_m(v; s_0) = \begin{cases} n_0 & \text{for } n_0 \leq m \\ m & \text{for } n_0 > m. \end{cases}$$
We denote by $\mu_m(v; s_0)$ the number of indices $i \in \{0, \dots, m-1\}$ for which $v^{(i)}(s_0) = 0$ (if the set of such indices is empty, then we put $\mu_m(v; s_0) = 0$).

If the function v has a finite number of distinct zeros s_i $(i = 1, \dots, k)$ in the interval I, then
$$\lambda_m(v; I) = \sum_{i=1}^k \lambda_m(v; s_i).$$

Moreover, we set

$$\nu_{*m}(v; a, b) = \lambda_m(v;]a, b]) + \mu_m(v; a).$$

It is obvious that $\nu_{*n}(u; t_0, t^0) \geq n + m + 1$. We consider two possibilities.

1) Suppose the number of zeros of all $u^{(i)}$ $(i = 1, \ldots, n-1)$ in the interval $[t_0, t^0]$ is finite. Note that

$$\nu_{*n-i}(u^{(i)}; t_0, t^0) \geq \nu_{*n-i+1}(u^{(i-1)}; t_0, t^0) - 1 \qquad (i = 1, \ldots, n-1). \qquad (1.26)$$

Indeed, if $u^{(i-1)}(t_0) \neq 0$, then $\lambda_{n-i}(u^{(i)};]t_0, t^0]) \geq \lambda_{n-i+1}(u^{(i-1)};]t_0, t^0]) - 1$ and $\mu_{n-i}(u^{(i)}; t_0) = \mu_{n-i+1}(u^{(i-1)}; t_0)$, while if $u^{(i-1)}(t_0) = 0$, then $\lambda_{n-i}(u^{(i)};]t_0, t^0]) \geq \lambda_{n-i+1}(u^{(i-1)};]t_0, t^0])$ and $\mu_{n-i}(u^{(i)}; t_0) = \mu_{n-i+1}(u^{(i-1)}; t_0) - 1$.

From (1.26) we get

$$\nu_{*1}(u^{(n-1)}; t_0, t^0) \geq \nu_{*n}(u; t_0, t^0) - (n-1) \geq m + 2.$$

Therefore $u^{(n-1)}$ has at least $m + 2$ distinct zeros on $[t_0, t^0]$. So, for at least one j_0, the interval I_{j_0} contains two zeros of the function $u^{(n-1)}$, which is impossible, because $u^{(n)}$ does not change sign in this interval.

2) Suppose now that $u^{(i)}$, with $i \in \{1, \ldots, n-1\}$, has infinitely many zeros in the interval $[t_0, t^0]$. By (1.2), for every bounded interval $I \subset \mathbf{R}_+$ there exists a finite set of intervals T_{ij} (which can degenerate into points) such that $I \cap T_{ij} \neq \emptyset$, $u^{(i)}(t) \equiv 0$ on T_{ij} and $u^{(i)}(t) \neq 0$ on $T_{ij}^\epsilon \setminus T_{ij}$, where T_{ij}^ϵ is an ϵ-neighborhood of T_{ij}. Note that T_{0j} are points. Furthermore, if $T_{i_1 j_1}$ and $T_{i_2 j_2}$ are not points, then either $T_{i_1 j_1} = T_{i_2 j_2}$ or $T_{i_1 j_1} \cap T_{i_2 j_2} = \emptyset$.

Put

$$\bar{\nu}_{*n-i}(u^{(i)}; t_0, t^0) = \sum_{j=1}^{k_i} \bar{\lambda}_{n-i}(u^{(i)}; T_{ij}) + \mu_{n-i}(u^{(i)}; t_0),$$

where $\{T_{i1}, \ldots, T_{ik}\}$ is the set of all intervals having nonempty intersection with $]t_0, t^0]$ and

$$\bar{\lambda}_{n-i}(u^{(i)}; T_{ij}) = \begin{cases} \lambda_{n-i}(u^{(i)}; t_*) & \text{if } T_{ij} = \{t_*\} \\ n - i & \text{otherwise.} \end{cases}$$

As in the first case, we obtain $\bar{\nu}_{*1}(u^{(n-1)}; t_0, t^0) \geq m + 2$. Therefore, for at least one j_0 there exist points t_* and t^* in the interval I_{j_0} such that

$$u^{(n-1)}(t_*) = u^{(n-1)}(t^*) = 0, \qquad u^{(n-1)}(t) \neq 0 \qquad \text{for } t \in]t_*, t^*[.$$

But this is impossible, since $u^{(n)}$ does not change sign in $[t_*, t^*]$. $\quad\square$

REMARK 1.4. Let $u: [t_0, t^0] \to \mathbf{R}$ be absolutely continuous together with $u^{(i)}$ $(i = 1, \ldots, n-1)$,

$$u(t_0) = u(t^0) = 0,$$

$$u(t) \not\equiv 0, \qquad u(t) \geq 0, \qquad u^{(n)}(t) \leq 0 \qquad \text{on } [t_0, t^0].$$

It then follows from the proof of Lemma 1.4 that $\lambda_n(u; [t_0, t^0]) \leq n$; moreover, if $\lambda_n(u; [t_0, t^0]) = n$, then $\lambda_n(u; t_0)$ is an odd number.

Now let $0 \leq t_0 < t_1 < \ldots < t_k < t^0 < +\infty$, $n^0, n_i \in \{1, \ldots, n-1\}$ $(i = 0, \ldots, k)$, $n_0 + n_1 + \ldots + n_k + n^0 = n$, $1 \leq i_1 < i_2 < \ldots < i_{n^0} \leq n$. Consider the equation (1.1) under the conditions

$$u^{(j-1)}(t_i) = 0 \qquad (j = 1, \ldots, n_i; \ i = 0, \ldots, k),$$

$$(1.27)$$

$$u^{(i_j-1)}(t^0) = 0 \qquad (j = 1, \ldots, n^0).$$

LEMMA 1.5. If the inequality (1.3) holds, n^0 is odd and n_i $(i = 1, \ldots, k)$ are even, then the problem (1.1), (1.27) has only the trivial solution.

PROOF. If n is odd, then by the change of the independent variable $s = -t$ this lemma turns into the previous one. Therefore we assume that n is even.

Let the problem (1.1), (1.27) have a nontrivial solution u. As above, we consider two possibilities.

1) Suppose the number of zeros of all $u^{(i)}$ $(i = 1, \ldots, n-1)$ in the interval $[t_0, t^0]$ is finite and u changes sign exactly m times in this interval, i.e. there exist $m+1$ intervals I_j $(j = 0, \ldots, m)$, $\cup_{j=0}^{m} I_j = [t_0, t^0]$, such that $u^{(n)}$ does not change sign in each of them. In addition to the notation introduced in the proof of Lemma 1.4, we set

$$\nu_m^*(v; a, b) = \lambda_m(v; [a, b[) + \mu_m(v; b).$$

Clearly,

$$\mu_n(u; t^0) \geq n^0, \qquad \nu_n^*(u; t_0, t^0) \geq n + m,$$

$$\nu_{n-i}^*(u^{(i)}; t_0, t^0) \geq \nu_{n-i+1}^*(u^{(i-1)}; t_0, t^0) - 1 \qquad (i = 1, \ldots, n-1).$$

If at least one of these inequalities is strict, then $\nu_1^*(u^{(n-1)}; t_0, t^0) \geq m+2$, i.e. $u^{(n-1)}$ has at least $m+2$ distinct zeros in the interval $[t_0, t^0]$. Hence, for a certain j_0 the function $u^{(n-1)}$ has at least two distinct zeros in the interval I_{j_0}, which is impossible. So

$$\mu_n(u; t^0) = n^0, \qquad \nu_n^*(u; t_0, t^0) = n + m,$$

$$\nu_{n-i}^*(u^{(i)}; t_0, t^0) = \nu_{n-i+1}^*(u^{(i-1)}; t_0, t^0) - 1 \qquad (i = 1, \ldots, n-1),$$

and $u^{(n-1)}$ has exactly $m+1$ distinct zeros in $[t_0, t^0]$.

Denote by $t_*^{(i)}$ $(i = 0, \ldots, n-1)$ the smallest zero of the function $u^{(i)}$ in the interval $]t_0, t^0]$. It is obvious that

$$t_0 < t_*^{(n_0)} < t_*^{(n_0-1)} < \ldots < t_*^{(0)}, \qquad t_*^{(n_0)} < t_*^{(n_0+1)} < \ldots < t_*^{(n-1)}.$$

If $t_*^{(n-1)} \notin [t_0, t_*^{(0)}]$, then for at least one j_0 the function $u^{(n-1)}$ has two distinct zeros in the interval I_{j_0}, which is impossible. Thus $t_*^{(n-1)} \in]t_0, t_*^{(0)}]$. But in this case, assuming

$$u(t) > 0 \qquad \text{for } t \in]t_0, t_*^{(0)}[, \qquad (1.28)$$

by (1.3) we obtain

$$(-1)^{n-i+1} u^{(i-1)}(t) > 0 \qquad \text{for } t \in [t_0, t_*^{(i-1)}[\qquad (i = n_0 + 1, \ldots, n),$$

$$(-1)^{n-n_0} u^{(i-1)}(t) > 0 \qquad \text{for } t \in]t_0, t_*^{(i-1)}[\qquad (i = 1, \ldots, n_0).$$

Since n is even and n_0 is odd, this contradicts the inequality (1.28).

2) The case when, for a certain $i \in \{1, \dots, n-1\}$, $u^{(i)}$ has infinitely many zeros in the interval $[t_0, t^0]$ can be treated in the same way as in the proof of Lemma 1.4. $\square$

REMARK 1.5. As above, we note that if $u \colon [t_0, t^0] \to \mathbf{R}$ is absolutely continuous together with $u^{(i)}$ $(i = 1, \dots, n-1)$,

$$u(t_0) = u(t^0) = 0,$$

$$u(t) \not\equiv 0, \qquad u(t) \geq 0, \qquad u^{(n)}(t) \geq 0 \qquad \text{on } [t_0, t^0],$$

then $\lambda_n(u; [t_0, t^0]) \leq n$. If, moreover, $\lambda_n(u; [t_0, t^0]) = n$, then the constant $\lambda_n(u; t^0)$ is even.

DEFINITION 1.3. Let $l \in \{1, \dots, n-1\}$. The equation (1.1) is said to be $(l, n-l)$ conjugate in the interval $I \subset \mathbf{R}_+$ if there exist $t_1, t_2 \in I$, $t_2 > t_1$, and a nontrivial solution u of this equation such that

$$u^{(i)}(t_1) = 0 \qquad (i = 0, \dots, l-1),$$

$$\tag{1.29_l}$$

$$u^{(i)}(t_2) = 0 \qquad (i = 0, \dots, n-l-1).$$

Otherwise the equation (1.1) is said to be $(l, n-l)$ disconjugate in I.

DEFINITION 1.4. The equation (1.1) is said to be conjugate in the interval $I \subset \mathbf{R}_+$ if there exists a nontrivial solution of this equation having n zeros (taking into account multiplicity) in I. Otherwise (1.1) is said to be disconjugate in I.

Lemma 1.9, stated below, implies that (1.1) is conjugate in an interval I if and only if it is $(l, n-l)$ conjugate in this interval for a certain $l \in \{1, \dots, n-1\}$.

LEMMA 1.6. Let the inequalities (1.2) and (1.6) (the inequalities (1.3) and (1.6)) hold, $l \in \{1, \dots, n-1\}$, and let $l+n$ be odd (even). Then the equation (1.1) under the condition (1.7_l) is solvable if and only if it is $(l, n-l)$ disconjugate in a certain neighborhood of $+\infty$.

PROOF. Necessity. Suppose there exists a solution u_1 of (1.1) satisfying (1.7_l) and for some $t_2 > t_1 \geq t_0$ the boundary value problem (1.1), (1.29_l) has a nontrivial solution u_2. Let u_2 vanish at m distinct points in the interval $]t_1, t_2[$ and let s_0 be the smallest zero of u_2 in $]t_1, t_2[$. Without loss of generality we assume that $u_2(t) > 0$ for $t \in]t_1, s_0[$.

Since u_2 satisfies (1.29_l), $u_2^{(n-1)}$ has at least $m+1$ zeros $s_{n-1}^1 < \dots < s_{n-1}^{m+1}$ in the interval $]t_1, t_2[$. Furthermore,

$$u_2^{(n-1)}(t) \not\equiv 0 \qquad \text{on } [s_{n-1}^j, s_{n-1}^{j+1}] \qquad (j = 1, \dots, m),$$

and thus $s_{n-1}^1 \in]t_1, s_0[$. It is easy to verify that there exist $s_i \in]t_1, t_2[$ $(i = 1, \dots, n-1)$ such that

$$s_l \leq s_{l-1} \leq \dots \leq s_1 \leq s_0, \qquad s_l < s_{l+1} < \dots < s_{n-1} < s_0,$$

$$u_2^{(i)}(s_i) = 0 \qquad (i = 1, \dots, n-1),$$

$$u_2^{(i)}(t) \leq 0 \qquad \text{for } s_i \leq t \leq s_{i-1} \qquad (i = 1, \dots, l).$$

Consider the solution $v_\epsilon = u_1 - \epsilon u_2$. Let $\epsilon = \epsilon_0$ be the largest constant for which

$$v_\epsilon^{(i)}(t) \geq 0 \quad \text{on } [t_1, s_i] \quad (i = 0, \ldots, l-1),$$

$$(-1)^{i+l} v_\epsilon^{(i)}(t) \geq 0 \quad \text{on } [t_1, s_i] \quad (i = l, \ldots, n-1).$$

(1.30)

Then for $\epsilon = \epsilon_0$ at least one of the inequalities (1.30) is nonstrict. On the other hand, since

$$v_\epsilon^{(i)}(t_1) > 0 \quad (i = 0, \ldots, l-1), \qquad (-1)^{i+l} v_\epsilon^{(i)}(s_i) > 0 \quad (i = l, \ldots, n-1),$$

$$(-1)^{n+l} v_\epsilon^{(n)}(t) \geq 0 \quad \text{for } t \in [t_1, s_0],$$

we obtain

$$v_\epsilon^{(i)}(t) > 0 \quad \text{for } t \in [t_1, s_i] \quad (i = 0, \ldots, l-1),$$

$$(-1)^{i+l} v_\epsilon^{(i)}(t) > 0 \quad \text{for } t \in [t_1, s_i] \quad (i = l, \ldots, n-1).$$

This contradiction proves the necessity part.

Sufficiency. Suppose the equation (1.1) is $(l, n-l)$ disconjugate in any neighborhood of $+\infty$, i.e. for any $t_2 > t_1 \geq t_0$ the boundary value problem (1.1),(1.29$_l$) has only the trivial solution. Let $s > t_0$ and let u_s be the solution of (1.1) under the conditions

$$u_s^{(i)}(t_0) = 0 \quad (i = 0, \ldots, l-1), \qquad u_s^{(i)}(s) = 0 \quad (i = 0, \ldots, n-l-2),$$

(1.31)

$$\sum_{i=0}^{n-1} |u_s^{(i)}(t_0)| = 1, \qquad u_s(t) > 0 \quad \text{for } t \in]t_0, t_0 + \delta[,$$

where δ is a sufficiently small constant. Assume that for a certain $s > t_0$, u_s has at least n zeros in the interval $[t_0, s]$. If S is the set of all such s and $\alpha = \inf S$, then we can readily verify that $\alpha \in S$ and $u_\alpha(t) \geq 0$ for $t \in [t_0, \alpha]$. On the other hand, since the boundary value problem (1.1), (1.29$_l$) with $t_1 = t_0$, $t_2 = \alpha$ has only the trivial solution, $u_\alpha^{(n-l-1)}(\alpha) \neq 0$ and by Lemma 1.4 (Lemma 1.5),

$$u_\alpha(t) > 0 \quad \text{for } t \in [t_0, \alpha], \qquad u_\alpha^{(l)}(t_0) \neq 0.$$

Therefore $\alpha \notin S$. In view of this contradiction, for any $s > t_0$, u_s has $n-1$ zeros in $[t_0, s]$. So

$$u_s(t) > 0 \quad \text{for } t \in]t_0, s[.$$

Clearly, we can find a sequence $(s_k)_{k=1}^{+\infty}$ such that $t_0 < s_1 < s_2 < \ldots$, $\lim_{k \to +\infty} s_k = +\infty$ and $\lim_{k \to +\infty} u_{s_k} = u$ uniformly on each bounded set. Hence u is a nontrivial solution of the equation (1.1), and $u(t) \geq 0$ for $t \geq t_0$.

By Lemma 1.1 there exist $l_0 \in \{0, \ldots, n\}$ and $t_1 \geq t_0$ such that $l_0 + n$ is odd (even) and

$$u^{(i)}(t) > 0 \quad \text{for } t \geq t_1 \quad (i = 0, \ldots, l_0 - 1),$$

$$(-1)^{i+l_0} u^{(i)}(t) > 0 \quad \text{for } t \geq t_1 \quad (i = l_0, \ldots, n-1).$$

Thus, for a sufficiently large k_0 we have $s_{k_0} > t_1$ and

$$u_{k_0}^{(i)}(t_1) > 0 \qquad (i = 0, \dots, l_0 - 1),$$

$$(-1)^{i+l_0} u_{k_0}^{(i)}(t_1) > 0 \qquad (i = l_0, \dots, n - 1).$$

$$(1.32)$$

Suppose $l_0 \neq l$. Since $l_0 + l$ is even, either $l_0 \leq l - 2$ or $l_0 \geq l + 2$.

If v is a solution of the equation (1.1), then by the inequality (1.2) (the inequality (1.3)), for every finite interval $I \subset \mathbf{R}_+$ there exists a finite set of intervals T_{ij} (some of which can degenerate into points) such that $I \cap T_{ij} \neq \emptyset$, $v^{(i)}(t) \equiv 0$ on T_{ij} and $v^{(i)}(t) \neq 0$ on $T_{ij}^\epsilon \setminus T_{ij}$, where T_{ij}^ϵ is an ϵ-neighborhood of T_{ij}. We denote by $\nu_i(v; I)$, $i \in \{0, \dots, n-1\}$, the number of such intervals.

Let $l_0 \leq l - 2$ and let u_{k_0} have m distinct zeros in the interval $]t_0, t_1[$. So $\nu_0(u_{k_0};]t_0, t_1[) = m$ and there exist $m+1$ intervals I_j $(j = 0, \dots, m)$, $\cup_{j=0}^m I_j = [t_0, t_1]$, such that $u_{k_0}^{(n)}$ does not change sign in each I_j. Then (1.31) and (1.32) yield

$$\nu_i(u_{k_0};]t_0, t_1[) \geq m \qquad (i = 0, \dots, l_0),$$

$$\nu_i(u_{k_0};]t_0, t_1[) \geq m + i - l_0 \qquad (i = l_0 + 1, \dots, l),$$

$$\nu_i(u_{k_0};]t_0, t_1[) \geq m + l - l_0 \qquad (i = l + 1, \dots, n - 1).$$

This implies $\nu_{n-1}(u_{k_0};]0, t_1[) \geq m+2$. Hence, at least one interval I_{j_0}, $j_0 \in \{0, \dots, m\}$, contains points t_* and t^* such that

$$u_{k_0}^{(n-1)}(t_*) = u_{k_0}^{(n-1)}(t^*) = 0, \qquad u_{k_0}^{(n-1)}(t) \neq 0 \qquad \text{for } t \in]t_*, t^*[.$$

But this is impossible, because $u_{k_0}^{(n)}$ does not change sign in the interval $[t_*, t^*]$.

Now let $l_0 \geq l + 2$ and let u_{k_0} have m distinct zeros in the interval $]t_1, s_{k_0}[$. Then from (1.31) and (1.32) we obtain

$$\nu_i(u_{k_0};]t_1, s_{k_0}[) \geq m \qquad (i = 0, \dots, l),$$

$$\nu_i(u_{k_0};]t_1, s_{k_0}[) \geq m + i - l \qquad (i = l + 1, \dots, l_0),$$

$$\nu_i(u_{k_0};]t_1, s_{k_0}[) \geq m + l_0 - l \qquad (i = l_0 + 1, \dots, n - 1).$$

As above, applying the inequality $\nu_{n-i}(u_{k_0};]t_1, s_{k_0}[) \geq m + 2$, we arrive at a contradiction. Thus $l_0 = l$ and u is a solution of the equation (1.1) under the condition (1.7_l). $\square$

Since the boundary value problem (1.1), (1.29_l) has only the trivial solution if and only if the adjoint problem

$$u^{(n)} = (-1)^n p(t) u, \qquad\qquad (1.33)$$

$$u^{(i)}(t_1) = 0 \quad (i = 0, \dots, n - l - 1), \qquad u^{(i)}(t_2) = 0 \quad (i = 0, \dots, l - 1)$$

has only the trivial solution, Lemma 1.6 implies

COROLLARY 1.1. Let the inequality (1.2) (the inequality (1.3)) hold, $l \in \{1, \dots, n-1\}$, and let $l + n$ be odd (even). Then the equation (1.1) under the condition (1.7_l) is solvable if and only if the equation (1.33) under the condition (1.7_{n-l}) is solvable.

LEMMA 1.7. Let (1.2) hold, $l \in \{1, \ldots, n-1\}$, $k \in \{0, \ldots, l-1\}$, $c_0 \in \,]0, +\infty[$, and let there exist a continuous function $v: [t_1, +\infty[\to \mathbf{R}_+$ such that

$$v(t) \geq c_0(t - t_1)^k - \frac{1}{(l-1)!\,(n-l-1)!} \int_{t_1}^t (t-s)^{l-1} \int_s^{+\infty} (\xi - s)^{n-l-1} p(\xi) v(\xi) \, d\xi \, ds$$

in the interval $[t_1, +\infty[$. Then the equation

$$u^{(n)} = (-1)^{n-l-1} p(t) u$$

has a solution u satisfying the conditions (1.7_l) and $u(t_1) = \ldots = u^{(k-1)}(t_1) = 0$.

PROOF. Denote by S the set of continuous functions $x: [t_1, +\infty[\to \mathbf{R}_+$ satisfying the inequalities

$$c_0(t - t_1)^k \leq x(t) \leq v(t) \qquad \text{for } t \in [t_1, +\infty[,$$

and by $F: S \to S$ the operator for which

$$F(x)(t) = c_0(t-t_1)^k - \frac{1}{(l-1)!\,(n-l-1)!} \int_{t_1}^t (t-s)^{l-1} \int_s^{+\infty} (\xi-s)^{n-l-1} p(\xi) x(\xi) \, d\xi \, ds.$$

Consider the sequence $(u_j)_{j=1}^{+\infty}$, where

$$u_1(t) = c_0(t - t_1)^k, \qquad u_{j+1} = F(u_j) \qquad (j = 1, 2, \ldots).$$

Clearly,

$$u_{j+1}(t) \geq u_j(t) \qquad \text{for } t \in [t_1, +\infty[\qquad (j-1, 2, \ldots).$$

The inequality

$$|(F(u_j))'| \leq (F(v))'(t) \qquad \text{for } t \geq t_1$$

implies that the sequence $(u_j)_{j=1}^{+\infty}$ converges uniformly on each interval in $[t_1, +\infty[$. It is obvious that $u = \lim_{j \to +\infty} u_j$ satisfies the assertion of the lemma. $\square$

LEMMA 1.8. Let $n \geq 4$ and let the inequality (1.2) (the inequality (1.3)) hold. Furthermore, suppose that $l \in \{2, \ldots, [n/2]\}$, $l + n$ is odd (even) and the equation (1.1) has a solution satisfying the condition (1.7_l). Then the equation

$$u^{(n)} = -p(t)u \tag{1.34}$$

has a solution satisfying the condition (1.7_{l-1}).

PROOF. We assume that the condition (1.8) holds, because otherwise the equation (1.34) is nonoscillatory and for any $k \in \{1, \ldots, n-1\}$ such that $k+n$ is even (odd) it has a solution satisfying (1.7_k) and, thus, (1.7_{l-1}). Let u be a solution of the equation (1.1) under the condition (1.7_l). Since $n \geq 2l$, by Lemma 1.3 (from the inequalities (1.16) with $i = 1$, (1.17) and (1.20) with $i = l$) we obtain for $t \geq t_1$,

$$u'(t) \geq u'(t_1) + \frac{(-1)^{n-l}}{(l-2)!\,(n-l)!} \int_{t_1}^t (t-s)^{l-2} \int_s^{+\infty} (\tau - s)^{n-l} p(\tau) u'(\tau) \, d\tau \, ds.$$

Hence Lemma 1.7 implies that the equation (1.34) has a solution satisfying the condition (1.7_{l-1}). $\square$

1.2. Conjugate points of the equation (1.1). Let $k \in \mathbf{N}$, $t_0 \in \mathbf{R}_+$, and let $F_n^k(t_0; p)$ be the set of $t_1 > t_0$ such that a certain nontrivial solution of the equation (1.1) has at least $n + k - 1$ zeros in the interval $[t_0, t_1]$. Put

$$\eta_n^k(t_0; p) = \inf F_n^k(t_0; p);$$

as usual, $\eta_n^k(t_0; p) = +\infty$ if $F_n^k(t_0; p) = \emptyset$).

DEFINITION 1.5. $\eta_n^k(t_0; p)$ is called the k-*th conjugate point* for t_0 with respect to the equation (1.1). Furthermore, $\eta_n^1(t_0; p)$ is simply denoted by $\eta_n(t_0; p)$ and is called the *conjugate point* for t_0 with respect to the equation (1.1).

It is readily verified that $\eta_n^k(t_0; p) > t_0$, and if $\eta_n^k(t_0; p) < +\infty$, then there exists a solution of the equation (1.1) which has at least $n + k - 1$ zeros in the interval $[t_0, \eta_n^k(t_0; p)]$.

First we will state two auxiliary propositions, taken from [259] (Theorem 3.3) and [99] (Theorem 1).

LEMMA 1.9. Let $t^0 = \eta_n(t_0; p) < +\infty$. Then there exist a solution u of the equation (1.1) and $l \in \{1, \dots, n - 1\}$ such that

$$u(t) > 0 \qquad \text{for } t \in \ ]t_0, t^0[,$$

$$u^{(i-1)}(t_0) = 0 \quad (i = 1, \dots, l), \qquad u^{(i-1)}(t^0) = 0 \quad (i = 1, \dots, n - l).$$

REMARK 1.6. Lemma 1.4 (Lemma 1.5) implies that if p is nonpositive (nonnegative), then $n - l$ is odd (even).

LEMMA 1.10. Let either (1.2) or (1.3) hold and $t^0 = \eta_n^k(t_0; p) < +\infty$. Then there exists a solution u of the equation (1.1) having exactly $n + k - 1$ zeros (taking into account multiplicity) in the interval $[t_0, t^0]$ and, moreover, the interval $]t_0, t^0[$ contains exactly $k - 1$ zeros, each being of odd multiplicity.

Let $t_0 \in \mathbf{R}_+$. Define the sets $E_n(t_0; p)$, $E_n^*(t_0; p)$ and $E_n^{**}(t_0; p)$ as follows: $t_1 \in E_n(t_0; p)$ if $t_1 > t_0$ and the equation (1.1) has a solution satisfying the conditions

$$u(t_0) = u(t_1) = 0, \qquad u(t) > 0 \qquad \text{for } t \in \ ]t_0, t_1[;$$

$t_1 \in E_n^*(t_0; p)$ if $t_1 > t_0$ and the equation (1.1) has a nontrivial solution satisfying the conditions

$$u(t_0) = u(t_1) = 0, \qquad u(t) \geq 0 \qquad \text{for } t \in [t_0, t_1];$$

$t_1 \in E_n^{**}(t_0; p)$ if $t_1 > t_0$ and the equation (1.1) has a nontrivial solution satisfying the conditions

$$u(t_0) = 0, \qquad u(t) \geq 0 \qquad \text{for } t \in [t_0, t_1],$$

$$u^{(i)}(t_1) \geq 0 \quad (i = 0, \dots, n - 1), \qquad \mu_n(u; t_1) > 0,$$

with $\mu_n(u, t_1)$ being defined in the proof of Lemma 1.4. Set

$$\tau_n(t_0; p) = \sup E_n(t_0; p);$$

$$\tau_n^*(t_0; p) = \sup E_n^*(t_0; p); \qquad \tau_n^{**}(t_0; p) = \sup E_n^{**}(t_0; p).$$

It is easy to conclude that if $\tau_n^*(t_0; p) < +\infty$ ($\tau_n^{**}(t_0; p) < +\infty$), then $\tau_n^*(t_0; p) \in E_n^*(t_0; p)$ ($\tau_n^{**}(t_0; p) \in E_n^{**}(t_0; p)$).

LEMMA 1.11. If the inequality (1.2) (the inequality (1.3)) holds, then $\tau_n(t_0; p) = \tau_n^*(t_0; p)$.

PROOF. Clearly, $E_n(t_0; p) \subset E_n^*(t_0; p)$, i.e. $\tau_n(t_0; p) \leq \tau_n^*(t_0; p)$. Therefore, if $\tau_n(t_0; p) = +\infty$, then the lemma is obvious.

Suppose $\tau_n(t_0; p) < +\infty$ and $\tau_n(t_0; p) < \tau_n^*(t_0; p)$. Then there exist $t^0 \in \,]\tau_n(t_0; p), \tau_n^*(t_0; p)[$ and a solution u_1 of (1.1) such that

$$u_1(t_0) = u_1(t^0) = 0, \qquad u_1(t) \geq 0 \qquad \text{for } t \in [t_0, t^0].$$

Let $n_0, n_1, \dots, n_k, n^0$ be the respective multiplicities of the zeros $t_0 < t_1 < \dots < t_k < t^0$ of u_1. Then the integers n_i ($i = 1, \dots, k$) are even, $n_0 + n_1 + \dots + n_k + n^0 \leq n$, and according to Lemma 1.4 (Lemma 1.5), if n^0 is even (odd), the last inequality is strict. By Lemma 1.4 (Lemma 1.5) the equation (1.1) has a unique solution u_0 for which

$$u_0^{(j-1)}(t_i) = \delta_{1j} \qquad (j = 1, \dots, n_i; \; i = 1, \dots, k),$$
$$u_0^{(j-1)}(t_0) = 0 \qquad \left(j = 1, \dots, n - \sum_{i=1}^k n_i - n^*\right),$$
$$u_0^{(j-1)}(t^0) = 0 \qquad (j = 1, \dots, n^*),$$

where δ_{ij} is the Kronecker delta and

$$n^* = \begin{cases} n^0 & \text{if } n^0 \text{ is even (odd)} \\ n^0 - 1 & \text{if } n^0 \text{ is odd (even)}. \end{cases}$$

Let $\delta \in \,]0, t^0 - \tau_n(t_0; p)[$. The inequality $n - \sum_{i=1}^k n_i - n^* > n^0$ implies that

$$u(t) > 0 \qquad \text{for } t \in \,]t_0, t^0 - \delta] \qquad (1.35)$$

whenever $u = u_1 + \epsilon u_0$ and $\epsilon > 0$ is a sufficiently small constant. Moreover, if $n^* \neq 0$, then $u(t^0) = 0$, and if $n^* = 0$ (which is impossible unless the inequality (1.3) holds and $n^0 = 1$), then $u_1(t^0 + \delta) < 0$ and

$$u(t^0 + \delta) = u_1(t^0 + \delta) + \epsilon u_0(t^0 + \delta) < 0.$$

Hence, in both cases there exists $t^* > t^0 - \delta > \tau_n(t_0; p)$ such that

$$u(t) > 0 \qquad \text{for } t \in \,]t_0, t^*[, \qquad u(t^*) = 0,$$

and we arrive at a contradiction. $\square$

LEMMA 1.12. Let the inequality (1.2) (the inequality (1.3)) hold and $\tau_n(t_0; p) < +\infty$. Then the equation (1.1) possesses a solution u which has n zeros in the interval $[t_0, \tau_n(t_0; p)]$ and satisfies the conditions

$$u(t_0) = u(\tau_n(t_0; p)) = 0, \qquad u(t) \geq 0 \qquad \text{for } t \in [t_0, \tau_n(t_0; p)].$$

PROOF. Since $\tau_n(t_0; p) = \tau_n^*(t_0; p) \in E_n^*(t_0; p)$, according to the definition of the set $E_n^*(t_0; p)$ there exists a solution of the equation (1.1) such that

$$u(t_0) = u(t^0) = 0, \qquad u(t) \geq 0 \qquad \text{for } t \in [t_0, t^0],$$

where $t^0 = \tau_n(t_0; p)$. Choose u so that $\lambda_n(u; [t_0, t^0])$ is maximal. Lemma 1.4 (Lemma 1.5) yields $\lambda_n(u; [t_0, t^0]) \leq n$. (Here $\lambda_n(u; [t_0, t^0])$ is the number of zeros of u in the interval $[t_0, t^0]$ taking into account multiplicities.) Our aim is to prove that $\lambda_n(u; [t_0, t^0]) = n$.

Assume the contrary, i.e. $\lambda_n(u; [t_0, t^0]) < n$. Let $t_0 < t_1 < \ldots < t_k < t^0$ be all the zeros of u in the interval $[t_0, t^0]$ and let $n_0, n_1, \ldots, n_k, n^0$ be their respective multiplicities. Clearly, n_i $(i = 1, \ldots, k)$ are even.

Consider a sequence $(v_m)_{m=1}^{+\infty}$ of nontrivial solutions of the equation (1.1) satisfying the conditions

$$v_m^{(j-1)}(t_i) = 0 \qquad (j = 1, \ldots, n_i;\ i = 0, \ldots, k),$$

$$v_m^{(j-1)}\left(t^0 + \frac{1}{m}\right) = 0 \qquad (j = 1, \ldots, n^0).$$

Such solutions exist, because each one is defined by $\lambda_n(u; [t_0, t^0]) < n$ homogeneous conditions.

Since $t^0 = \tau_n(t_0; p) = \sup E_n^*(t_0; p)$, v_m changes sign in the interval $]t_0, t^0 + 1/m[$. If the function v_m changes sign at a point which differs from t_i, then it has at least $\lambda_n(u; [t_0, t^0]) + 1$ zeros in $[t_0, t^0 + 1/m]$, and if the sign changes at t_i, then, since n_i is even, the multiplicity of the zero t_i is at least $n_i + 1$. So

$$\lambda_n(v_m; [t_0, t^0 + 1/m]) \geq \lambda_n(u; [t_0, t^0]) + 1.$$

Without loss of generality we may assume that $\sum_{j=1}^{m} |v_m^{(j-1)}(t_0)| = 1$ and that the sequence $(v_m)_{m=1}^{+\infty}$ converges uniformly on each finite interval in $\mathbf{R}_+$. Let $v_0 = \lim_{m \to +\infty} v_m$. It is readily verified that

$$\lambda_n(v_0; [t_0, t^0]) \geq \lambda_n(v_m; [t_0, t^0 + 1/m]) \geq \lambda_n(u; [t_0, t^0]) + 1.$$

Hence, u and v_0 are linearly independent. Besides, owing to the maximality of $\lambda_n(u; [t_0, t^0])$, v_0 changes sign in the interval $[t_0, t^0]$.

Let $u_\epsilon = u - \epsilon v_0$. Since v_0 vanishes at the zeros of u, we have

$$u_\epsilon(t) \geq 0 \qquad \text{for } t \in [t_0, t^0] \tag{1.36}$$

whenever $\epsilon > 0$ is sufficiently small. For sufficiently large ϵ this inequality is violated. Thus, there exists a largest $\epsilon = \epsilon_0$ such that (1.36) remains valid. We have $\lambda_n(u_{\epsilon_0}; [t_0, t^0]) > \lambda_n(u; [t_0, t^0])$. But this contradicts the maximality of $\lambda_n(u; [t_0, t^0])$. $\square$

REMARK 1.7. It follows from Lemma 1.2 that if either (1.2) or (1.3) holds, then $\eta_n(t_0; p) \leq \tau_n(t_0; p)$.

LEMMA 1.13. Let the inequality (1.2) be fulfilled and $\tau_n(t_0; p) < +\infty$. Then every solution of the equation (1.1) vanishing at t_0 has a zero in the interval $]t_0, \tau_n(t_0; p)]$.

PROOF. If the lemma is false, then the equation (1.1) has a solution u such that

$$u(t_0) = 0, \qquad u(t) > 0 \qquad \text{for } t \in]t_0, \tau_n(t_0; p)].$$

According to Lemma 1.12, there exists a solution v of (1.1) having n zeros in the interval $[t_0, \tau_n(t_0; p)]$ and satisfying the conditions

$$v(t_0) = v(\tau_n(t_0; p)) = 0, \qquad v(t) \geq 0 \qquad \text{for } t \in [t_0, \tau_n(t_0; p)].$$

By Lemma 1.4, $\tau_n(t_0; p)$ is a zero of odd multiplicity. Hence we can choose $t_1 > \tau_n(t_0; p)$ and $\epsilon_0 > 0$ so that

$$u(t_1) + \epsilon_0 v(t_1) = 0, \qquad u(t) + \epsilon_0 v(t) > 0 \qquad \text{for } t \in \,]t_0, t_1[.$$

This yields $t_1 \in E_n(t_0; p)$, and we arrive at a contradiction. $\square$

LEMMA 1.14. Let the inequality (1.3) be fulfilled, $\tau_n^{**}(t_0; p) < +\infty$, and let u be a solution of the equation (1.1) vanishing at t_0. Then either u has a zero in the interval $]t_0, \tau_n^{**}(t_0; p)]$ or

$$u^{(i-1)}(\tau_n^{**}(t_0; p))u(\tau_n^{**}(t_0; p)) \geq 0 \qquad (i = 1, \ldots, n).$$

PROOF. Assume the contrary. Then there exist $t_1 > \tau_n^{**}(t_0; p)$, $i_0 \in \{1, \ldots, n\}$ and solutions u_0 and u_1 such that

$$u_0(t_0) = 0, \qquad u_0^{(i_0-1)}(t_1) < 0, \qquad u_0(t) > 0 \qquad \text{for } t \in \,]t_0, t_1],$$

$$u_1(t_0) = 0, \qquad u_1(t) \geq 0 \qquad \text{for } t \in [t_0, \tau_n^{**}(t_0; p)],$$

$$u_1^{(i-1)}(\tau_n^{**}(t_0; p)) \geq 0 \quad (i = 1, \ldots, n), \qquad \mu_n(u_1; \tau_n^{**}(t_0; p)) > 0$$

(for the definition of $\mu_n(u, t)$ see the proof of Lemma 1.4). Clearly,

$$u_1^{(i-1)}(t) > 0 \qquad \text{for } t \in \,]\tau_n^{**}(t_0; p), +\infty[\qquad (i = 1, \ldots, n).$$

Therefore we can choose $\epsilon > 0$ so that $v = u_0 + \epsilon u_1$ satisfies the inequalities

$$v(t) \geq 0 \qquad \text{for } t \in [t_0, t_1],$$

$$v^{(i-1)}(t_1) \geq 0 \quad (i = 1, \ldots, n), \qquad \mu_n(v; t_1) > 0.$$

Hence $t_1 \in E_n^{**}(t_0; p)$. But this is impossible. $\square$

LEMMA 1.15. If the inequality (1.3) holds, then $\tau_n(t_0; p) = \tau_n^{**}(t_0; p)$.
PROOF. First consider the case when $\tau_n^{**}(t_0; p) < +\infty$. Then Lemma 1.14 implies $\tau_n(t_0; p) \leq \tau_n^{**}(t_0; p)$. Suppose $t^0 = \tau_n^{**}(t_0; p) > \tau_n(t_0; p)$. So, there exists a solution v_1 such that

$$v_1(t_0) = 0, \qquad \mu_n(v_1; t^0) > 0, \qquad v_1(t) \geq 0 \qquad \text{for } t \in [t_0, t^0]$$

and

$$v_1^{(i-1)}(t) > 0 \qquad \text{for } t \in \,]t^0, +\infty[\qquad (i = 1, \ldots, n). \tag{1.37}$$

Let $n_0, n_1, \ldots, n_k$ be the respective multiplicities of the zeros $t_0 < t_1 < \ldots < t_k$ of v_1 in the interval $[t_0, t^0[$, $\mu_n(v_1; t^0) = n^0$ and $v_1^{(i_0-1)}(t^0) = 0$. Clearly, n_i $(i = 1, \ldots, k)$ are even and by Lemma 1.5 $n_0 + n_1 + \ldots + n_k + n^0 \leq n$; moreover, if n^0 is odd, then the last inequality is strict.

In view of Lemma 1.5, the equation (1.1) under the conditions

$$u^{(j-1)}(t_i) = \delta_{1j} \qquad (j = 1, \ldots, n_i;\ i = 1, \ldots, k),$$

$$u^{(j-1)}(t_0) = 0 \qquad \left(j = 1, \ldots, n - \sum_{i=1}^{k} n_i - 1\right),$$

$$u^{(i_0-1)}(t^0) = -1$$

has a unique solution v_0.

If $\delta \in \,]0, t^0 - \tau_n(t_0; p)[$, then, by applying the inequality $n - \sum_{i=1}^{k} n_i - 1 > n_0$, we can easily conclude that for sufficiently small $\epsilon_0 > 0$, (1.35) with $u = v_1 + \epsilon v_0$ is valid. Since $t^0 - \delta > \tau_n(t_0; p)$, we have $u(t) > 0$ for $t \in \,]t_0, +\infty[$. Besides,

$$u^{(i_0-1)}(t^0) = v_1^{(i_0-1)}(t^0) + \epsilon_0 v_0^{(i_0-1)}(t^0) = -\epsilon_0 < 0.$$

Hence there exists $t^* > t^0$ such that

$$u^{(i_0-1)}(t^*) < 0. \tag{1.38}$$

Put $v = u + \epsilon v_1$. Then for any $\epsilon > 0$, $v(t) > 0$ on $]t_0, t^*]$ and according to (1.37) and (1.38) we can find $\epsilon > 0$ so that

$$v^{(i-1)}(t^*) \geq 0 \qquad (i = 1, \dots, n), \qquad \mu_n(v; t^*) > 0.$$

Thus, $t^* \in E_n^{**}(t_0; p)$. But $t^* > \tau_n^{**}(t_0; p)$. This contradiction completes the proof in the case under consideration.

Now let $\tau_n^{**}(t_0; p) = +\infty$. Assume $\tau_n(t_0; p) < +\infty$. Then there exists $t_1 \in E_n^{**}(t_0; p)$ such that $t_1 > \tau_n(t_0; p)$. Set

$$p^*(t) = \begin{cases} p(t) & \text{for } t \in [0, t_1] \\ p(t) + 1 & \text{for } t \in \,]t_1, +\infty[. \end{cases}$$

It is easy to show that $t_1 \leq t^0 = \tau_n^{**}(t_0; p^*) < +\infty$. As already proved, $t^0 = \tau_n(t_0; p^*)$. Hence, in view of Lemmas 1.11 and 1.12, the equation

$$u^{(n)} = p^*(t)u$$

has a solution u_0 satisfying the conditions

$$u_0(t_0) = u_0(t^0) = 0, \qquad u_0(t) \geq 0 \qquad \text{for } t \in [t_0, t^0],$$

$$\lambda_n(u_0; [t_0, t^0]) = n.$$

Denote by $t_0 < t_1 < \dots < t_k < t^0$ the zeros of u_0 and by $n_0, n_1, \dots, n_k, n^0$ their respective multiplicities. Clearly, $n_1, \dots, n_k$ are even, and in view of Lemma 1.5 so is n^0. Consider a nontrivial solution v of the equation (1.1) such that

$$v^{(j-1)}(t_i) = 0 \qquad (j = 1, \dots, n_i; \ i = 0, \dots, k),$$

$$v^{(j-1)}(t^0) = 0 \qquad (j = 1, \dots, n^0 - 1).$$

Since $t^0 > \tau_n(t_0; p)$, v changes sign in the interval $[t_0, t^0]$. Assume that $v(t) > 0$ for $t \in \,]t^0 - \delta, t^0[$. Then for sufficiently small $\epsilon > 0$ the inequality (1.36) with $u_\epsilon = u_0 + \epsilon v$ holds. On the other hand, for sufficiently large ϵ this inequality is violated. Let $\epsilon = \epsilon_0$ be the largest constant for which (1.36) remains valid. Then

$$\lambda_n(u_{\epsilon_0}; [t_0, t^0]) > \lambda_n(u_\epsilon; [t_0, t^0]) \qquad \text{for } 0 < \epsilon < \epsilon_0$$

and

$$u_{\epsilon_0}^{(n)}(t) = p(t)u_{\epsilon_0}(t) + (p^*(t) - p(t))u_0(t) \geq 0 \qquad \text{for } t \in [t_0, t^0].$$

Hence, by Remark 1.5, $u_{\epsilon_0}^{(n_0-1)}(t^0) = 0$, i.e. $v^{(n_0-1)}(t^0) = 0$, which leads to the contradiction

$$n \geq \lambda_n(u_{\epsilon_0}; [t_0, t^0]) > \lambda_n(u_\epsilon; [t_0, t^0]) \geq n.$$

So $\tau_n(t_0; p) = +\infty$. $\square$

PROBLEM 1.6. If $c \neq 0$, then

$$\tau_n(t_0; c) = t_0 + \tau_n(0; c) = \frac{1}{\sqrt[n]{|c|}} \tau_n(t_0; \operatorname{sign} c).$$

PROBLEM 1.7. If u is a solution of the equation

$$u^{(n)} = ct^{-n}u, \tag{1.39}$$

$\alpha > 0$, $u_\alpha(t) = u(\alpha t)$, and $v(t) = t^{n-1}u(t^{-1})$, then u_α is a solution of the equation (1.39), while v is a solution of the equation

$$v^{(n)} = (-1)^n ct^{-n}v.$$

PROBLEM 1.8. If $p(t) = ct^{-n}$ for $t \in]0, +\infty[$, then

$$\tau_n(t_0; p) = t_0 \tau_n(1; p) \qquad \text{for } t_0 > 0,$$

$$\tau_n(1; p) = \tau_n(1; -p) \qquad \text{for } n \text{ odd.}$$

Together with the equation (1.1), consider the equation

$$v^{(n)} = q(t)v, \tag{1.40}$$

where $q \in L_{\text{loo}}(\mathbf{R}_+)$.

LEMMA 1.16. If either

$$p(t) \leq q(t) \leq 0 \qquad \text{for } t \in \mathbf{R}_+ \tag{1.41}$$

or

$$p(t) \geq q(t) \geq 0 \qquad \text{for } t \in \mathbf{R}_+, \tag{1.42}$$

then for any $t_0 \in \mathbf{R}_+$ we have

$$\tau_n(t_0; p) \leq \tau_n(t_0; q). \tag{1.43}$$

PROOF. Suppose the inequality (1.41) (the inequality (1.42)) holds. First we assume that $\tau_n(t_0; p) < +\infty$. Then by Lemma 1.12 the equation (1.1) possesses a solution u having n zeros in the interval $[t_0, t^0]$ with $t^0 = \tau_n(t_0; p)$ and satisfying

$$u(t_0) = u(t^0) = 0, \qquad u(t) \geq 0 \qquad \text{for } t \in [t_0, t^0].$$

Let $n_0, n_1, \ldots, n_k, n^0$ be the respective multiplicities of the zeros $t_0 < t_1 < \ldots < t_k < t^0$ of u in the interval $[t_0, t^0]$. It follows from Lemma 1.4 (Lemma 1.5) that $n_0 + n_1 + \ldots + n_k + n^0 = n$ and n^0 is odd (even).

Consider a nontrivial solution v of the equation (1.40) under the conditions

$$v^{(j-1)}(t_i) = 0 \qquad (j = 1, \ldots, n_i; \ i = 0, \ldots, k),$$

$$v^{(j-1)}(t^0) = 0 \qquad (j = 1, \ldots, n^0 - 1).$$

If v changes sign in the interval $[t_0, t^0]$, then, assuming v is positive in $]t^0 - \delta, t^0[$, we easily conclude that

$$u_\epsilon(t) \geq 0 \qquad \text{for } [t_0, t^0], \tag{1.44}$$

where $u_\epsilon = u + \epsilon v$ and $\epsilon > 0$ is sufficiently small. For sufficiently large ϵ this inequality is violated. Let $\epsilon_0 > 0$ be the largest constant for which (1.44) is true. Applying the equality

$$u_{\epsilon_0}^{(n)}(t) = (p(t) - q(t))u(t) + q(t)u_{\epsilon_0}(t),$$

by (1.41) and (1.44) (by (1.42) and (1.44)) we obtain

$$u_{\epsilon_0}^{(n)}(t) \leq 0 \, (\geq 0) \qquad \text{for } t \in [t_0, t^0].$$

Clearly, $u_{\epsilon_0}(t) \not\equiv 0$ and $\lambda_n(u_{\epsilon_0}; [t_0, t^0]) > n - 1$. Hence, according to Remark 1.4 (Remark 1.5), $\lambda_n(u_{\epsilon_0}; [t_0, t^0]) = n$ and $u_{\epsilon_0}^{(n^0-1)}(t^0) = 0$. It is readily verified that this contradicts the definition of ϵ_0, and so v does not change sign in the interval $[t_0, t^0]$.

Thus, if $v(t^0) = 0$, then $\tau_n(t_0; q) \geq t^0 = \tau_n(t_0; p)$, while if $v(t^0) \neq 0$ (which is impossible unless (1.41) holds and $n^0 = 1$) and $\tau_n(t_0; q) < t^0$, then without loss of generality we may assume that $k = 0$, i.e. $v(t) > 0$ for $t \in]t_0, t^0]$. Indeed, otherwise we consider a solution w of the equation (1.40) satisfying the conditions

$$w^{(j-1)}(t_i) = \delta_{ij} \qquad (j = 1, \ldots, n_i; \ i = 1, \ldots, k),$$

$$w^{(j-1)}(t_0) = 0 \qquad (j = 1, \ldots, n_0 + 1)$$

and note that for sufficiently small $\epsilon > 0$ the solution $v + \epsilon w$ satisfies the inequality in question.

On the other hand, by Lemmas 1.4 and 1.12 the equation (1.40) has a solution v_0 such that

$$v_0(t) \geq 0 \qquad \text{for } t \in [t_0, \tau_n(t_0; q)], \qquad v_0(t_0) = v_0(\tau_n(t_0; q)) = 0,$$

$\tau_n(t_0; q)$ being a zero of odd multiplicity. Then for certain $t^* \in]\tau_n(t_0; q), t^0[$ and $c > 0$,

$$v(t^*) + cv_0(t^*) = 0, \qquad v(t) + cv_0(t) > 0 \qquad \text{for } t \in]t_0, t^*[,$$

i.e. $t^* \in E_n(t_0; q)$, which is impossible. So (1.43) holds, irrespective of whether $v(t^0) = 0$ or $v(t^0) \neq 0$. This completes the proof in case $\tau_n(t_0; p) < +\infty$.

Now suppose $\tau_n(t_0; p) = +\infty$ and (1.43) is violated, i.e. $\tau_n(t_0; q) < +\infty$. Then there exist $t^0 > \tau_n(t_0; q)$ and a nontrivial solution u_0 of the equation (1.1) such that

$$u_0(t_0) = u_0(t^0) = 0, \qquad u_0(t) \geq 0 \qquad \text{for } t \in [t_0, t^0]. \tag{1.45}$$

If u_0 has n zeros in the interval $[t_0, t^0]$, the proof can be carried out as above. Let $\lambda_n(u_0; [t_0, t^0]) < n$ and let $n_0, n_1, \ldots, n_k, n^0$ $(n_0 + n_1 + \ldots + n_k + n^0 < n)$ be the respective multiplicities of the zeros $t_0 < t_1 < \ldots < t_k < t^0$ of u_0. Consider a nontrivial solution v_0 of the equation (1.40) satisfying the conditions

$$v_0^{(j-1)}(t_i) = 0 \qquad (j = 1, \ldots, n_i; \ i = 0, \ldots, k),$$

$$v_0^{(j-1)}(t^0) = 0 \qquad (j = 1, \ldots, n^0).$$

Since $t^0 > \tau_n(t_0; q)$, v_0 changes sign in the interval $[t_0, t^0]$.

Denote by ϵ_0 the largest $\epsilon > 0$ for which $u_\epsilon(t) \geq 0$ on $[t_0, t^0]$, where $u_\epsilon = u_0 + \epsilon v_0$. It is readily shown that

$$\lambda_n(u_0; [t_0, t^0]) < \lambda_n(u_{\epsilon_0}; [t_0, t^0]) \leq n.$$

By repeating this procedure, we obtain a function $u = u_0 + v_*$, where u_0 is a solution of the equation (1.1) under the condition (1.45), v_* is a solution of the equation (1.40), and u also satisfies the condition (1.45) and has n zeros in the interval $[t_0, t^0]$. Now, arguing as in the previous case, we arrive at a contradiction. $\square$

PROBLEM 1.9. There exists a number $a > 0$ such that if $|p(t)| \geq c > 0$ for $t \geq t_0$ and if $t_2 > t_1 \geq t_0$ are consecutive zeros of a solution of the equation (1.1), then $\sqrt[n]{|c|}(t_2 - t_1) \leq a$. In particular, if $\lim_{t \to +\infty} |p(t)| = +\infty$ and $(t_k)_{k=1}^{+\infty}$ is the increasing sequence of all zeros of a solution of the equation (1.1), then $\lim_{k \to +\infty}(t_{k+1} - t_k) = 0$.

LEMMA 1.17. Let M_n^* be the largest local maximum of the polynomial (1.9), $c < -M_n^*$, $t_0 > 0$, and $p(t) = ct^{-n}$ for $t \in {]}0, +\infty[$. Then $\tau_n(t_0; p) < +\infty$ and, moreover, any solution of the equation (1.39) has a zero in the interval $[t_0, \tau_n(t_0; p)[$ if n is even, and either has a zero in this interval or satisfies the inequalities

$$(-1)^{i-1}u^{(i-1)}(t_0)u(t_0) > 0 \qquad (i = 1, \ldots, n)$$

if n is odd.

PROOF. Since the equation (1.39) with $c < -M_n^*$ has property A on $[t_0, +\infty[$, in view of Theorem 1.1, proved below, $\tau_n(t_0; p) < +\infty$. Besides, in the case under consideration $\tau_n(t_0; p) = t_0\tau_n(1; p)$ (see Problem 1.8) and so it suffices to show that any solution of the equation (1.39) has a zero in $[1, \tau_n(1; p)[$ for n even, and either has a zero in this interval or satisfies the inequalities

$$(-1)^{i-1}u^{(i-1)}(1)u(1) > 0 \qquad (i = 1, \ldots, n) \tag{1.46}$$

for n odd.

First, let n be even. The function $t^{n-1}u(t^{-1})$ is a solution of the equation (1.39) together with $u(t)$. Hence an arbitrary solution of this equation has infinitely many zeros in any neighborhood of $+\infty$ as well as of zero. So, if $u(t) \neq 0$ for $t \in [1, \tau_n(1; p)[$, there exist $t_0 \in {]}0, 1[$ and $t_1 \in [\tau_n(1; p), +\infty[$ such that

$$u(t_0) = u(t_1) = 0, \qquad u(t) \neq 0 \qquad \text{for } t \in {]}t_0, t_1[.$$

Therefore $u_0(t) = u(t_0 t)$ is a solution of the equation (1.39) satisfying the conditions

$$u_0(1) = u_0(t_1/t_0) = 0, \qquad u_0(t) \neq 0 \qquad \text{for } t \in {]}1, t_1/t_0[.$$

Since $t_1/t_0 > \tau_n(1; p)$, this contradicts the definition of $\tau_n(1; p)$. Hence, for n even, any solution of the equation (1.39) has a zero in $[1, \tau_n(1; p)[$.

Now assume n to be odd. Then $\tau_n(1; p) = \tau_n(1; -p)$ (see Problem 1.8).

Arguing as in the proof of Lemma 1.14, we can show that any solution u of the equation

$$u^{(n)} = -ct^{-n}u \tag{1.47}$$

vanishing at $\tau_n(t_0; -p)$ either has a zero in the interval $[t_0, \tau_n(t_0; -p)[$ or satisfies the inequalities

$$(-1)^{i-1}u^{(i-1)}(t_0)u(t_0) \geq 0 \qquad (i = 1, \dots, n).$$

Let u be a solution of the equation (1.39) which does not vanish in the interval $[1, \tau_n(1; p)]$. Recall that the equation (1.39) has property A in the interval $[1, +\infty[$. Thus, any solution of this equation either has infinitely many zeros in $[1, +\infty[$ or $(-1)^{i-1}v^{(i-1)}(t)v(t) > 0$ for $t > 0$ $(i = 1, \dots, n)$. So, if the solution u has no zero in $[1, +\infty[$, it satisfies the inequalities (1.46). Furthermore, if $u(t_0) = 0$ for a certain $t_0 > \tau_n(1; p) = t_1$, then we consider the solution $u_0(t) = u((t_0/t_1)t)$. For it,

$$u_0(t_1) = 0, \qquad u_0(t) \neq 0 \qquad \text{for } t \in [t_1/t_0, t_1[$$

and

$$(-1)^{i-1}u_0^{(i-1)}(t)u_0(t) > 0 \qquad \text{for } t \in]0, 1] \qquad (i = 1, \dots, n).$$

This yields

$$(-1)^{i-1}u^{(i-1)}(t)u(t) > 0 \qquad \text{for } t \in]0, t_0/t_1] \qquad (i = 1, \dots, n),$$

and since $t_0/t_1 > 1$, (1.46) holds. $\square$

COROLLARY 1.2. For an arbitrary $c < -M_n^*$ there exists $\gamma > 1$ such that if $\alpha, \beta > 0$ and $\beta \geq \alpha\gamma$, then any solution of the equation (1.39) has a zero in the interval $]\alpha, \beta[$ for n even, and either has a zero in $]\alpha, \beta[$ or satisfies the inequalities

$$(-1)^{i-1}u^{(i-1)}(\alpha)u(\alpha) > 0 \qquad (i = 1, \dots, n)$$

for n odd.

1.3. Necessary and sufficient conditions for the equation (1.1) to have properties A and B.

The assertions proved in this subsection express a relation of properties A and B with other asymptotic properties of solutions of the equation (1.1).

THEOREM 1.1. Let the inequality (1.2) hold. Then the following assertions are equivalent:

   (i) the equation (1.1) has property A;
  (ii) for no $l \in \{1, \dots, n-1\}$ such that $l+n$ is odd the equation (1.1) has a solution satisfying the condition (1.7_l);
 (iii) the equation (1.1) has no solution satisfying the condition (1.7_{n-1});
 (iv) every solution of the equation (1.1) vanishing at least once is oscillatory;
  (v) $\tau_n(t_0; p) < +\infty$ for any $t_0 \in \mathbf{R}_+$.

PROOF. (i)$\Rightarrow$(v). Suppose that the equation (1.1) has property A, but $\tau_n(t_0; p) = +\infty$ for a certain $t_0 \in \mathbf{R}_+$. Then there exist sequences of points $(t_k)_{k=1}^{+\infty}$ and of solutions of (1.1) $(u_k)_{k=1}^{+\infty}$ such that

$$t_0 < t_1 < t_2 < \dots, \qquad \lim_{k \to +\infty} t_k = +\infty,$$

$$(1.48)$$

$$u_k(t_0) = u_k(t_k) = 0, \qquad u_k(t) > 0 \qquad \text{for } t \in]t_0, t_k[.$$

Without loss of generality we may assume that $\sum_{i=0}^{n-1}|u_k^{(i)}(t_0)| = 1$ and that the sequence $(u_k)_{k=1}^{+\infty}$ converges uniformly on each finite interval in $\mathbf{R}_+$. Set $u = \lim_{k\to+\infty} u_k$. Clearly, u is a nonoscillatory solution of the equation (1.1). If n is even, this contradicts the definition of property A. If n is odd, for a certain $t_* > t_0$ the inequalities

$$(-1)^i u^{(i)}(t) > 0 \quad \text{for } t \in [t_*, +\infty[\quad (i = 0, \dots, n-1)$$

are fulfilled. So, taking into account (1.2) and the oddness of n, we get (see Problem 1.2)

$$(-1)^i u^{(i)}(t) > 0 \quad \text{for } t \in \mathbf{R}_+ \quad (i = 0, \dots, n-1).$$

But this is impossible, because $u(t_0) = 0$.

(v)$\Rightarrow$(iv). Follows from Lemma 1.13.

(iv)$\Rightarrow$(iii). Suppose that the equation (1.1) has a solution satisfying the condition (1.7_{n-1}). Then we easily obtain for $t \geq t_1$,

$$u(t) \geq u^{(n-2)}(t_1)\frac{(t-t_1)^{n-2}}{(n-2)!} + \frac{1}{(n-2)!}\int_{t_1}^t (t-s)^{n-2}\int_s^{+\infty} p(\xi)u(\xi)\,d\xi\,ds.$$

By Lemma 1.7, this implies the existence of a nonoscillatory solution of the equation (1.1) vanishing at t_1, which contradicts (iv).

(iii)$\Rightarrow$(ii). Let the equation (1.1) have no solution satisfying the condition (1.7_{n-1}). Then Remark 1.1 yields (1.8). Hence, if there exists $l \in \{1, \dots, n-3\}$ such that $l+n$ is odd and the equation (1.1) under the condition (1.7_l) is solvable, then (1.18) holds and from (1.20), applying Lemma 1.3, we obtain

$$u(t) \geq \frac{t^{l-1}}{l!}u^{(l-1)}(t).$$

Thus, (1.17) implies

$$u^{(l-1)}(t) \geq u^{(l-1)}(t_1) - \frac{1}{(n-2)!}\int_{t_2}^t\int_s^{+\infty}(\xi-s)^{n-2}p(\xi)u^{(l-1)}(\xi)\,d\xi\,ds \quad \text{for } t \geq t_2.$$

So, by Lemma 1.7 there exists a solution v of the equation (1.33) satisfying the condition (1.7_1), which is impossible by Corollary 1.1.

(ii)$\Rightarrow$(i). It follows from (ii) and Lemma 1.1 that if n is even, then the equation (1.1) has no nonoscillatory solution and if n is odd, then any nonoscillatory solution of (1.1) satisfies the condition (1.7_0). In view of Remark 1.1 this solution satisfies the condition (1.4) as well. Therefore, (1.1) has property A. $\square$

Theorem 1.1 and Lemma 1.6 imply

COROLLARY 1.3. Let the inequality (1.2) hold. Then either of the following two conditions is necessary and sufficient for the equation (1.1) to have property A:

(i) for an arbitrary $l \in \{1, \dots, n-1\}$ such that $l+n$ is odd the equation (1.1) is $(l, n-l)$ conjugate in any neighborhood of $+\infty$;

(ii) the equation (1.1) is $(n-1, n)$ conjugate in any neighborhood of $+\infty$.

THEOREM 1.2. Let the inequality (1.3) hold. Then the following assertions are equivalent:

    (i) the equation (1.1) has property B;

    (ii) for no $l \in \{1, \ldots, n-2\}$ such that $l+n$ is even the equation (1.1) has a solution satisfying the condition (1.7_l);

    (iii) the equation (1.1) has no solution satisfying the condition (1.7_1) (in case n odd) and the condition (1.7_2) (in case n even);

    (iv) every solution of the equation (1.1) vanishing at least once either is oscillatory or satisfies the condition (1.5);

    (v) $\tau_n(t_0; p) < +\infty$ for any $t_0 \in \mathbf{R}_+$.

PROOF. (i)$\Rightarrow$(v). Let the equation (1.1) have property B, but $\tau_n(t_0; p) = +\infty$ for a certain $t_0 \in \mathbf{R}_+$. Then there exist sequences of points $(t_k)_{k=1}^{+\infty}$ and of solutions $(u_k)_{k=1}^{+\infty}$ of (1.1) satisfying the condition (1.48). Without loss of generality we may assume that $\sum_{i=0}^{n-1} |u_k^{(i)}(t_0)| = 1$ and $\lim_{k \to +\infty} u_k = u$, where u is a nonoscillatory solution of (1.1). Since $u(t_0) = 0$, as in the proof of Theorem 1.1 we conclude that u does not satisfy the condition (1.4). So

$$\lim_{t \to +\infty} u^{(i)}(t) = +\infty \qquad (i = 0, \ldots, n-1).$$

Hence, for a certain $t_* > t_0$ we have

$$u^{(i)}(t_*) \geq 2c > 0 \qquad (i = 0, \ldots, n-1).$$

Consequently, if k_0 is sufficiently large, then $t_{k_0} > t_*$ and

$$u_{k_0}^{(i)}(t_*) \geq 0 \qquad (i = 0, \ldots, n-1).$$

By (1.3) this yields

$$u_{k_0}^{(i)}(t) \geq c \qquad \text{for } t \geq t_* \qquad (i = 0, \ldots, n-1),$$

which contradicts the condition (1.48).

    (v)$\Rightarrow$(iv). Follows from Lemma 1.14 and Remark 1.1.

    (iv)$\Rightarrow$(iii). If n is even and the equation (1.1) under the condition (1.7_2) is solvable, then, as in the proof of Theorem 1.1 ((iv)$\Rightarrow$(iii)), we can establish the existence of a nonoscillatory solution of (1.1) which does not satisfy the condition (1.5), but vanishes at least once.

If n is odd and the equation (1.1) has a solution satisfying (1.7_1):

$$u(t) > 0, \qquad (-1)^{i+1} u^{(i)}(t) > 0 \qquad \text{for } t \geq t_0 \qquad (i = 1, \ldots, n-1),$$

then (1.1) under the condition (1.7_1) has a nonoscillatory solution v such that $v(t_0) = 0$.

Indeed, first note that for any $t^* > t_0$ the solution v_0 of the equation (1.1) under the initial conditions

$$v_0(t^*) = 1, \qquad v_0^{(i)}(t^*) = 0 \qquad (i = 1, \ldots, n-1)$$

satisfies the inequalities

$$v_0(t) > 0, \qquad (-1)^{i+1} v_0^{(i)}(t) > 0 \qquad (i = 1, \ldots, n-1) \tag{1.49}$$

for $t \in [t_0, t^*[$. Assume the contrary. Then by (1.3) there exists $t_* \in [t_0, t^*[$ such that $v_0(t_*) = 0$ and (1.49) holds in the interval $]t_*, t^*[$. Consider the solution $u_\epsilon = u - \epsilon v_0$. Let $\epsilon = \epsilon_0$ be the largest constant for which the inequalities

$$u_\epsilon(t) \geq 0, \qquad (-1)^{i+1} u_\epsilon^{(i)}(t) \geq 0 \qquad \text{for } t \in [t_*, t^*] \qquad (i = 1, \dots, n-1) \qquad (1.50)$$

are fulfilled. Clearly, for $\epsilon = \epsilon_0$ at least one of these inequalities is nonstrict. On the other hand, since

$$u_{\epsilon_0}(t_*) > 0, \qquad (-1)^{i+1} u_{\epsilon_0}^{(i)}(t^*) > 0 \qquad (i = 1, \dots, n-1),$$

$$u_{\epsilon_0}^{(n)}(t) \geq 0 \qquad \text{for } t \in [t_*, t^*],$$

the inequalities (1.50) should be strict for $\epsilon = \epsilon_0$. This contradiction shows that (1.49) is valid in the interval $[t_0, t^*[$.

Now let u_0 be a solution of the equation (1.1) satisfying the conditions

$$u_0(t_0) = 0, \qquad u_0'(t_0) > 0, \qquad u_0^{(i)}(t^*) = 0 \qquad (i = 2, \dots, n-1).$$

By Lemma 1.5 such a solution exists. Moreover, the inequality

$$v_0(t_0) u_0'(t_0) - u_0(t_0) v_0'(t_0) > 0$$

and the same lemma yield

$$v_0(t) u_0'(t) - u_0(t) v_0'(t) > 0 \qquad \text{for } t \in [t_0, t^*], \qquad (1.51)$$

because otherwise for a certain $t_1 \in]t_0, t^*[$ there would exist a nontrivial solution of the equation (1.1) under the boundary conditions

$$u(t_1) = u'(t_1) = 0, \qquad u^{(i)}(t^*) = 0 \qquad (i = 2, \dots, n-1).$$

From (1.51) we obtain $u_0'(t) > 0$ for $t \in [t_0, t^*]$. Thus (1.3) implies

$$u_0(t) > 0, \qquad (-1)^{i+1} u_0^{(i)}(t) \geq 0 \qquad \text{for } t \in]t_0, t^*] \qquad (i = 1, \dots, n-1).$$

Finally, let $(t_k)_{k=1}^{+\infty}$ be an unbounded increasing sequence $(t_k > t_0)$, and let u_k be solutions of the equation (1.1) under the conditions

$$u_k(t_0) = 0, \qquad u_k^{(i)}(t_k) > 0 \qquad (i = 2, \dots, n-1),$$

$$u_k'(t_0) > 0, \qquad \sum_{j=0}^{n-1} |u_k^{(j)}(t_0)| = 1.$$

As already proved, such solutions exist and

$$u_k(t) > 0, \qquad (-1)^{i+1} u_k^{(i)}(t) \geq 0 \qquad \text{for } t \in]t_0, t_k] \qquad (i = 1, \dots, n-1).$$

Assume $\lim_{k \to +\infty} u_k = v$. Then v is a solution of the equation (1.1) and

$$v(t_0) = 0, \qquad v(t) > 0 \qquad \text{for } t > t_0,$$

$$(-1)^{i+1} v^{(i)}(t) \geq 0 \qquad \text{for } t \geq t_0 \qquad (i = 1, \dots, n-1).$$

Hence, v vanishes for $t = t_0$, is nonoscillatory and does not satisfy the condition (1.5), which contradicts (iv).

(iii)$\Rightarrow$(ii). Suppose that the equation (1.1) with n even has no solution satisfying the condition (1.7$_2$), but has a solution satisfying (1.7$_l$) for a certain $l \in \{2, \ldots, n-1\}$. Corollary 1.1 yields $l \in \{4, \ldots, n-4\}$. By Lemma 1.3 (the validity of the condition (1.18) follows from Remark 1.1),

$$u^{(l-2)}(t) \geq u^{(l-2)}(t_2) + \frac{1}{(n-l)!} \int_{t_2}^{t} (t-s) \int_{s}^{+\infty} (\xi - s)^{n-l-1} p(\xi) u(\xi) \, d\xi \, ds,$$

$$u(t) \geq \frac{2}{l!} t^{l-2} u^{(l-2)}(t) \qquad \text{for } t \geq t_2.$$

Since $(n-l-1)! \, l! \leq 2! \, (n-3)!$, these inequalities and Lemma 1.7 imply the existence of a solution of the equation (1.1) under the condition (1.7$_2$), which contradicts (iii).

For n odd the proof can be carried out similarly.

(ii)$\Rightarrow$(i). Follows from Remark 1.3. $\square$

As above we obtain

COROLLARY 1.4. If the inequality (1.3) holds, then either of the following two conditions is necessary and sufficient for the equation (1.1) to have property B:

(i) for an arbitrary $l \in \{1, \ldots, n-1\}$ such that $l+n$ is even the equation (1.1) is $(l, n-l)$ conjugate in any neighborhood of $+\infty$;

(ii) the equation (1.1) is $(1, n-1)$ conjugate (in case n odd) and $(2, n-2)$ conjugate (in case n even) in any neighborhood of $+\infty$.

Theorems 1.1 ((i)$\Rightarrow$(iii)) and 1.2 ((i)$\Rightarrow$(iii)) together with Lemma 1.6 immediately imply

THEOREM 1.3. Let n be odd and let the inequality (1.2) hold. Then the equation (1.1) has property A if and only if the equation (1.34) has property B.

THEOREM 1.3'. Let $n \geq 4$ be even and let the inequality (1.3) hold. Then the equation (1.1) has property B if the equation (1.34) has property A.

PROOF. If the equation (1.34) has property A, then according to Theorem 1.1 ((i)$\Rightarrow$(iii)) it has no solution satisfying (1.7$_1$). Thus, by Lemma 1.8 the equation (1.1) has no solution satisfying (1.7$_2$). Consequently, Theorem 1.2 ((i)$\Rightarrow$(iii)) implies that the equation (1.1) has property B. $\square$

Note that the converse is not true. Indeed, for $n \geq 4$ even we have $M_n^* > M_{*n}$. Hence, if $c \in \,]M_{*n}, M_n^*]$ and $p(t) = ct^{-n}$ for $t \geq 1$, then the equation (1.1) has property B, but the equation (1.34) does not have property A.

PROBLEM 1.10. Let p be nonpositive (nonnegative and $n \geq 3$). If the equation (1.1) has property A (property B), then the solutions of the equation

$$v'' + \frac{1}{(n-2)!} t^{n-2} |p(t)| v = 0$$

are oscillatory.

PROBLEM 1.11. Let p be nonpositive (nonnegative and $n \geq 3$). If the solutions of the equation

$$v'' + \frac{1}{(n-1)!}t^{n-2}|p(t)|v = 0$$

are oscillatory, then the equation (1.1) has property A (property B).

1.4. Integral criteria for the equation (1.1) to have properties A and B.
First of all we state two integral comparison theorems concerning properties A and B.

THEOREM 1.4. Let the inequalities

$$p(t) \leq 0, \qquad q(t) \leq 0 \qquad \text{for } t \in \mathbf{R}_+, \tag{1.52}$$

$$\int_t^{+\infty} s^{n-2}|p(s)|\, ds \geq \int_t^{+\infty} s^{n-2}|q(s)|\, ds \qquad \text{for } t \in \mathbf{R}_+$$

hold, and let the equation (1.40) have property A. Then the equation (1.1) also has this property.

THEOREM 1.5. Let the inequalities

$$p(t) \geq 0, \qquad q(t) \geq 0 \qquad \text{for } t \in \mathbf{R}_+, \tag{1.53}$$

$$\int_t^{+\infty} s^{\mu}p(s)\, ds \geq \int_t^{+\infty} s^{\mu}q(s)\, ds \qquad \text{for } t \in \mathbf{R}_+,$$

where $\mu = n-2$ in case n odd and $\mu = n-3$ in case n even, hold, and let the equation (1.40) have property B. Then the equation (1.1) also has this property.

The proofs of these propositions are based on the following lemmas.

LEMMA 1.18. Let, in any neighborhood of $+\infty$, the function q be distinct from zero on a set of positive measure, and let $l + n$, $l \in \{1, \dots, n-1\}$, be odd. Suppose, in addition, that the inequalities (1.52) and

$$\int_t^{+\infty} s^{l-1}|p(s)|\, ds \geq \int_t^{+\infty} s^{l-1}|q(s)|\, ds \qquad \text{for } t \in \mathbf{R}_+ \tag{1.54}$$

hold and that the equation (1.1) has a solution satisfying the condition (1.7_l). Then the equation (1.40) also has such a solution.

PROOF. In view of Remark 1.1 we may assume that

$$\int_0^{+\infty} t^{n-1}|q(t)|\, dt = +\infty.$$

So (1.54) yields (1.8).

Let u be a solution of the equation (1.1) under the condition (1.7_l). From Lemma 1.3 we obtain

$$tu'(t) \geq (l-1)u(t) \qquad \text{for } t \geq t_2.$$

Hence the function $t^{1-l}u(t)$ does not decrease for $t \geq t_2$. First we will show that (1.54) implies the inequality

$$\int_t^{+\infty} |p(s)|u(s)\, ds \geq \int_t^{+\infty} |q(s)|u(s)\, ds \qquad \text{for } t \geq t_2. \tag{1.55}$$

Indeed, otherwise there would exist $s_2 > s_1 \geq t_2$ such that

$$\int_{s_1}^{s_2} |q(s)||u(s)|\, ds > \int_{s_1}^{+\infty} |p(s)||u(s)|\, ds.$$

Thus, by the second mean value theorem, for a certain $\tau \in [s_1, s_2]$ we obtain

$$s_2^{1-l} u(s_2) \int_{s_2}^{+\infty} s^{l-1} |p(s)|\, ds \leq \int_{s_2}^{+\infty} |p(s)||u(s)|\, ds$$

$$< \int_{s_1}^{s_2} (|q(s)| - |p(s)|) u(s)\, ds = s_2^{1-l} u(s_2) \int_{\tau}^{s_2} s^{l-1} (|q(s)| - |p(s)|)\, ds,$$

i.e.

$$\int_{\tau}^{+\infty} s^{l-1} |p(s)|\, ds < \int_{\tau}^{s_2} s^{l-1} |q(s)|\, ds,$$

which contradicts the inequality (1.54). Hence (1.55) is proved.

In view of Lemma 1.3,

$$u(t) \geq u(t_2) + \frac{1}{(l-1)!\,(n-l-1)!} \int_{t_2}^{t} (t-s)^{l-1} \int_{s}^{+\infty} (\tau - s)^{n-l-1} |p(\tau)||u(\tau)|\, d\tau\, ds$$

for $t \geq t_2$.

So, taking into account (1.55), we get

$$u(t) \geq u(t_2) + \frac{1}{(l-1)!\,(n-l-1)!} \int_{t_2}^{t} (t-s)^{l-1} \int_{s}^{+\infty} (\tau - s)^{n-l-1} |q(\tau)||u(\tau)|\, d\tau\, ds.$$

By Lemma 1.7 this implies the solvability of the equation (1.40) under the condition (1.7_l). $\square$

The proof of the following assertion is quite similar.

LEMMA 1.19. Let, in any neighborhood of $+\infty$, the function q be distinct from zero on a set of positive measure, and let $l + n$, $l \in \{1, \ldots, n-1\}$, be even. Suppose, in addition, that the inequalities (1.53) and (1.54) hold and that the equation (1.1) has a solution satisfying the condition (1.7_l). Then the equation (1.40) also has such a solution.

It is obvious that Theorem 1.4 immediately follows from Theorem 1.1 ((i)$\Rightarrow$(iii)) and Lemma 1.18, while Theorem 1.5 is a consequence of Theorems 1.3 and 1.4 in case n odd and of Theorem 1.2 ((i)$\Rightarrow$(iii)) and Lemma 1.19 in case n even.

COROLLARY 1.5. If the inequalities (1.41) hold and the equation (1.40) has property A, then the equation (1.1) also has this property.

COROLLARY 1.6. If the inequalities (1.42) hold and the equation (1.40) has property B, then the equation (1.1) also has this property.

THEOREM 1.6. Let the inequality (1.2) hold. Then the condition

$$\lim_{t \to +\infty} \sup t \int_{t}^{+\infty} s^{n-2} |p(s)|\, ds \geq M_n^*, \tag{1.56}$$

where M_n^* is the largest local maximum of the polynomial (1.9), is necessary, while either of the conditions

$$\lim_{t\to+\infty} \inf t \int_t^{+\infty} s^{n-2}|p(s)|\,ds > M_n^* \tag{1.57}$$

and

$$\lim_{t\to+\infty} \sup t \int_t^{+\infty} s^{n-2}|p(s)|\,ds > (n-1)! \tag{1.58}$$

is sufficient for the equation (1.1) to have property A.

PROOF. The necessity of (1.56) and the sufficiency of (1.57) follow from Theorem 1.4, because the Euler equation (1.11) for $c < -M_n^*$ has, and for $-M_n^* \leq c < 0$ does not have, property A.

It remains to prove the sufficiency of the condition (1.58). If the equation (1.1) has a solution satisfying the condition (1.7$_l$) with $l \in \{1,\dots,n-1\}$ and $l+n$ odd, then by Lemma 1.3 (see the inequalities (1.14) and (1.15)) we obtain

$$u^{(l-1)}(t) \geq \frac{t}{(n-1)!} \int_t^{+\infty} s^{n-2}|p(s)||u^{(l-1)}(s)|\,ds \qquad \text{for } t \geq t_2.$$

But since $u^{(l-1)}$ is nondecreasing, this contradicts (1.58). Therefore, for any $l \in \{1,\dots,n-1\}$ such that $l+n$ is odd, the equation (1.1) under the condition (1.7$_l$) is not solvable. So Theorem 1.1 ((i)$\Rightarrow$(ii)) implies that the equation (1.1) has property A. $\square$

THEOREM 1.7. Let the inequality (1.3) hold, and let n be odd. Then the condition

$$\lim_{t\to+\infty} \sup t \int_t^{+\infty} s^{n-2}p(s)\,ds \geq M_{*n},$$

where M_{*n} is the largest local maximum of the polynomial (1.10), is necessary, while either of the conditions

$$\lim_{t\to+\infty} \inf t \int_t^{+\infty} s^{n-2}p(s)\,ds > M_{*n}$$

and

$$\lim_{t\to+\infty} \sup t \int_t^{+\infty} s^{n-2}p(s)\,ds > (n-1)!$$

is sufficient for the equation (1.1) to have property B.

PROOF. Since $M_n^* = M_{*n}$ for n odd, this theorem immediately follows from Theorems 1.3 and 1.6. $\square$

THEOREM 1.7′. Let the inequality (1.3) hold, and let $n \geq 4$ be even. Then the condition

$$\lim_{t\to+\infty} \sup t^2 \int_t^{+\infty} s^{n-3}p(s)\,ds \geq \frac{1}{2}M_{*n}$$

is necessary, while either of the conditions

$$\lim_{t\to+\infty} \inf t^2 \int_t^{+\infty} s^{n-3}p(s)\,ds > \frac{1}{2}M_{*n}$$

and

$$\lim_{t \to +\infty} \sup t \int_t^{+\infty} s^{n-2} p(s) \, ds > 2(n-2)!$$

is sufficient for the equation (1.1) to have property B.

The proof is similar to that of Theorem 1.6, but instead of Theorems 1.1 and 1.4 one should apply Theorems 1.2 and 1.5 together with the following fact: the equation (1.11) for $c > M_{*n}$ has, and for $0 < c \leq M_{*n}$ does not have, property B.

REMARK 1.8. In a manner similar to the proof of the sufficiency of the condition (1.58) we can show that if the inequality (1.2) (the inequality (1.3)) holds, $l \in \{1, \ldots, n-1\}$, $l + n$ is odd (even), and

$$\lim_{t \to +\infty} \sup t \int_t^{+\infty} s^{n-2} |p(s)| \, ds > l! \, (n-l)!,$$

then the equation (1.1) has no solution satisfying the condition (1.7_l).

Theorems 1.6, 1.7 and 1.7′ imply

COROLLARY 1.7. If p is nonpositive (nonnegative) and

$$\lim_{t \to +\infty} \inf t^n |p(t)| > M_n^* \qquad (> M_{*n}),$$

then the equation (1.1) has property A (property B).

COROLLARY 1.8. Let the inequality (1.2) (the inequality (1.3)) hold, and let there exist a continuous nondecreasing function $\omega \colon \mathbf{R}_+ \to \,]0, +\infty[$ such that

$$\int_1^{+\infty} \frac{dt}{t\omega(t)} < +\infty, \qquad \int_0^{+\infty} \frac{t^{n-1}}{\omega(t)} |p(t)| \, dt = +\infty. \qquad (1.59)$$

Then the equation (1.1) has property A (property B).
PROOF. First we will show that (1.59) yields

$$\lim_{t \to +\infty} \sup t^{n-1} \int_t^{+\infty} |p(s)| \, ds = +\infty. \qquad (1.60)$$

Indeed, assuming that

$$t^{n-1} \int_t^{+\infty} |p(s)| \, ds \leq c_0 \qquad \text{for } t \geq 0,$$

we obtain

$$\int_1^t \frac{s^{n-1}}{\omega(s)} |p(s)| \, ds = -\int_1^t s^{n-1} \, d \left[\int_s^{+\infty} \frac{|p(\tau)|}{\omega(\tau)} \, d\tau \right]$$

$$\leq \int_1^{+\infty} \frac{|p(\tau)|}{\omega(\tau)} \, d\tau + (n-1) \int_1^t s^{n-2} \int_s^{+\infty} \frac{|p(\tau)|}{\omega(\tau)} \, d\tau \, ds$$

$$\leq \frac{c_0}{\omega(1)} + (n-1)c_0 \int_1^t \frac{ds}{s\omega(s)} \qquad \text{for } t \geq 1,$$

which contradicts (1.59). Consequently, (1.60) is valid. Hence, by Theorem 1.6 (by Theorems 1.7 and 1.7′), the equation (1.1) has property A (property B). □
From Corollary 1.8, setting $\omega(t) = (1+t)^\epsilon$, $\epsilon > 0$, we obtain

COROLLARY 1.9. If the inequality (1.2) (the inequality (1.3)) holds and

$$\int_0^{+\infty} t^{n-1-\epsilon}|p(t)|\,dt = +\infty$$

for a certain $\epsilon \in \,]0, n-1]$, then the equation (1.1) has property A (property B).

PROBLEM 1.12. If the inequality (1.2) holds and

$$\lim_{t\to+\infty} \sup \frac{1}{\ln t} \int_0^t s^{n-1}|p(s)|\,ds > (n-1)!,$$

then the equation (1.1) has property A. Furthermore, if the inequality (1.3) holds, n is odd (even) and

$$\lim_{t\to+\infty} \sup \frac{1}{\ln t} \int_0^t s^{n-1}|p(s)|\,ds > (n-1)! \qquad (> 2(n-2)!),$$

then the equation (1.1) has property B.

PROBLEM 1.13. Let the inequality (1.2) (the inequality (1.3)) hold. Then either of the conditions

$$\int_0^{+\infty} |p(t)|^{1/\gamma}\,dt = +\infty, \qquad \gamma \in \,]1, n[,$$

and

$$\lim_{t\to+\infty} \sup \frac{1}{\ln t} \int_0^t |p(s)|^{1/n}\,ds > \sqrt[n]{(n-1)!}$$

is sufficient for the equation (1.1) to have property A (property B).

PROBLEM 1.14*. Suppose that p is nonpositive (nonnegative) and that M_n^* (M_{*n}) is the largest local maximum of the polynomial (1.9) (the polynomial (1.10)). Is the condition

$$\int_1^{+\infty} t^{n-1}[-p(t) - M_n^* t^{-n}]\,dt = +\infty$$

$$\left(\int_1^{+\infty} t^{n-1}[-p(t) - M_{*n} t^{-n}]\,dt = +\infty \right)$$

sufficient for the equation (1.1) to have property A (property B)?

REMARK 1.9. It is well known that for $n = 2$ the condition

$$\int_0^{+\infty} p(t)\,dt = \lim_{t\to+\infty} \int_0^t p(\tau)\,d\tau = -\infty \tag{1.61}$$

is sufficient for the oscillation of all solutions of the equation (1.1), without the assumption that p is of constant sign. But for higher order equations the analogous statement is not true.

Indeed, let $n = 4m$, $\epsilon > 0$, $\ln((1 - \sqrt{2}/2 + \epsilon)/\epsilon) > 1/(1 + \epsilon)$ and $p(t) = -(1 - \cos t + \epsilon)^{-1}\cos t$. Then

$$\int_0^{\pi/2} \frac{\cos t\,dt}{1 - \cos t + \epsilon} \geq \int_0^{\pi/4} \frac{\cos t\,dt}{1 - \cos t + \epsilon} \geq \int_0^{\pi/4} \frac{\sin t\,dt}{1 - \cos t + \epsilon} = \ln \frac{1 - \sqrt{2}/2 + \epsilon}{\epsilon},$$

$$\left| \int_{\pi/2}^\pi \frac{\cos t\,dt}{1 - \cos t + \epsilon} \right| \leq \frac{1}{1+\epsilon} \left| \int_{\pi/2}^\pi \cos t\,dt \right| \leq \frac{1}{1+\epsilon},$$

and so

$$-\int_0^{2\pi} p(t)\,dt = 2\int_0^{\pi} \frac{\cos t\,dt}{1 - \cos t + \epsilon} \geq 2\left(\ln\frac{1 - \sqrt{2}/2 + \epsilon}{\epsilon} - \frac{1}{1+\epsilon}\right).$$

Hence the condition (1.61) is satisfied. But $u(t) = 1 + \cos t + \epsilon$ is a nonoscillatory solution of the equation (1.1).

PROBLEM 1.15*. Is the condition (1.61) sufficient for the existence of at least one oscillatory solution of the equation (1.1)?

1.5. On general equations. Consider a general linear differential equation

$$u^{(n)} = \sum_{i=1}^{n} p_i(t)u^{(i-1)} \tag{1.62}$$

with $p_i \in L_{\mathrm{loc}}(\mathbf{R}_+)$ $(i = 1, \dots, n)$.

DEFINITION 1.1'. The equation (1.62) has *property A* if every nontrivial solution of this equation for n even is oscillatory and for n odd either is oscillatory or satisfies the condition

$$u(t)u'(t) < 0 \qquad \text{for } t \geq t_0, \tag{1.63}$$

where t_0 is a positive number depending on the solution.

REMARK 1.10. It follows from Lemma 1.1 and Remark 1.3 that for equations of the form (1.1) with nonpositive coefficients p, Definitions 1.1 and 1.1' are equivalent.

Below, in Theorems 1.8–1.14, we assume that

$$p_i = p_{i1} + p_{i2}, \qquad p_{i1}, p_{i2} \in L_{\mathrm{loc}}(\mathbf{R}_+) \qquad (i = 1, \dots, n).$$

THEOREM 1.8. Suppose there exists a locally absolutely continuous function $\omega\colon \mathbf{R}_+ \to\,]0, +\infty[$ such that

$$\lim_{t\to+\infty} \sup |\omega'(t)| < +\infty,$$

$$\omega^{n-i} p_{i2} \in L(\mathbf{R}_+) \qquad (i = 1, \dots, n), \tag{1.64}$$

$$\lim_{t\to+\infty} \sup \omega^{n-i+1}(t)|p_{i1}(t)| < +\infty \qquad (i = 2, \dots, n),$$

$$\lim_{t\to+\infty} \omega^n(t)p_{11}(t) = -\infty. \tag{1.65}$$

Then the equation (1.62) has property A.

THEOREM 1.9. Suppose there exists a locally absolutely continuous function $\omega\colon \mathbf{R}_+ \to\,]0, +\infty[$ such that (1.64) holds together with

$$\lim_{t\to+\infty} \omega'(t) = 0, \tag{1.66}$$

$$\lim_{t\to+\infty} \omega^{n-i+1}(t)p_{i1}(t) = 0 \qquad (i = 2, \dots, n), \tag{1.67}$$

$$\lim_{t\to+\infty} \sup \omega^n(t)p_{11}(t) < 0.$$

Then the equation (1.62) has property A.

THEOREM 1.10. *If there exists a locally absolutely continuous function $\omega\colon \mathbf{R}_+ \to$ $]0, +\infty[$ satisfying the conditions (1.64)–(1.67), then the equation (1.62) has property A.*

This assertion follows from Theorem 1.8 as well as from Theorem 1.9. We will show that the converse is also true. For this purpose we need

LEMMA 1.20. *If $f \in L_{\text{loc}}(\mathbf{R}_+)$ and $\lim_{t \to +\infty} f(t) = +\infty$, then there exists a positive nondecreasing unbounded function $\varphi \in \tilde{C}_{\text{loc}}(\mathbf{R}_+)$ such that*

$$\lim_{t \to +\infty} \frac{f(t)}{\varphi(t)} = +\infty, \qquad \lim_{t \to +\infty} t\varphi'(t) = 0.$$

PROOF. Without loss of generality we may assume that f is nondecreasing and $f(t) > 1$ for $t \geq 0$. Let $(t_k)_{k=0}^{+\infty}$ and $(\tau_k)_{k=1}^{+\infty}$ be sequences of real numbers such that

$$t_0 = 0, \qquad f(\tau_k) \geq f(t_{k-1}) + 1, \qquad t_k = 3\tau_k \qquad (k = 1, 2, \dots).$$

Clearly, $t_{k-1} < \tau_k < t_k$ and $\lim_{k \to +\infty} \tau_k = +\infty$. So there exists a continuous function $\psi_k\colon [\tau_k, t_k] \to \mathbf{R}_+$ satisfying the conditions

$$\psi_k(\tau_k) = \psi_k(t_k) = 0, \qquad t\psi_k(t) \leq 1 \qquad \text{for } t \in [\tau_k, t_k],$$

$$\int_{\tau_k}^{t_k} \psi_k(t)\, dt = 1.$$

Set

$$\varphi_0(t) = \begin{cases} f(0) & \text{for } t \in [0, \tau_1] \\ \int_{\tau_k}^{t} \psi_k(s)\, ds + \varphi_0(t_{k-1}) & \text{for } t \in [\tau_k, t_k[\\ 1 + \varphi_0(t_{k-1}) & \text{for } t \in [t_k, \tau_{k+1}[. \end{cases}$$

Since φ_0 is continuously differentiable and unbounded and, in addition,

$$f(t) \geq \varphi_0(t) > 1, \qquad \varphi_0'(t) \geq 0, \qquad t\varphi_0'(t) \leq 1 \qquad \text{for } t \geq 0,$$

the function $\varphi(t) = \ln \varphi_0(t)$ satisfies the assertion of the lemma. $\square$

COROLLARY 1.10. *If $g \in L(\mathbf{R}_+)$, then there exists a positive nondecreasing unbounded function $\varphi \in \tilde{C}_{\text{loc}}(\mathbf{R}_+)$ such that $g\varphi \in L(\mathbf{R}_+)$ and $\lim_{t \to +\infty} t\varphi'(t) = 0$.*
PROOF. Let $\alpha \in]0, 1[$ and $f(t) = (\int_t^{+\infty} |g(s)|\, ds)^{-\alpha}$. Then the function φ chosen for f by Lemma 1.20 satisfies all the requirements. $\square$

Now we can prove that Theorems 1.8 and 1.9 are consequences of Theorem 1.10. Indeed, let the hypotheses of Theorem 1.8 be satisfied. Suppose that φ is a function chosen for $f = \omega|p_{11}|^{1/n}$ by Lemma 1.20, and $\omega_* = \omega/\varphi$. Then

$$\lim_{t \to +\infty} \omega_*'(t) = 0, \qquad \lim_{t \to +\infty} \omega_*^n(t)p_{11}(t) = -\infty,$$

$$\omega_*^{n-k}p_{k2} \in L(\mathbf{R}_+) \qquad (k = 1, \dots, n), \tag{1.68}$$

$$\lim_{t \to +\infty} \omega_*^{n-k+1}(t)p_{k1}(t) = 0 \qquad (k = 1, \dots, n-1).$$

Hence, in view of Theorem 1.10 the equation (1.62) has property A. Similarly, under the hypotheses of Theorem 1.9 there exists a locally integrable function $\epsilon: \mathbf{R}_+ \to]0, +\infty[$ such that

$$\lim_{t \to +\infty} \epsilon(t) = 0,$$

$$|\omega'(t)| \leq \epsilon(t), \qquad \omega(t)|p_{k1}(t)|^{1/(n-k+1)} \leq \epsilon(t) \quad (k = 1, \ldots, n-1) \qquad \text{for } t \geq 0.$$

Let $\varphi = \min\{\varphi_0, \varphi_1, \ldots, \varphi_n\}$, where φ_0 and φ_k are functions chosen by Lemma 1.20 for $f = 1/\epsilon$ and by Corollary 1.10 for $g = \omega^{n-k}p_{2k}$, respectively. Then

$$\lim_{t \to +\infty} \varphi(t) = +\infty, \qquad \lim_{t \to +\infty} \epsilon(t)\varphi(t) = \lim_{t \to +\infty} t\varphi'(t) = 0,$$

and the function $\omega_* = \omega\varphi$ satisfies the conditions (1.68). Thus, by Theorem 1.10 the equation (1.62) has property A.

Consequently, Theorems 1.8, 1.9 and 1.10 are equivalent.

In order to prove Theorem 1.10 we need

LEMMA 1.21. *For any $\epsilon \in]0, 1[$ there exists $\delta > 0$ such that if $b > a \geq 0$ and*

$$(b-a)^{n-k} \int_a^b |p_k(t)|\, dt \leq \delta \qquad (k = 1, \ldots, n),$$

then an arbitrary $u \in \tilde{C}^{n-1}([a,b])$ has the representation

$$u^{(n)} - \sum_{k=1}^n p_k(t)u^{(k-1)} = D^n(u; a_0, \ldots, a_n), \tag{1.69}$$

where $a_k \in \tilde{C}^{n-k-1}([a,b])$ $(k = 0, \ldots, n-1)$, $a_n \in \tilde{C}([a,b])$,

$$a_0'(t) \leq 0 \qquad \text{for } t \in [a, b], \tag{1.70}$$

$$a_k(t) \geq 1 - \epsilon \qquad \text{for } t \in [a, b] \quad (k = 0, \ldots, n), \tag{1.71}$$

$$D^0(u; a_0) = \frac{u}{a_0},$$

$$D^k(u; a_0, \ldots, a_k) = \frac{1}{a_0}(D^{k-1}(u; a_0, \ldots, a_{k-1}))' \qquad (k = 1, \ldots, n). \tag{1.72}$$

PROOF. For $\epsilon \in]0, 1[$ we choose $\epsilon_1 \in]0, 1[, \epsilon_2 > 0$ and $\delta > 0$ so that

$$(1 - \epsilon_1)^2 \geq (1 + \epsilon_1)^2(1 - \epsilon), \tag{1.73}$$

$$n!\, n\epsilon_2(1 + \epsilon_2)^{n-1} \leq \epsilon_1, \tag{1.74}$$

$$n(1 + \epsilon_2)\delta < \epsilon_2. \tag{1.75}$$

Let $v_1(t) = 1 - \epsilon_2(t - a)/(b - a)$, $v_k(t) = (t - a)^{k-1}/(k-1)!$ $(k = 2, \ldots, n)$, $\gamma_{jk} = k - j$ if $k > j$ and $\gamma_{jk} = 0$ if $k \leq j$. We claim that if the functions $\delta_{jk} \in C([a,b])$ $(j, k = 1, \ldots, n)$ satisfy the inequalities

$$|\delta_{jk}(t)| \leq \epsilon_2(b - a)^{-\gamma_{kj}}(t - a)^{\gamma_{jk}} \qquad \text{for } t \in [a, b],$$

then

$$|\Delta(t) - 1| \leq \epsilon_1 \qquad \text{for } t \in [a, b], \tag{1.76}$$

where Δ is the determinant of order n with entries $\Delta_{jk} = v_k^{(j-1)} + \delta_{jk}$.

The determinant Δ can be represented as the sum of $n+1$ determinants, one of which is the Wronskian of the system of functions $v_1, \ldots, v_n$ (i.e. equals 1) and the others have at least one column consisting of only the functions δ_{jk}. Since for all $j, k \in \{1, \ldots, n\}$ the estimates

$$|\delta_{jk}(t)| \le \epsilon_2 (b-a)^{k-j}, \qquad |\Delta_{jk}(t)| \le (1+\epsilon_2)(b-a)^{k-j}$$

hold in the interval $[a, b]$, we have

$$|\Delta(t) - 1| \le n! \, n\epsilon_2 (1+\epsilon_2)^{n-1} \qquad \text{for } t \in [a, b].$$

So (1.74) yields the inequality (1.76).

Now we will show that if u_k is the solution of the equation (1.62) under the initial condition $u_k^{(j-1)}(a) = v_k^{(j-1)}(a)$ $(j = 1, \ldots, n)$, then for $t \in \,]a, b]$,

$$|u_k^{(j-1)}(t) - v_k^{(j-1)}(t)| < \epsilon_2 (b-a)^{-\gamma_{kj}}(t-a)^{\gamma_{jk}} \qquad (j = 1, \ldots, n). \tag{1.77}$$

Indeed, otherwise there would exist $j_0 \in \{1, \ldots, n\}$ and $t^* \in \,]a, b]$ such that the inequalities (1.77) hold in $\,]a, t^*[$ and

$$|u_k^{(j_0-1)}(t^*) - v_k^{(j_0-1)}(t^*)| = \epsilon_2 (b-a)^{-\gamma_{kj_0}}(t^* - a)^{\gamma_{j_0k}}.$$

On the other hand, if $k \in \{2, \ldots, n\}$, then the equality

$$u_k^{(j-1)}(t) = v_k^{(j-1)}(t) + \int_a^t \frac{(t-s)^{n-j}}{(n-j)!} u_k^{(n)}(s)\, ds$$

implies

$$|u_k^{(j_0-1)}(t^*) - v_k^{(j_0-1)}(t^*)| \le \sum_{i=1}^k \int_a^{t^*} |p_i(s)|\, ds \frac{(t^* - a)^{n-i-j_0+k}}{(n-j_0)!\,(k-i)!}$$

$$+ \sum_{i=1}^n \int_a^{t^*} |p_i(s)|\, ds \frac{\epsilon_2}{(n-j_0)!}(b-a)^{-\gamma_{ki}}(t^* - a)^{n-j_0+\gamma_{ik}}.$$

Hence

$$\epsilon_2 \le \sum_{i=1}^k (b-a)^{n-i} \int_a^b |p_i(s)|\, ds + \sum_{i=1}^n (b-a)^{n-i}\epsilon_2 \int_a^b |p_i(s)|\, ds \le \delta n(1+\epsilon_2),$$

which contradicts (1.75). Similarly, if $k = 1$, then

$$|u_1^{(j_0-)1}(t^*) - v_1^{(j_0-1)}(t^*)| \le (b-a)^{n-j_0} \int_a^b |p_1(s)|\, ds$$

$$+\epsilon_2 (b-a)^{n-j_0-1} \int_a^b |p_2(s)|\, ds + \sum_{i=1}^n \int_a^b |p_i(s)|\, ds\, \epsilon_2 (b-a)^{n-i-j_0-1}.$$

So

$$\epsilon_2 \le \delta(1 + \epsilon_2(n+1)) \le \delta n(1+\epsilon_2),$$

and (1.77) is proved. Consequently,

$$0 < 1 - \epsilon_1 \le w_k(t) \le 1 + \epsilon_1 \qquad \text{for } t \in [a, b] \qquad (k = 1, \ldots, n), \tag{1.78}$$

where w_k is the Wronskian of the system $u_1, \ldots, u_k$. Applying the Frobenius factorization (see e.g. [127]), we conclude that for any $u \in \tilde{C}^{n-1}([a, b])$ the representation

(1.69) with $w_0 = w_{-1} = 1$, $w_{n+1} = w_n$, $a_k = w_{k-1}w_{k+1}/w_k^2$ $(k = 0, \ldots, n)$ is valid. In view of (1.73) and (1.78) the inequalities (1.71) hold. Moreover, $a_0 = w_1 = u_1$, and (1.77) yields

$$\left| u_1'(t) + \frac{\epsilon_2}{b-a} \right| < \frac{\epsilon_2}{b-a} \qquad \text{for } t \in [a,b].$$

Thus, (1.70) is fulfilled. $\square$

PROOF OF THEOREM 1.10. Let $c_0 \in]0,1[$, $c_1 > 0$, $\epsilon \in]0,1[$, and $t_0 \geq 0$ be such that

$$(1+c_0)^{n(n-1)+3} \leq 2, \tag{1.79}$$

$$(1-\epsilon)^{n+1} c_0^n (1-c_0)^{n-1} c_1 \geq (n-1)! \, (1+c_0)^n, \tag{1.80}$$

$$\omega^n(t) p_{11}(t) \leq -c_1 \qquad \text{for } t \geq t_0. \tag{1.81}$$

Choose $\delta > 0$ for ϵ by Lemma 1.21. Moreover, we assume that

$$2\omega^{n-k-1}(t)|p_{k1}(t)| \leq \delta(1-\epsilon)^{n-k+1} \qquad (k = 2, \ldots, n),$$

$$2 \int_{t_0}^{+\infty} \omega^{n-k}(t)|p_{k2}(t)| \, dt \leq \delta(1-\epsilon)^{n-k} \qquad (k = 1, \ldots, n), \tag{1.82}$$

$$|\omega'(t)| \leq \epsilon \qquad \text{for } t \geq t_0.$$

Suppose that u is a nonoscillatory solution of the equation (1.62), $u(t) \neq 0$ for $t \geq t_0$, $a \in [t_0, +\infty[$ and $b = a + \omega(a)$. According to (1.82), for $t \in [a,b]$ we have

$$\omega(t) = \omega(a) + \omega'(\xi)(t-a) \geq (b-a)(1-\epsilon),$$

$$(b-a)^{n-k} \int_a^b |p_k(t)| \, dt \leq \frac{1}{(b-a)(1-\epsilon)^{n-k}} \int_a^b \omega^{n-k+1}(t)|p_{k1}(t)| \, dt$$

$$+\frac{1}{(1-\epsilon)^{n-k}} \int_a^b \omega^{n-k}(t)|p_{k2}(t)| \, dt \leq \delta \qquad (k = 2, \ldots, n),$$

$$(b-a)^{n-1} \int_a^b |p_{12}(t)| \, dt \leq \delta.$$

Therefore, by Lemma 1.21 there exist functions $a_k \in C([a,b])$ $(k = 0, \ldots, n)$ such that

$$a_k(t) \geq 1 - \epsilon \quad (k = 0, \ldots, n-1), \qquad a_n(t) \geq (1-\epsilon)^2 |p_{11}(t)|, \qquad a_0'(t) \leq 0,$$

and

$$x_k'(t) = a_k(t) x_{k+1}(t) \qquad (k = 1, \ldots, n-1),$$

$$\tag{1.83}$$

$$x_n'(t) = -a_n(t) x_1(t)$$

for $t \in [a,b]$, where $x_k = D^{k-1}(u; a_0, \ldots, a_{k-1})$ $(k = 1, \ldots, n)$.

Since $x_1(t) \neq 0$ and $a_k(t) > 0$ $(k = 1, \ldots, n)$ on $[a,b]$, for some $m \in \{0, \ldots, n(n-1)/2\}$, $\tau_j \in [a,b]$ $(j = 0, \ldots, m+1)$, $a = \tau_0 < \tau_1 < \ldots < \tau_{m+1} = b$, and $\nu_{kj} \in \{0,1\}$ we have

$$(-1)^{\nu_{kj}} x_k(t) > 0 \qquad \text{for } t \in]\tau_j, \tau_{j+1}[\qquad (k = 1, \ldots, n; \, j = 0, \ldots, m).$$

Set $m^* = n(n-1)+2$, $\tau_0^* = a$, $\tau_{j+1}^* = \tau_j^* + c_0\omega(\tau_j^*)$ $(j = 0, \dots, m^*)$. By induction with respect to j we can show that $\omega(\tau_j^*) \leq (1+c_0)^j\omega(\tau_0^*)$, $\tau_j^* \leq \tau_0^* + \omega(\tau_0^*)((1+c_0)^j - 1)$ $(j = 0, \dots, m^* + 1)$. Hence (1.79) yields $\tau_{m^*+1}^* \leq a + \omega(a) = b$. According to the inequality $m^* + 1 > 2(m+1)$ this implies the existence of $j_0 \in \{0, \dots, m\}$ and $j_1 \in \{0, \dots, m^* - 1\}$ for which $\tau_{j_0} \leq \tau_{j_1}^* < \tau_{j_1+2} \leq \tau_{j_0+1}$.

Let $\tau_{j_1}^* = \alpha$, $\tau_{j_1+2}^* = \beta$, $\tau_{j_1+1}^* = \gamma$, $\nu_{kj_0} = \nu_k$, $\mu_k = \nu_{k+1}$ $(k = 1, \dots, n - 1)$, $\mu_n + \nu_1 = 1$. Then

$$(-1)^{\nu_k}x_k(t) > 0, \qquad (-1)^{\mu_k}x_k'(t) > 0 \qquad \text{for } t \in]\alpha, \beta[\qquad (k = 1, \dots, n). \qquad (1.84)$$

Moreover, in view of (1.83), $\prod_{k=1}^n x_k'(t)x_k(t) < 0$ for $t \in]\alpha, \beta[$. Thus, the constant $\sum_{k=1}^n (\mu_k + \nu_k)$ is odd and $\mu_k \neq \nu_k$ for an odd number of indices $k \in \{1, \dots, n\}$.

Assume that $\mu_k = \nu_k$ for some k. Then there exist $k_*, k_j \in \{1, \dots, n\}$ $(j = 1, \dots, 2q)$, $0 = k_0 < k_1 < \dots < k_{2q} = n$, such that $l = k_2 - k_1 + \dots + k_{2q} - k_{2q-1}$ is odd and

$$\nu_{n-k_*+j} = \mu_{n-k_*+j} \qquad (j = j_1, \dots, j_l),$$
$$\nu_{n-k_*+j} \neq \mu_{n-k_*+j} \qquad (j = j_{l+1}, \dots, j_n),$$

where $\nu_{n+j} = \nu_j$, $\mu_{n+j} = \mu_j$, $(j_1, \dots, j_l) = (k_1 + 1, \dots, k_2, \dots, k_{2q-1} + 1, \dots, k_{2q})$, $(j_{l+1}, \dots, j_n) = (1, \dots, k_1, \dots, k_{2q-2} + 1, \dots, k_{2q-1})$.

Put

$$y_j = (-1)^{\nu_{n-k_*+j}}x_{n-k_*+j} \qquad (j = 1, \dots, n; \; x_{n+j} = x_j),$$
$$b_j = (-1)^{\nu_{n-k_*+j} + \mu_{n-k_*+1}}a_{n-k_*+j} \qquad (j = 1, \dots, n; \; a_{n+j} = a_j).$$

Then

$$y_j'(t) = b_j(t)y_{j+1}(t) \qquad (j = 1, \dots, n; \; y_{n+1} = y_1), \qquad (1.85)$$
$$b_j(t) < 0 \quad (j = j_1, \dots, j_l), \qquad b_j(t) > 0 \quad (j = j_{l+1}, \dots, j_n), \qquad (1.86)$$
$$|b_j(t)| \geq 1 - \epsilon \qquad (j = 1, \dots, n; \; j \neq k_*), \qquad (1.87)$$
$$|b_{k_*}(t)| \geq (1 - \epsilon)^2|p_{11}(t)| \qquad \text{for } t \in [\alpha, \beta], \qquad (1.88)$$
$$y_j(t) > 0 \qquad \text{for } t \in]\alpha, \beta[\qquad (j = 1, \dots, n). \qquad (1.89)$$

The relations (1.85) yield

$$y_i(t) = \sum_{j=i}^k (-1)^{j-i}y_j(s)I^{j-i}(s, t; b_{j-1}, \dots, b_i)$$

$$+(-1)^{k-i-1}\int_t^s I^{k-i}(\xi, t; b_{k-1}, \dots, b_i)y_k'(\xi)\,d\xi$$

$$(i = 1, \dots, k; \; k = 1, \dots, n), \qquad (1.90)$$

where $I^0 = 1$,

$$I^j(t, s; b_1, \dots, b_j) = \int_t^s b_1(\xi)I^{j-1}(\xi, s; b_2, \dots, b_j)\,d\xi \qquad (j = 1, \dots, n).$$

Applying (1.86), (1.89) and the equality

$$I^j(t, s; b_1, \dots, b_j) = (-1)^j I^j(s, t; b_j, \dots, b_1)$$

to (1.90) with $s = \alpha$, $i = k_{2i-2} + 1$, $k = k_{2i-1}$, we obtain

$$y_{k_{2i-2}+1}(t) \geq \int_{\alpha}^{t} I^{k_{2i-1}-k_{2i-2}-1}(t, \xi; b_{k_{2i-2}+1}, \ldots, b_{k_{2i-1}-1}) b_{k_{2i-1}}(\xi) y_{k_{2i-1}+1}(\xi)\, d\xi$$

$$> I^{k_{2i-1}-k_{2i-2}}(t, \alpha; b_{k_{2i-2}+1}, \ldots, b_{k_{2i-1}}) y_{k_{2i-1}+1}(t)$$

$$\text{for } t \in \,]\alpha, \beta] \qquad (i = 1, \ldots, q). \tag{1.91}$$

Similarly, from (1.90) with $s = \beta$, $i = k_{2i-1} + 1$ and $k = k_{2i}$, we derive

$$y_{k_{2i-1}+1}(t)$$

$$\geq (-1)^{k_{2i}-k_{2i-1}} \int_{t}^{\beta} I^{k_{2i}-k_{2i-1}-1}(\xi, t; b_{k_{2i-1}}, \ldots, b_{k_{2i-1}+1}) b_{k_{2i}}(\xi) y_{k_{2i}+1}(\xi)\, d\xi$$

$$> I^{k_{2i}-k_{2i-1}}(\beta, t; |b_{k_{2i}}|, \ldots, |b_{k_{2i-1}+1}|) y_{k_{2i}+1}(t)$$

$$\text{for } t \in [\alpha, \beta[\qquad (i = 1, \ldots, q;\ y_{n+1} = y_1). \tag{1.92}$$

According to (1.91) and (1.92),

$$1 > I^{n-l}(t, \alpha; b_{j_{l+1}}, \ldots, b_{j_n}) I^l(\beta, t; |b_{j_l}|, \ldots, |b_{j_1}|) \qquad \text{for } t \in [\alpha, \beta]. \tag{1.93}$$

If $k_* \in \{j_{l+1}, \ldots, j_n\}$, then, taking into account (1.81), (1.87), (1.88) and the relations

$$\gamma = \alpha + c_0 \omega(\alpha), \qquad \beta = \gamma + c_0 \omega(\gamma),$$

$$(1 - c_0)\omega(\alpha) \leq \omega(\gamma) \leq \omega(\alpha)(1 + c_0),$$

$$\omega(t) \leq \omega(\alpha)(1 + c_0) \qquad \text{for } t \in [\alpha, \gamma],$$

we obtain

$$I^l(\beta, \gamma; |b_{j_l}|, \ldots, |b_{j_1}|) \geq \frac{(1 - \epsilon)^l}{l!} c_0^l (1 - c_0)^l [\omega(\alpha)]^l,$$

$$I^{n-l}(\gamma, \alpha; b_{j_{l+1}}, \ldots, b_{j_n}) \geq \frac{c_1}{(n-l)!}(1 - \epsilon)^{n-l+1} c_0^{n-l}(1 + c_0)^{-n}[\omega(\alpha)]^{-l}.$$

This, together with (1.93), implies

$$l!\,(n-l)!\,(1 + c_0)^n > c_1(1 - \epsilon)^{n+1} c_0^n (1 - c_0)^l,$$

which contradicts (1.80).

Furthermore, if $k_* \in \{j_1, \ldots, j_l\}$, we similarly get

$$I^{n-l}(\gamma, \alpha; b_{j_{l+1}}, \ldots, b_{j_n}) \geq \frac{(1 - \epsilon)^{n-l} c_0^{n-l}}{(n-l)!\,(1 + c_0)^{n-l}}[\omega(\gamma)]^{n-l},$$

$$I^l(\beta, \gamma; |b_{j_l}|, \ldots, |b_{j_1}|) \geq \frac{c_1(1 - \epsilon)^{l+1} c_0^l}{l!\,(1 + c_0)^n}[\omega(\gamma)]^{l-n},$$

and, thus,

$$l!\,(n-l)!\,(1 + c_0)^{n-l} > c_1(1 - \epsilon)^{n+1} c_0^n,$$

which also contradicts (1.80). So $\mu_k \neq \nu_k$ $(k = 1, \ldots, n)$.

On the other hand, we have established above that the number of indices $k \in \{1, \ldots, n\}$ for which $\mu_k \neq \nu_k$ is odd. Hence, in case n even the equation (1.62) does not have a nonoscillatory solution. In case n odd the inequalities

$$(-1)^{\nu_k} x_k(t) > 0, \qquad (-1)^{\nu_k} x_k'(t) < 0 \qquad (k = 1, \ldots, n)$$

hold in the interval $]\alpha, \beta[$ (see (1.84)). By (1.83) we easily conclude that these inequalities remain true in the interval $[a, \beta[$. Therefore, $u(t)u'(t) < 0$ for $t \in [a, \beta[$ and, since $a \geq t_0$ is arbitrary, we get (1.63). $\square$

For $\omega = |p_{11}|^{-1/n}$ Theorem 1.9 yields

THEOREM 1.11. Let $p_{11} \in \tilde{C}_{\mathrm{loc}}(\mathbf{R}_+)$ be negative,

$$\lim_{t \to +\infty} |p_{11}(t)|^{-1-1/n} p'_{11}(t) = 0,$$

$$p_{i2}|p_{11}|^{-(n-i)/n} \in L(\mathbf{R}_+) \qquad (i = 1, \dots, n),$$
$$\lim_{t \to +\infty} p_{i1}(t)|p_{11}(t)|^{-(n-i+1)/n} = 0 \qquad (i = 2, \dots, n).$$

Then the equation (1.62) has property A.

Setting $\omega(t) = t^{\sigma/n}$ in Theorems 1.8 and 1.9, we obtain the following assertions.

THEOREM 1.12. Let, for a certain $\sigma \leq n$,

$$\int_0^{+\infty} t^{(n-i)\sigma/n} |p_{i2}(t)| \, dt < +\infty \qquad (i = 1, \dots, n), \tag{1.94}$$

$$\lim_{t \to +\infty} \sup t^{(n-i+1)\sigma/n} |p_{i1}(t)| < +\infty \qquad (i = 2, \dots, n),$$

$$\lim_{t \to +\infty} t^\sigma p_{11}(t) = -\infty. \tag{1.95}$$

Then the equation (1.62) has property A.

THEOREM 1.13. Let, for a certain $\sigma < n$, the condition (1.94) hold,

$$\lim_{t \to +\infty} t^{(n-i+1)\sigma/n} p_{i1}(t) = 0 \qquad (i = 2, \dots, n),$$

and

$$\lim_{t \to +\infty} \sup t^\sigma p_{11}(t) < 0. \tag{1.96}$$

Then the equation (1.62) has property A.

Note that we cannot replace (1.95) by (1.96) in Theorem 1.12. Indeed, consider the equation

$$u'' = p_1(t)u' + p(t)u \qquad (t \in [1, +\infty[), \tag{1.97}$$

where $p_1(t) = c_1 t^{-\sigma/2}$, $p(t) = ct^{-\sigma}$, and

$$0 < -c < \frac{c_1^2}{4} \quad \text{if } \sigma < 2, \qquad 0 < -c < \frac{c_1^2 - 2c_1}{4} \quad \text{if } \sigma = 2. \tag{1.98}$$

Setting

$$u = v \exp\left(\frac{1}{2} \int_1^t p_1(s) \, ds\right),$$

we transform (1.97) into the equation

$$v'' = q(t)v, \tag{1.99}$$

with

$$q(t) = t^{-\sigma}\left(c + \frac{c_1^2}{4} + \frac{c_1\sigma}{4} t^{\sigma/2-1}\right).$$

In view of (1.98) the equation (1.99) and, hence, the equation (1.97) are nonoscillatory. So they do not have property A.

Corollary 6.4 implies that in Theorems 1.12 and 1.13 the estimates of the parameter σ are best possible. Moreover, for the estimate of Theorem 1.13 the strict inequality cannot be replaced by a nonstrict one. However, the following theorem is true.

THEOREM 1.14. Let

$$\lim_{t \to +\infty} t^{n-i+1} p_{i1}(t) = 0 \qquad (i = 2, \ldots, n),$$

$$\int_0^{+\infty} t^{n-i} |p_{i2}(t)| \, dt < +\infty \qquad (i = 1, \ldots, n),$$

$$\lim_{t \to +\infty} \sup t^n p_{11}(t) < -M_n^*,$$

where M_n^* is the largest local maximum of the polynomial (1.9). Then the equation (1.62) has property A.

Before starting the proof of this theorem, we have to justify the following lemma.

LEMMA 1.22. Let $p, q, a_k, b_k \in L([\alpha, \beta])$ $(k = 0, \ldots, n)$,

$$a_k(t) \geq b_k(t) > 0 \qquad \text{for } t \in [\alpha, \beta] \qquad (k = 1, \ldots, n-1),$$

$$\tag{1.100}$$

$$a_0(t) a_n(t) p(t) \leq b_0(t) b_n(t) q(t) \leq 0 \qquad \text{for } t \in [\alpha, \beta],$$

$\nu_k \in \{0, 1\}$ $(k = 1, \ldots, n)$, and let the equation

$$D^n(u; a_0, a_1, \ldots, a_n) = p(t) u$$

have a solution u such that

$$(-1)^{\nu_k} D^{k-1}(u; a_0, \ldots, a_{k-1})(t) > 0 \qquad \text{for } t \in]\alpha, \beta[\qquad (k = 1, \ldots, n).$$

Then the equation

$$D^n(v; b_0, \ldots, b_n) = q(t) v \tag{1.101}$$

has a solution v such that

$$(-1)^{\nu_k} D^{k-1}(v; b_0, \ldots, b_{k-1})(t) > 0 \qquad \text{for } t \in]\alpha, \beta[\qquad (k = 1, \ldots, n). \tag{1.102}$$

PROOF. Suppose $\alpha < \alpha_j < \beta_j < \beta$, $\lim_{j \to +\infty} \alpha_j = \alpha$ and $\lim_{j \to +\infty} \beta_j = \beta$. Fix a positive integer j and set $\gamma_k = \alpha_j$ if $\nu_k = \nu_{k+1}$ and $\gamma_k = \beta_j$ if $\nu_k \neq \nu_{k+1}$ (here $\nu_{n+1} = 1 - \nu_1$). Then

$$|D^{k-1}(u; a_0, \ldots, a_{k-1})(t)| = |D^{k-1}(u; a_0, \ldots, a_{k-1})(\gamma_k)|$$

$$+ \left| \int_{\gamma_k}^t a_k(s) D^k(u; a_0, \ldots, a_k)(s) \, ds \right| \qquad \text{for } t \in [\alpha_j, \beta_j].$$

Therefore

$$|u(t)| \geq \frac{|u(\gamma_1)|}{a_0(\gamma_1)} a_0(t)$$

$$+ a_0(t) \left| \int_{\gamma_1}^t a_1(s_1) \int_{\gamma_2}^{s_1} \cdots \int_{\gamma_n}^{s_{n-1}} a_n(s_n) p(s_n) u(s_n) \, ds_n \ldots ds_1 \right| \qquad \text{for } t \in [\alpha_j, \beta_j].$$

By (1.100) this yields

$$|u(t)| \geq \frac{|u(\gamma_1)|}{a_0(\gamma_1)} a_0(t)$$

$$+ a_0(t) \left| \int_{\gamma_1}^t b_1(s_1) \int_{\gamma_2}^{s_1} \cdots \int_{\gamma_n}^{s_{n-1}} \frac{b_0(s_n) b_n(s_n)}{a_0(s_n)} q(s_n) u(s_n) \, ds_n \ldots ds_1 \right| \qquad \text{for } t \in [\alpha_j, \beta_j].$$

From the last estimate, applying the method of successive approximations, we easily conclude that there exists a continuous function $v_j : [\alpha_j, b_j] \to \mathbf{R}$ such that

$$\frac{|u(\gamma_1)|}{a_0(\gamma_1)} b_0(t) \leq (-1)^{\nu_1} v_j(t) \leq \frac{b_0(t)|u(t)|}{a_0(t)} \qquad \text{for } t \in [\alpha_j, \beta_j]$$

and

$$v_j(t) = \frac{|u(\gamma_1)|}{a_0(\gamma_1)} b_0(t)$$

$$+ (-1)^{\nu_1} b_0(t) \left| \int_{\gamma_1}^t b_1(s_1) \int_{\gamma_2}^{s_1} \cdots \int_{\gamma_n}^{s_{n-1}} b_n(s_n) q(s_n) v_j(s_n) \, ds_n \ldots ds_1 \right| \qquad \text{for } t \in [\alpha_j, \beta_j].$$

It is easily verified that v_j is a solution of the equation (1.101) satisfying the condition

$$(-1)^{\nu_k} D^{k-1}(v_j; b_0, \ldots, b_{k-1})(t) > 0 \qquad \text{for } t \in [\alpha_j, \beta_j] \qquad (k = 1, \ldots, n). \qquad (1.103)$$

Extend v_j to the whole of $[\alpha, \beta]$ as a solution of (1.101). Then we may assume that

$$\sum_{k=1}^n |D^{k-1}(v_j; b_0, \ldots, b_{k-1})(\alpha)| = 1$$

and that the sequence $(v_j)_{j=1}^{+\infty}$ converges uniformly on $[\alpha, \beta]$. Hence $v = \lim_{j \to +\infty} v_j$ is a nontrivial solution of the equation (1.101), and by (1.103),

$$(-1)^{\nu_k} D^{k-1}(v; b_0, \ldots, b_{k-1})(t) \geq 0 \qquad \text{for } t \in [\alpha, \beta] \qquad (k = 1, \ldots, n). \qquad (1.104)$$

Let $D^{k_0-1}(v; b_0, \ldots, b_{k_0-1})(t_*) = 0$ for some $k_0 \in \{1, \ldots, n\}$ and $t_* \in \,]\alpha, \beta[$. In view of (1.104) we have

$$D^{k-1}(v; b_0, \ldots, b_{k-1})(t) \equiv 0 \qquad \text{on } [\alpha, t_*] \qquad (k = 1, \ldots, n)$$

if $\nu_{k_0} = \nu_{k_0+1}$, and

$$D^{k-1}(v; b_0, \ldots, b_{k-1})(t) \equiv 0 \qquad \text{on } [t_*, \beta] \qquad (k = 1, \ldots, n)$$

if $\nu_{k_0} \neq \nu_{k_0+1}$. This is impossible, and so v satisfies (1.102). $\square$

PROOF OF THEOREM 1.14. Let u be a nonoscillatory solution of the equation (1.62), and let $\epsilon \in \,]0, 1[$, $c < -M_n^*$ and $t_0 > 0$ be such that

$$u(t) \neq 0 \qquad \text{for } t \geq t_0$$

and

$$(1 - \epsilon)^{n+1} p_{11}(t) \leq c t^{-n} \qquad \text{for } t \geq t_0. \qquad (1.105)$$

Choose $\gamma > 1$ for c by Corollary 1.2 and $\delta > 0$ for ϵ by Lemma 1.21. Furthermore, assume t_0 to be so large that

$$2(\gamma^{n(n-1)/2+1} - 1)^{n-k+1} t^{n-k+1} |p_{k1}(t)| \leq \delta \qquad \text{for } t \geq t_0 \qquad (k = 2, \ldots, n)$$

and

$$2(\gamma^{n(n-1)/2+1} - 1)^{n-k} \int_{t_0}^{+\infty} t^{n-k}|p_{k2}(t)|\, dt \le \delta \qquad (k = 1, \dots, n).$$

Now suppose that $a \ge t_0$ is an arbitrary constant and $b = \gamma^{n(n-1)/2+1}a$. Then

$$(b-a)^{n-k} \int_a^b |p_{k1}(t)|\, dt \le \frac{1}{(b-a)}(\gamma^{n(n-1)/2+1} - 1)^{n-k} \int_a^b t^{n-k}|p_{k1}(t)|\, dt \le \frac{\delta}{2},$$

$$(b-a)^{n-k} \int_a^b |p_{k2}(t)|\, dt \le (\gamma^{n(n-1)/2+1} - 1)^{n-k} \int_a^b t^{n-k}|p_{k2}(t)|\, dt \le \frac{\delta}{2},$$

and via Lemma 1.21 we can write the equation (1.62) on $[a, b]$ as

$$D^n(u; a_0, \dots, a_n) = p_{11}(t)u,$$

where $a_k \in \tilde{C}^{n-k-1}([a,b])$ $(k = 0, \dots, n-1)$, $a_n \in \tilde{C}([a,b])$ and

$$a_0'(t) \le 0, \qquad a_k(t) \ge 1 - \epsilon \qquad \text{for } t \in [a,b] \qquad (k = 1, \dots, n). \tag{1.106}$$

The solution u and, thus, the function $D^n(u; a_0, \dots, a_n)$ have no zeros in the interval $[a, b]$. Therefore, $D^{n-k}(u; a_0, \dots, a_{n-k})$ has at most k zeros $(k = 1, \dots, n-1)$. If m is the number of distinct zeros of all $D^{n-k}(u; a_0, \dots, a_{n-k})$ $(k = 1, \dots, n)$, then $m \le n(n-1)/2$.

Set $\tau_0 = a$, $\tau_j = \tau_{j-1}\gamma$ $(j = 1, \dots, m+1)$. Clearly, $\tau_{m+1} = \tau_0\gamma^{m+1} \le b$. The interval $[a, b]$ contains $m + 1$ intervals $]\tau_{j-1}, \tau_j[$ altogether, and hence for at least one of these intervals (denote it by $]\alpha, \beta[$) we obtain

$$D^{k-1}(u; a_0, \dots, a_{k-1})(t) \ne 0 \qquad \text{for } t \in]\alpha, \beta[\qquad (k = 1, \dots, n).$$

Moreover, since $D^n(u; a_0, \dots, a_n)(t)u(t) < 0$ on $[a, b]$, for an odd number of indices $k \in \{1, \dots, n\}$ we get $D^{k-1}(u; a_0, \dots, a_{k-1})(t)D^k(u; a_0, \dots, a_k)(t) < 0$ in $]\alpha, \beta[$.

Let $D^{k_0-1}(u; a_0, \dots, a_{k_0-1})(t)D^{k_0}(u; a_0, a_1, \dots, a_{k_0})(t) > 0$ in $]\alpha, \beta[$ for at least one $k_0 \in \{1, \dots, n\}$. Then by the inequalities (1.105), (1.106) and Lemma 1.22, the equation (1.39) has a solution satisfying the conditions

$$v(t) \ne 0, \qquad v^{(k_0-1)}(t)v^{(k_0)}(t) > 0 \qquad \text{for } t \in]\alpha, \beta[.$$

As $\beta = \gamma\alpha$, this contradicts Corollary 1.2. So,

$$(-1)^{k-1}D^{k-1}(u; a_0, a_1, \dots, a_{k-1})(t)u(t) > 0 \qquad \text{for } t \in]\alpha, \beta[\qquad (k = 1, \dots, n).$$

On the other hand, these inequalities should hold for an odd number of indices $k \in \{1, \dots, n\}$. Hence, the proof is completed if n is even. If n is odd, the last inequalities yield

$$(-1)^{k-1}D^{k-1}(u; a_0, a_1, \dots, a_{k-1})(a)u(a) > 0 \qquad (k = 1, \dots, n).$$

But $a_0'(t) \le 0$ for $t \in [a, b]$. Therefore $u'(a)u(a) < 0$, which, since $a \ge t_0$ is arbitrary, implies the validity of the theorem for odd n. $\square$

PROBLEM 1.16. If the equation $u^{(n)} = \sum_{i=2}^n p_i(t)u^{(i-1)}$ is disconjugate in $[t_0, +\infty[$ and

$$(-1)^n p_1(t) \ge 0 \qquad \text{for } t \ge t_0, \tag{1.107}$$

then the equation (1.62) has a solution satisfying the inequalities

$$u(t) > 0, \qquad u'(t) \leq 0 \qquad \text{for } t \geq t_0. \tag{1.108}$$

Let n be odd. As the proofs of Theorems 1.10 and 1.14 imply, under the hypotheses of Theorems 1.8–1.14 we may assume that t_0 appearing in Definition 1.1' is independent of the solution. Therefore, under the hypotheses of these theorems the equation (1.62) has oscillatory solutions.

PROBLEM 1.17*. Does the equation (1.62) with n odd have a solution of the type (1.63) under the hypotheses of Theorems 1.8–1.14?

PROBLEM 1.18. Let the inequalities

$$\eta p_1(t) \leq p_k(t) \leq 0 \qquad (k = 2, \dots, n-2),$$

$$-\eta \leq p_{n-1}(t) \leq 0, \qquad \eta p_1(t) \leq p_n(t) \leq \delta p_1(t),$$

where δ and η are positive constants, hold on $[t_0, +\infty[$,

$$\int_{t_0}^{+\infty} p_n(t)\, dt = -\infty,$$

and let u be a solution of the equation (1.62). Then either $u^{(n-2)}$ has a sequence of zeros converging to $+\infty$ or $u^{(i)}(t) \to 0$ as $t \to +\infty$ $(i = 0, \dots, n-1)$.

Notes. The investigation of oscillatory properties of solutions of second order linear differential equations was begun by C. Sturm [341]. Quite detailed surveys of the results obtained in this field are contained in [305, 370]. Oscillation criteria for solutions of higher order linear differential equations were first established by A. Kneser [208]. Subsequently they were generalized and improved in [1, 67, 68, 74, 113, 172, 176, 188, 210–213, 269, 364].

Properties A and B were introduced in [213] and [185], respectively. For Problem 1.5 see [176, 213].

For the class of functions defined by the inequalities (1.7_l) see [102, 172, 238].

Lemma 1.4 for $i_j = j$ $(j = 1, \dots, n_0)$ was proved in [270] and in the general case in [74].

For $(n/2, n/2)$ disconjugacy of even order equations see [93, 118, 236, 261, 283, 312]. We also refer to [11, 258], where $(l, n - l)$ disconjugacy was studied on finite intervals. Lemma 1.6 with $p(t) \neq 0$ for $t \geq 0$ was proved in [101] and in the general case in [82].

Conjugate points of general equations were investigated in [99, 160, 257, 322]. The modifications $\tau_n^*(t_0; p)$, $\tau_n^{**}(t_0; p)$ and $\tau_n(t_0; p)$ of the conjugate point were introduced in [213], [57] and [74], respectively. For the case of nonpositive p and $n \in \{3, 4\}$ see [270].

For Theorems 1.1, 1.2 and Corollaries 1.3, 1.4 see [74, 82].

A comparison theorem concerning property A (Corollary 1.5) was first proved by V.A. Kondratyev [211, 213]; for Corollary 1.6 see [57, 61].

The integral comparison theorems (Theorems 1.4 and 1.5) were established in [74, 78]. They extend to higher order equations the comparison theorems of E. Hille [139] for second order equations. For the case of $n \in \{3, 4\}$ see also [104, 105].

Corollary 1.7, which generalizes the results of A. Kneser [208] and W.B. Fite [113], is due to V.A. Kondratyev [213]. Corollary 1.8 was proved by W.B. Fite [113] for $\omega(t) = (1 + t)^{n-1}$, by J.G. Mikusiński [269] for $\omega(t) = (1 + t)^{\epsilon}$ and even n, and by I.T. Kiguradze [172, 176] in the form given here. Theorems 1.6, 1.7 and 1.7' were established by T.A. Chanturia [67, 74, 78].

The propositions in subsection 1.5 are due to T.A. Chanturia [73, 84]. Theorem 1.13 with even n and $\sigma = 0$ implies the result in [358] on the oscillation of all solutions of the equation (1.62). Theorem 1.14 for $p_i = 0$ $(i = 2, \dots, n)$, $p_{12} = 0$ was proved in [213].

For Problem 1.18 see [202].

§2. Oscillatory and Nonoscillatory Equations

As already mentioned (see Problem 1.5), if p is nonpositive (nonnegative), $n \geq 3$ and the equation (1.1) has property A (property B), then this equation possesses an oscillatory solution. In general, the converse is not true. There exist functions satisfying the inequality (1.2) (the inequality (1.3)) for which the equation (1.1) possesses an oscillatory solution, but does not have property A (property B). Indeed, let $n \in \{5, 6, \dots\}$ (either $n = 5$ or $n \in \{7, 8, \dots\}$). Then $m_n^* < M_n^*$ $(m_{*n} < M_{*n})$, where m_n^* (m_{*n}) is the smallest and M_n^* (M_{*n}) is the largest local maximum of the polynomial (1.9) (the polynomial (1.10)). Now if $c \in \ ] - M_n^*, -m_n^*[$ $(c \in \ ]m_{*n}, M_{*n}[)$, then the equation $P_n^*(x) + c = 0$ $(P_{*n}(x) - c = 0)$ has real roots belonging to the interval $]0, n - 1[$, as well as complex roots. Hence in the case under consideration, (1.39) does not have property A (property B), but possesses an oscillatory solution. So, for higher order equations the problem on whether the equation has property A or B does not coincide with the problem on the existence of an oscillatory solution.

DEFINITION 2.1. The equation (1.1) is said to be *nonoscillatory* if its all solutions are nonoscillatory, and *oscillatory* if at least one solution is oscillatory.

REMARK 2.1. Suppose that p is nonnegative (nonpositive) and that the equation (1.1) does not have property A (property B), but possesses an oscillatory solution. Then, by Theorem 1.1 (Theorem 1.2), $\tau_n(t_0; p) = +\infty$ for a certain $t_0 \in \mathbf{R}_+$. Thus there exist sequences of points $(t_k)_{k=1}^{+\infty}$ and of solutions $(u_k)_{k=1}^{+\infty}$ such that

$$t_0 < t_k, \qquad \lim_{k \to +\infty} t_k = +\infty,$$

$$u_k(t_0) = u_k(t_k) = 0, \qquad u_k(t) > 0 \qquad \text{for } t \in \ ]t_0, t_k[.$$

Since in this case the equation (1.1) has oscillatory solutions, for any positive integer m the interval between two successive zeros of a solution can contain at least m zeros of another solution of the same equation.

2.1. Some auxiliary assertions. Below we use certain asymptotic properties of solutions of the equation

$$D^n(u; a_0, \dots, a_n) = p^*(t)u, \qquad (2.1)$$

where $p^* \in L_{\text{loc}}(\mathbf{R}_+)$, $a_i \in L_{\text{loc}}(\mathbf{R}_+)$, $a_i(t) > 0$ for $t \geq 0$ $(i = 0, \dots, n)$,

$$\int_0^{+\infty} a_i(t)\, dt = +\infty \qquad (i = 1, \dots, n-1), \qquad (2.2)$$

and $D^n(u; a_0, \dots, a_n)$ is defined by (1.72). Moreover, we assume that either

$$p^*(t) \leq 0 \qquad \text{for } t \in \mathbf{R}_+ \qquad (2.3)$$

or

$$p^*(t) \geq 0 \qquad \text{for } t \in \mathbf{R}_+. \qquad (2.4)$$

A function $u \colon \mathbf{R}_+ \to \mathbf{R}$ is said to be a solution of the equation (2.1) if it is locally absolutely continuous together with $D^i(u; a_0, \dots, a_i)$ $(i = 1, \dots, n-1)$ and satisfies (2.1) for almost all $t \in \mathbf{R}_+$.

The definitions we have introduced for the equation (1.1) can be naturally extended to the equation (2.1).

LEMMA 2.1. *Let the inequality (2.3) (the inequality (2.4)) hold and let in any neighborhood of $+\infty$ the function p^* be distinct from zero on a set of positive measure. If u is a solution of the equation (2.1) satisfying the condition*

$$u(t) > 0 \qquad \text{for } t \geq t_0,$$

then there exist $l \in \{0, \dots, n\}$ and $t_1 \geq t_0$ such that $l + n$ is odd (even) and

$$D^i(u; a_0, \dots, a_i)(t) > 0 \qquad \text{for } t \geq t_1 \qquad (i = 0, \dots, l-1),$$

$$(2.5_l)$$

$$(-1)^{i+l} D^i(u; a_0, \dots, a_i)(t) > 0 \qquad \text{for } t \geq t_1 \qquad (i = l, \dots, n-1).$$

The proof of this lemma is similar to that of Lemma 1.1.

LEMMA 2.2. *Let the inequality (2.3) (the inequality (2.4)) hold and let in any neighborhood of $+\infty$ the function p^* be distinct from zero on a set of positive measure. Then the equation (2.1) is nonoscillatory if and only if for any $l \in \{1, \dots, n-1\}$ such that $l + n$ is odd (even) this equation has a solution satisfying (2.5_l).*

PROOF. Necessity. Let v_k be a solution of (2.1) under the conditions

$$D^i(v_k; a_0, \dots, a_i)(0) = 0 \qquad (i = 0, \dots, l-1),$$

$$D^i(v_k; a_0, \dots, a_i)(k) = 0 \qquad (i = l, \dots, n-2),$$

$$\sum_{j=0}^{n-1} \left| D^j(v_k; a_0, \dots, a_j)(0) \right| = 1.$$

Without loss of generality we may assume that the sequence $(v_k)_{k=1}^{+\infty}$ converges uniformly on every finite interval in $\mathbf{R}_+$ and $v(t) > 0$ for $t \geq t_0$, where $v = \lim_{k \to +\infty} v_k$. Then, as in the proof of sufficiency in Lemma 1.6, we can show that v is a solution of the equation (2.1) satisfying (2.5_l).

Sufficiency. Suppose that for any $l \in \{1, \ldots, n-1\}$ such that $l+n$ is odd (even) the equation (2.1) under the condition (2.5_l) has an oscillatory solution. Let $t_* > t_1$ be a zero of some oscillatory solution of this equation. We will show that there exist $t^* > t_*$, $l_0 \in \{1, \ldots, n-1\}$ and a solution v of (2.1) satisfying the relations

$$D^j(v; a_0, \ldots, a_j)(t_*) = 0 \qquad (j = 0, \ldots, l_0 - 1), \tag{2.6}$$

$$D^j(v; a_0, \ldots, a_j)(t^*) = 0 \qquad (j = l_0, \ldots, n - 1), \tag{2.7}$$

$$D^j(v; a_0, \ldots, a_j)(t) \neq 0 \qquad \text{for } t \in \,]t_*, t^*[\qquad (j = 0, \ldots, n - 1). \tag{2.8}$$

Since the equation (2.1) has an oscillatory solution vanishing at t_*, we can find $k \in \{1, \ldots, n-1\}$, $t_{1*} > t_*$ and a solution v_0 so that t_* is a k-fold zero[3] of $D^0(v_0; a_0)$ and $D^k(v_0; a_0, \ldots, a_k)$ has $n - k$ zeros (taking into account multiplicity) in $]t_*, t_{1*}[$. Denote by T_{1*} the set of all such t_{1*} and put $t^* = \inf T_{1*}$. Clearly, $t^* > t_*$ and $t^* \in T_{1*}$.

Let v be a solution of the equation (2.1) under the condition (2.6) and let $D^{l_0}(v; a_0, \ldots, a_{l_0})$ have $n - l_0$ zeros in $]t_*, t^*]$, l_0 being the largest possible value. Then

$$D^{l_0}(v; a_0, \ldots, a_{l_0})(t) \neq 0 \qquad \text{for } t \in [t_*, t^*[.$$

Hence (2.8) for $j = 0, \ldots, l_0$ and (2.7) are fulfilled. For other j, (2.8) follows from the inequality (2.3) (the inequality (2.4)). Applying (2.6)–(2.8) and assuming v to be positive in $]t_*, t^*]$, we easily conclude that $l_0 + n$ is odd (even) and

$$D^j(v; a_0, \ldots, a_j)(t) > 0 \qquad \text{for } t \in \,]t_*, t^*] \qquad (j = 0, \ldots, l_0 - 1),$$

$$(-1)^{j+l_0} D^j(v; a_0, \ldots, a_j)(t) > 0 \qquad \text{for } t \in [t_*, t^*[\qquad (j = l_0, \ldots, n - 1).$$

Now, let u be a solution of the equation (2.1) satisfying the condition (2.5_{l_0}). Consider the solution $w_\epsilon = u - \epsilon v$. Denote by ϵ_0 the largest ϵ for which

$$D^j(w_\epsilon; a_0, \ldots, a_j)(t) \geq 0 \qquad \text{for } t \in [t_*, t^*] \qquad (j = 0, \ldots, l_0 - 1),$$

$$\tag{2.9}$$

$$(-1)^{j+l_0} D^j(w_\epsilon; a_0, \ldots, a_j)(t) \geq 0 \qquad \text{for } t \in [t_*, t^*] \qquad (j = l_0, \ldots, n - 1).$$

Then for $\epsilon = \epsilon_0$ at least one of the inequalities (2.9) is nonstrict. On the other hand, (2.9) together with the relations

$$(-1)^{n+l_0} D^n(w_\epsilon; a_0, \ldots, a_n)(t) \geq 0 \qquad \text{for } t \in [t_*, t^*],$$

$$D^j(w_\epsilon; a_0, \ldots, a_j)(t_*) > 0 \qquad (j = 0, \ldots, l_0 - 1),$$

$$(-1)^{j+l_0} D^j(w_\epsilon; a_0, \ldots, a_j)(t^*) > 0 \qquad (j = l_0, \ldots, n - 1)$$

yields

$$D^j(w_{\epsilon_0}; a_0, \ldots, a_j)(t) > 0 \qquad \text{for } t \in [t_*, t^*] \qquad (j = 0, \ldots, l_0 - 1),$$

$$(-1)^{j+l_0} D^j(w_{\epsilon_0}; a_0, \ldots, a_j)(t) > 0 \qquad \text{for } t \in [t_*, t^*] \qquad (j = l_0, \ldots, n - 1).$$

This contradiction shows that the equation (2.1) is nonoscillatory. $\square$

---

[3] A zero of a function $D^i(v; a_0, \ldots, a_i)$ is regarded as being k-fold, $k \in \{0, \ldots, n-i-1\}$, if $D^j(v; a_0, \ldots, a_j)(t_0) = 0$ $(j = i, \ldots, i+k-1)$ and, moreover, $D^{k+i}(v; a_0, \ldots, a_{k+i})(t_0) \neq 0$ for $k + i \neq n$.

COROLLARY 2.1. If either (2.3) or (2.4) holds and the equation (2.1) is nonoscillatory, then there exists $t_0 \geq 0$ such that this equation is disconjugate in $[t_0, +\infty[$.

COROLLARY 2.2. Let the inequality (1.2) (the inequality (1.3)) hold and let in any neighborhood of $+\infty$ the function p be distinct from zero on a set of positive measure. Then the equation (1.1) is nonoscillatory if and only if for any $l \in \{1, \ldots, n-1\}$ such that $l + n$ is odd (even) this equation has a solution satisfying (1.7_l).

COROLLARY 2.3. If either (1.2) or (1.3) holds and the equation (1.1) is nonoscillatory, then there exists $t_0 \geq 0$ such that this equation is disconjugate in $[t_0, +\infty[$.

Obviously, for second order equations Corollary 2.3 is valid without the hypotheses (1.2) and (1.3).

PROBLEM 2.1*. Is the assertion of Corollary 2.3 true when $n \geq 3$ and the function p changes sign?

PROBLEM 2.2. There exists a nonoscillatory third order equation which is conjugate in any neighborhood of $+\infty$.

PROBLEM 2.3. Let $p_*, p^*, p \in L_{\text{loc}}(\mathbf{R}_+)$,

$$p_*(t) \leq p(t) \leq p^*(t) \qquad \text{for } t \geq 0,$$

and let the equations $u^{(n)} = p_*(t)u$ and $u^{(n)} = p^*(t)u$ be disconjugate in $\mathbf{R}_+$. Then the equation (1.1) is also disconjugate in $\mathbf{R}_+$.

We claim that

$$I^{j-l}(t; a_{j-1}, \ldots, a_l) I^{n-l-1}(t; a_{n-1}, \ldots, a_{l+1})$$

$$\geq I^{j-l-1}(t; a_{j-1}, \ldots, a_{l+1}) I^{n-l}(t; a_{n-1}, \ldots, a_l)$$

$$\text{for } t \geq 0 \qquad (j = l+1, \ldots, n), \tag{2.10}$$

where

$$I^0 = 1, \qquad I^k(t; a_1, \ldots, a_k) = \int_0^t a_1(s) I^{k-1}(s; a_2, \ldots, a_k) \, ds \quad (k = 1, \ldots, n).$$

The proof will be carried out by induction with respect to $j + n$. If $j + n = 2n + 2$, i.e. $j = l+1$ and $n = l+1$, then (2.10) is clear. Assume that $k \geq 2l + 2$ and that (2.10) holds for all j and n such that $j + n \leq k$. Let $j + n = k + 1$. It is easy to verify that

$$r'_{jn}(t) = a_{j-1}(t) r_{j-1n}(t) + a_{n-1}(t) r_{jn-1}(t) \qquad \text{for } t \geq 0, \tag{2.11}$$

where $r_{ln}(t) \equiv 1$, while if $l + 1 \leq j \leq n$, then

$$r_{jn}(t) = I^{j-l}(t; a_{j-1}, \ldots, a_l) I^{n-l-1}(t; a_{n-1}, \ldots, a_{l+1})$$

$$- I^{j-l-1}(t; a_{j-1}, \ldots, a_{l+1}) I^{n-l}(t; a_{n-1}, \ldots, a_l).$$

According to our assumption, $r_{j-1n}(t) \geq 0$, $r_{jn-1}(t) \geq 0$ for $t \geq 0$ and, since $r_{jn}(0) = 0$, (2.11) implies (2.10) with $j + n = k + 1$. So (2.10) is proved.

LEMMA 2.3. Let the inequality (2.3) (the inequality (2.4)) hold, $l \in \{1,\dots,n-1\}$, $l+n$ being odd (even), and

$$\lim_{t \to +\infty} \sup \frac{I^{n-l}(t; a_{n-1},\dots,a_l)}{I^{n-l-1}(t; a_{n-1},\dots,a_{l+1})} \int_t^{+\infty} |q_l(s)| \, ds > 1, \tag{2.12}$$

where

$$q_l(t) = I^{n-l-1}(t; a_{n-1},\dots,a_{l+1}) I^l(t; a_1,\dots,a_l) \frac{a_0(t) a_n(t)}{I^1(t; a_l)} p^*(t).$$

Then the equation (2.1) has no solution satisfying (2.5_l).

PROOF. By (2.10) with $j = n-1$, the function

$$\frac{I^{n-l}(\cdot; a_{n-1},\dots,a_l)}{I^{n-l-1}(\cdot; a_{n-1},\dots,a_{l+1})}$$

does not decrease in $]0,+\infty[$. Moreover, since

$$I^l(t; a_1,\dots,a_l) \le I^{l-1}(t; a_1,\dots,a_{l-1}) I^1(t; a_l) \qquad \text{for } t \ge 0,$$

in view of (2.12) we obtain

$$\int_0^{+\infty} I^{n-l}(s; a_{n-1},\dots,a_l) I^{l-1}(s; a_1,\dots,a_{l-1}) a_0(s) a_n(s) |p^*(s)| \, ds = +\infty. \tag{2.13}$$

Now assume that the lemma is false and that u is a solution of the equation (2.1) under the condition (2.5_l).

Integration by parts yields

$$\sum_{j=i}^n (-1)^{j-1} I^{j-1}(t; a_{j-1},\dots,a_i) D^{j-1}(u; a_0,\dots,a_{j-1})(t)$$

$$= \sum_{j=i}^n (-1)^{j-1} I^{j-1}(s; a_{j-1},\dots,a_i) D^{j-1}(u; a_0,\dots,a_{j-1})(s)$$

$$+ (-1)^{n-1} \int_s^t I^{n-1}(\xi; a_{n-1},\dots,a_i) a_n(\xi) D^n(u; a_0,\dots,a_n)(\xi) \, d\xi. \tag{2.14_i}$$

Setting $i = l$, $s = t_0$ and applying (2.5), (2.13) and the inequality

$$u(t) \ge c_0 a_0(t) I^{l-1}(t; a_1,\dots,a_{l-1}) \qquad \text{for } t \ge t_0,$$

where c_0 is a positive constant, we get

$$\lim_{t \to +\infty} \left(D^{l-1}(u; a_0,\dots,a_{l-1})(t) - I^1(t; a_l) D^l(u; a_0,\dots,a_l)(t) \right) = +\infty$$

and

$$D^{l-1}(u; a_0,\dots,a_{l-1})(t)$$

$$\ge \sum_{j=l+1}^n (-1)^{j-l+1} I^{j-l}(t; a_{j-1},\dots,a_l) D^{j-1}(u; a_0,\dots,a_{j-1})(t)$$

$$\text{for } t \ge t_2, \tag{2.15}$$

where t_2 is a sufficiently large constant. According to (2.14_{l+1}),

$$\sum_{j=l+1}^{n} (-1)^{j-l+1} I^{j-l-1}(t; a_{j-1}, \dots, a_{l+1}) D^{j-1}(u; a_0, \dots, a_{j-1})(t)$$

$$\geq (-1)^{n-l} \int_{t}^{+\infty} I^{n-l-1}(\xi; a_{n-1}, \dots, a_{l+1}) a_n(\xi) D^n(u; a_0, \dots, a_n)(\xi) \, d\xi$$

$$\text{for } t \geq t_1. \tag{2.16}$$

The inequalities (2.10),(2.15) and (2.16) imply

$$D^{l-1}(u; a_0, \dots, a_{l-1})(t) \geq \frac{I^{n-l}(t; a_{n-1}, \dots, a_l)}{I^{n-l-1}(t; a_{n-1}, \dots, a_{l+1})}$$

$$\times \int_{t}^{+\infty} I^{n-l-1}(\xi; a_{n-1}, \dots, a_{l-1}) a_n(\xi) p^*(\xi) u(\xi) \, d\xi \qquad \text{for } t \geq t_2. \tag{2.17}$$

Set

$$\rho_{ij}(t) = I^{l-j+1}(t; a_j, \dots, a_l) D^{i-1}(u; a_0, \dots, a_{i-1})(t)$$

$$-I^{l-i+1}(t; a_i, \dots, a_l) D^{j-1}(u; a_0, \dots, a_{j-1})(t)$$

$$(i = 1, \dots, j; \; j = 1, \dots, l+1).$$

Clearly,

$$\rho_{il+1}'(t) = a_i(t) \rho_{i+1l+1}(t)$$

$$-a_{l+1}(t) I^{l-i+1}(t; a_i, \dots, a_l) D^{l+1}(u; a_0, \dots, a_{l+1})(t) \qquad (i = 1, \dots, l)$$

and

$$\rho_{ij}'(t) = a_j(t) \rho_{ij+1}(t) + a_i(t) \rho_{i+1j}(t) \qquad (i = 1, \dots, j-1; \; j = 2, \dots, l).$$

As $\lim_{t \to +\infty} \rho_{ll+1}(t) = +\infty$, by (2.2) and (2.5) these equalities yield

$$\lim_{t \to +\infty} \rho_{ij}(t) = +\infty \qquad (i = 1, \dots, j-1; \; j = 2, \dots, l+1).$$

Therefore we may assume t_2 to be so large that

$$I^1(t; a_l) u(t) \geq a_0(t) I^l(t; a_1, \dots, a_l) D^{l-1}(u; a_0, \dots, a_{l-1})(t) \qquad \text{for } t \geq t_2. \tag{2.18}$$

Since $D^{l-1}(u; a_0, \dots, a_{l-1})$ does not decrease in $[t_2, +\infty[$, the inequalities (2.17) and (2.18) contradict (2.12). $\square$

Let m_n^* be the smallest local maximum of the polynomial (1.9) and let $\alpha_1 \leq \alpha_2 \leq \dots \leq \alpha_n$ be the roots of the equation $P_n^*(x) = m_n^*$. It is easy to verify that for $n \neq 4m$,

$$\alpha_i < \alpha_{i+1} \qquad (i = 1, \dots, n-1; \; i \neq l_n^*), \qquad \alpha_{l_n^*} = \alpha_{l_n^*+1},$$

while for $n = 4m$,

$$\alpha_i < \alpha_{i+1} \qquad (i = 1, \dots, n-1; \; i \neq l_n^* - 2, l_n^*),$$

$$\alpha_{l_n^*-2} = \alpha_{l_n^*-1}, \qquad \alpha_{l_n^*} = \alpha_{l_n^*+1},$$

where

$$l_n^* = \begin{cases} n/2 + 1 & \text{if } n = 4m \\ n/2 & \text{if } n = 4m + 2 \\ (n-1)/2 & \text{if } n = 4m + 1 \\ (n+1)/2 & \text{if } n = 4m + 3. \end{cases} \tag{2.19}$$

Put

$$v_i(t) = t^{\alpha_i} \quad \text{for } t \in [1, +\infty[\quad (i = 1, \dots, n;\ i \neq l_n^* + 1),$$
$$v_{l_n^*+1}(t) = t^{\alpha_{l_n^*}} \ln t \quad \text{for } t \in [1, +\infty[$$

if $n \neq 4m$, and

$$v_i(t) = t^{\alpha_i} \quad \text{for } t \in [1, +\infty[\quad (i = 1, \dots, n;\ i \neq l_n^* - 1, l_n^* + 1),$$
$$v_i(t) = t^{\alpha_i - 1} \ln t \quad \text{for } t \in [1, +\infty[\quad (i = l_n^* - 1, l_n^* + 1)$$

if $n = 4m$. It is clear that $v_1, \dots, v_n$ is a fundamental system of solutions of the equation

$$v^{(n)} = -m_n^* t^{-n} v \quad (t \in [1, +\infty[).$$

Denote by $W_k[v_{i_1}, \dots, v_{i_k}]$ the Wronskian of solutions $v_{i_1}, \dots, v_{i_k}$. Then

$$W_n[v_1, \dots, v_n](t) \equiv c_0 \neq 0,$$

$$|W_{n-1}[v_1, \dots, v_{i-1}, v_{i+1}, \dots, v_n](t)|$$
$$\leq c t^{n-1-\alpha_i-\alpha_{i+1}} v_{i+1}(t) \quad \text{for } t \geq 1 \quad (i = 1, \dots, n), \tag{2.20}$$

where $\alpha_{n+1} = \alpha_n$, $v_{n+1}(t) = t^{\alpha_n}$.

LEMMA 2.4. *If the function $b: [1, +\infty[\to \mathbf{R}$ is locally integrable and*

$$\int_1^{+\infty} t^{n-1} \ln t |b(t)| \, dt < +\infty, \tag{2.21}$$

then the equation

$$u^{(n)} = (-m_n^* t^{-n} + b(t))u \tag{2.22}$$

has a fundamental system of solutions $u_1, \dots, u_n$ such that

$$u_k^{(j-1)}(t) = v_k^{(j-1)}(t)(1 + o(1)) \quad (j, k = 1, \dots, n).$$

PROOF. Fix $k \in \{1, \dots, n\}$ and choose $t_0 > 1$ so that

$$\theta = \frac{2cn}{|c_0|} \int_{t_0}^{+\infty} t^{n-1} \ln t |b(t)| \, dt < 1.$$

Let $C_{v_k}([t_0, +\infty[)$ be the Banach space of continuous functions $u: [t_0, +\infty[\to \mathbf{R}$ satisfying the condition

$$\sup\left\{ \frac{|u(t)|}{v_k(t)} : t \in [t_0, +\infty[\right\} < +\infty$$

with the norm

$$\|u\|_{v_k} = \sup\left\{ \frac{|u(t)|}{v_k(t)} : t \in [t_0, +\infty[\right\}.$$

Moreover, let $B \subset C_{v_k}([t_0, +\infty[)$ be the set of functions for which $\|u\|_{v_k} \leq 2$. Define the operator $G \colon B \to C_{v_k}([t_0, +\infty[)$ by the equality

$$G(u)(t) = v_k(t)$$

$$+ \sum_{i=1}^{k-1} (-1)^{n-i} \frac{v_i(t)}{c_0} \int_{t_0}^{t} W_{n-1}[v_1, \ldots, v_{i-1}, v_{i+1}, \ldots, v_n](s)b(s)u(s)\,ds$$

$$- \sum_{i=k}^{n} (-1)^{n-i} \frac{v_i(t)}{c_0} \int_{t}^{+\infty} W_{n-1}[v_1, \ldots, v_{i-1}, v_{i+1}, \ldots, v_n](s)b(s)u(s)\,ds.$$

Since v_k/v_i does not decrease for $i \in \{1, \ldots, k-1\}$ and does not increase for $i \in \{k, \ldots, n\}$, in view of (2.20) we get

$$|G(u)(t)| \leq v_k(t) \left[1 + \sum_{i=1}^{k-1} \frac{c\|u\|_{v_k}}{|c_0|} \frac{v_i(t)}{v_k(t)} \int_{t_0}^{t} \frac{v_k(s)}{v_i(s)} s^{n-1} \ln s \, |b(s)|\,ds \right.$$

$$\left. + \sum_{i=k}^{n} \frac{c\|u\|_{v_k}}{|c_0|} \frac{v_i(t)}{v_k(t)} \int_{t}^{+\infty} \frac{v_k(s)}{v_i(s)} s^{n-1} \ln s \, |b(s)|\,ds \right]$$

$$\leq v_k(t) \left[1 + \frac{2cn}{|c_0|} \int_{t_0}^{+\infty} s^{n-1} \ln s \, |b(s)|\,ds \right] \qquad \text{for } t \geq t_0,$$

i.e. $\|G(u)\|_{v_k} \leq 2$ and so $G(B) \subset B$. We can similarly show that if $u, v \in B$, then

$$\|G(u) - G(v)\|_{v_k} \leq \theta \|u - v\|_{v_k}.$$

Thus, the operator G is contractive and by the well-known Banach theorem there exists $u_k \in B$ such that $u_k = G(u_k)$. Clearly, u_k satisfies the equation (2.22). As above we can show that for any $t_1 \geq t_0$ the estimate

$$\left| \frac{u_k(t)}{v_k(t)} - 1 \right| \leq \sum_{i=1}^{k-1} \frac{2c}{|c_0|} \frac{v_i(t)}{v_k(t)} \int_{t_0}^{t_1} \frac{v_k(s)}{v_i(s)} s^{n-1} \ln s \, |b(s)|\,ds$$

$$+ \sum_{i=k}^{n} \frac{2c}{|c_0|} \int_{t_1}^{+\infty} s^{n-1} \ln s \, |b(s)|\,ds \qquad \text{for } t \geq t_1$$

is true. This, together with the arbitrariness of t_1 and the equalities

$$\lim_{t \to +\infty} \frac{v_i(t)}{v_k(t)} = 0 \qquad (i = 1, \ldots, k-1),$$

implies

$$\lim_{t \to +\infty} \frac{u_k(t)}{v_k(t)} = 1.$$

Analogously,

$$\lim_{t \to +\infty} \frac{u_k^{(i-1)}(t)}{v_k^{(i-1)}(t)} = 1 \qquad (i = 1, \ldots, n). \qquad \square$$

REMARK 2.2. If (2.21) is replaced by the condition

$$\int_1^{+\infty} t^{n-1} \ln^2 t \, |b(t)| \, dt < +\infty, \tag{2.23}$$

then the asymptotic representations

$$u_k^{(j-1)}(t) = v_k^{(j-1)}(t)(1 + \ln^{-1} t \, o(1)) \qquad (j, k = 1, \dots, n),$$

which are more precise, hold.

The validity of this remark can be verified similarly to the proof of Lemma 2.4, provided that if $k = l_n^* + 1$ and $n \neq 4m$ or $k \in \{l_n^* - 1, l_n^* + 1\}$ and $n = 4m$, one considers the operator

$$G(u)(t) = v_k(t)$$

$$+ \sum_{i=1}^{k-2} (-1)^{n-i} \frac{v_i(t)}{c_0} \int_{t_0}^t W_{n-1}[v_1, \dots, v_{i-1}, v_{i+1}, \dots, v_n](s) b(s) u(s) \, ds$$

$$- \sum_{i=k-1}^n (-1)^{n-i} \frac{v_i(t)}{c_0} \int_t^{+\infty} W_{n-1}[v_1, \dots, v_{i-1}, v_{i+1}, \dots, v_n](s) b(s) u(s) \, ds.$$

LEMMA 2.5. If the function $b: [1, +\infty[\to \mathbf{R}$ is locally integrable and (2.23) holds, then there exist $t_0 \geq 1$ and positive $a_i \in C^{n-i}([t_0, +\infty[)$ $(i = 0, \dots, n)$ such that

$$u^{(n)} + (m_n^* t^{-n} - b(t))u = D^n(u; a_0, \dots, a_n) \tag{2.24}$$

for any $u \in \tilde{C}_{\text{loc}}^{n-1}([t_0, +\infty[)$, where $D^n(u; a_0, \dots, a_n)$ is defined by (1.72) and, moreover,

$$a_0(t) = t^{\alpha_1}(1 + o(1)), \qquad a_n(t) = t^{n-1-\alpha_n}(1 + o(1)),$$

$$\tag{2.25}$$

$$a_i(t) = t^{\alpha_{i+1}-\alpha_i-1}(1 + o(1)) \qquad (i = 1, \dots, n-1).$$

PROOF. Let $u_1, \dots, u_n$ be a fundamental system of solutions of the equation (2.22) constructed by Lemma 2.4 and Remark 2.2. If w_k denotes the Wronskian of the solutions $u_1, \dots, u_k$, then

$$w_k(t) = t^{\sum_{i=1}^k \alpha_i - k(k-1)/2}(c_k + o(1)), \tag{2.26}$$

where $c_k \neq 0$ $(k = 1, \dots, n)$. Choose $t_0 \geq 1$ so large that $w_k(t) \neq 0$ for $t \geq t_0$ $(k = 1, \dots, n-1)$. Then the Frobenius factorization yields (2.24) with

$$a_i = \frac{w_{i-1} w_{i+1}}{w_i^2} \qquad (i = 0 \dots, n), \qquad w_{-1} = w_0 = 1, \qquad w_{n+1} = w_n.$$

Without loss of generality we may assume that $c_k = 1$ $(k = 1, \dots, n)$. Hence (2.26) implies (2.25). $\square$

REMARK 2.3. It follows from the proof that the lemma remains true if in case of $n = 3$, (2.23) is replaced by

$$\int_1^{+\infty} t^2 \ln t \, |b(t)| \, dt < +\infty.$$

Let m_{*n} be the smallest local maximum of the polynomial (1.10), and let $\beta_1 \le \beta_2 \le \ldots \le \beta_n$ be the roots of the equation $P_{*n}(x) = m_{*n}$. We can easily show that

$$\beta_i < \beta_{i+1} \quad (i = 1, \ldots, n-1; \ i \ne l_{*n}), \qquad \beta_{l_{*n}} = \beta_{l_{*n}+1}$$

for $n \ne 4m + 2$, and

$$\beta_i < \beta_{i+1} \quad (i = 1, \ldots, n-1; \ i \ne l_{*n} - 2, l_{*n}),$$

$$\beta_{l_{*n}-2} = \beta_{l_{*n}-1}, \qquad \beta_{l_{*n}} = \beta_{l_{*n}+1}$$

for $n = 4m + 2$, where

$$l_{*n} = \begin{cases} n/2 & \text{if } n = 4m \\ n/2 + 1 & \text{if } n = 4m + 2 \\ (n+1)/2 & \text{if } n = 4m + 1 \\ (n-1)/2 & \text{if } n = 4m + 3. \end{cases} \tag{2.27}$$

Set

$$v_i(t) = t^{\beta_i} \quad \text{for } t \ge 1 \quad (i = 1, \ldots, n; \ i \ne l_{*n} + 1),$$

$$v_{l_{*n}+1}(t) = t^{\beta_{l_{*n}}} \ln t \quad \text{for } t \ge 1$$

if $n \ne 4m + 2$, and

$$v_i(t) = t^{\beta_i} \quad \text{for } t \ge 1 \quad (i = 1, \ldots, n; \ i \ne l_{*n} - 1, l_{*n} + 1),$$

$$v_i(t) = t^{\beta_i - 1} \ln t \quad \text{for } t \ge 1 \quad (i = l_{*n} - 1, l_{*n} + 1)$$

if $n = 4m + 2$. Obviously, $v_1, \ldots, v_n$ is a fundamental system of solutions of the equation

$$v^{(n)} = m_{*n} t^{-n} v \quad (t \in [1, +\infty[).$$

As above we can justify the following assertions.

LEMMA 2.6. *If the function* $b: [1, +\infty[\to \mathbf{R}$ *is locally integrable and (2.21) holds, then the equation*

$$u^{(n)} = (m_{*n} t^{-n} + b(t)) u$$

has a fundamental system of solutions $u_1, \ldots, u_n$ *such that*

$$u_k^{(j-1)}(t) = v_k^{(j-1)}(t)(1 + o(1)) \quad (j, k = 1, \ldots, n).$$

LEMMA 2.7. *If the function* $b: [1, +\infty[\to \mathbf{R}$ *is locally integrable and (2.23) holds, then there exist* $t_0 \ge 1$ *and positive* $a_i \in C^{n-i}([t_0, +\infty[)$ $(i = 0, \ldots, n)$ *such that*

$$u^{(n)} - (m_{*n} t^{-n} + b(t)) u = D^n(u; a_0, \ldots, a_n)$$

for any $u \in \tilde{C}_{\text{loc}}^{m-1}([t_0, +\infty[)$, *where* $D^n(u; a_0, \ldots, a_n)$ *is defined by (1.72) and, moreover,*

$$a_0(t) = t^{\beta_1}(1 + o(1)), \qquad a_n(t) = t^{n-1-\beta_n}(1 + o(1)),$$

$$a_i(t) = t^{\beta_{i+1}-\beta_i-1}(1 + o(1)) \quad (i = 1, \ldots, n-1).$$

2.2. Oscillation and nonoscillation criteria. First we establish necessary and sufficient conditions for the equation (1.1) to be oscillatory.

THEOREM 2.1. Let the inequality (1.2) (the inequality (1.3)) hold. Then the following statements are equivalent:

  (i) the equation (1.1) is oscillatory;
  (ii) the equation (1.1) has no solution satisfying the condition (1.7_l) with $l = l_n^*$ $(l = l_{*n})$;
  (iii) the equation (1.1) has no solution satisfying the condition (1.7_l) for some $l \in \{1, \ldots, n-1\}$ such that $l + n$ is odd (even);
  (iv) $\eta_n(t_0; p) < +\infty$ for any $t_0 \in \mathbf{R}_+$.

PROOF. (i) $\Rightarrow$ (iv) is obvious.

(iv) $\Rightarrow$ (iii). According to Lemma 1.9 and Remark 1.6, there exists $l \in \{1, \ldots, n-1\}$ such that $l + n$ is odd (even) and the equation (1.1) is $(l, n - l)$ conjugate. Hence by Lemma 1.6 this equation does not have a solution satisfying (1.7_l).

(iii) $\Rightarrow$ (ii). If the equation (1.1) has a solution satisfying (1.7_l) with $l = l_n^*$ $(l = l_{*n})$, then via Corollary 1.1 and Lemma 1.8 we can show that this equation has a solution of the type (1.7_l) for any $l \in \{1, \ldots, n-1\}$ such that $l + n$ is odd (even). But this contradicts (iii).

(ii) $\Rightarrow$ (i) follows from Corollary 2.2. $\square$

PROBLEM 2.4*. Is the condition (iv) of Theorem 2.1 sufficient for the equation (1.1) to be oscillatory in case p changes sign?

Theorem 2.1 ((i) $\Leftrightarrow$ (ii)) and Lemma 1.6 imply

COROLLARY 2.4. Let the inequality (1.2) (the inequality (1.3)) hold. Then the equation (1.1) is oscillatory if and only if it is $(l_n^*, n - l_n^*)$ conjugate $((l_{*n}, n - l_{*n})$ conjugate) in any neighborhood of $+\infty$.

Since $l_n^* + l_{*n} = n$ for odd n, by Corollary 1.1 and Theorem 2.1 ((i) $\Leftrightarrow$ (ii)) we can derive the following theorem.

THEOREM 2.2. Let n be odd and (1.2) hold. Then the equation (1.1) is oscillatory if and only if the equation (1.34) is oscillatory.

Furthermore, Theorem 2.1 ((i) $\Leftrightarrow$ (ii)) and Lemma 1.8 yield

THEOREM 2.2'. Let $n = 4m$ $(n = 4m + 2)$, $m \in \mathbf{N}$, and let p be nonpositive (nonnegative). If the equation (1.34) is oscillatory, then the equation (1.1) is also oscillatory.

Note that the converse is not true. Indeed, in the case under consideration, $m_n^* > m_{*n}$ $(m_{*n} > m_n^*)$. Hence, if $p(t) = ct^{-n}$ for $t \geq 1$, where $c \in [-m_n^*, -m_{*n}[$ $(c \in]m_n^*, m_{*n}])$, then the equation (1.34) is oscillatory, but the equation (1.1) is nonoscillatory.

THEOREM 2.3. Let [4]

$$\lim_{t \to +\infty} \sup \ln t \int_t^{+\infty} s^{n-1}[p(s) + m_n^* s^{-n}]_- \, ds = +\infty \tag{2.28}$$

and

$$\int_1^{+\infty} t^{n-1} \ln^2 t \, [p(t) + m_n^* t^{-n}]_+ \, dt < +\infty, \tag{2.29}$$

where m_n^* is the smallest local maximum of the polynomial (1.9). Then the equation (1.1) is oscillatory.

PROOF. For $t \geq 1$ we have

$$p(t) = -m_n^* t^{-n} + [p(t) + m_n^* t^{-n}]_+ - [p(t) + m_n^* t^{-n}]_-.$$

Therefore, by (2.29) and Lemma 2.5 there exists $t_0 \geq 1$ such that in $[t_0, +\infty[$ the equation (1.1) can be written in the form (2.1) with $a_i \in C^{n-i}([t_0, +\infty[)$ $(i = 0, \dots, n)$ satisfying (2.25) and

$$p^*(t) = -[p(t) + m_n^* t^{-n}]_-.$$

In view of (2.25),

$$\lim_{t \to +\infty} \frac{I^{n-l_n^*}(t; a_{n-1}, \dots, a_{l_n^*})}{t^{\alpha_n - \alpha_{l_n^*}+1} \ln t} = c_0,$$

$$\lim_{t \to +\infty} \frac{I^{n-l_n^*-1}(t; a_{n-1}, \dots, a_{l_n^*+1})}{t^{\alpha_n - \alpha_{l_n^*}+1}} = c_0,$$

$$\lim_{t \to +\infty} \frac{I^{l_n^*}(t; a_1, \dots, a_{l_n^*})}{t^{\alpha_{l_n^*} - \alpha_1} \ln t} = c_1,$$

where

$$c_0 = \begin{cases} (\alpha_n - \alpha_{l_n^*+1}) \dots (\alpha_{l_n^*+2} - \alpha_{l_n^*+1}) & \text{if } l_n^* \in \{1, \dots, n-2\} \\ 1 & \text{if } l_n^* = n-1, \end{cases}$$

$$c_1 = (\alpha_{l_n^*} - \alpha_1) \dots (\alpha_{l_n^*} - \alpha_{l_n^*-1}).$$

Consequently, (2.28) is equivalent to the condition

$$\lim_{t \to +\infty} \sup \frac{I^{n-l_n^*}(t; a_{n-1}, \dots, a_{l_n^*})}{I^{n-l_n^*-1}(t; a_{n-1}, \dots, a_{l_n^*+1})} \int_t^{+\infty} |q_{l_n^*}(s)| \, ds = +\infty,$$

with q_l being defined in Lemma 2.3. Taking into account that $l_n^* + n$ is odd, by Lemma 2.3 we conclude that the equation (2.1) does not have solutions satisfying $(2.5_{l_n^*})$. Therefore, Lemma 2.2 implies that the equation (2.1) and, hence, the equation (1.1) have oscillatory solutions. □

COROLLARY 2.5. Let (2.29) hold and let there exist a continuous nondecreasing function $\omega: [1, +\infty[\to]0, +\infty[$ such that

$$\int_2^{+\infty} \frac{dt}{t\omega(t) \ln t} < +\infty, \tag{2.30}$$

$$\int_1^{+\infty} \frac{t^{n-1} \ln t}{\omega(t)} [p(t) + m_n^* t^{-n}]_- \, dt = +\infty. \tag{2.31}$$

---

[4]Here and below, $[x(t)]_+ = \max\{x(t), 0\}$, $[x(t)]_- = \max\{-x(t), 0\}$.

Then the equation (1.1) is oscillatory.

PROOF. It suffices to show that (2.30) and (2.31) imply (2.28). Assume, on the contrary, that

$$\ln t \int_t^{+\infty} s^{n-1}[p(s) + m_n^* s^{-n}]_- \, ds \le c_0 \qquad \text{for } t \ge 1.$$

Then

$$\int_1^t \frac{s^{n-1} \ln s}{\omega(s)} [p(s) + m_n^* s^{-n}]_- \, ds$$

$$= -\int_1^t \ln s \, d \left(\int_s^{+\infty} \frac{\tau^{n-1}}{\omega(\tau)} [p(\tau) + m_n^* \tau^{-n}]_- \, d\tau \right)$$

$$\le \int_1^t \frac{1}{s} \int_s^{+\infty} \frac{\tau^{n-1}}{\omega(\tau)} [p(\tau) + m_n^* \tau^{-n}]_- \, d\tau \, ds$$

$$\le c_1 + c_0 \int_{t_0}^t \frac{ds}{s\omega(s)\ln s} \qquad \text{for } t \ge t_0,$$

where $t_0 > 0$ and $c_1 > 0$ are certain constants. The last inequality contradicts (2.30), (2.31). □

The following proposition can be proved similarly to Theorem 2.3, provided that instead of Lemma 2.5 one applies Lemma 2.7.

THEOREM 2.4. Let

$$\lim_{t \to +\infty} \sup \ln t \int_t^{+\infty} s^{n-1}[p(s) - m_{*n} s^{-n}]_+ \, ds = +\infty$$

and

$$\int_1^{+\infty} t^{n-1} \ln^2 t [p(t) - m_{*n} t^{-n}]_- \, dt < +\infty, \qquad (2.32)$$

where m_{*n} is the smallest local maximum of the polynomial (1.10). Then the equation (1.1) is oscillatory.

COROLLARY 2.6. Let (2.32) hold and let there exist a continuous nondecreasing function $\omega: [1, +\infty[\to]0, +\infty[$ satisfying the conditions (2.30) and

$$\int_1^{+\infty} \frac{t^{n-1} \ln t}{\omega(t)} [p(t) - m_{*n} t^{-n}]_+ \, dt = +\infty.$$

Then the equation (1.1) is oscillatory.

THEOREM 2.5. Let

$$\int_1^{+\infty} t^{n-1} \ln t \, [p(t) + m_n^* t^{-n}]_- \, dt < +\infty$$

and

$$\int_1^{+\infty} t^{n-1} \ln t \, [p(t) - m_{*n} t^{-n}]_+ \, dt < +\infty,$$

where m_n^* and m_{*n} are the smallest local maxima of the polynomials (1.9) and (1.10), respectively. Then the equation (1.1) is nonoscillatory.

PROOF. Clearly,

$$-m_n^* t^{-n} - [p(t) + m_n^* t^{-n}]_- \leq p(t) \leq m_{*n} t^{-n} + [p(t) - m_{*n} t^{-n}]_+ \qquad \text{for } t \geq 1.$$

According to Lemmas 2.5 and 2.6 the equations

$$u^{(n)} = (-m_n^* t^{-n} - [p(t) + m_n^* t^{-n}]_-)u$$

and

$$u^{(n)} = (m_{*n} t^{-n} + [p(t) - m_{*n} t^{-n}]_+)u$$

are nonoscillatory. By Corollary 4.3 there exists $t_0 \geq 1$ such that these equations are disconjugate in $[t_0, +\infty[$. Hence the equation (1.1) is also disconjugate in $[t_0, +\infty[$ (see Problem 2.3). Therefore the equation (1.1) is nonoscillatory. □

In Theorems 2.6–2.10, l_n^* and l_{*n} are the constants defined by (2.19) and (2.27).

Lemmas 1.18 and 1.19 and Theorem 2.1 ((i)⟺(ii)) imply the following comparison theorems.

THEOREM 2.6. Let the inequalities (1.52) hold and

$$\int_t^{+\infty} s^{l_n^*-1} |p(s)| \, ds \geq \int_t^{+\infty} s^{l_n^*-1} |q(s)| \, ds \qquad \text{for } t \geq 0.$$

If the equation (1.40) is oscillatory, then the equation (1.1) is also oscillatory.

THEOREM 2.7. Let the inequalities (1.53) hold and

$$\int_t^{+\infty} s^{l_{*n}-1} p(s) \, ds \geq \int_t^{+\infty} s^{l_{*n}-1} q(s) \, ds \qquad \text{for } t \geq 0.$$

If the equation (1.40) is oscillatory, then the equation (1.1) is also oscillatory.

COROLLARY 2.7. If either (1.41) or (1.42) holds and if the equation (1.40) is oscillatory, then the equation (1.1) is also oscillatory.

THEOREM 2.8. If the inequality (1.2) holds, then either of the conditions

$$\lim_{t \to +\infty} \inf t^{n-l_n^*} \int_t^{+\infty} s^{l_n^*-1} |p(s)| \, ds > \frac{m_n^*}{n - l_n^*} \tag{2.33}$$

and

$$\lim_{t \to +\infty} \sup t \int_t^{+\infty} s^{n-2} |p(s)| \, ds > \mu_n^*, \tag{2.34}$$

where m_n^* denotes the smallest local maximum of the polynomial (1.9) and

$$\mu_n^* = \min\{l! \, (n-l)! : l \in \{1, \dots, n-1\}, l+n \text{ odd}\},$$

is sufficient for the equation (1.1) to be oscillatory.

PROOF. The sufficiency of (2.34) follows from Remark 1.7 and Corollary 2.2, while the sufficiency of (2.33) is a consequence of Theorem 2.6 and the fact that the equation (1.11) with $c < -m_n^*$ is oscillatory. □

THEOREM 2.9. If the inequality (1.3) holds, then either of the conditions

$$\lim_{t \to +\infty} \inf t^{n-l_{*n}} \int_t^{+\infty} s^{l_{*n}-1} p(s) \, ds > \frac{m_{*n}}{n - l_{*n}}$$

and

$$\lim_{t \to +\infty} \sup t \int_t^{+\infty} s^{n-2} p(s)\, ds > \mu_{*n},$$

where m_{*n} denotes the smallest local maximum of the polynomial (1.10) and

$$\mu_{*n} = \min\{l!\,(n-l)! : l \in \{1, \ldots, n-1\}, l+n \text{ even}\},$$

is sufficient for the equation (1.1) to be oscillatory.

THEOREM 2.10. Let $t_0 \in \mathbf{R}_+$,

$$t^{n-l_n^*} \int_t^{+\infty} s^{l_n^*-1} [p(s)]_-\, ds \leq \frac{m_n^*}{n - l_n^*} \qquad \text{for } t \geq t_0$$

and

$$t^{n-l_{*n}} \int_t^{+\infty} s^{l_{*n}-1} [p(s)]_+\, ds \leq \frac{m_{*n}}{n - l_{*n}} \qquad \text{for } t \geq t_0,$$

where m_n^* and m_{*n} denote the smallest local maxima of the polynomials (1.9) and (1.10), respectively. Then the equation (1.1) is nonoscillatory.

The proofs of Theorems 2.9 and 2.10 are similar to those of Theorems 2.8 and 2.5, respectively.

COROLLARY 2.8. If either

$$\lim_{t \to +\infty} \sup t^n p(t) < -m_n^*$$

or

$$\lim_{t \to +\infty} \inf t^n p(t) > m_{*n},$$

then the equation (1.1) is oscillatory.

COROLLARY 2.9. If $t_0 \geq 0$ and

$$-m_n^* \leq t^n p(t) \leq m_{*n} \qquad \text{for } t \geq t_0,$$

then the equation (1.1) is nonoscillatory.

REMARK 2.4. Note that

$$\mu_n^* = l_n^*!\,(n - l_n^*)!, \qquad \mu_{*n} = l_{*n}!\,(n - l_{*n})!.$$

REMARK 2.5. If p is nonnegative (nonpositive) and $n \in \{3, 4\}$ ($n \in \{3, 4, 6\}$), then the sufficiency conditions for the equation (1.1) to be oscillatory coincide with the sufficiency conditions under which this equation has property A (property B) (see Theorems 1.6, 1.7, 1.7', 2.8, and 2.9). This is not accidental, because, as Lemma 2.8', established below, implies, in these cases the equation (1.1) has property A (property B) if and only if it is oscillatory.

2.3. On third, fourth and sixth order equations. In some cases existence of at least one oscillatory solution implies that the equation has property A or B, just as for second order equations.

LEMMA 2.8. If the equation (1.1) with $n \in \{3,4\}$ ($n \in \{3,4,6\}$) and nonpositive (nonnegative) p is oscillatory, then it has property A (property B).

PROOF. For $n = 3$ ($n \in \{3,4\}$) the lemma immediately follows from Theorem 1.1 (Theorem 1.2) ((i)$\Leftrightarrow$(ii)) and Corollary 2.2. For $n = 4$ ($n = 6$) one should additionally take into account Corollary 1.1. $\square$

In view of Problem 1.5 we can restate Lemma 2.8 as follows.

LEMMA 2.8'. Let $n \in \{3,4\}$ ($n \in \{3,4,6\}$) and let the inequality (1.2) (the inequality (1.3)) hold. Then the equation (1.1) has property A (property B) if and only if it is oscillatory.

Below we study third, fourth and sixth order equations separately.

First we consider the equation

$$u''' = p(t)u. \tag{2.35}$$

THEOREM 2.11. (i) If the inequality (1.2) holds,

$$\lim_{t \to +\infty} \sup \ln t \int_t^{+\infty} s^2 \left[p(s) + \frac{2\sqrt{3}}{9s^3} \right]_- ds = +\infty$$

and

$$\int_1^{+\infty} t^2 \ln t \left[p(t) + \frac{2\sqrt{3}}{9t^3} \right]_+ dt < +\infty, \tag{2.36}$$

then the equation (2.35) has property A;

(ii) if

$$\int_1^{+\infty} t^2 \ln t \left[p(t) + \frac{2\sqrt{3}}{9t^3} \right]_- dt < +\infty$$

and

$$\int_1^{+\infty} t^2 \ln t \left[p(t) - \frac{2\sqrt{3}}{9t^3} \right]_+ dt < +\infty,$$

then the equation (2.35) is nonoscillatory;

(iii) if the inequality (1.3) holds,

$$\lim_{t \to +\infty} \sup \ln t \int_t^{+\infty} s^2 \left[p(s) - \frac{2\sqrt{3}}{9s^3} \right]_+ ds = +\infty$$

and

$$\int_1^{+\infty} t^2 \ln t \left[p(t) - \frac{2\sqrt{3}}{9t^3} \right]_- dt < +\infty, \tag{2.37}$$

then the equation (2.35) has property B.

PROOF. Assertion (ii) is a consequence of Theorem 2.5. By Theorem 1.3 it suffices to prove either (i) or (iii). Consider (i). Taking into account the proof of Theorem 2.3, Remark 2.3 and the equality $m_3^* = 2\sqrt{3}/9$, we conclude that the equation (2.35) is oscillatory. Hence, by Lemma 2.8, it has property A. $\square$

Arguing as in the proof of Corollary 2.5, we can verify

COROLLARY 2.10. If p is nonnegative, (2.36) holds and

$$\int_1^{+\infty} t^2 \ln^{1-\epsilon} t \left[p(t) + \frac{2\sqrt{3}}{9t^3} \right]_- dt = +\infty$$

for a certain $\epsilon \in \,]0,1]$, then the equation (2.35) has property A. Furthermore, if p is nonpositive, (2.37) holds and

$$\int_1^{+\infty} t^2 \ln^{1-\epsilon} t \left[p(t) - \frac{2\sqrt{3}}{9t^3} \right]_+ dt = +\infty$$

for a certain $\epsilon \in \,]0,1]$, then the equation (2.35) has property B.

Since $M_3^* = m_3^* = M_{*3} = m_{*3} = 2\sqrt{3}/9$, Theorems 2.2, 2.8, 2.10, and Lemma 2.8 imply

THEOREM 2.12. (i) If p is nonpositive and either

$$\lim_{t \to +\infty} \inf t \int_t^{+\infty} s|p(s)|\, ds > \frac{2\sqrt{3}}{9}$$

or

$$\lim_{t \to +\infty} \sup t \int_t^{+\infty} s|p(s)|\, ds > 2,$$

then the equation (2.35) has property A;

(ii) if there exists $t_0 \geq 0$ such that

$$t \int_t^{+\infty} s[p(s)]_-\, ds \leq \frac{2\sqrt{3}}{9}, \qquad t \int_t^{+\infty} s[p(s)]_+\, ds \leq \frac{2\sqrt{3}}{9} \qquad \text{for } t \geq t_0,$$

then the equation (2.35) is nonoscillatory;

(iii) if p is nonnegative and either

$$\lim_{t \to +\infty} \inf t \int_t^{+\infty} sp(s)\, ds > \frac{2\sqrt{3}}{9}$$

or

$$\lim_{t \to +\infty} \sup t \int_t^{+\infty} sp(s)\, ds > 2,$$

then the equation (2.35) has property B.

LEMMA 2.9. If either (1.2) or (1.3) holds, then

$$\tau_3(t_0; p) = \eta_3(t_0; p).$$

PROOF. Assuming the contrary, by Remark 1.7 we get $\eta_3(t_0; p) < \tau_3(t_0; p)$. According to the definition of $\tau_3(t_0; p)$, there exist $t_1 \in]\eta_3(t_0; p), \tau_3(t_0; p)[$ and a solution v of the equation (2.35) such that

$$v(t_0) = v(t_1) = 0, \qquad v(t) > 0 \qquad \text{for } t \in]t_0, t_1[. \tag{2.38}$$

First let (1.2) hold. Then by Lemma 1.9 and Remark 1.6, the equation (2.35) under the conditions

$$u(t_0) = u'(t_0) = u(t^0) = 0, \qquad u(t) > 0 \qquad \text{for } t \in]t_0, t^0[, \tag{2.39}$$

where $t^0 = \eta_3(t_0; p)$, has a solution u. It follows from (2.38) and (2.39) that $u(t_2)v'(t_2) - v(t_2)u'(t_2) = 0$ for a certain $t_2 \in]t_0, t^0[$. Thus we can choose constants c_1 and c_2 so that $w(t_2) = w'(t_2) = 0$, where $w = c_1 u + c_2 v$. But this contradicts Lemma 1.4, because $w(t_0) = 0$.

Now suppose the inequality (1.3) holds. Then, in view of Lemmas 1.5 and 1.9 and Remark 1.6, there exists a solution u such that

$$u(t) > 0 \qquad \text{for } t \in]t_0, t^0[\cup]t^0, +\infty[,$$

$$u(t_0) = u(t^0) = u'(t^0) = 0, \qquad u'(t_0) > 0.$$

Consider the solution

$$w = v - \frac{v'(t_0)}{u'(t_0)} u$$

of the equation (2.35). Obviously, $w(t^0) > 0$ and $w(t_1) < 0$. So $w(t_2) = 0$ for some $t_2 \in]t_0, t_1[$. Since $w(t_0) = w'(t_0) = 0$, this contradicts Lemma 1.5. $\square$

Applying Lemma 2.9, we can efficiently determine the constant $\tau_3(0; 1) = \tau_3(0; -1)$. In fact, it is the smallest positive zero of the solution of the equation $v''' = -v$ under the conditions $v(0) = v'(0) = 0$, $v''(0) = 1$, i.e. the smallest positive root of the equation $e^{-t} + 2e^{t/2}\sin(\sqrt{3}t/2 - \pi/6) = 0$. This, in particular, yields the estimate $\tau_3(0; 1) < 8\pi/3\sqrt{3}$.

Moreover, by means of Lemma 2.9 we can prove the following comparison theorem.

THEOREM 2.13. If either (1.41) or (1.42) holds, then the interval between any two consecutive zeros of a solution of the equation (2.35) contains at most two zeros of a solution of the equation

$$v''' = q(t)v. \tag{2.40}$$

PROOF. Assume the contrary. Then there exist solutions u and v of the equations (2.35) and (2.40), respectively, such that

$$u(t_0) = u(t_1) = 0, \qquad u(t) > 0 \qquad \text{for } t \in]t_0, t_1[$$

and v has three zeros in the interval $]t_0, t_1[$. Hence $\eta_3(t_0; q) < t_1$ and $t_1 \leq \tau_3(t_0; p)$. So, by Lemmas 1.16 and 2.9 we obtain a contradiction:

$$t_1 \leq \tau_3(t_0; p) \leq \tau_3(t_0; q) = \eta_3(t_0; q) < t_1. \qquad \square$$

PROBLEM 2.5. Let p and q be of constant sign and $|p(t)| \geq |q(t)|$ for $t \geq 0$. Then the interval between any two consecutive zeros of a solution of the equation (2.35) contains at most two zeros of a solution of the equation (2.40).

Now we consider the fourth order equation

$$u^{(4)} = p(t)u. \tag{2.41}$$

Taking into account the equalities $M_4^* = m_4^* = 1$ and $M_{*4} = m_{*4} = 9/16$, as in the proofs of Theorems 2.11 and 2.12 we can derive the following propositions.

THEOREM 2.14. (i) If the inequality (1.2) holds,

$$\lim_{t \to +\infty} \sup \ln t \int_t^{+\infty} s^3 \left[p(s) + \frac{1}{s^4} \right]_- ds = +\infty$$

and

$$\int_1^{+\infty} t^3 \ln^2 t \left[p(t) + \frac{1}{t^4} \right]_+ dt < +\infty,$$

then the equation (2.41) has property A;
(ii) if

$$\int_1^{+\infty} t^3 \ln t \left[p(t) + \frac{1}{t^4} \right]_- dt < +\infty$$

and

$$\int_1^{+\infty} t^3 \ln t \left[p(t) - \frac{9}{16t^4} \right]_+ dt < +\infty,$$

then the equation (2.41) is nonoscillatory;
(iii) if the inequality (1.3) holds,

$$\lim_{t \to +\infty} \sup \ln t \int_t^{+\infty} s^3 \left[p(s) - \frac{9}{16s^4} \right]_+ ds = +\infty$$

and

$$\int_1^{+\infty} t^3 \ln^2 t \left[p(t) - \frac{9}{16t^4} \right]_- dt < +\infty,$$

then the equation (2.41) has property B.

THEOREM 2.15. (i) If p is nonpositive and either

$$\lim_{t \to +\infty} \inf t \int_t^{+\infty} s^2 |p(s)| ds > 1$$

or

$$\lim_{t \to +\infty} \sup t \int_t^{+\infty} s^2 |p(s)| ds > 6,$$

then the equation (2.41) has property A;
(ii) if there exists $t_0 \in \mathbf{R}_+$ such that

$$t \int_t^{+\infty} s^2 [p(s)]_- ds \leq 1, \qquad t^2 \int_t^{+\infty} s[p(s)]_+ ds \leq \frac{9}{32} \qquad \text{for } t \geq t_0,$$

then the equation (2.41) is nonoscillatory;

(iii) if p is nonnegative and either

$$\liminf_{t\to+\infty} t^2 \int_t^{+\infty} sp(s)\,ds > \frac{9}{32}$$

or

$$\limsup_{t\to+\infty} t \int_t^{+\infty} s^2 p(s)\,ds > 4,$$

then the equation (2.41) has property B.

LEMMA 2.10. If (1.3) holds, then

$$\tau_4(t_0; p) = \eta_4(t_0; p). \tag{2.42}$$

PROOF. Assume the contrary. Then, in view of Remark 1.7, $\tau_4(t_0; p) > \eta_4(t_0; p)$. According to Lemmas 1.5 and 1.9 and Remark 1.6 there exists a solution u of the equation (2.41) satisfying the conditions

$$u(t_0) = u'(t_0) = u(t^0) = u'(t^0) = 0, \qquad u''(t_0) > 0,$$

$$u(t) > 0 \qquad \text{for } t \in]t_0, t^0[\,\cup\,]t^0, +\infty[,$$

where $t^0 = \eta_4(t_0; p)$. By the definition of $\tau_4(t_0; p)$ we can find $t_1 \in]t^0, \tau_4(t_0; p)]$ and a solution v_0 of the equation (2.41) so that

$$v_0(t_0) = v_0(t_1) = 0, \qquad v_0(t) > 0 \qquad \text{for } t \in]t_0, t_1[.$$

Let v be a nontrivial solution of the equation (2.41) under the conditions

$$v(t_0) = v'(t_0) = v(t_1) = 0, \qquad v''(t_0) \geq 0.$$

Lemma 1.5 yields $v''(t_0) > 0$. First we assume that $v(t^0) > 0$ and consider the solution

$$w = u - \frac{u''(t_0)}{v''(t_0)} v.$$

Clearly, $w(t^0) < 0$ and $w(t_1) > 0$. Thus $w(t_2) = 0$ for a certain $t_2 \in]t^0, t_1[$. Because of the equalities $w(t_0) = w'(t_0) = w''(t_0) = 0$, this contradicts Lemma 1.5. So $v(t^0) \leq 0$ and, since

$$\frac{v(t)}{v_0'(t)} \downarrow 0 \qquad \text{as } t \downarrow t_0,$$

we obtain

$$v(t_*)v_0'(t_*) - v_0(t_*)v_0(t_*) = 0$$

for some $t_* \in]t_0, t^0[$, i.e. there exist c_1 and c_2 such that $v_1(t_*) = v_1'(t_*) = 0$, where $v_1 = c_1 v + c_2 v_0$ is a nontrivial solution of the equation (2.41). Moreover, $v_1(t_0) = v_1(t_1) = 0$, which also contradicts Lemma 1.5. $\square$

Note that, as the following proposition implies, if the inequality (1.2) holds, then, in general, (2.42) is violated.

LEMMA 2.11. We have

$$\eta_4(0; -1) < 2\pi\sqrt{2} = \tau_4(0; -1).$$

PROOF. By Lemma 1.9 and Remark 1.6, $\eta_4(0; -1)$ is the smallest positive zero of a solution of the equation

$$v^{(4)} = -v \tag{2.43}$$

under the conditions

$$v(0) = v'(0) = v''(0) = 0, \qquad v'''(0) > 0.$$

We can immediately show that

$$v(t) = \sin\left(\frac{\sqrt{2}}{2}t - \frac{\pi}{4}\right)\exp\left(\frac{\sqrt{2}}{2}t\right) + \sin\left(\frac{\sqrt{2}}{2}t + \frac{\pi}{4}\right)\exp\left(-\frac{\sqrt{2}}{2}t\right)$$

is such a solution. Since $v(2\pi\sqrt{2}) < 0$, we get $\eta_4(0; -1) < 2\pi\sqrt{2}$.

On the other hand, the function

$$u(t) = \sin\left(\frac{\sqrt{2}}{2}t\right)\left(\exp\left(\frac{\sqrt{2}}{2}t - \pi\right) - \exp\left(-\frac{\sqrt{2}}{2}t + \pi\right)\right)$$

is a solution of the equation (2.43) satisfying the conditions

$$u(0) = u(2\pi\sqrt{2}) = 0, \qquad u(t) \geq 0 \qquad \text{for } t \in [0, 2\pi\sqrt{2}].$$

Also, $u(\pi\sqrt{2}) = 0$ and

$$u(\pi\sqrt{2} - t) = u(\pi\sqrt{2} + t) \qquad \text{for } t \in [0, \pi\sqrt{2}]. \tag{2.44}$$

Therefore Lemma 1.11 yields $\tau_4(0; 1) \geq 2\pi\sqrt{2}$.

Suppose $\tau_4(0; -1) > 2\pi\sqrt{2}$. Then there exist $t_0 > \pi\sqrt{2}$ and a solution u_0 of the equation (2.43) for which

$$u_0(-t_0 + \pi\sqrt{2}) = u_0(t_0 + \pi\sqrt{2}) = 0,$$

$$u_0(t) > 0 \qquad \text{for } t \in]\pi\sqrt{2} - t_0, \pi\sqrt{2} + t_0[.$$

Without loss of generality we may assume that

$$u_0(\pi\sqrt{2} - t) = u_0(\pi\sqrt{2} + t) \qquad \text{for } t \in [0, \pi\sqrt{2}], \tag{2.45}$$

because otherwise we can consider the solution $u_0(t) + u_0(2\pi\sqrt{2} - t)$.

Clearly, u_0 and u are linearly independent. Thus, for certain $c > 0$ and $t_* \in]0, \pi\sqrt{2}[$ we get $w(t_*) = w'(t_*) = 0$, where $w = u_0 - cu$. By (2.44) and (2.45), $w(2\pi\sqrt{2} - t_*) = w'(2\pi\sqrt{2} - t_*) = 0$. So w is a nontrivial solution of the equation (2.43) and has two double zeros. This contradicts Lemma 1.4. $\square$

In particular, Lemma 2.11 implies that if p is nonpositive, then the interval between two consecutive zeros of a solution of the equation (2.43) can contain four zeros of another solution of the same equation (see [212], p. 270).

To derive an upper bound for $\tau_4(t_0; p)$ when (1.2) holds, we will use the second conjugate number (see Definition 1.5).

LEMMA 2.12. If the inequality (1.2) holds, then

$$\tau_4(t_0; p) \leq \eta_4^2(t_0; p).$$

PROOF. Assume the contrary. Then by the definition of $\tau_4(t_0; p)$ and Lemma 1.10 there exist $s_1 > t_1 = \eta_4^2(t_0; p)$ and solutions u_0 and v_0 such that

$$v_0(t_0) = v_0(s_1) = 0, \qquad v_0(t) > 0 \qquad \text{for } t \in \,]t_0, s_1[,$$

$$u_0(t_0) = u_0'(t_0) = u_0''(t_0) = u_0(t^0) = u_0(t_1) = 0,$$

$$u_0(t) < 0 \qquad \text{for } t \in \,]t_0, t_1[, \qquad u_0(t) > 0 \qquad \text{for } t \in \,]t^0, t_1[,$$

where $t^0 = \eta_4(t_0; p)$.

Consider the solution $v_\epsilon = v_0 - \epsilon u_0$. Let $\epsilon = \epsilon_0$ be the largest constant for which $v_\epsilon(t) \geq 0$ on $[t^0, t_1]$. It is easily to shown that we can choose $t^* \in \,]t^0, t_1[$ so that

$$v_{\epsilon_0}(t_0) = v_{\epsilon_0}(t^*) = v_{\epsilon_0}'(t^*) = 0, \qquad v_{\epsilon_0}(t) > 0 \qquad \text{for } t \in \,]t_0, t^*[.$$

Furthermore, since $\eta_4(\cdot; p)$ is an increasing function, $\eta_4(t_2; p) = t^*$ for a certain $t_2 \in \,]t_0, t^*[$. Hence, taking into account Lemma 1.9, we can prove that there exists a solution u satisfying the conditions

$$u(t_2) = u(t^*) = u'(t^*) = u''(t^*) = 0, \qquad u(t) > 0 \qquad \text{for } t \in \,]t_2, t^*[.$$

Now consider the solution $u_\epsilon = v_{\epsilon_0} - \epsilon u$. If $\epsilon = \epsilon_1$ is the largest constant such that $u_\epsilon(t) \geq 0$ on $[t_2, t^*]$, then $u_{\epsilon_1}(t_*) = u_{\epsilon_1}'(t_*) = 0$ for a certain $t_* \in \,]t_2, t^*[$. But since $u_{\epsilon_0}(t^*) = u_{\epsilon_1}'(t^*) = 0$, this contradicts Lemma 1.4. $\square$

By means of Lemmas 2.10 and 2.12 we can prove the following comparison theorem.

THEOREM 2.16. If the inequality (1.41) (the inequality (1.42)) holds, then the interval between any two consecutive zeros of a solution of the equation (2.41) contains at most four (three) zeros of a solution of the equation

$$v^{(4)} = q(t)v. \tag{2.46}$$

PROOF. Assume the contrary. Then there exist solutions u and v of the equations (2.41) and (2.46), respectively, such that

$$u(t_0) = u(t_1) = 0, \qquad u(t) > 0 \qquad \text{for } t \in \,]t_0, t_1[,$$

and v has at least five (four) zeros in the interval $]t_0, t_1[$. Hence $\eta_4^2(t_0; q) < t_1$ $(\eta_4(t_0; q) < t_1)$ and $t_1 \leq \tau_4(t_0; p)$. So, by (2.40) and Lemmas 1.16 and 2.12 we obtain a contradiction: $t_1 \leq \tau_4(t_0; p) \leq \tau_4(t_0; q) < t_1$. $\square$

Finally, we consider the sixth order equation

$$u^{(6)} = p(t)u. \tag{2.47}$$

By Lemma 2.8, if p is nonnegative and the equation (2.47) is oscillatory, then this equation has property B. (As we mentioned at the beginning of §2, for nonpositive p oscillation of the equation (2.47) does not imply that this equation has property A.) Furthermore, note that $m_6^* = 225/64$, $M_6^* = 16(7\sqrt{7} + 10)/27$, $m_{*6} = M_{*6} = 16(7\sqrt{7} - 10)/27$. So the following assertions are true.

THEOREM 2.17. If the inequality (1.3) holds,

$$\lim_{t \to +\infty} \sup \ln t \int_t^{+\infty} s^5 \left[p(s) - \frac{16(7\sqrt{7} - 10)}{27s^6} \right]_+ ds = +\infty$$

and

$$\int_1^{+\infty} t^5 \ln^2 t \left[p(t) - \frac{16(7\sqrt{7} - 10)}{27t^6} \right]_- dt < +\infty,$$

then the equation (2.47) has property B.

THEOREM 2.18. If p is nonnegative and either

$$\lim_{t \to +\infty} \inf t^2 \int_t^{+\infty} s^3 p(s)\, ds > \frac{32(7\sqrt{7} - 10)}{135}$$

or

$$\lim_{t \to +\infty} \sup t \int_t^{+\infty} s^4 p(s)\, ds > 48,$$

then the equation (2.47) has property B.

LEMMA 2.13. Let the inequality (1.3) hold. Then

$$\tau_6(t_0; p) < \eta_6^6(t_0; p) \tag{2.48}$$

if $\tau_6(t_0; p) < +\infty$, and $\eta_6^6(t_0; p) = +\infty$ if $\tau_6(t_0; p) = +\infty$.

PROOF. First suppose $\tau_6(t_0; p) < +\infty$, but $\tau_6(t_0; p) \geq \eta_6^6(t_0; p)$. Then by the definition of $\eta_6^6(t_0; p)$ there exists a solution of the equation (2.47) with 11 zeros in the interval $[t_0, \eta_6^6(t_0; p)]$ and by Lemmas 1.5 and 1.12 there exists a solution u of the same equation which has exactly 6 zeros in the interval $[t_0, t^0]$, $t^0 = \tau_6(t_0; p)$, and satisfies the conditions

$$u(t_0) = u(t^0) = 0, \qquad u(t) \geq 0 \qquad \text{for } t \in [t_0, t^0],$$

$$u(t) > 0 \qquad \text{for } t \in [0, t_0[\,\cup\,]t^0, +\infty[,$$

t_0 and t^0 being zeros of even multiplicity.

If $u(t) > 0$ for $t \in\,]t_0, t^0[$, then either

$$\lambda_6(u; t_0) = 4, \qquad \lambda_6(u; t^0) = 2 \tag{2.49}$$

or

$$\lambda_6(u; t_0) = 2, \qquad \lambda_6(u; t^0) = 4 \tag{2.50}$$

(recall that $\lambda_6(u; t_0)$ denotes the multiplicity of a zero t_0 of a solution u).

Let (2.49) hold. It is easy to verify that

$$\eta_6(t_0; p) < \eta_6^6(t_0; p) \leq t_0 = \tau_6(t_0; p).$$

By Lemma 1.9 the equation (2.47) has a solution v satisfying the conditions

$$v(t) > 0 \qquad \text{for } t \neq t_0, \, t \neq \eta_6(t_0; p), \qquad v(t_0) = v(\eta_6(t_0; p)) = 0,$$

and we may assume that

$$\lambda_6(v; t_0) = 4, \qquad \lambda_6(v; \eta_6(t_0; p)) = 2.$$

Consider the solution

$$w = u - \frac{u^{(4)}(t_0)}{v^{(4)}(t_0)}v.$$

Then

$$w^{(i)}(t_0) = 0 \qquad (i = 0, \dots, 4), \qquad w(t_*) = 0$$

for a certain $t_* \in \,]\eta_6(t_0; p), t^0[$. But this contradicts Lemma 1.5.

We can similarly show that the case (2.50) is also impossible. Hence there exists $t_1 \in \,]t_0, t^0[$ for which $u(t_1) = 0$, $\lambda_6(u; t_0) = \lambda_6(u; t_1) = \lambda_6(u; t^0) = 2$, and

$$u(t) > 0 \qquad \text{on }]t_0, t_1[\,\cup\,]t_1, t^0[.$$

Since the equation (2.47) has a solution with 11 zeros in $[t_0, t^0]$, either the interval $[t_0, t_1]$ or the interval $[t_1, t^0]$ contains 6 zeros of this solution. For the sake of being specific we assume that such is $[t_1, t^0]$. Then $\eta_6(t_1; p) \leq t^0$. In view of Lemma 1.9 the equation (2.47) has a solution u_1 such that

$$u_1^{(i)}(t_1) = 0 \qquad (i = 0, \dots, 4), \qquad u_1(\eta_6(t_1; p)) = u_1'(\eta_6(t_1; p)) = 0,$$

$$u_1(t) > 0 \qquad \text{for } t \neq t_1,\, t \neq \eta_6(t_1; p).$$

If $\eta_6(t_1; p) = t^0$, then the solution

$$w_1 = u - \frac{u''(t^0)}{u_1''(t^0)}u_1$$

satisfies the equalities

$$w_1(t_1) = w_1'(t_1) = 0, \qquad w_1(t^0) = w_1'(t^0) = w_1''(t^0) = 0,$$

and $w_1(t_*) = 0$ for some $t_* \in \,]t_0, t_1[$. But this contradicts Lemma 1.5.

Furthermore, if $\eta_6(t_1; p) < t^0$, then there exists $c > 0$ such that the solution $w_2 = u - cu_1$ is subject to the condition

$$w_2(t_*) = w_2(t^*) = w_2'(t^*) = w_2'(t^{**}) = 0$$

for certain $t_* \in \,]t_0, t_1[$, $t^* \in \,]t_1, t^0[$ and $t^{**} \in \,]t^*, t^0[$. Moreover, $w_2(t_1) = w_2'(t_1) = 0$, which contradicts Lemma 1.5 as well.

Consequently, if $\tau_6(t_0; p) < +\infty$, then (2.48) is true.

Now let $\tau_6(t_0; p) = +\infty$ and $\eta_6^6(t_0; p) < +\infty$. Then there exist $t^0 > \eta_6(t_0; p)$ and a solution u of the equation (2.47) under the conditions

$$u(t_0) = u(t^0) = 0, \qquad u(t) > 0 \qquad \text{for } t \in \,]t_0, t^0[.$$

Consider the equation

$$u^{(6)} = p_*(t)u \tag{2.51}$$

with

$$p_*(t) = \begin{cases} p(t) & \text{for } t \in [0, t^0] \\ p(t) + 1 & \text{for } t \in \,]t^0, +\infty[. \end{cases}$$

Since the equation (2.51) has property B, $\tau_6(t_0; p_*) < +\infty$. On the other hand, $\eta_6^6(t_0; p_*) = \eta_6^6(t_0; p) < t_0$ and $\tau_6(t_0; p_*) \geq t_0$, which contradicts the already proven part of the lemma. $\square$

THEOREM 2.19. If the inequality (1.42) holds, then the interval between any two consecutive zeros of a solution of the equation (2.47) contains at most ten zeros of a solution of the equation

$$v^{(6)} = q(t)v.$$

PROOF. Assume the contrary. Then

$$t_1 \leq \tau_6(t_0; p), \qquad \eta_6^6(t_0; q) \leq t_1$$

for certain $t_0, t_1 \in \mathbf{R}_+$. Hence, by Lemmas 1.16 and 2.13 we arrive at a contradiction:

$$t_1 \leq \tau_6(t_0; p) \leq \tau_6(t_0; q) < \eta_6^6(t_0; q) \leq t_1. \qquad \square$$

As mentioned in Remark 2.1, in case of the inequality (1.41) the analogue of this theorem is not true. Namely, there exists a function p satisfying (1.2) for which the interval between two consecutive zeros of a solution of the equation (2.47) may contain any number of zeros of another solution of this equation.

REMARK 2.6. Note that $\eta_6(0; 1) < \tau_6(0; 1)$. Indeed, consider the set of solutions

$$u_c(t) = e^t + e^{-t} - (e^{t/2} + e^{-t/2}) \left(\cos \frac{\sqrt{3}}{2} t - c \sin \frac{\sqrt{3}}{2} t \right)$$

of the equation

$$u^{(6)} = u. \tag{2.52}$$

Clearly, $u_c(0) = u_c'(0) = 0$ and for sufficiently small $c > 0$ we obtain $u_c(t) > 0$ if $t \neq 0$. Set $c_0 = \sup\{c > 0 : u_c(t) > 0 \text{ for } t \neq 0\}$. Then u_{c_0} has three double zeros: $-t_0$, 0 and t_0, where $t_0 \in \,]2\sqrt{3}\pi/3, \sqrt{3}\pi[$, and $u_{c_0}(t) > 0$ for $t \notin \{-t_0, 0, t_0\}$. Thus, according to Lemma 1.11 and Definition 1.5, $\tau_6(0; 1) \geq 2t_0 \geq \eta_6(0; 1)$. If $\eta_6(0; 1) = 2t_0$, then by Lemma 1.9 and Remark 1.6 there exists a solution v of the equation (2.52) such that

$$v(-t_0) = v'(-t_0) = v''(-t_0) = v'''(-t_0) = v(t_0) = v'(t_0) = 0,$$

$$v(t) > 0 \qquad \text{for } t \in \,]-t_0, t_0[.$$

Let $u(t) = v(t) + v(-t)$. Then the function u is even, $u(-t_0) = u'(-t_0) = u(t_0) = u'(t_0) = 0$, and $u(t) > 0$ for $t \in \,]-t_0, t_0[$. Consider the set of solutions $v_c = u - c u_{c_0}$ of the equation (2.52). Put $c_1 = \sup\{c > 0 : v_c(t) \geq 0 \text{ for } t \in [-t_0, t_0]\}$. Then v_{c_1} has either two triple or four double zeros in the interval $[-t_0, t_0]$, which contradicts Lemma 1.5. So $\eta_6(0; 1) < 2t_0 \leq \tau_6(0; 1)$.

If for any nonpositive (nonnegative) function p the maximal number of zeros of a nontrivial solution of the equation (1.1) which can lie between two consecutive zeros of another solution of this equation is finite; we denote that number by z_n^* (z_{*n}). Otherwise we set $z_n^* = +\infty$ ($z_{*n} = +\infty$).

By Remark 2.1, $z_n^* = +\infty$ for $n \notin \{3, 4\}$ and $z_{*n} = +\infty$ for $n \notin \{3, 4, 6\}$. Theorem 2.13 yields $z_3^* = z_{*3} = 2$. It follows from Lemma 2.11 and Theorem 2.16 that $z_4^* = 4$ and $z_{*4} = 3$. Theorem 2.19 and Remark 2.6 imply $z_{*6} \in \{6, \ldots, 10\}$.

PROBLEM 2.6*. What is the value of z_{*6}?

PROBLEM 2.7. For any $n \geq 3$ and $m \in \mathbf{N}$ there exists a continuous function $p \colon \mathbf{R}_+ \to \mathbf{R}$ which changes sign and is such that at least m zeros of a solution of the equation (1.1) lie between two consecutive zeros of another solution of this equation.

Notes. An example similar to the one in Remark 2.1 was first constructed by V.A. Kondratyev [213]. For $n \in \{3,4\}$ he also showed that if p changes sign, then the interval between two consecutive zeros of a solution of the equation (1.1) can contain arbitrarily many zeros of another solution [212].

When $p^*(t) \neq 0$ for $t \geq 0$, Lemma 2.1 can be derived from results in [100, 101]. Corollary 2.1 under the same restriction on p^* was proved in [100, 257, 292]. For Problems 2.2 and 2.3 see [259] and [213], respectively. Theorem 2.1 ((i)⇔(iv)) was proved in [100].

In case of either nonpositive p and $n = 4m + 2$ or nonnegative p and $n = 4m$, Theorems 2.3, 2.4, 2.8 and 2.9 improve certain results of I.M. Glazman [118] (Theorems 9, 11 and 12). In order to verify this one should take into account that for those above-mentioned p and n, by Corollary 2.4 oscillation of the equation (1.1) is equivalent to $(n/2, n/2)$ conjugacy in any neighborhood of $+\infty$ (oscillation by I.M. Glazman).

For nonpositive p, Lemma 2.8 with $n = 3$ was proved by V.A. Kondratyev [212] and with $n = 4$ by V.A. Kondratyev [212], W. Leighton and Z. Nehari [257] and M. Švec [342].

Corollary 2.10 sharpens a result in [212]. Theorems 2.13 and 2.16 with $p = q$ were proved in [212].

For the other results of this section see [68, 74].

§3. Oscillation Properties of Solutions of Equations with Strongly Oscillating Coefficients

3.1. Zeros of solutions on a finite interval. As is well known, the classical Sturm comparison theorem states that if $n = 2$ and $p(t) \le q(t)$ for $t \in [a, b]$, then the interval between any two consecutive zeros of a solution of the equation (1.1) contains at most one zero of a solution of the equation (1.40). The example constructed in Remark 2.1 shows that, in general, for $n \ne 2$ an arbitrary number of zeros of a solution of the equation (1.1) can lie between two consecutive zeros of another solution of this equation.

However, the Sturm theorem has the following, equivalent, form: if $n = 2$, $p(t) \le q(t)$ for $t \in [a, b]$ and every solution of the equation (1.40) has a zero in the interval $[a, b]$, then every solution of the equation (1.1) also has a zero in this interval. It was found that in this form the theorem remains true in the case of higher order equations, both for even and odd n (with some natural changes for odd n).

THEOREM 3.1. Let n be even,

$$p(t) \le q(t) \le 0 \qquad \text{for } t \in [a, b], \tag{3.1}$$

and let every solution of the equation (1.40) have a zero in the interval $[a, b]$. Then every solution of the equation (1.1) also has a zero in this interval.

PROOF. Assume the contrary. Then the equation (1.1) has a solution u such that

$$u(t) > 0 \qquad \text{for } t \in [a, b]. \tag{3.2}$$

Suppose p, q and the solution u are defined on $\mathbf{R}$ and

$$p(t) \le q(t) \le 0 \qquad \text{for } t \in \mathbf{R}, \qquad -q(t) \ge c > 0 \qquad \text{for } |t| \ge t_*,$$

where t_* is a sufficiently large constant. Theorem 1.6 implies that the equation (1.1) has property A in the interval $\mathbf{R}_+$, i.e. any solution of it is oscillatory in this interval. Via the transformation of the independent variable $t = -s$ we can similarly conclude that any solution of the equation (1.1) is oscillatory in $\mathbf{R}_-$ as well. Hence, there exist $t_0 < a$ and $t_1 > b$ for which

$$u(t_0) = u(t_1) = 0, \qquad u(t) > 0 \qquad \text{in }]t_0, t_1[.$$

Consequently, $\tau_n(t_0; p) \ge t_1$, and by Lemma 1.16, $\tau_n(t_0; q) \ge \tau_n(t_0; p) \ge t_1$. So we can find $t_2 \in]b, \tau_n(t_0; q)[$ and a solution v of the equation (1.40) for which

$$v(t_0) = v(t_2) = 0, \qquad v(t) > 0 \qquad \text{for } t \in]t_0, t_2[.$$

But this contradicts the hypotheses of the theorem. $\square$

THEOREM 3.2. Let n be odd and let the inequality (3.1) hold. Suppose that every solution of the equation (1.40) either has a zero in the interval $[a, b]$ or satisfies the condition

$$(-1)^{i-1} v^{(i-1)}(a) v(a) > 0 \qquad (i = 1, \dots, n).$$

Then every solution of the equation (1.1) either has a zero in this interval or satisfies the condition

$$(-1)^{i-1} u^{(i-1)}(a) u(a) > 0 \qquad (i = 1, \dots, n). \tag{3.3}$$

By applying the transformation of the independent variable $s = b + a - t$, we reduce Theorem 3.2 to the following assertion.

THEOREM 3.3. Let n be odd,

$$p(t) \geq q(t) \geq 0 \qquad \text{for } t \in [a, b], \tag{3.4}$$

and let every solution of the equation (1.40) either have a zero in the interval $[a, b]$ or satisfy the condition

$$v^{(i-1)}(b)v(b) > 0 \qquad (i = 1, \dots, n).$$

Then every solution of the equation (1.1) either has a zero in this interval or satisfies the condition

$$u^{(i-1)}(b)u(b) > 0 \qquad (i = 1, \dots, n). \tag{3.5}$$

PROOF. Suppose the equation (1.1) has a solution u satisfying the inequality (3.2) and $u^{(i_0-1)}(b) < 0$ for a certain $i_0 \in \{1, \dots, n\}$. We assume that p, q and the solution u are defined on $\mathbf{R}$ so that

$$p(t) \geq q(t) \geq 0 \qquad \text{for } t \in \mathbf{R}, \qquad q(t) \geq c > 0 \qquad \text{for } |t| \geq t_*,$$

where t_* is a sufficiently large constant.

Set

$$w(t) = u(-t), \qquad p_*(t) = -p(-t) \qquad \text{for } t \geq -b.$$

Then w is a solution of the equation

$$w^{(n)} + p_*(t)w = 0. \tag{3.6}$$

Since $p_*(t) \geq c > 0$ for $t \geq t_*$, the equation (3.6) has property A. If w is nonoscillatory, we have

$$(-1)^{i-1}w^{(i-1)}(t) > 0 \qquad \text{for } t \geq -b \qquad (i = 1, \dots, n).$$

But $(-1)^{i_0-1}w^{(i_0-1)}(b) = u^{(i_0-1)}(b) < 0$. Hence w is an oscillatory solution of the equation (3.6) and there exists $t_0 < a$ such that

$$w(-t_0) = 0, \qquad w(t) > 0 \qquad \text{for } t \in [-b, -t_0[.$$

This yields

$$u(t_0) = 0, \qquad u(t) > 0 \qquad \text{for } t \in]t_0, b].$$

The equation (1.1) has property B in the interval $[t_0, +\infty[$. Thus, by Theorem 1.2 ((i)$\leftrightarrow$(v)), $\tau_n(t_0; p) < +\infty$, and in view of Lemmas 1.14–1.16, $\tau_n(t_0; q) \geq \tau_n(t_0; p) > b$. Therefore, according to the definition of $\tau_n(t_0; q)$ there exist $t_1 \in]b, \tau_n(t_0; q)]$ and a solution v of the equation (1.40) satisfying the conditions

$$v(t_0) = v(t_1) = 0, \qquad v(t) > 0 \qquad \text{for } t \in]t_0, t_1[. \tag{3.7}$$

The hypotheses of the theorem imply $v^{(i-1)}(b) > 0$ $(i = 1, \dots, n)$. So

$$v^{(i-1)}(t) > 0 \qquad \text{for } t \in [b, +\infty[\qquad (i = 1, \dots, n),$$

which contradicts the condition (3.7). $\square$

PROBLEM 3.1. If either (3.1) or (3.4) holds and the equation (1.40) is conjugate in $[a, b]$, then the equation (1.1) is also conjugate in $[a, b]$.

Set

$$\gamma_n = [\tau_n(0; -1)]^n. \tag{3.8}$$

Recall that $\tau_n(0; -1)$ is the least upper bound of those $t_0 > 0$ for which there exists a solution v of the equation

$$v^{(n)} = -v \tag{3.9}$$

under the conditions

$$v(0) = v(t_0) = 0, \qquad v(t) > 0 \qquad \text{for } t \in \,]0, t_0[.$$

Furthermore, note that $\tau_n(0; 1) = \tau_n(0; -1)$ for n odd, because if v is a solution of (3.9), then $v(t_0 - t)$ is a solution of the equation

$$v^{(n)} = v. \tag{3.10}$$

THEOREM 3.4. Let n be even and

$$(b - a)^n p(t) \le -\gamma_n \qquad \text{for } t \in [a, b]. \tag{3.11}$$

Then every solution of the equation (1.1) has a zero in the interval $[a, b]$.

PROOF. According to Theorem 3.1 it suffices to show that every solution of the equation

$$w^{(n)} = -\frac{\gamma_n}{(b - a)^n} w \tag{3.12}$$

has a zero in the interval $[a, b]$. Assume the contrary. Then there exists a solution w of the equation (3.12) which is positive on $[a, b]$. Since any solution of (3.12) is oscillatory both in $\mathbf{R}_+$ and in $\mathbf{R}_-$, we can choose $t_0 < a$ and $t_1 > b$ so that

$$w(t_0) = w(t_1) = 0, \qquad w(t) > 0 \qquad \text{for } t \in \,]t_0, t_1[.$$

Clearly, the function

$$v(s) = w(t), \qquad s = \frac{\sqrt[n]{\gamma_n}}{b - a}(t - t_0),$$

is a solution of the equation (3.8). Moreover,

$$v(0) = v(s_0) = 0, \qquad v(s) > 0 \qquad \text{for } s \in \,]0, s_0[,$$

where

$$s_0 = \frac{\sqrt[n]{\gamma_n}}{b - a}(t_1 - t_0) > \sqrt[n]{\gamma_n} = \tau_n(0; 1).$$

But this contradicts the definition of $\tau_n(0; 1)$. $\square$

THEOREM 3.5. Let n be odd. If the inequality (3.11) holds, then every solution of the equation (1.1) which does not satisfy the condition (3.3) has a zero in $[a, b]$. Furthermore, if

$$(b - a)^n p(t) \ge \gamma_n \qquad \text{on } [a, b], \tag{3.13}$$

then the same is true for every solution of the equation (1.1) which does not satisfy the condition (3.5).

PROOF. As already mentioned, via a transformation of the independent variable the case of (3.11) can be reduced to the case of (3.13). So we assume that (3.13) holds. By Theorem 3.3 it suffices to prove that any solution of the equation

$$w^{(n)} = \frac{\gamma_n}{(b-a)^n} w \qquad (3.14)$$

either has a zero in $[a, b]$ or satisfies the inequalities

$$w^{(i-1)}(b)w(b) > 0 \qquad (i = 1, \dots, n). \qquad (3.15)$$

Let w be a solution of the equation (3.14). If

$$w_*(t) = w(b - t) \qquad \text{for } t \in \mathbf{R}_+, \qquad (3.16)$$

then we easily conclude that w_* satisfies the equation (3.12). Since this equation has property A in $\mathbf{R}_+$, either

$$(-1)^{i-1} w_*^{(i-1)}(t) w_*(t) > 0 \qquad \text{for } t \in \mathbf{R}_+ \qquad (i = 1, \dots, n)$$

or w_* is oscillatory, i.e., in view of (3.16), either

$$w^{(i-1)}(t)w(t) > 0 \qquad \text{for } t \le b \qquad (i = 1, \dots, n)$$

or w has a zero in $]-\infty, b[$. In the former case we have (3.15). In the latter we let t_0 be the largest zero of w in $]-\infty, b]$ and assume that $t_0 < a$. The function

$$v(s) = w(t), \qquad s = \frac{\sqrt[n]{\gamma_n}}{b-a}(t - t_0),$$

is a solution of the equation (3.10). Moreover, $v(0) = w(t_0) = 0$ and

$$s_0 = \frac{\sqrt[n]{\gamma_n}}{b-a}(b - t_0) > \tau_n(0; 1) = \tau_n(0; -1).$$

Thus, by Lemmas 1.14 and 1.15, $v^{(i-1)}(s_0)v(s_0) > 0$ $(i = 1, \dots, n)$, i.e. (3.15) is fulfilled. $\square$

REMARK 3.1. Observe that there is no $\mu_n = \mu_n(a, b) > 0$ such that we can replace (3.11) (and hence (3.13)) in Theorems 3.4 and 3.5 by the integral condition

$$\int_a^b |p(t)|\, dt \ge \mu_n.$$

Indeed, consider the equation

$$u^{(n)} = -c(t - a + \epsilon)^{-n} u \qquad (3.17)$$

with $c \in]0, m_n^*]$ and

$$\epsilon \in \left]0, \left(\frac{(n-1)\mu_n}{c} + (b-a)^{-n+1}\right)^{1/(1-n)}\right].$$

It is obvious that this equation is discongugate in $]a - \epsilon, +\infty[$. Hence the solution of (3.17) under the initial condition

$$u^{(i)}\left(a - \frac{\epsilon}{2}\right) = 0 \qquad (i = 0, \dots, n-2), \qquad u^{(n-1)}\left(a - \frac{\epsilon}{2}\right) = 1$$

does not have a zero in the interval $[a, b]$ and for n odd does not satisfy (3.3). But

$$\int_a^b c(t - a + \epsilon)^{-n} dt \geq \frac{c}{n-1} \left(\epsilon^{1-n} - (b-a)^{1-n} \right) \geq \mu_n.$$

PROBLEM 3.2. (i) Let $n = 2$ and let p be nonpositive. If there exist $\alpha \in \,]a, b[$ and $\beta \in \,]\alpha, b[$ such that

$$\int_\alpha^\beta |p(t)| \, dt > \frac{1}{\alpha - a} + \frac{1}{b - \beta}, \tag{3.18}$$

then every solution of the equation (1.1) has a zero in $]a, b[$;
  (ii) the condition (3.18) cannot be replaced by the condition

$$\int_\alpha^\beta |p(t)| \, dt > \frac{c}{\alpha - a} + \frac{c}{b - \beta}$$

with $c \in \,]0, 1[$.

PROBLEM 3.3. If $n = 2$, p is nonpositive, and

$$\frac{1}{t-a} \int_a^t (\tau - a)^2 (b - \tau) |p(\tau)| \, d\tau + \frac{1}{b-t} \int_t^b (\tau - a)(b - \tau)^2 |p(\tau)| \, d\tau \geq b - a$$

$$\text{for } a < t < b,$$

then every solution of the equation (1.1) has a zero in $]a, b[$.

THEOREM 3.6. Let n be odd, $a = a_1 < b_1 < c < a_2 < b_2 = b$,

$$p(t) \geq 0 \quad \text{for } t \in [a, c], \qquad p(t) \leq 0 \quad \text{for } t \in [c, b],$$

and

$$(b_k - a_k)^n |p(t)| \geq \gamma_n \quad \text{for } t \in [a_k, b_k] \quad (k = 1, 2).$$

Then every solution of the equation (1.1) has a zero in $[a, b]$.

PROOF. If a solution u of the equation (1.1) does not have a zero in the interval $[a, b]$, then by Theorem 3.5,

$$(-1)^{i-1} u^{(i-1)}(a_2) u(a_2) > 0 \quad (i = 1, \ldots, n),$$

$$u^{(i-1)}(b_1) u(b_1) > 0 \quad (i = 1, \ldots, n).$$

This yields (see Problem 1.2)

$$(-1)^{i-1} u^{(i-1)}(t) u(t) > 0 \quad \text{for } t \in [c, a_2] \quad (i = 1, \ldots, n),$$

$$u^{(i-1)}(t) u(t) > 0 \quad \text{for } t \in [b_1, c] \quad (i = 1, \ldots, n).$$

When $t = c$, these inequalities contradict each other. $\square$

Note that the estimates given in Theorems 3.4 and 3.5 are best possible. However, the constant γ_n which appears in these estimates is defined implicitly. So, to obtain effective sufficient conditions we must derive an upper bound for it.

LEMMA 3.1. The estimate

$$\gamma_n \leq \left(n + \frac{n^2(n-1)}{2} \right)^n \gamma_n^*$$

with

$$\gamma_n^* = \begin{cases} \left(\frac{n-1}{2}!\right)^2 \left(\frac{\sqrt{n^2-1}}{2}\right)^{1-n} & \text{if } n = 2m+1 \\ \left(\frac{n}{2}!\right)^2 \left(\frac{n}{2}\right)^{-n} & \text{if } n = 4m+2 \\ \frac{n-2}{n}\left(\frac{n}{2}!\right)^2 \left(\frac{\sqrt{n^2-4}}{2}\right)^{-n} & \text{if } n = 4m \end{cases}$$

holds.

PROOF. Assume the contrary, i.e.

$$\left(n + \frac{n^2(n-1)}{2}\right)\sqrt[n]{\gamma^*} < \tau_n(0; -1).$$

Then for $a, b \in \mathbf{R}$ such that

$$b - a = \left(n + \frac{n^2(n-1)}{2}\right)\sqrt[n]{\gamma^*}$$

the equation (3.9) has a solution v which satisfies the inequality $v(t) > 0$ for $t \in [a, b]$ and does not satisfy the inequalities $(-1)^{i-1}v^{(i-1)}(a) > 0$ $(i = 1, \dots, n)$. As $v^{(n)}(t) \neq 0$ for $t \in [a, b]$, the function $v^{(i)}$ $(i = 1, \dots, n-1)$ can have at most $n - i$ zeros in $[a, b]$. Let $t_1 < t_2 < \dots < t_k$ be these zeros. Clearly, $k \leq n(n-1)/2$, and there exists $]a_0, b_0[\subset [a, b]$ such that

$$b_0 - a_0 \geq \frac{b - a}{1 + n(n-1)/2} = n\sqrt[n]{\gamma^*}, \tag{3.19}$$

$$v^{(i-1)}(t) \neq 0 \quad \text{for } t \in]a_0, b_0[\quad (i = 1, \dots, n),$$

and for at least one $i_* \in \{1, \dots, n\}$,

$$(-1)^{i_*}v^{(i_*-1)}(t)v(t) > 0 \quad \text{in }]a_0, b_0[.$$

Since $v^{(i)}$ is also a solution of the equation (3.9), without loss of generality we can assume the existence of $0 < i_1 < \dots < i_m < n$ (m being odd) for which

$$v^{(i)}(t)v(t) > 0 \quad (i = 0, \dots, i_1),$$
$$(-1)^{i-i_1}v^{(i)}(t)v^{(i_1)}(t) > 0 \quad (i = i_1, \dots, i_2),$$
$$v^{(i)}(t)v^{(i_2)}(t) > 0 \quad (i = i_2, \dots, i_3),$$
$$\dots\dots\dots\dots\dots$$
$$(-1)^{i-i_m}v^{(i)}(t)v^{(i_m)}(t) > 0 \quad (i = i_m, \dots, n)$$

in $]a_0, b_0[$. It is easy to verify that $l = i_2 - i_1 + \dots + n - i_m$ is odd.

The equality

$$v^{(i)}(t) = \sum_{j=i}^{k-1} \frac{v^{(j)}(s)}{(j-i)!}(t-s)^{j-i} + \frac{1}{(k-i-1)!}\int_s^t (t-\tau)^{k-i-1}v^{(k)}(\tau)\,d\tau \tag{3.20}$$

with $i = 0$, $k = i_1$ and $s = a_0$ implies

$$|v(t)| \geq \frac{1}{(i_1-1)!}\int_{a_0}^t (t-\tau)^{i_1-1}v^{(i_1)}(\tau)\,\text{sign}\,v(\tau)\,d\tau > |v^{(i_1)}(t)|\frac{(t-a_0)^{i_1}}{i_1!}$$

$$\text{for } t \in]a_0, b_0].$$

Setting $i = i_1$, $k = i_2$ and $s = b_0$, we similarly obtain

$$|v^{(i_1)}(t)| \geq \frac{(-1)^{i_2-i_1}}{(i_2 - i_1 - 1)!} \int_t^{b_0} (\tau - t)^{i_2-i_1-1} v^{(i_2)}(\tau) \operatorname{sign} v^{(i_1)}(\tau) \, d\tau$$

$$> |v^{(i_2)}(t)| \frac{(b_0 - t)^{i_2-i}}{(i_2 - i_1)!} \qquad \text{for } t \in [a_0, b_0[,$$

etc. Finally, (3.20) with $i = i_m$, $k = n$ and $s = b_0$ yields

$$|v^{(i_m)}(t)| \geq \frac{(-1)^{n-i_m}}{(n - i_m - 1)!} \int_t^{b_0} (\tau - t)^{n-i_m-1} v^{(n)}(\tau) \operatorname{sign} v^{(i_m)}(\tau) \, d\tau$$

$$= \frac{1}{(n - i_m - 1)!} \int_t^{b_0} (\tau - t)^{n-i_m-1} |v(\tau)| \, d\tau > |v(t)| \frac{(b_0 - t)^{n-i_m}}{(n - i_m)!}$$

$$\text{for } t \in [a_0, b_0[.$$

From these relations we obtain

$$\frac{(t - a_0)^{n-l}(b_0 - t)^l}{(n - l)!\, l!} < 1 \qquad \text{for } t \in [a_0, b_0],$$

which for

$$t = \frac{a_0 l + (n - l) b_0}{n}$$

becomes

$$(b_0 - a_0)^n < n^n \frac{l!\,(n - l)!}{l^l(n - l)^{n-l}}.$$

Since

$$\gamma_n^* = \max \left\{ \frac{l!\,(n - l)!}{l^l(n - l)^{n-l}} : l \in \{1, \dots, n - 1\}, l \text{ odd} \right\},$$

the last inequality contradicts (1.19). $\square$

3.2. Oscillation of all solutions of equations with strongly oscillating coefficients. Theorems 3.4 and 3.6 immediately imply the following assertions.

THEOREM 3.7. Let n be even and let there exist sequences $(a_k)_{k=1}^{+\infty}$ and $(b_k)_{k=1}^{+\infty}$ such that

$$0 \leq a_k < b_k < a_{k+1}, \qquad \lim_{k \to +\infty} a_k = +\infty, \tag{3.21}$$

$$(b_k - a_k)^n p(t) \leq -\gamma_n \qquad \text{for } t \in [a_k, b_k] \tag{3.22}$$

(γ_n being defined in (3.8)). Then every solution of the equation (1.1) is oscillatory.

COROLLARY 3.1. Let n be even and let there exist sequences $(a_k)_{k=1}^{+\infty}$ and $(b_k)_{k=1}^{+\infty}$ such that (3.21) holds and

$$\lim_{k \to +\infty} (b_k - a_k)^n p_k = +\infty, \tag{3.23}$$

where

$$p_k = \operatorname{ess\,inf}\{p(t) : t \in [a_k, b_k]\}.$$

Then every solution of the equation (1.1) is oscillatory.

THEOREM 3.8. Let n be odd and let there exist sequences $(a_k)_{k=1}^{+\infty}$, $(b_k)_{k=1}^{+\infty}$ and $(c_k)_{k=1}^{+\infty}$ such that

$$0 \le a_{2k-1} < b_{2k-1} < c_k < a_{2k} < b_{2k} < a_{2k+1}, \qquad \lim_{k \to +\infty} a_k = +\infty, \qquad (3.24)$$

$$p(t) \ge 0 \quad \text{for } t \in [a_{2k+1}, c_k], \qquad p(t) \le 0 \quad \text{for } t \in [c_k, b_{2k}], \qquad (3.25)$$

and

$$(b_k - a_k)^n |p(t)| \ge \gamma_n \quad \text{for } t \in [a_k, b_k].$$

Then every solution of the equation (1.1) is oscillatory.

COROLLARY 3.2. Let n be odd and let there exist sequences $(a_k)_{k=1}^{+\infty}$, $(b_k)_{k=1}^{+\infty}$ and $(c_k)_{k=1}^{+\infty}$ such that the conditions (3.23)–(3.25) hold, where

$$p_k = \text{ess inf}\{|p(t)| : t \in [a_k, b_k]\}.$$

Then every solution of the equation (1.1) is oscillatory.

EXAMPLES. 1) All the solutions of the equation

$$u^{(n)} + ct^\sigma \sin t^\mu u = 0,$$

where $\mu \in]0,1]$, $c \ne 0$ if $\sigma + n(1-\mu) > 0$ and

$$|c| \ge \left(\frac{\mu(n+1)}{\pi n}\right)^n \frac{\gamma_n}{\sin(\pi/2(n+1))}$$

if $\sigma + n(1-\mu) = 0$, are oscillatory.

Indeed, let n be even,

$$a_k^\mu = 2k\pi + \frac{\pi}{2(n+1)}, \qquad b_k^\mu = (2k+1)\pi - \frac{\pi}{2(n+1)} \qquad \text{for } c > 0$$

and

$$a_k^\mu = (2k-1)\pi - \frac{\pi}{2(n+1)}, \qquad b_k^\mu = 2k\pi - \frac{\pi}{2(n+1)} \qquad \text{for } c < 0.$$

Then the hypotheses of Theorem 3.7 are satisfied.

If n is odd,

$$a_k^\mu = k\pi + \frac{\pi}{2(n+1)}, \qquad b_k^\mu = (k+1)\pi - \frac{\pi}{2(n+1)},$$

$$c_k^\mu = 2k\pi \qquad \text{for } c > 0$$

and

$$a_k^\mu = (k+1)\pi + \frac{\pi}{2(n+1)}, \qquad b_k^\mu = (k+2)\pi - \frac{\pi}{2(n+1)},$$

$$c_k^\mu = (2k+1)\pi \qquad \text{for } c < 0,$$

then the hypotheses of Theorem 3.8 are fulfilled.

2) Similarly, all the solutions of the equation

$$u^{(n)} + ct^\sigma \sin \ln tu = 0, \qquad (3.26)$$

where $c \neq 0$ if $\sigma + n > 0$ and

$$|c| \geq \left(\frac{\exp \frac{\pi n}{n+1}}{\exp \frac{\pi n}{n+1} - 1} \right)^n \frac{\gamma_n}{\sin \frac{\pi}{2(n+1)}} \qquad \text{if } \sigma + n = 0,$$

are oscillatory.

Note that if either $\sigma < -n$ and c is arbitrary or $\sigma = -n$ and $|c| \leq \min\{m_n^*, m_{*n}\}$, then the equation (3.26) is nonoscillatory.

Clearly, an odd order linear differential equation with constant coefficients always has a nonoscillatory solution. The same is true for equations (1.1) if p is of constant sign. As the above examples indicate, this is not the case if p changes sign.

Notes. The assertions in this section are due to T.A. Chanturia [65, 69, 74]. For Problems 3.1, 3.2 and 3.3 see [213], [231, 232] and [263], respectively. For third and fourth order equations, Corollaries 3.1 and 3.2 improve results of V.A. Kondratyev [212]. Theorem 3.7 implies a result of D.V. Izyumova [150].

The first examples of third order equations all solutions of which are oscillatory were constructed by G. Ascoli [6] and G. Sansone [315]. Other examples of similar type are given in [95, 120, 293]. Some statements concerning the asymptotic behavior of solutions of third order linear differential equations are contained in the monograph by M. Greguš [121] and in the survey paper by J.H. Barrett [12].

§4. The Subspace of Solutions Vanishing at Infinity

This section deals with criteria for differential equations

$$u^{(n)} = \sum_{k=0}^{n-1} p_k(t) u^{(k)} \tag{4.1}$$

and

$$u^{(n)} = p(t) u \tag{4.2}$$

to have a nontrivial subspace of solutions vanishing at infinity.

Below, unless otherwise stipulated,

$$p, p_0 \in L_{\text{loc}}(\mathbf{R}_+), \qquad p_k \in \tilde{C}_{\text{loc}}^{k-1}(\mathbf{R}_+) \qquad (k = 1, \dots, n-1). \tag{4.3}$$

Besides, n_0 denotes the integral part of $n/2$, $V^{(n,\sigma)}(p_0, \dots, p_{n-1})$ and $V^{(n,\sigma)}(p)$ are the sets of solutions u of the equations (4.1) and (4.2), respectively, satisfying the conditions

$$\int^{+\infty} t^{2i-\sigma} |u^{(i)}(t)|^2 \, dt < +\infty \qquad (i = 0, \dots, n_0), \tag{4.4}$$

and $Z^{(n)}(p)$ is the set of solutions u of the equation (4.2) for which

$$\lim_{t \to +\infty} u(t) = 0.$$

By μ_i^k $(i = 0, 1, \dots; k = 2i, 2i+1, \dots)$ we denote the positive constants such that

$$\mu_0^{i+1} = \frac{1}{2}, \qquad \mu_i^{2i} = 1, \qquad \mu_{i+1}^k = \mu_{i+1}^{k-1} + \mu_i^{k-2}$$

$$(i = 0, 1, \dots; k = 2i+3, 2i+4, \dots) \tag{4.5}$$

and by $\dim X$ the dimension of a linear space X.

The first three subsections contain auxiliary information. The main results are given in subsections 4.4–4.6, in which, in particular, we prove theorems on the dimensions of the spaces $V^{(n,\sigma)}(p_0, \ldots, p_{n-1})$ and $V^{(n,\sigma)}(p)$ as well as Milloux and Armellini–Tonelli–Sansone types theorems on the dimension of the space $Z^{(2)}(p)$.

4.1. Some integral identities and inequalities. Together with the above-mentioned constants μ_i^k, we consider the constants μ_{ij}^k $(i = 0, 1, \ldots; \, j = i, i+1, \ldots; \, k = i+j+1, i+j+2, \ldots)$ defined by the equalities

$$\mu_{00}^k = \frac{1}{2}, \qquad \mu_{0j}^k = 1 \qquad (k = 1, 2, \ldots; \, j = 1, \ldots, k-1),$$

$$\mu_{ii}^{2i+1} = \frac{1}{2}, \qquad \mu_{ik-i-1}^k = 1 \qquad (i = 1, 2, \ldots; \, k = 2i+2, \ldots), \qquad (4.6)$$

$$\mu_{ij}^k = \mu_{ij}^{k-1} + \mu_{i-1j-1}^{k-2} \qquad (i = 1, 2, \ldots; \, j = i, i+1, \ldots; \, k = i+j+2, \ldots).$$

LEMMA 4.1. Let $-\infty < a < b < +\infty$ and $u, v \in \tilde{C}^{k-1}([a,b])$, where k is a positive integer. Then

$$\int_a^b v(t)u(t)u^{(k)}(t)\, dt$$

$$= \sum_{i=0}^{\left[\frac{k-1}{2}\right]} \sum_{j=i}^{1-i} (-1)^{k-1-j} \mu_{ij}^k \left(v^{(k-1-i-j)}(b)u^{(i)}(b)u^{(j)}(b) - v^{(k-1-i-j)}(a)u^{(i)}(a)u^{(j)}(a) \right)$$

$$+ \sum_{i=0}^{\left[\frac{k}{2}\right]} (-1)^{k-i} \mu_i^k \int_a^b v^{(k-2i)}(t)|u^{(i)}(t)|^2\, dt. \qquad (4.7)$$

PROOF. For $k = 1$ the equality (4.7) is obvious. Assume that it is true for any $k \in \{1, \ldots, m\}$, with m a positive integer. Then

$$\int_a^b v(t)u(t)u^{(m+1)}(t)\, dt = v(b)u(b)u^{(m)}(b) - v(a)u(a)u^{(m)}(a)$$

$$- \int_a^b v'(t)u(t)u^{(m)}(t)\, dt - \int_a^b v(t)u'(t)u^{(m)}(t)\, dt, \qquad (4.8)$$

whatever $u, v \in \tilde{C}^m([a,b])$. According to our assumption,

$$\int_a^b v'(t)u(t)u^{(m)}(t)\, dt$$

$$= \sum_{i=0}^{\left[\frac{m-1}{2}\right]} \sum_{j=i}^{m-1-i} (-1)^{m-1-j} \mu_{ij}^m \left(v^{(m-i-j)}(b)u^{(i)}(b)u^{(j)}(b) \right.$$

$$\left. - v^{(m-i-j)}(a)u^{(i)}(a)u^{(j)}(a) \right)$$

$$+ \sum_{i=0}^{\left[\frac{m}{2}\right]} (-1)^{m-i} \mu_i^m \int_a^b v^{(m+1-2i)}(t)|u^{(i)}(t)|^2\, dt \qquad (4.9)$$

and

$$\int_a^b v(t)u'(t)u^{(m)}(t)\,dt$$

$$= \sum_{i=0}^{\left[\frac{m-2}{2}\right]} \sum_{j=i}^{m-2-i} (-1)^{m-2-j}\mu_{ij}^{m-1}$$

$$\times \left(v^{(m-2-i-j)}(b)u^{(1+i)}(b)u^{(1+j)}(b) - v^{(m-2-i-j)}(a)u^{(1+i)}(a)u^{(1+j)}(a) \right)$$

$$+ \sum_{i=0}^{\left[\frac{m-1}{2}\right]} (-1)^{m-1-i}\mu_i^{m-1} \int_a^b v^{(m-1-2i)}(t)|u^{(1+i)}(t)|^2\,dt$$

$$= \sum_{i=1}^{\left[\frac{m}{2}\right]} \sum_{j=i}^{m-i} (-1)^{m-1-j}\mu_{i-1j-1}^{m-1} \left(v^{(m-i-j)}(b)u^{(i)}(b)u^{(j)}(b) - v^{(m-i-j)}(a)u^{(i)}(a)u^{(j)}(a) \right)$$

$$+ \sum_{i=1}^{\left[\frac{m+1}{2}\right]} (-1)^{m-i}\mu_{i-1}^{m-1} \int_a^b v^{(m+1-2i)}(t)|u^{(i)}(t)|^2\,dt. \tag{4.10}$$

From (4.8)–(4.10), applying (4.5) and (4.6), we derive (4.7) with $k = m + 1$. $\quad\square$

LEMMA 4.2. Let $k \geq 2$ be a positive integer, $0 < a < b < +\infty$, $\sigma \geq 0$, $c_0 \geq 0$, and let the function $u\colon [a, b] \to \mathbf{R}$ be k times continuously differentiable and satisfy the inequality

$$b^{2i-\sigma}u^{(i)}(b)u^{(i-1)}(b) - \left(i - \frac{\sigma}{2} \right) b^{2i-\sigma-1}|u^{(i-1)}(b)|^2$$

$$-a^{2i-\sigma}u^{(i)}(a)u^{(i-1)}(a) + \left(i - \frac{\sigma}{2} \right) a^{2i-\sigma-1}|u^{(i-1)}(a)|^2 \leq c_0$$

$$(i = 1, \dots, k - 1). \tag{4.11}$$

Then for any $\lambda > (k-2)(4k^2 - k + 3 + 3\sigma^2)/3$ we have

$$\int_a^b t^{2i-\sigma}|u^{(i)}(t)|^2\,dt \leq 2(\lambda + \sigma(2k^2 - 3k - 2))^{k-2}c_0 + \alpha_i(\lambda, k) \int_a^b t^{-\sigma}|u(t)|^2\,dt$$

$$+\beta_i(\lambda, k) \int_a^b t^{2k-\sigma}|u^{(k)}(t)|^2\,dt \qquad (i = 1, \dots, k-1), \tag{4.12}$$

where

$$\alpha_1(\lambda, k) = (k-1)\left(\left(1 - \frac{\sigma}{2} \right)(1 - \sigma) + \frac{\lambda}{4} \right),$$

$$\alpha_i(\lambda, k) = \frac{k-i}{k-1}\alpha_1(\lambda, k)$$

$$\times \prod_{j=1}^{i-1} \left(\lambda - \frac{(j-1)(4j^2 + 7j + 6)}{3} + \sigma(j-1)(2j + 3 - \sigma) \right) \tag{4.13}$$

$$(i = 2, \dots, k-1),$$

$$\beta_i(\lambda, k) = \prod_{j=i}^{k-1} \left(\lambda - \frac{(j-1)(4j^2 + 7j + 6)}{3} + \sigma(j-1)(2j + 3 - \sigma) \right)^{-1}$$

$$(i = 1, \ldots, k-1).$$

PROOF. Set

$$\gamma_0 = \left(1 - \frac{\sigma}{2}\right)(1 - \sigma) + \frac{\lambda}{4},$$

$$\gamma_i = \lambda - \frac{(i-1)(4i^2 + 7i + 6)}{3} + \sigma(i-1)(2i + 3 - \sigma) \qquad (i = 1, \ldots, k-1),$$

and

$$\rho_i = \int_a^b t^{2i-\sigma} |u^{(i)}(t)|^2 \, dt \qquad (i = 0, \ldots, k-1).$$

Then by (4.11),

$$\rho_i = b^{2i-\sigma} u^{(i)}(b) u^{(i-1)}(b) - \left(i - \frac{\sigma}{2}\right) b^{2i-\sigma-1} |u^{(i-1)}(b)|^2$$

$$- a^{2i-\sigma} u^{(i)}(a) u^{(i-1)}(a) + \left(i - \frac{\sigma}{2}\right) a^{2i-\sigma-1} |u^{(i-1)}(a)|^2$$

$$+ \left(i - \frac{\sigma}{2}\right)(2i - \sigma - 1)\rho_{i-1} - \int_a^b t^{2i-\sigma} u^{(i+1)}(t) u^{(i-1)}(t) \, dt$$

$$\leq c_0 + \left(i - \frac{\sigma}{2}\right)(2i - \sigma - 1)\rho_{i-1} + \int_a^b t^{2i-\sigma} |u^{(i+1)}(t) u^{(i-1)}(t)| \, dt$$

for any $i \in \{1, \ldots, k-1\}$. Hence, taking into account the relations

$$t^{2-\sigma} |u(t) u''(t)| \leq \frac{\lambda}{4} t^{-\sigma} |u(t)|^2 + \frac{1}{\lambda} t^{4-\sigma} |u''(t)|^2,$$

$$t^{2i-\sigma} |u^{(i+1)}(t) u^{(i-1)}(t)|$$

$$\leq \frac{\gamma_i}{2} t^{2i-2-\sigma} |u^{(i-1)}(t)|^2 + \frac{1}{2\gamma_i} t^{2i+2-\sigma} |u^{(i+1)}(t)|^2,$$

and

$$\left(i - \frac{\sigma}{2}\right)(2i - \sigma - 1) + \frac{\gamma_i}{2} = \frac{\gamma_{i-1}}{2} \qquad (i = 2, \ldots, k-1),$$

we get

$$\rho_1 \leq c_0 + \gamma_0 \rho_0 + \frac{1}{\gamma_1} \rho_2, \tag{4.14}$$

$$\rho_i \leq c_0 + \frac{\gamma_{i-1}}{2} \rho_{i-1} + \frac{1}{2\gamma_i} \rho_{i+1} \qquad (i = 2, \ldots, k-1). \tag{4.15}$$

If $k = 2$, then the inequality (4.14) coincides with (4.12). So it remains to consider the case $k > 2$. In this case $0 < \gamma_i < \lambda_0 - 2$ $(i = 2, \ldots, k-1)$, where $\lambda_0 = \lambda + \sigma(2k^2 - 3k - 2)$. Consequently, (4.14) and (4.15) imply

$$\rho_i \leq \lambda_0^{i-1} c_0 + \frac{\alpha_i(\lambda, k)}{k-i} \rho_0 + \frac{1}{\gamma_i} \rho_{i+1} \qquad (i = 1, \ldots, k-1). \tag{4.16}$$

Now we will show that for any $i \in \{1, \ldots, k-1\}$,

$$\rho_i \leq 2(1 - 2^{i-k})\lambda_0^{k-2} c_0 + \alpha_i(\lambda, k)\rho_0 + \beta_i(\lambda, k)\rho_k. \tag{4.17}$$

For $i = k - 1$ this estimate is the same as (4.16). Assume (4.17) to be true for a certain $i \in \{2, \ldots, k - 1\}$. Then by (4.16),

$$\rho_{i-1} \le \lambda_0^{i-2} c_0 + \frac{\alpha_{i-1}(\lambda, k)}{k - i + 1} \rho_0 + \frac{2}{\gamma_{i-1}} (1 - 2^{i-k}) \lambda_0^{k-2} c_0 + \frac{\alpha_i(\lambda, k)}{\gamma_{i-1}} \rho_0 + \frac{\beta_i(\lambda, k)}{\gamma_{i-1}} \rho_k. \quad (4.18)$$

The equalities (4.13) imply

$$\frac{\alpha_{i-1}(\lambda, k)}{k - i + 1} + \frac{\alpha_i(\lambda, k)}{\gamma_{i-1}} = \alpha_{i-1}(\lambda, k), \qquad \frac{\beta_i(\lambda, k)}{\gamma_{i-1}} = \beta_{i-1}(\lambda, k).$$

On the other hand,

$$\lambda_0^{i-2} \le \lambda_0^{k-2}, \qquad 1 + \frac{2}{\gamma_{i-1}} (1 - 2^{i-k}) \le 2(1 - 2^{i-1-k}).$$

Thus, from (4.18) we obtain

$$\rho_{i-1} \le 2(1 - 2^{i-1-k}) \lambda_0^{k-2} c_0 + \alpha_{i-1}(\lambda, k) \rho_0 + \beta_{i-1}(\lambda, k) \rho_k,$$

and so (4.17) is proved for any $i \in \{1, \ldots, k - 1\}$ by induction. This yields (4.12). $\square$

LEMMA 4.3. Let $k \ge 2$ be a positive integer, $a > 0$, $\sigma \ge 0$, $c_0 \ge 0$, and let the function $u: [a, +\infty[\to \mathbf{R}$ be k times continuously differentiable and satisfy the conditions

$$\int_a^{+\infty} t^{2i-\sigma} |u^{(i)}(t)|^2 \, dt < +\infty \qquad (i = 0, \ldots, k) \quad (4.19)$$

and

$$\left(i - \frac{\sigma}{2}\right) a^{2i-\sigma-1} |u^{(i-1)}(a)|^2 - a^{2i-\sigma} u^{(i)}(a) u^{(i-1)}(a) \le c_0$$

$$(i = 1, \ldots, k - 1). \quad (4.20)$$

Then for any $\lambda > (k - 2)(4k^2 - k + 3 + 3\sigma^2)/3$ the inequalities

$$\int_a^{+\infty} t^{2i-\sigma} |u^{(i)}(t)|^2 \, dt \le 2(\lambda + \sigma(2k^2 - 3k - 2))^{k-2} c_0 + \alpha_i(\lambda, k) \int_a^{+\infty} t^{-\sigma} |u(t)|^2 \, dt$$

$$+ \beta_i(\lambda, k) \int_a^{+\infty} t^{2k-\sigma} |u^{(k)}(t)|^2 \, dt \qquad (i = 1, \ldots, k - 1), \quad (4.21)$$

with the constants $\alpha_i(\lambda, k)$ and $\beta_i(\lambda, k)$ defined by (4.13), hold.

PROOF. According to (4.19),

$$\int_a^{+\infty} t^{2i-\sigma-1} |u^{(i)}(t) u^{(i-1)}(t)| \, dt$$

$$\le \left(\int_a^{+\infty} t^{2i-\sigma} |u^{(i)}(t)|^2 \, dt\right)^{1/2} \left(\int_a^{+\infty} t^{2i-\sigma-2} |u^{(i-1)}(t)|^2 \, dt\right)^{1/2} < +\infty$$

$$(i = 1, \ldots, k).$$

Thus there exist sequences $(b_{im})_{m=1}^{+\infty} \subset [a, +\infty[$ $(i = 1, \ldots, k - 1)$ converging to $+\infty$ so that

$$b_{im}^{2i-\sigma} |u^{(i)}(b_{im}) u^{(i-1)}(b_{im})| \le \frac{1}{m} \qquad (i = 1, \ldots, k - 1; \; m = 1, 2, \ldots). \quad (4.22)$$

For any positive integer m, Lemma 4.2 and the inequalities (4.20) and (4.22) imply

$$\int_a^{b_{im}} t^{2i-\sigma}|u^{(i)}(t)|^2\,dt \leq 2(\lambda + \sigma(2k^2 - 3k - 2))^{k-2}\left(c_0 + \frac{1}{m}\right)$$

$$+\alpha_i(\lambda, k)\int_a^{b_{im}} t^{-\sigma}|u(t)|^2\,dt + \beta_i(\lambda, k)\int_a^{b_{im}} t^{2k-\sigma}|u^{(k)}(t)|^2\,dt$$

$$(i = 1, \ldots, k-1).$$

From these relations, by passing m to $+\infty$, we obtain (4.21). $\square$

4.2. On functions satisfying the conditions (4.19).

LEMMA 4.4. Let m be a positive integer, $a > 0, \sigma \geq 0$, and let k denote the integral part of $m/2$. Furthermore, suppose that the function $u: [a, +\infty[\to \mathbf{R}$ is $m - 1$ times continuously differentiable and satisfies the conditions (4.19). Then for any constants c_{ij} $(i = 0, \ldots, k; j = i, \ldots, m_i)$ the function

$$w(t) = \sum_{i=0}^{k}\sum_{j=i}^{m_i} c_{ij}t^{i+j+1-\sigma}u^{(i)}(t)u^{(j)}(t),$$

where $m_0 = m - 1$ and $m_i = m - i$ if $i \neq 0$, satisfies the condition

$$\lim_{t \to +\infty} \inf |w(t)| = 0. \tag{4.23}$$

PROOF. Assume, on the contrary, that for certain c_{ij} $(i = 0, \ldots, k; j = i, \ldots, m_i)$ (4.23) is violated. Then, without loss of generality,

$$\sum_{i=0}^{k}\sum_{j=i}^{m_i} c_{ij}t^{i+j-\sigma}u^{(i)}(t)u^{(j)}(t) \geq \frac{\delta}{t} \qquad \text{for } t \geq t_1,$$

where the constants $\delta > 0$ and $t_1 > a$ are sufficiently small and sufficiently large, respectively. By Lemma 4.1, integration of both parts of this inequality over $[t_1, t]$ leads to

$$\sum_{i=0}^{[\frac{m-1}{2}]}\sum_{j=i}^{m-1-i} c_{ij}^{(1)}t^{i+j+1-\sigma}u^{(i)}(t)u^{(j)}(t) + \sum_{i=0}^{k} c_i \int_{t_1}^{t} \tau^{2i-\sigma}|u^{(i)}(\tau)|^2\,d\tau \geq c + \delta\ln\frac{t}{t}$$

$$\text{for } t \geq t_1,$$

where $c_{ij}^{(1)}$, c_i and c are certain constants. Hence, in view of (4.19) there exists $t_2 > t_1$ such that

$$\sum_{i=0}^{[\frac{m-1}{2}]}\sum_{j=i}^{m-1-i} c_{ij}^{(1)}t^{i+j-\sigma}u^{(i)}(t)u^{(j)}(t) \geq \frac{1}{t} \qquad \text{for } t \geq t_2.$$

From the last inequality, taking into account Lemma 4.1, we easily conclude that for any $\nu \in \{1, \ldots, m\}$,

$$\sum_{i=0}^{[\frac{m-\nu}{2}]}\sum_{j=i}^{m-\nu-i} c_{ij}^{(\nu)}t^{i+j-\sigma}u^{(i)}(t)u^{(j)}(t) \geq \frac{1}{t} \qquad \text{for } t \geq t_\nu,$$

where $c_{ij}^{(\nu)} = \text{const}$ and $t_\nu > t_{\nu-1} > \ldots > t_1$. Therefore,

$$c_{00}^{(m)} t^{-\sigma} |u(t)|^2 \geq \frac{1}{t} \qquad \text{for } t \geq t_m,$$

which contradicts (4.19). $\square$

LEMMA 4.5. Let k be a positive integer, $\sigma \neq 2i+1$ $(i = 0, \ldots, k-1)$, and let the function $u \colon [a, +\infty[\to \mathbf{R}$ be k times continuously differentiable. Then the condition (4.19) is fulfilled if and only if

$$\int_a^{+\infty} t^{2k-\sigma} |u^{(k)}(t)|^2 \, dt < +\infty \tag{4.24}$$

and

$$\lim_{t \to +\infty} t^{i+(1-\sigma)/2} u^{(i)}(t) = 0 \qquad (i = 0, \ldots, k-1). \tag{4.25}$$

PROOF. Suppose that (4.24) and (4.25) hold. Then the identities

$$\int_a^t \tau^{2i-\sigma} |u^{(i)}(\tau)|^2 \, d\tau \equiv \frac{2}{2i+1-\sigma} (t^{2i+1-\sigma} |u^{(i)}(t)|^2 - a^{2i+1-\sigma} |u^{(i)}(a)|^2)$$

$$+ \frac{4}{(2i+1-\sigma)^2} \int_a^t \tau^{2i+2-\sigma} |u^{(i+1)}(\tau)|^2 \, d\tau$$

$$- \int_a^t \left(\frac{2}{2i+1-\sigma} \tau^{i+1-\sigma/2} u^{(i+1)}(\tau) + \tau^{i-\sigma/2} u^{(i)}(\tau) \right)^2 \, d\tau$$

$$(i = 0, \ldots, k-1) \tag{4.26}$$

imply the conditions (4.19).

Let now the conditions (4.19) be satisfied. Then

$$\liminf_{t \to +\infty} t^{i+(1-\sigma)/2} |u^{(i)}(t)| = 0 \qquad (i = 0, \ldots, k-1).$$

On the other hand, it follows from (4.19) and (4.26) that the limit of each function $t^{2i+1-\sigma} |u^{(i)}(t)|^2$ as $t \to +\infty$ exists. So the equalities (4.25) are valid. $\square$

Moreover, the identities (4.26) yield

LEMMA 4.6. Let k be a positive integer, $\sigma > 2k-1$, and let the function $u \colon [a, +\infty[\to \mathbf{R}$ be k times continuously differentiable and satisfy (4.24). Then the condition (4.19) holds.

4.3. On a related problem. Let $p_k \in L_{\text{loc}}(\mathbf{R}_+)$ $(k = 0, \ldots, n-1)$, $m_0 \in \{0, \ldots, n-2\}$, $m \in \{m_0+1, \ldots, n-1\}$, $a \in \mathbf{R}_+$, $c_i \in \mathbf{R}$ $(i = 0, \ldots, m-m_0-1)$, and let the functions $g_k \in L_{\text{loc}}([a, +\infty[)$ $(k = 0, \ldots, n_0)$ be nonnegative, g_0 differing from zero on a set of positive measure. Consider the problem of solving the differential equation (4.1) under the conditions

$$u^{(i+m_0)}(a) = c_i \qquad (i = 0, \ldots, m-m_0-1),$$

$$\int_a^{+\infty} g_k(t) |u^{(k)}(t)|^2 \, dt < +\infty \qquad (k = 0, \ldots, n_0). \tag{4.27}$$

LEMMA 4.7. Let there exist a continuous function $r: \mathbf{R}_+ \rightarrow \mathbf{R}_+$ such that for any sufficiently large $b \in]a, +\infty[$ every solution of the equation (4.1) satisfies the inequalities

$$\int_a^b g_k(t)|u^{(k)}(t)|^2 \, dt \leq r(a) \sum_{i=0}^{n-1} |u^{(i)}(a)| \sum_{i=m_0}^{m-1} |u^{(i)}(a)|$$

$$+ r(b) \sum_{i=0}^{n-1} |u^{(i)}(b)| \sum_{i=0}^{n-m+m_0-1} |u^{(i)}(b)| \qquad (k = 0, \dots, n_0). \tag{4.28}$$

Then the problem (4.1), (4.27) has at least one solution.

PROOF. Let $b_0 \in]a, +\infty[$ be so large that for any $b > b_0$ the solutions of (4.1) satisfy (4.28). Set $b_j = b_0 + j$, and for each positive integer j consider the equation (4.1) under the boundary conditions

$$u^{(i+m_0)}(a) = c_i \qquad (i = 0, \dots, m - m_0 - 1),$$

$$\tag{4.29}$$

$$u^{(k)}(b_j) = 0 \qquad (k = 0, \dots, n - m + m_0 - 1).$$

It is well known (see e.g. [127]) that the problem (4.1), (4.29) is uniquely solvable if and only if the equation (4.1) does not have a nonzero solution satisfying the homogeneous boundary conditions

$$u^{(i+m_0)}(a) = 0 \qquad (i = 0, \dots, m - m_0 - 1),$$

$$\tag{4.29_0}$$

$$u^{(k)}(b_j) = 0 \qquad (k = 0, \dots, n - m + m_0 - 1).$$

Suppose that u is a solution of the problem (4.1), (4.29_0). Then, in view of (4.28),

$$\int_a^{b_j} g_0(t)|u(t)|^2 \, dt = 0,$$

and, thus, $u(t) \equiv 0$. So the solution of the problem (4.1), (4.29) is unique. Denote it by u_j.

According to (4.28) and (4.29), for any positive integer j,

$$\int_a^{b_j} g_k(t)|u_j^{(k)}(t)|^2 \, dt \leq r_0 \rho_j \qquad (k = 0, \dots, n_0), \tag{4.30}$$

where

$$r_0 = r(a) \sum_{i=0}^{m-m_0-1} |c_i|, \qquad \rho_j = \sum_{i=0}^{n-1} |u_j^{(i)}(a)|.$$

First we establish the boundedness of the sequence $(\rho_j)_{j=1}^{+\infty}$. Assume the contrary. Then there exists a subsequence $(\rho_{j_\nu})_{\nu=1}^{+\infty}$ such that $\rho_{j_\nu} > \nu$ $(\nu = 1, 2, \dots)$.

Put

$$v_\nu(t) = \frac{u_{j_\nu}(t)}{\rho_{j_\nu}}.$$

Clearly,

$$\sum_{i=0}^{n-1} |v_\nu^{(i)}(a)| = 1 \qquad (\nu = 1, 2, \dots)$$

and

$$\int_a^{b_0} g_0(t) |v_\nu(t)|^2 \, dt \leq \frac{r_0}{\nu} \qquad (\nu = 1, 2, \dots). \tag{4.31}$$

Without loss of generality we may assume that the sequences $(v_\nu^{(i)}(a))_{\nu=1}^{+\infty}$ $(i = 0, \dots, n-1)$ converge. Then

$$\lim_{\nu \to +\infty} v_\nu(t) = v(t)$$

uniformly on $[a, b_0]$, where v is a solution of the equation (4.1) under the condition

$$\sum_{i=0}^{n-1} |v^{(i)}(a)| = 1.$$

On the other hand, by (4.31), $v(t) \equiv 0$. This contradiction proves the boundedness of $(\rho_j)_{j=1}^{+\infty}$.

Hence the sequences $(u_j^{(i)}(a))_{j=1}^{+\infty}$ $(i = 0, \dots, n-1)$ contain converging subsequences $(u_{j_\nu}^{(i)}(a))_{\nu=1}^{+\infty}$ $(i = 0, \dots, n-1)$. So

$$\lim_{\nu \to +\infty} u_{j_\nu}^{(i)}(t) = u^{(i)}(t) \qquad (i = 0, \dots, n-1)$$

uniformly on each finite interval in $\mathbf{R}_+$, where u is a solution of the equation (4.1). According to (4.29) and (4.30), u satisfies the conditions (4.27). $\square$

4.4. Theorems on the dimension of the space $V^{(n,\sigma)}(p_0, \dots, p_{n-1})$.

THEOREM 4.1. Let

$$\lim_{t \to +\infty} \sum_{k=0}^{n-1} (-1)^{n-n_0-k-1} \mu_0^k (t^n p_k(t))^{(k)} = +\infty, \tag{4.32}$$

$$\lim_{t \to +\infty} \sup t^{-2i} \sum_{k=2i}^{n-1} (-1)^{n-n_0-i-k} \mu_i^k (t^n p_k(t))^{(k-2i)} < +\infty$$

$$(i = 1, \dots, n_0 - 1), \tag{4.33}$$

and let, for n odd, in addition,

$$\lim_{t \to +\infty} \inf t p_{n-1}(t) > -\frac{n^2}{2}. \tag{4.34}$$

Then

$$\dim V^{(n,0)}(p_0, \dots, p_{n-1}) \geq n_0. \tag{4.35}$$

PROOF. Set $g_k(t) = t^{2k}$ $(k = 0, \dots, n_0)$, $m_0 = 0$ and $m = n_0$. The inequality (4.35) will be proved if for a certain $a \in]0, +\infty[$ we establish the solvability of the problem (4.1), (4.27) with arbitrary $c_i \in \mathbf{R}$ $(i = 0, \dots, n_0 - 1)$. By Lemma 4.7, this problem is solvable whenever for each sufficiently large $b \in]a, +\infty[$ every solution u of the equation (4.1) satisfies the estimates (4.28), where $r \colon \mathbf{R}_+ \to \mathbf{R}_+$ is a continuous function which does not depend on u and b.

Applying Lemma 4.1, multiplying both sides of (4.1) by $(-1)^{n-n_0}t^n u(t)$ and integrating from a to $b > a$, we get

$$\sum_{i=0}^{n_0}\int_a^b h_i(t)|u^{(i)}(t)|^2\,dt = l(u)(b) - l(u)(a),\qquad(4.36)$$

where $h(t) = t^n$,

$$l(u)(t) = \sum_{i=0}^{n-n_0-1}\sum_{j=i}^{n-1-i}(-1)^{n_0-j}\mu_{ij}^n h^{(n-1-i-j)}(t)u^{(i)}(t)u^{(j)}(t)$$

$$+\sum_{k=1}^{n-1}\sum_{i=0}^{\left[\frac{k-1}{2}\right]}\sum_{j=i}^{k-1-i}(-1)^{n-n_0+k-1-j}\mu_{ij}^k(h(t)p_k(t))^{(k-1-i-j)}u^{(i)}(t)u^{(j)}(t),\qquad(4.37)$$

$$h_i(t) = \sum_{k=2i}^{n-1}(-1)^{n-n_0-1-i-k}\mu_i^k(h(t)p_k(t))^{(k-2i)} + (-1)^{n_0-i}\mu_i^n h^{(n-2i)}(t)$$

$$(i = 0,\ldots,n_0-1),\qquad(4.38)$$

$$h_{n_0}(t) = h(t)\qquad\text{for } n = 2n_0,\qquad(4.39)$$

$$h_{n_0}(t) = h(t)p_{n-1}(t) + \frac{n}{2}h'(t)\qquad\text{for } n = 2n_0 + 1.$$

Put

$$r_0(t) = \sum_{i=0}^{n-n_0-1}\sum_{j=i}^{n-1-i}\mu_{ij}^n|h^{(n-1-i-j)}(t)|$$

$$+\sum_{k=1}^{n-1}\sum_{i=0}^{\left[\frac{k-1}{2}\right]}\sum_{j=i}^{k-1-i}\mu_{ij}^k|(h(t)p_k(t))^{(k-1-i-j)}|.\qquad(4.40)$$

From (4.37) we obtain

$$-r_0(t)\sum_{i=0}^{n-1}|u^{(i)}(t)|\sum_{i=0}^{n_0-1}|u^{(i)}(t)| \le l(u)(t) \le r_0(t)\sum_{i=0}^{n-1}|u^{(i)}(t)|\sum_{i=0}^{n-n_0-1}|u^{(i)}(t)|.\qquad(4.41)$$

By the equalities (4.13) and the conditions (4.32)–(4.34), there exist positive constants a, γ, γ_0, and λ such that $\lambda > n^{3,5}$

$$2\gamma_0\sum_{i=1}^{n_0-1}\beta_i(\lambda,n_0) < \gamma,\qquad(4.42)$$

$$h_{n_0}(t) > \gamma t^{2n_0},\qquad h_i(t) > -\gamma_0 t^{2i}\qquad\text{for } t \ge a\qquad(i = 1,\ldots,n_0-1),\qquad(4.43)$$

$$h_0(t) > 1 + \gamma_1\qquad\text{for } t \ge a,\qquad(4.44)$$

where $\gamma_1 = \gamma_0\sum_{i=1}^{n_0-1}\alpha_i(\lambda,n_0)$.

---

[5]When $n_0 = 1$, here and below we assume the sum $\sum_{i=0}^{n_0-1}$ to be zero.

According to (4.41), (4.43) and (4.44) the equality (4.36) yields

$$(1+\gamma_1)\int_a^b |u(t)|^2\,dt + \gamma\int_a^b t^{2n_0}|u^{(n_0)}(t)|^2\,dt - \gamma_0\sum_{i=1}^{n_0-1}\int_a^b t^{2i}|u^{(i)}(t)|^2\,dt$$

$$\le r_0(b)\sum_{i=0}^{n-1}|u^{(i)}(b)|\sum_{i=0}^{n-n_0-1}|u^{(i)}(b)| + r_0(a)\sum_{i=0}^{n-1}|u^{(i)}(a)|\sum_{i=0}^{n_0-1}|u^{(i)}(a)|.$$

On the other hand, by Lemma 4.2 and the inequality (4.42) we have

$$\gamma_0\sum_{i=0}^{n_0-1}\int_a^b t^{2i}|u^{(i)}(t)|^2\,dt \le r_1(a)\left(\sum_{i=0}^{n_0-1}|u^{(i)}(a)|\right)^2$$

$$+r_1(b)\left(\sum_{i=0}^{n_0-1}|u^{(i)}(b)|\right)^2 + \gamma_1\int_a^b |u(t)|^2\,dt + \frac{\gamma}{2}\int_a^b t^{2n_0}|u^{(n_0)}(t)|^2\,dt,$$

with $r_1(t) = n\gamma_0\lambda^n(1+t)^n$. Consequently, the inequalities (4.28) hold, where the function

$$r(t) = \left(\frac{2+\gamma_1}{\gamma_0} + \frac{2}{\gamma} + 1\right)(r_0(t)+r_1(t))$$

does not depend on b and u. □

By Lemma 4.5, Theorem 4.1 implies

COROLLARY 4.1. Under the hypotheses of Theorem 4.1, the equation (4.1) has an n_0-dimensional subspace of solutions satisfying the conditions

$$\lim_{t\to+\infty} t^{i+1/2}u^{(i)}(t) = 0 \qquad (i=0,\dots,n_0-1). \tag{4.45}$$

PROBLEM 4.1. Let the hypotheses of Theorem 4.1 be fulfilled, $q \in L^2_{loc}(\mathbf{R}_+)$ and

$$\int^{+\infty} \frac{t^{2n}q^2(t)}{g_0(t)}\,dt < +\infty,$$

where $g_0(t) = \sum_{k=0}^{n-1}(-1)^{n-n_0-k-1}\mu_0^k(t^n p_k(t))^{(k)}$.

Then there exists $a_0 \in \mathbf{R}_+$ such that for any $a > a_0$ and $c_i \in \mathbf{R}$ $(i=0,\dots,n_0-1)$ the equation

$$u^{(n)} = \sum_{k=0}^{n-1} p_k(t)u^{(k)} + q(t) \tag{4.46}$$

has at least one solution satisfying the conditions

$$u^{(i)}(a) = c_i \qquad (i=0,\dots,n_0-1),$$

$$\int_a^{+\infty} t^{2k}|u^{(k)}(t)|^2\,dt < \infty \qquad (k=0,\dots,n_0).$$

THEOREM 4.2. Let $n = 2n_0 + 1$, $\sigma > n$,

$$\lim_{t\to+\infty} t^\sigma g_0(t) = +\infty, \tag{4.47}$$

where $g_0(t) = \sum_{k=0}^{n-1} (-1)^{n_0+1-k} \mu_0^k (t^{n-\sigma} p_k(t))^{(k)}$, and let

$$\lim_{t \to +\infty} \sup t^{\sigma-2i} \sum_{k=2i}^{n-1} (-1)^{n_0-i-k} \mu_i^k (t^{n-\sigma} p_k(t))^{(k-2i)} < +\infty$$

$$(i = 1, \ldots, n_0 - 1), \qquad (4.48)$$

$$\lim_{t \to +\infty} \sup t p_{n-1}(t) < \frac{n(\sigma - n)}{2}.$$

Then

$$\dim V^{(n,\sigma)}(p_0, \ldots, p_{n-1}) \geq n_0 + 1; \qquad (4.49)$$

moreover, $V^{(n,\sigma)}(p_0, \ldots, p_{n-1})$ contains an $(n_0 + 1)$-dimensional subspace of solutions u for which

$$\int_1^{+\infty} g_0(t) |u(t)|^2 \, dt < +\infty. \qquad (4.50)$$

PROOF. Suppose l, h_i $(i = 0, \ldots, n_0)$ and r_0 are the operator and functions defined by (4.37)–(4.40) with $h(t) = t^{n-\sigma}$. According to (4.13), (4.47) and (4.48), we can find constants $a > 1$, $\gamma_0 > 1$, $\gamma > 0$, and $\lambda > (n + \sigma)^3$ so that the inequality (4.42) holds and

$$h_0(t) < -\frac{1}{2} g_0(t) < -2\gamma_0 \sum_{i=1}^{n_0-1} \alpha_i(\lambda, n_0) t^{-\sigma} \qquad \text{for } t \geq a, \qquad (4.51)$$

$$h_i(t) \leq \gamma_0 t^{2i-\gamma} \qquad (i = 1, \ldots, n_0 - 1),$$

$$\qquad (4.52)$$

$$h_{n_0}(t) = -t^{2n_0-\sigma} \left(\frac{n(\sigma - n)}{2} - t p_{n-1}(t) \right) \leq -\gamma t^{2n_0-\sigma} \qquad \text{for } t \geq a.$$

Let $b \in \,]a, +\infty[$ and let u be a solution of the equation (4.1). Then (4.36) and (4.41) are valid. So, taking into account (4.52), we obtain

$$\frac{1}{2} \int_a^b g_0(t) |u(t)|^2 \, dt + \gamma \int_a^b t^{2n_0-\sigma} |u^{(n_0)}(t)|^2 \, dt$$

$$\leq r_0(a) \sum_{i=0}^{n-1} |u^{(i)}(a)| \sum_{i=0}^{n_0} |u^{(i)}(a)| + r_0(b) \sum_{i=0}^{n-1} |u^{(i)}(b)| \sum_{i=0}^{n_0-1} |u^{(i)}(b)|$$

$$+ \gamma_0 \sum_{i=0}^{n_0-1} \int_a^b t^{2i-\sigma} |u^{(i)}(t)|^2 \, dt.$$

On the other hand, in view of Lemma 4.2 and conditions (4.42) and (4.51),

$$\gamma_0 \sum_{i=1}^{n_0-1} \int_a^b t^{2i-\sigma} |u^{(i)}(t)|^2 \, dt \leq \gamma_1 \left(\sum_{i=0}^{n_0-1} |u^{(i)}(a)| \right)^2 + \gamma_1 \left(\sum_{i=0}^{n_0-1} |u^{(i)}(b)| \right)^2$$

$$+ \frac{1}{4} \int_a^b g_0(t) |u(t)|^2 \, dt + \frac{\gamma}{2} \int_a^b t^{2n_0-\sigma} |u^{(n_0)}(t)|^2 \, dt,$$

with $\gamma_1 = 2\sigma(\lambda + n^2\sigma)^n \gamma_0$. Hence the inequalities (4.28) are fulfilled, where $m_0 = 0$, $m = m_0 + 1$, $g_k(t) = t^{2k-\sigma}$ $(k = 1, \ldots, n_0)$, and the function $r(t) = (4 + 2/\gamma)(r_0(t) + \gamma_1)$ does not depend on b and u. By Lemma 4.7, for any $c_i \in \mathbf{R}$ $(i = 0, \ldots, n_0)$ the problem (4.1), (4.27) has at least one solution and by (4.47) each solution of this problem belongs to $V^{(n,\sigma)}(p_0, \dots, p_{n-1})$. So (4.49) is true. $\square$

COROLLARY 4.2. Let $n = 2n_0 + 1$, $\sigma > n$,

$$\lim_{t \to +\infty} \sup t^{2-\sigma} \sum_{k=0}^{n-1} (-1)^{n_0+1-k} \mu_0^k (t^{n-\sigma} p_k(t))^{(k)} > 0, \qquad (4.53)$$

and let the conditions (4.48) hold. Then $V^{(n,\sigma)}(p_0, \dots, p_{n-1})$ contains an $(n_0 + 1)$-dimensional subspace of solutions vanishing at infinity.

PROOF. Since all the hypotheses of Theorem 4.2 are fulfilled, $V^{(n,\sigma)}(p_0, \dots, p_{n-1})$ contains an $(n_0 + 1)$-dimensional subspace of solutions satisfying (4.50). Let u be a solution in this subspace. Then by (4.4) and (4.53),

$$\int_1^{+\infty} t^{\sigma-2} |u(t)|^2 \, dt < +\infty, \qquad \int_1^{+\infty} t^{2-\sigma} |u'(t)|^2 \, dt < +\infty.$$

So

$$|u(t)|^2 \le 2 \left(\int_t^{+\infty} t^{\sigma-2} |u(\tau)|^2 \, d\tau \right)^{1/2} \left(\int_t^{+\infty} t^{2-\sigma} |u'(\tau)|^2 \, d\tau \right)^{1/2} \to 0$$

as $t \to +\infty$. $\square$

PROBLEM 4.2. Let $n = 2n_0 + 1$, $\sigma > n$, and let (4.47) and (4.48) hold. Furthermore, suppose that $q \in L^2_{\text{loc}}(\mathbf{R}_+)$ and

$$\int^{+\infty} \frac{t^{2n-2\sigma} |q(t)|^2}{g_0(t)} \, dt < +\infty.$$

Then there exists $a_0 \in \mathbf{R}_+$ such that for $a > a_0$ and $c_i \in \mathbf{R}$ ($i = 0, \dots, n_0 - 1$) the equation (4.46) has at least one solution satisfying the conditions

$$u^{(i)}(a) = c_i \qquad (i = 0, \dots, n_0),$$

$$\int_a^{+\infty} t^{2k-\sigma} |u^{(k)}(t)|^2 \, dt < +\infty \qquad (k = 0, \dots, n_0).$$

THEOREM 4.3. Let $n = 2n_0 \ge 4$, $p_0 \in \tilde{C}_{\text{loc}}(\mathbf{R}_+)$, $p_1 \in L_{\text{loc}}(\mathbf{R}_+)$, $p_k \in \tilde{C}^{k-2}_{\text{loc}}(\mathbf{R}_+)$ ($k = 2, \dots, n-1$),

$$\lim_{t \to +\infty} (-1)^{n_0} (t^{n+1} p_0(t))' = +\infty$$

and

$$\lim_{t \to +\infty} \sup t^{-2i} \sum_{k=2i-1}^{n-1} (-1)^{k+n_0-i} \mu_{i-1}^{k-1} (t^{n+1} p_k(t))^{(k+1-2i)} < +\infty$$

$$(i = 1, \dots, n).$$

Then

$$\dim V^{(n,0)}(p_0, \dots, p_{n-1}) \ge n_0 - 1.$$

The proof of this theorem is omitted, because it repeats that of Theorem 4.1 with the following differences: in the equality (4.36), instead of (4.37)–(4.39) we have $h(t) = t^{n+1}$,

$$l(u)(t) = \frac{(-1)^{n_0}}{2} p_0(t) h(t) |u(t)|^2$$

$$+ \sum_{i=1}^{n_0} \sum_{j=i}^{n-1} (-1)^{n_0-j} \mu_{i-1j-1}^{n-1} h^{(n-i-j)}(t) u^{(i)}(t) u^{(j)}(t)$$

$$+ \sum_{k=2}^{n-1} \sum_{i=1}^{[\frac{k}{2}]} \sum_{j=i}^{k-i} (-1)^{n_0-k-j+1} \mu_{i-1j-1}^{k-1} (h(t) p_k(t))^{(k-i-j)} u^{(i)}(t) u^{(j)}(t),$$

$$h_0(t) = \frac{(-1)^{n_0}}{2} (p_0(t) h(t))',$$

$$h_i(t) = \sum_{k=2i-1}^{n-1} (-1)^{n_0-i-k+1} \mu_{i-1}^{k-1} (h(t) p_k(t))^{(k+1-2i)} + (-1)^{n_0-i} \mu_{i-1}^{n-1} h^{(n+1-2i)}(t)$$

$$(i = 1, \dots, n_0),$$

and the solvability of the problem (4.1), (4.27) should be proved for $m = n_0$, $m_0 = 1$.

In view of Lemma 4.5, Theorem 4.3 implies

COROLLARY 4.3. Under the hypotheses of Theorem 4.3 the equation (4.1) has an $(n_0 - 1)$-dimensional subspace of solutions satisfying (4.45).

PROBLEM 4.3. Let the hypotheses of Theorem 4.3 be fulfilled, $q \in L^2_{loc}(\mathbf{R}_+)$ and

$$\int^{+\infty} t^{2n} |q(t)|^2 \, dt < +\infty.$$

Then there exists $a_0 \in \mathbf{R}_+$ such that for any $a > a_0$ and $c_i \in \mathbf{R}$ $(i = 0, \dots, n_0 - 2)$ the equation (4.46) has at least one solution satisfying the conditions

$$u^{(i+1)}(a) = c_i \qquad (i = 0, \dots, n_0 - 2),$$

$$\int_a^{+\infty} t^{2k} |u^{(k)}(t)|^2 \, dt < \infty \qquad (k = 0, \dots, n_0).$$

4.5. Theorems on the dimension of the space $V^{(n,\sigma)}(p)$.

THEOREM 4.4. If for large t,

$$(-1)^{n-n_0-1} p(t) \geq 0, \tag{4.54}$$

then

$$\dim V^{(n,2n_0)}(p) = n_0. \tag{4.55}$$

Moreover, if

$$\lim_{t \to +\infty} (-1)^{n-n_0-1} t^n p(t) = +\infty, \tag{4.56}$$

then together with (4.55) we have

$$V^{(n,0)}(p) = V^{(n,2n_0)}(p). \tag{4.57}$$

PROOF. Let $a \in \mathbf{R}_+$ be so large that (4.54) holds on $[a, +\infty[$. By Lemma 4.6, in order to prove (4.55) it suffices to show that for any $c_i \in \mathbf{R}$ $(i = 0, \ldots, n_0 - 1)$ the equation (4.2) has one and only one solution satisfying the conditions

$$u^{(i)}(a) = c_i \qquad (i = 0, \ldots, n_0 - 1), \qquad \int_a^{+\infty} |u^{(n_0)}(t)|^2 \, dt < +\infty. \qquad (4.58)$$

If $p(t) = 0$ for $t \geq a$, then the function $u(t) = \sum_{i=0}^{n_0-1} c_i (t-a)^i / i!$ is the unique solution of the problem (4.2), (4.58). Hence it remains to consider the case when p is distinct from zero on a set of positive measure contained in $[a, +\infty[$.

Let $b \in \,]a, +\infty[$ and let u be a solution of equation (4.2). Multiplying both sides of (4.2) by $(-1)^{n-n_0} t^{n-2n_0} u(t)$ and integrating over $[a, b]$, by Lemma 4.1 and the inequality (4.54) we get

$$\mu_{n_0}^n \int_a^b |u^{(n_0)}(t)|^2 \, dt + \int_a^b t^{n-2n_0} |p(t)| \, |u(t)|^2 \, dt = l(u)(b) - l(u)(a),$$

with

$$l(u)(t) = \sum_{i=0}^{n-n_0-2} (-1)^{n-n_0-i} \mu_{in-i-2}^n (t^{n-2n_0})' u^{(i)}(t) u^{(n-2-i)}(t)$$

$$+ \sum_{i=0}^{n-n_0-1} (-1)^{n-n_0-1-i} t^{n-2n_0} u^{(i)}(t) u^{(n-1-i)}(t).$$

Hence the inequalities (4.28) hold, where $m_0 = 0$, $m = n_0$, $g_0(t) = t^{n-2n_0} |p(t)|$, $g_n(t) = 1$ and $g_k(t) = 0$ $(k = 1, \ldots, n_0 - 1)$. Therefore, by Lemma 4.7 the problem (4.2), (4.58) is solvable.

We will show that this problem has at most one solution, i.e. that the equation (4.2) under the homogeneous boundary conditions

$$u^{(i)}(a) = 0 \qquad (i = 0, \ldots, n_0 - 1), \qquad \int_a^{+\infty} |u^{(n_0)}(t)|^2 \, dt < +\infty \qquad (4.58_0)$$

has only the trivial solution.

Assume that, on the contrary, the problem (4.2), (4.58$_0$) has a nontrivial solution u. Clearly,

$$\lim_{b \to +\infty} b^{2i-n} \int_a^b |u^{(i)}(t)|^2 \, dt = 0 \qquad (i = 0, \ldots, n - n_0 - 1). \qquad (4.59)$$

Set

$$w(t) = \sum_{i=0}^{n_0-1} (-1)^{n-n_0-1-i} u^{(n-1-i)}(t) u^{(i)}(t)$$

for n even, and

$$w(t) = \sum_{i=0}^{n_0-1} (-1)^{n-n_0-1-i} u^{(n-1-i)}(t) u^{(i)}(t) + \frac{1}{2} |u^{(n_0)}(t)|^2$$

for n odd. Then

$$w(a) \geq 0, \qquad w'(t) \geq (-1)^{n-n_0-1} p(t) |u(t)|^2 \geq 0 \qquad \text{for } t \geq a,$$

and w' is distinct from zero on a set of positive measure. So there exist $t_0 > a$ and $\gamma > 0$ such that

$$w(t) > \gamma \qquad \text{for } t \geq t_0. \tag{4.60}$$

By Lemma 4.1 and the conditions (4.58_0),

$$\int_a^b (b-t)^{n-1} w(t)\, dt = \sum_{i=0}^{n-n_0-1} l_i \int_a^b (b-t)^{2i} |u^{(i)}(t)|^2\, dt$$

for any $b > a$, where the constants l_i $(i = 0, \ldots, n - n_0 - 1)$ do not depend on b. Dividing both parts of this equality by b^n and passing b to $+\infty$, from (4.59) and (4.60) we obtain a contradiction: $0 < \gamma/n \leq 0$, which proves the unique solvability of the problem (4.2), (4.58). Thus (4.55) is true.

Now let the condition (4.56) hold. Then Theorem 4.1 yields $\dim V^{(n,0)}(p) \geq n_0$. On the other hand, as already proven, (4.55) is fulfilled. Consequently, the inclusion $V^{(n,0)}(p) \subset V^{(n,2n_0)}(p)$ implies (4.57). $\square$

REMARK 4.1. The condition (4.56) cannot be replaced by the condition

$$\lim_{t \to +\infty} \inf (-1)^{n-n_0-1} t^n p(t) > 0. \tag{4.61}$$

Indeed, if $n_0 - 2 < \alpha < n_0 - 1$, $l = \alpha(\alpha - 1) \ldots (\alpha - n + 1)$,

$$p(t) = \frac{l}{(1+t)^n}, \qquad u(t) = (1+t)^\alpha,$$

then (4.61) holds, $u \in V^{(n,2n_0)}(p)$, but $u \notin V^{(n,0)}$, i.e. (4.57) is violated.

According to Lemma 4.5, every solution $u \in V^{(n,0)}(p)$ satisfies the conditions (4.45). Hence, if (4.56) holds, then $\dim Z^{(n)}(p) \geq n_0$.

PROBLEM 4.4*. Does the condition (4.56) imply

$$\dim Z^{(n)}(p) = n_0?$$

THEOREM 4.5. If $n = 2n_0 + 1$ and

$$\lim_{t \to +\infty} (-1)^{n_0+1} t^n p(t) = +\infty, \tag{4.62}$$

then for any $\sigma > n$,

$$\dim V^{(n,\sigma)}(p) = n_0 + 1. \tag{4.63}$$

Furthermore, if

$$\lim_{t \to +\infty} \inf (-1)^{n_0+1} t^{2-2\sigma+n} p(t) > 0 \tag{4.64}$$

for a certain $\sigma > n$, then

$$V^{(n,\sigma)}(p) \subset Z^{(n)}(p). \tag{4.65}$$

PROOF. By Theorem 4.2, $\dim V^{(n,\sigma)}(p) \geq n_0 + 1$. So the equality (4.63) will be established if we show that for a certain $a \in \mathbf{R}_+$ the equation (4.2) under the homogeneous boundary conditions

$$u^{(i)}(a) = 0 \qquad (i = 0, \ldots, n_0),$$

$$\int_a^{+\infty} t^{2k-\sigma} |u^{(k)}(t)|^2 \, dt < +\infty \qquad (k = 0, \ldots, n_0)$$

(4.66)

has only the trivial solution.

Set

$$\nu_i = \mu_i^n \prod_{j=i}^{n-2} (\sigma - n + j - 1) \qquad (i = 0, \ldots, n_0),$$

and choose $\lambda > (n + \sigma)^3$ and $a > 0$ so that

$$\sum_{i=0}^{n_0-1} \nu_i \beta_i(\lambda, n_0) < \nu_{n_0},$$

(4.67)

$$(-1)^{n_0+1} t^n p(t) > \sum_{i=0}^{n_0-1} \nu_i \alpha_i(\lambda, n_0) + 1 \qquad \text{for } t \geq a,$$

where $\alpha_0(\lambda, n_0) = 1$, $\beta_0(\lambda, n_0) = 0$, and the constants $\alpha_i(\lambda, n_0)$ and $\beta_i(\lambda, n_0)$ ($i = 1, \ldots, n_0 - 1$) are defined by (4.13).

Let u be a solution of the problem (4.2), (4.66). Multiplying both sides of (4.2) by $(-1)^{n_0+1} t^{n-\sigma} u(t)$ and integrating from a to t, we obtain

$$w(t) + \sum_{i=0}^{n_0-1} (-1)^{n_0-i-1} \nu_i \int_a^t \tau^{2i-\sigma} |u^{(i)}(\tau)|^2 \, d\tau$$

$$= (-1)^{n_0+1} \int_a^t \tau^{n-\sigma} p(\tau) |u(\tau)|^2 \, d\tau + \nu_{n_0} \int_a^t \tau^{2n_0-\sigma} |u^{(n_0)}(\tau)|^2 \, d\tau$$

with $w(t) = \sum_{i=0}^{n_0} \sum_{j=i}^{n-1-i} c_{ij} t^{i+j+1-\sigma} u^{(i)}(t) u^{(j)}(t)$, c_{ij}=const. Since according to Lemma 4.4,

$$\lim_{t \to +\infty} \inf |w(t)| = 0,$$

the last equality yields

$$\sum_{i=0}^{n_0-1} (-1)^{n_0-i-1} \nu_i \int_a^{+\infty} t^{2i-\sigma} |u^{(i)}(t)|^2 \, dt = (-1)^{n_0+1} \int_a^{+\infty} t^{n-\sigma} p(t) |u(t)|^2 \, dt$$

$$+ \nu_{n_0} \int_a^{+\infty} t^{2n_0-\sigma} |u^{(n_0)}(t)|^2 \, dt.$$

(4.68)

By Lemma 4.3 and the inequalities (4.67),

$$\sum_{i=0}^{n_0-1} \nu_i \int_a^{+\infty} t^{2i-\sigma} |u^{(i)}(t)|^2 \, dt \leq \int_a^{+\infty} ((-1)^{n_0+1} t^{n-\sigma} p(t) - 1) |u(t)|^2 \, dt$$

$$+\nu_{n_0}\int_a^{+\infty} t^{2n_0-\sigma}|u^{(n_0)}(t)|^2\,dt.$$

Thus from (4.68) we get

$$\int_a^{+\infty}|u(t)|^2\,dt = 0,$$

i.e. $u(t) \equiv 0$, which shows the validity of (4.63). $\square$

Now suppose that for a certain $\sigma > n$ (4.64) holds. Then by Corollary 4.2 and the equality (4.63), every solution $u \in V^{(n,\sigma)}(p)$ vanishes at infinity. So the inclusion (4.65) is true. $\square$

REMARK 4.2. The condition (4.62) cannot be replaced by the conditions

$$\lim_{t\to+\infty}\inf(-1)^{n_0+1}t^n p(t) > 0. \tag{4.69}$$

Indeed, let $n_0 + 1 < \alpha < n_0 + 2$, $l = \alpha(\alpha-1)\ldots(\alpha-n+1)$,

$$p(t) = \frac{l}{(1+t)^n}, \qquad u(t) = (1+t)^\alpha.$$

Then (4.69) holds, $u \in V^{(n,n+4)}(p)$, but $u \notin V^{(n,n+1)}(p)$. Hence the equality (4.63) is violated for either $\sigma = n + 1$ or $\sigma = n + 4$.

PROBLEM 4.5*. Let $n = 2n_0 + 1$. Does the condition (4.62) imply

$$\dim Z^{(n)}(p) = n_0 + 1?$$

THEOREM 4.6. If $n = 2n_0 \geq 4$, $p \in \tilde{C}_{\text{loc}}(\mathbf{R}_+)$ and

$$\lim_{t\to+\infty}(-1)^{n_0}(t^{n+1}p(t))' = +\infty, \tag{4.70}$$

then

$$\dim V^{(n,0)}(p) = n_0 - 1. \tag{4.71}$$

PROOF. Theorem 4.3 yields $\dim V^{(n,0)}(p) \geq n_0 - 1$. So it suffices to show that the equation (4.2) under the homogeneous boundary conditions

$$u^{(i)}(a) = 0 \qquad (i = 0,\ldots,n_0 - 1), \tag{4.72}$$

$$\int_a^{+\infty} t^{2k}|u^{(k)}(t)|^2\,dt < +\infty \qquad (k = 0,\ldots,n_0)$$

has only the trivial solution whenever a is sufficiently large.

Choose $a > 0$ and $\lambda > n^3$ so that

$$(-1)^{n_0}t^n p(t) > \nu_1 \qquad \text{for } t \geq a, \tag{4.73}$$

and

$$\sum_{i=1}^{n_0-1} \nu_i \beta_i(\lambda, n_0) < 1,$$

(4.74)

$$(-1)^{n_0}(t^{n+1}p(t))' > 2 + \sum_{i=1}^{n_0-1} \nu_i \alpha_i(\lambda, n_0) \qquad \text{for } t \geq a,$$

where $\nu_i = 2(n+1)!\mu_{i-1}^{n-1}/(2i)!$ and the constants $\alpha_i(\lambda, n_0)$ and $\beta_i(\lambda, n_0)$ are defined by (4.13).

Let the problem (4.2), (4.72) have a nontrivial solution u. By Lemmas 4.1 and 4.4,

$$w(t) - w(a) + \sum_{i=0}^{n_0}(-1)^{n_0-i}\frac{n!}{(2i)!}\mu_i^n\int_a^t \tau^{2i}|u^{(i)}(\tau)|^2\,d\tau$$

$$= (-1)^{n_0}\int_a^t \tau^n p(\tau)|u(\tau)|^2\,d\tau$$

and

$$\lim_{t\to+\infty}\inf|w(t)| = 0,$$

where

$$w(t) = \sum_{i=0}^{n_0-1}\sum_{j=i}^{n-1-i}(-1)^{n_0-1-j}\frac{n!}{(i+j+1)!}\mu_{ij}^n t^{i+j+1}u^{(i)}(t)u^{(j)}(t).$$

According to (4.72) and (4.73) this yields

$$\int_a^{+\infty} t^n|p(t)|\,|u(t)|^2\,dt < +\infty.$$

(4.75)

Multiplying both sides of (4.2) by $(-1)^{n_0-1}2t^{n+1}u'(t)$, integrating from a to t and applying Lemma 4.1, we obtain

$$\sum_{i=1}^{n_0}\sum_{j=i}^{n-i}\nu_{ij}t^{i+j+1}u^{(i)}(t)u^{(j)}(t)$$

$$= (-1)^{n_0}a^{n+1}p(a)|u(a)|^2 + a^{n+1}|u^{(n_0)}(a)|^2 - (-1)^{n_0}t^{n+1}p(t)|u(t)|^2$$

$$+(-1)^{n_0}\int_a^t(\tau^{n+1}p(\tau))'|u(\tau)|^2\,d\tau + \nu_{n_0}\int_a^t \tau^{2n_0}|u^{(n_0)}(\tau)|^2\,d\tau$$

$$+ \sum_{i=1}^{n_0-1}(-1)^{n_0-i}\nu_i\int_a^t \tau^{2i}|u^{(i)}(\tau)|^2\,d\tau$$

(4.76)

with

$$\nu_{ij} = (-1)^{n_0-j}2\frac{(n+1)!}{(i+j+1)!}\mu_{i-1j-1}^{n-1}.$$

Lemma 4.3 and the inequalities (4.74) imply the existence of $a_0 > a$ such that

$$(-1)^{n_0}\int_a^t(\tau^{n+1}p(\tau))'|u(\tau)|^2\,d\tau + \nu_{n_0}\int_a^t \tau^{2n_0}|u^{(n_0)}(\tau)|^2\,d\tau$$

$$> \delta - \nu_1 a |u(a)|^2 + \sum_{i=1}^{n_0-1} \nu_i \int_a^t \tau^{2i} |u^{(i)}(\tau)|^2 \, d\tau \qquad \text{for } t \geq a_0, \qquad (4.77)$$

where $\delta = \int_a^{+\infty} |u(\tau)|^2 \, d\tau > 0$.

In view of (4.73), (4.76) and (4.77),

$$\sum_{i=1}^{n_0} \sum_{j=i}^{n-i} \nu_{ij} t^{i+j} u^{(i)}(t) u^{(j)}(t) \geq \frac{\delta}{t} - t^n |p(t)| \, |u(t)|^2 \qquad \text{for } t \geq a_0.$$

By Lemma 4.1 the integration of this inequality over $[a_0, t]$ gives

$$w_1(t) \geq w_1(a) + \delta \ln \frac{t}{a_0} - \int_a^t \tau^n |p(\tau)| \, |u(\tau)|^2 \, d\tau - \sum_{i=1}^{n_0} c_i \int_a^t \tau^{2i} |u^{(i)}(\tau)|^2 \, d\tau$$

$$\text{for } t \geq a_0, \qquad (4.78)$$

where

$$w_1(t) = \sum_{i=1}^{n_0-1} \sum_{j=i}^{n-1-i} c_{ij} t^{i+j+1} u^{(i)}(t) u^{(j)}(t)$$

and c_{ij} and c_i are certain constants.

From (4.78), applying (4.72) and (4.75), we get

$$\lim_{t \to +\infty} w_1(t) = +\infty.$$

On the other hand, by Lemma 4.4,

$$\lim_{t \to +\infty} \inf |w_1(t)| = 0.$$

Thus we arrive at a contradiction. $\square$

REMARK 4.3. The condition (4.70) cannot be replaced by the condition

$$\lim_{t \to +\infty} \inf (-1)^{n_0} (t^{n+1} p(t))' > 0. \qquad (4.79)$$

Indeed, let

$$0 < l < \frac{(2n-1)!!}{2^n}, \qquad p(t) = \frac{(-1)^{n_0} l}{(1+t)^n},$$

and let u be a solution in $V^{(n,0)}(p)$. Then, in view of (4.45),

$$\rho = \sup\{t^{1/2} |u(t)| : t \in \mathbf{R}_+\} < +\infty.$$

Besides, $\lim_{t \to +\infty} u^{(i)}(t) = 0$ $(i = 1, \ldots, n-1)$. So, after n successive integrations of the inequality

$$|u^{(n)}(t)| \leq \frac{l}{(1+t)^{n+1/2}} \rho \qquad \text{for } t \geq 0,$$

we obtain

$$|u(t)| \leq \frac{\epsilon}{(1+t)^{1/2}} \rho \qquad \text{for } t \geq 0,$$

and $\rho \leq \epsilon \rho$ with $\epsilon = 2^n l / (2n-1)!! < 1$. So $\rho = 0$ and, consequently, $\dim V^{(n,0)}(p) = 0$, although the condition (4.79) is fulfilled.

PROBLEM 4.6*. Let $n = 2n_0 \geq 4$, $p \in \tilde{C}_{\mathrm{loc}}(\mathbf{R}_+)$ and $(-1)^{n_0}p'(t) \geq 0$ for $t \geq 0$. Does the condition $\lim_{t \to +\infty}(-1)^{n_0}p(t) = +\infty$ imply

$$\dim Z^{(n)}(p) \geq n_0?$$

PROBLEM 4.7*. Let $n = 2n_0 \geq 4$, $p \in \tilde{C}_{\mathrm{loc}}(\mathbf{R}_+)$, $(-1)^{n_0}p'(t) \geq 0$ for $t \geq 0$ and $\lim_{t \to +\infty}(-1)^{n_0}p(t) = +\infty$. Does the validity of the condition

$$\int_I \frac{p'(t)}{p(t)}\,dt = +\infty$$

for any open set $I \subset \mathbf{R}_+$ such that

$$\mathrm{meas}\{I \cap\,]t, t+1[\} \to 1 \qquad \text{as } t \to +\infty$$

imply

$$\dim Z^{(n)}(p) = n_0 + 1?$$

4.6. On solutions of second order equations. This subsection is devoted to the equation

$$u'' = p(t)u \tag{4.80}$$

with $p \in L_{\mathrm{loc}}(\mathbf{R}_+)$.

THEOREM 4.7. Let

$$p(t) \geq 0 \tag{4.81}$$

for large t. Then

$$\dim V^{(2,2)}(p) = 1 \tag{4.82}$$

and, moreover, the equality

$$Z^{(2)}(p) = V^{(2,2)}(p) \tag{4.83}$$

holds if and only if

$$\int_0^{+\infty} tp(t)\,dt = +\infty. \tag{4.84}$$

Furthermore, if

$$\lim_{t \to +\infty} \inf t^2 p(t) > \frac{3}{4}, \tag{4.85}$$

then together with (4.83) we have

$$Z^{(2)}(p) = V^{(2,0)}(p). \tag{4.86}$$

PROOF. Theorem 4.4 yields (4.82).

Let $u \in V^{(2,2)}(p)$ be a nonzero solution. Then by (4.81), $a \in\,]0, +\infty[$ can be found so that $u(t) \neq 0$ and $u(t)u'(t) \leq 0$ for $t \geq a$. Without loss of generality we may assume that (4.81) is fulfilled on $[a, +\infty[$ and

$$u(t) > 0, \qquad u'(t) \leq 0 \qquad \text{for } t \geq a. \tag{4.87}$$

Thus, applying the equality

$$tu'(t) - u(t) + c_0 = \int_a^t \tau p(\tau)u(\tau)\,d\tau$$

with $c_0 = u(a) - au'(a)$, we obtain

$$c_0 \geq u(t) \int_a^t \tau p(\tau) \, d\tau \qquad \text{for } t \geq a.$$

If the condition (4.84) holds, then the last inequality implies $u \in Z^{(2)}(p)$. Consequently, $Z^{(2)}(p) \subset V^{(2,2)}(p)$. On the other hand, $\dim Z^{(2)}(p) \leq 1$, because any solution of the equation (4.80) under the conditions $u(a) > 0$, $u'(a) > 0$ is unbounded. So by (4.82), the validity of (4.83) becomes obvious.

Now we will show that the condition (4.84) is necessary for (4.83) to hold. Assume the contrary, i.e.

$$\int_0^{+\infty} tp(t) \, dt < +\infty, \qquad (4.88)$$

while (4.83) is true. Then there exists a function $u \in Z^{(2)}(p)$ satisfying the conditions (4.87) and

$$u(t) = \int_t^{+\infty} (\tau - t)p(\tau)u(\tau) \, d\tau \leq u(t) \int_t^{+\infty} \tau p(\tau) \, d\tau \qquad \text{for } t \geq a.$$

Hence

$$\int_t^{+\infty} \tau p(\tau) \, d\tau \geq 1 \qquad \text{for } t \geq a,$$

which contradicts (4.88). This shows the necessity of the condition (4.84).

Finally, consider the case when (4.85) is fulfilled. To establish (4.86) it suffices to prove that any solution of the equation (4.80) under the conditions (4.87) belongs to the set $V^{(2,0)}(p)$.

In view of (4.85), without loss of generality we may assume that

$$t^2 p(t) > \frac{3 + 5\epsilon}{4} \qquad \text{for } t \geq a$$

with a constant $\epsilon \in \,]0, 1[$. By (4.87) the identity

$$c_1 = t|u(t)|^2 - t^2 u'(t)u(t) + \int_a^t \tau^2 p(\tau)|u(\tau)|^2 \, d\tau + \int_a^t \tau^2 |u'(\tau)|^2 \, d\tau - \int_a^t |u(\tau)|^2 \, d\tau,$$

where $c_1 = a|u(a)|^2 - a^2 u'(a)u(a)$, yields

$$t|u(t)|^2 + \int_a^t (\tau^2 |u'(\tau)|^2 + \epsilon |u(\tau)|^2) \, d\tau - \frac{1-\epsilon}{4} \int_a^t |u(\tau)|^2 \, d\tau \leq c_1 \qquad \text{for } t \geq a.$$

From this inequality, since

$$\frac{1-\epsilon}{4} \int_a^t |u(\tau)|^2 \, d\tau = \frac{1-\epsilon}{2}(t|u(t)|^2 - a|u(a)|^2) + (1-\epsilon) \int_a^t \tau^2 |u'(\tau)|^2 \, d\tau$$

$$- \int_a^t (2\tau u'(\tau) + u(\tau))^2 \, d\tau \leq \frac{1-\epsilon}{2} t|u(t)|^2 + (1-\epsilon) \int_a^t \tau^2 |u'(\tau)|^2 \, d\tau \qquad \text{for } t \geq a,$$

we obtain

$$\int_a^{+\infty} (\tau^2 |u'(\tau)|^2 + |u(\tau)|^2) \, d\tau \leq \frac{c_1}{\epsilon}.$$

Therefore, $u \in V^{(2,0)}(p)$. $\square$

REMARK 4.4. The strict inequality in (4.85) cannot be replaced by a nonstrict inequality, because for $p(t) = 3/(4(1+t)^2)$ the function $u(t) = (1+t)^{-1/2}$ belongs to $Z^2(p)$, but not to $V^{(2,0)}(p)$.

THEOREM 4.8. Let

$$p(t) = -p_0(t) - p_1(t) - p_2(t), \qquad p_i \in \tilde{C}_{\text{loc}}(\mathbf{R}_+) \quad (i = 0, 1), \qquad p_2 \in L_{\text{loc}}(\mathbf{R}_+),$$

$$p_0(t) > 0, \qquad p'(t) \geq 0 \qquad \text{for } t \geq 0, \tag{4.89}$$

$$\int_0^{+\infty} \frac{|p_1'(t)|}{p_0(t)} \, dt < +\infty, \qquad \int_0^{+\infty} \frac{|p_2(t)|}{\sqrt{p_0(t)}} < +\infty$$

and

$$\liminf_{t \to +\infty} \frac{p_1(t)}{p_0(t)} > -1. \tag{4.90}$$

Then for any solution u of the equation (4.80) the function

$$\rho(u)(t) = \frac{|u'(t)|^2}{p_0(t) + p_1(t)} + |u(t)|^2 \tag{4.91}$$

tends to a finite limit as $t \to +\infty$.

PROOF. According to (4.90) there exist $a \in \mathbf{R}_+$ and $\epsilon \in \,]0, 1[$ such that

$$p_1(t) > -(1 - \epsilon)p_0(t) \qquad \text{for } t \geq a. \tag{4.92}$$

Suppose u is a nonzero solution of the equation (4.80). Then by (4.91) we have

$$\frac{\rho'(u)(t)}{\rho(u)(t)} = \alpha_0(t) + \alpha_1(t) + \alpha_2(t) \qquad \text{for } t \geq a,$$

where

$$\alpha_i(t) = -\frac{p_i'(t)}{p_0(t) + p_1(t)} \left(1 - \frac{|u'(t)|^2}{\rho(u)(t)} \right) \qquad (i = 0, 1)$$

and

$$\alpha_2(t) = -\frac{2p_2(t)}{p_0(t) + p_1(t)} \frac{u(t)u'(t)}{\rho(u)(t)}.$$

So

$$\rho(u)(t) = \rho(u)(a) \exp \left(\sum_{i=0}^{2} \int_a^t \alpha_i(\tau) \, d\tau \right). \tag{4.93}$$

In view of (4.89) and (4.92), $\alpha_0(t) \leq 0$ for $t \geq a$,

$$\int_a^{+\infty} |\alpha_1(\tau)| \, d\tau \leq \frac{1}{\epsilon} \int_a^{+\infty} \frac{|p_1'(\tau)|}{p_0(\tau)} \, d\tau < +\infty,$$

and

$$\int_a^{+\infty} |\alpha_2(\tau)| \, d\tau \leq \frac{1}{\sqrt{\epsilon}} \int_a^{+\infty} \frac{|p_2(\tau)|}{\sqrt{p_0(\tau)}} \, d\tau < +\infty.$$

Hence the representation (4.93) implies that $\rho(u)$ tends to a finite limit as $t \to +\infty$. $\square$

PROBLEM 4.8. If the conditions (4.89) and (4.92) hold, then for any solution u of the equation (4.80) the function $\rho(u)$ is of bounded variation on $[a, +\infty[$.

THEOREM 4.9. Let (4.89) be fulfilled and

$$\lim_{t \to +\infty} p_0(t) = +\infty. \tag{4.94}$$

Then $\dim Z^{(2)}(p) \geq 1$, i.e. the equation (4.80) has at least one nonzero solution vanishing at infinity.

PROOF. By (4.89) and (4.94), the inequality

$$\frac{|p_1(t)|}{p_0(t)} \leq \frac{|p_1(0)|}{p_0(t)} + \frac{1}{p_0(t)} \int_0^t |p_1'(\tau)| \, d\tau \qquad \text{for } t \geq 0$$

yields

$$\lim_{t \to +\infty} \frac{p_1(t)}{p_0(t)} = 0 \tag{4.95}$$

and

$$\lim_{t \to +\infty} (p_0(t) + p_1(t)) = +\infty. \tag{4.96}$$

So the hypotheses of Theorem 4.8 are satisfied.

Let u_1 and u_2 be linearly independent solutions of the equation (4.80). According to Theorem 4.8 there exist finite limits

$$\rho_i = \lim_{t \to +\infty} \rho(u_i)(t) \qquad (i = 1, 2). \tag{4.97}$$

If either ρ_1 or ρ_2 is equal to zero, then, clearly, $\dim Z^{(2)}(p) \geq 1$. Thus it remains to consider the case $\rho_i > 0$ $(i = 1, 2)$.

Since u_1 is bounded, there exists an increasing sequence of positive constants $(t_k)_{k=1}^{+\infty}$ for which

$$\lim_{k \to +\infty} u_1'(t_k) = 0.$$

From (4.91), by (4.96) and (4.97) we obtain

$$\lim_{k \to +\infty} |u_1(t_k)| = \rho_1.$$

Let u be a solution of the equation (4.80). Then

$$u(t)u_1'(t) - u'(t)u_1(t) = c \qquad \text{for } t \geq 0,$$

where $c = u(0)u_1'(0) - u'(0)u_1(0)$. Hence

$$\lim_{k \to +\infty} |u'(t_k)| = \frac{|c|}{\rho_1}.$$

From (4.91), applying the condition (4.96) once again, we get

$$\lim_{k \to +\infty} |u(t_k)| = \lim_{t \to +\infty} \rho(u)(t).$$

Without loss of generality we may assume that the sequences $(u_i(t_k))_{k=1}^{+\infty}$ $(i = 1, 2)$ are convergent. Set

$$\alpha_i = \lim_{k \to +\infty} u_i(t_k) \qquad (i = 1, 2), \qquad u_0(t) = \alpha_2 u_1(t) - \alpha_1 u_2(t).$$

Then

$$\lim_{k \to +\infty} u_0(t_k) = 0.$$

On the other hand, as proven above, $|\alpha_i| = \rho_i > 0$ $(i = 1, 2)$ and

$$\lim_{t \to +\infty} \rho(u_0)(t) = \lim_{k \to +\infty} u_0(t_k) = 0.$$

Consequently, u_0 is a nonzero solution of the equation (4.80) which vanishes at infinity. $\square$

THEOREM 4.10. Let $(m_k)_{k=1}^{+\infty}$ be a sequence of positive integers and $(r_k)_{k=1}^{+\infty}$ a nondecreasing sequence of positive constants such that

$$\lim_{k \to +\infty} r_k = +\infty, \qquad t_k = \pi \sum_{i=1}^{k} \frac{m_i}{r_i} \to +\infty \qquad \text{as } k \to +\infty.$$

Furthermore, assume

$$\sum_{k=1}^{+\infty} \frac{1}{r_k} \int_{t_k}^{t_{k+1}} |p(t) + r_k^2| \, dt < +\infty. \tag{4.98}$$

Then $\dim Z^{(2)}(p) = 1$.

To prove this theorem we need

LEMMA 4.8. Let $q \in L_{\text{loc}}(\mathbf{R}_+)$ and let the equation

$$v'' = q(t)v \tag{4.99}$$

have linearly independent solutions v_1 and v_2 satisfying the estimates

$$|v_k(t)| \le \varphi_k(t) \qquad \text{for } t \ge 0 \qquad (k = 1, 2), \tag{4.100}$$

where the functions $\varphi_k \colon \mathbf{R}_+ \to \,]0, +\infty[$ $(k = 1, 2)$ are measurable and bounded on each finite interval and the function φ_2/φ_1 does not increase. Suppose, in addition,

$$\int_0^{+\infty} \varphi_1(t)\varphi_2(t)|p(t) - q(t)| \, dt < +\infty. \tag{4.101}$$

Then the equation (4.80) has a fundamental system of solutions u_1, u_2 such that

$$u_k(t) = v_k(t) + o(\varphi_k(t)) \qquad \text{as } t \to +\infty \qquad (k = 1, 2). \tag{4.102}$$

PROOF. Without loss of generality we may assume that the Wronskian of the system v_1, v_2 equals 1.

By (4.101) we can choose a constant $a \in \mathbf{R}_+$ such that

$$\int_a^{+\infty} \varphi_1(t)\varphi_2(t)|p(t) - q(t)| \, dt < \frac{1}{4}. \tag{4.103}$$

Let $L^{+\infty}([a, +\infty[)$ be the Banach space of essentially bounded measurable functions $x \colon [a, +\infty[\, \to \mathbf{R}$ with the norm

$$\|x\| = \operatorname{ess\,sup}\{|x(t)| : t \in \mathbf{R}_+\}.$$

First we assume that

$$\lim_{t \to +\infty} \frac{\varphi_2(t)}{\varphi_1(t)} = 0 \qquad (4.104)$$

and consider the linear operators

$$A_1(x)(t) = \frac{v_1(t)}{\varphi_1(t)} + \int_t^{+\infty} \frac{v_1(t)v_2(\tau)}{\varphi_1(t)}\varphi_1(\tau)(p(\tau) - q(\tau))x(\tau)\,d\tau$$

$$+ \int_a^t \frac{v_1(\tau)v_2(t)}{\varphi_1(t)}\varphi_1(\tau)(p(\tau) - q(\tau))x(\tau)\,d\tau$$

and

$$A_2(x)(t) = \frac{v_2(t)}{\varphi_2(t)} + \int_t^{+\infty} \frac{v_1(t)v_2(\tau) - v_1(\tau)v_2(t)}{\varphi_2(t)}\varphi_2(\tau)(p(\tau) - q(\tau))x(\tau)\,d\tau.$$

In view of (4.100) and (4.103), $A_k \colon L^{+\infty}([a, +\infty[) \to L^{+\infty}([a, +\infty[)$ $(k = 1, 2)$ and for any $x, y \in L^{+\infty}([a, +\infty[)$,

$$\|A_k(x) - A_k(y)\| \le \frac{1}{2}\|x - y\|.$$

By the Banach theorem there exist functions $x_k \in L^{+\infty}([a, +\infty[)$ $(k = 1, 2)$ such that

$$x_k(t) = A_k(x_k)(t) \qquad \text{for } t \ge a \qquad (k = 1, 2).$$

According to (4.100), (4.101) and (4.104), $u_k(t) = \varphi_k(t)x_k(t)$ $(k = 1, 2)$ are linearly independent solutions of the equation (4.80) satisfying (4.102).

The case $\lim_{t \to +\infty} \varphi_2(t)/\varphi_1(t) > 0$ can be treated similarly. The only difference is that the operator A_1 should be defined as

$$A_1(x)(t) = \frac{v_1(t)}{\varphi_1(t)} + \int_t^{+\infty} \frac{v_1(t)v_2(\tau) - v_1(\tau)v_2(t)}{\varphi_1(t)}\varphi_1(\tau)(p(\tau) - q(\tau))x(\tau)\,d\tau. \qquad \square$$

PROOF OF THEOREM 4.10. Set $t_0 = 0$, $m_0 = 0$, $r_0 = m_1\pi/t_1$, $n_k = \sum_{i=0}^k m_i$ $(k = 0, 1, \ldots)$, and

$$q(t) = -r_k^2 \qquad \text{for } t_k \le t < t_{k+1} \qquad (k = 0, 1, \ldots).$$

Then the functions

$$v_1(t) = (-1)^{n_k} \cos r_k(t - t_k) \qquad \text{for } t_k \le t < t_{k+1} \qquad (k = 0, 1, \ldots)$$

and

$$v_2(t) = \frac{(-1)^{n_k}}{r_k} \sin r_k(t - t_k) \qquad \text{for } t_k \le t < t_{k+1} \qquad (k = 0, 1, \ldots)$$

are linearly independent solutions of the equation (4.80) and satisfy (4.100) with

$$\varphi_1(t) = 1, \qquad \varphi_2(t) = \frac{1}{r_k} \qquad \text{for } t_k \le t < t_{k+1} \qquad (k = 0, 1, \ldots).$$

On the other hand, (4.98) implies (4.101). Hence, by Lemma 4.8 the equation (4.80) has a fundamental system of solutions of the type (4.102). Clearly,

$$\lim_{t \to +\infty} \sup u_1(t) = 1, \qquad \lim_{t \to +\infty} u_2(t) = 0.$$

Therefore, $\dim Z^{(2)}(p) = 1$. $\square$

THEOREM 4.11. Let (4.89) and (4.94) hold,

$$-1 < \lim_{t \to +\infty} \inf \frac{p_2(t)}{p_0(t)} \leq \lim_{t \to +\infty} \sup \frac{p_2(t)}{p_0(t)} < +\infty, \tag{4.105}$$

and let there exist a positive constant ϵ such that for any sequence $(\tau_k)_{k=1}^{+\infty}$ satisfying the conditions

$$0 \leq \tau_k < \tau_{k+1} \quad (k = 1, 2, \dots), \qquad \lim_{k \to +\infty} \tau_k = +\infty, \tag{4.106}$$

$$\lim_{k \to +\infty} \int_{\tau_{2k}}^{\tau_{2k+1}} \sqrt{p_0(t)}\, dt \in]\pi - \epsilon, \pi[, \tag{4.107}$$

$$\lim_{k \to +\infty} \inf \sqrt{p_0(\tau_{2k})}(\tau_{2k} - \tau_{2k-1}) > 0, \tag{4.108}$$

$$\lim_{k \to +\infty} \sup \sqrt{p_0(\tau_{2k-1})}(\tau_{2k} - \tau_{2k-1}) < \epsilon \tag{4.109}$$

we have

$$\sum_{k=1}^{+\infty} (\ln p_0(\tau_{2k+1}) - \ln p_0(\tau_{2k})) = +\infty. \tag{4.110}$$

Then $\dim Z^{(2)}(p) = 2$.

PROOF. According to Theorem 4.8, for any solution u of the equation (4.80) the function $\rho(u)$ defined by (4.91) tends to a finite limit as $t \to +\infty$. Assume

$$\lim_{t \to +\infty} \rho(u)(t) > 0. \tag{4.111}$$

By (4.96) and (4.105) we can find $\delta \in]0, 1[$ and $a > 0$ so that

$$-p(t) \geq \delta p_0(t), \qquad p_0(t) + p_1(t) > 0 \qquad \text{for } t \geq. \tag{4.112}$$

Then in view of (4.89) and (4.95), the equality (4.93) yields

$$\int_a^{+\infty} \frac{|u'(t)|^2}{p_0(t) + p_1(t)} \frac{p_0'(t)}{p_0(t)}\, dt < +\infty. \tag{4.113}$$

Let $(t_{2k-1})_{k=1}^{+\infty}$ and $(t_{2k})_{k=1}^{+\infty}$ be the sequences of zeros of u and u', respectively, in the interval $[a, +\infty[$ and let, in addition, $t_k < t_{k+1}$ $(k = 1, 2, \dots)$. Choose $\nu \in]0, 1[$ so that $2 \arcsin \nu \in]\pi\sqrt{\delta\epsilon}, \pi[$. Since $|u(t_{2k-1})|^2 = 0$ and $|u(t_{2k})|^2 = \rho(u)(t_{2k})$, each interval $]t_{2k-1}, t_{2k+1}[$ contains two points τ_{2k-1}, τ_{2k} such that

$$\tau_{2k-1} < t_{2k} < \tau_{2k}, \qquad |u(\tau_k)|^2 = \nu^2 \rho(u)(\tau_k),$$

$$\tag{4.114}$$

$$|u(t)|^2 < \nu^2 \rho(u)(t) \qquad \text{for } t \in]\tau_{2k}, \tau_{2k+1}[.$$

By (4.114),

$$|u'(t)|^2 > (1 - \nu^2)(p_0(t) + p_1(t))\rho(u)(t) \qquad \text{for } t \in]\tau_{2k}, \tau_{2k+1}[.$$

So from (4.113), taking into account (4.111), we get

$$\sum_{k=1}^{+\infty}(\ln p_0(\tau_{2k+1}) - \ln p_0(\tau_{2k})) < +\infty \tag{4.115}$$

and

$$\lim_{k\to+\infty}\frac{p_0(\tau_{2k+1})}{p_0(\tau_{2k})} = 1. \tag{4.116}$$

In view of (4.89) and (4.114), the equality

$$\sqrt{p_0(t) + p_1(t)} = \left(\frac{u(t)}{\sqrt{\rho(u)(t)}}\right)'\left(1 - \frac{|u(t)|^2}{\rho(u)(t)}\right)^{-1/2}\operatorname{sign} u'(t)$$

$$-\frac{1}{2}\frac{u(t)\operatorname{sign} u'(t)}{\sqrt{\rho(u)(t)}}\left(1 - \frac{|u(t)|^2}{\rho(u)(t)}\right)^{1/2}\frac{p_0'(t) + p_1'(t)}{p_0(t) + p_1(t)}$$

$$-\frac{p_2(t)}{\sqrt{p_0(t) + p_1(t)}}\frac{|u(t)|^2}{\rho(u)(t)}$$

yields

$$\lim_{k\to+\infty}\int_{\tau_{2k}}^{\tau_{2k+1}}\sqrt{p_0(t) + p_1(t)}\,dt$$

$$= \lim_{k\to|\infty}\operatorname{sign} u'(t)\cdot\arcsin\frac{u(t)}{\sqrt{\rho(u)(t)}}\Bigg|_{\tau_{2k}}^{\tau_{2k+1}} = 2\arcsin\nu \in\,]\pi - \epsilon, \pi[.$$

Consequently, (4.107) holds.

From (4.80), by (4.112) we derive

$$u'^2(t) = -2\int_t^{t_{2k}} p(\tau)u(\tau)u'(\tau)\,d\tau$$

$$\geq \delta p_0(\tau_{2k-1})(|u(t_{2k})|^2 - |u(t)|^2)\qquad\text{for } t\in\,]\tau_{2k-1}, t_{2k}[.$$

Thus

$$\sqrt{p_0(\tau_{2k-1})}(t_{2k} - \tau_{2k-1}) \leq \frac{1}{\sqrt{\sigma}}\int_{\tau_{2k-1}}^{t_{2k}}\frac{|u'(t)|\,dt}{\sqrt{|u(t_{2k})|^2 - |u(t)|^2}}$$

$$= \frac{1}{\sqrt{\sigma}}\left(\frac{\pi}{2} - \arcsin\nu\sqrt{\frac{\rho(u)(\tau_{2k-1})}{\rho(u)(t_{2k})}}\right),$$

and according to (4.111),

$$\lim_{k\to+\infty}\sup\sqrt{p_0(\tau_{2k-1})}(t_{2k} - \tau_{2k-1}) < \frac{\epsilon}{2}.$$

Similarly,

$$\lim_{k\to+\infty}\sup\sqrt{p_0(\tau_{2k-1})}(\tau_{2k} - t_{2k}) < \frac{\epsilon}{2}.$$

Hence (4.109) is fulfilled. In the same way, applying the second part of (4.105), we can prove the validity of (4.108). Therefore, the sequence $(\tau_k)_{k=1}^{+\infty}$ satisfies the conditions (4.106)–(4.109) together with (4.115), which contradicts (4.110). □

REMARK 4.5. It follows from the proof that the theorem remains true if (4.105) is replaced by $\lim_{t \to +\infty} \inf p_2(t)/p_0(t) > -1$ while (4.108) is omitted.

REMARK 4.6. The estimate (4.107) in Theorem 4.11 is optimal in the following sense: there exists an infinitely differentiable nonincreasing unbounded function p such that for any sequence $(\tau_k)_{k=1}^{+\infty}$ satisfying the conditions (4.106),

$$\lim_{k \to +\infty} \int_{\tau_{2k}}^{\tau_{2k+1}} \sqrt{|p(t)|}\, dt > \pi \qquad (4.117)$$

and

$$\lim_{k \to +\infty} \sup \sqrt{p(\tau_{2k-1})}(\tau_{2k} - \tau_{2k-1}) < \frac{\pi}{4} \qquad (4.118)$$

we have

$$\sum_{k=1}^{+\infty} (\ln |p(\tau_{2k+1})| - \ln |p(\tau_{2k})|) = +\infty, \qquad (4.119)$$

but $\dim Z^{(2)}(p) = 1$.

Set

$$t_1 = 0, \qquad t_k = \pi \sum_{i=1}^{k-1} \frac{1}{i} \qquad (k = 2, 3, \dots), \qquad (4.120)$$

$$\varphi(t) = -\exp\left(-\frac{1}{t^2}\exp\left(-\frac{1}{1-t^2}\right)\right) \qquad \text{for } t \in\,]0, 1[, \qquad (4.121)$$

and

$$p(t) = -k^2 \qquad \text{for } t \in \left[t_k, t_k + \frac{(k-1)\pi}{k^2}\right],$$

$$p(t) = -k^2 - (2k+1)\varphi\left(\frac{k^2(t - t_k)}{\pi} - k + 1\right) \qquad \text{for } t \in\, \left]t_k + \frac{(k-1)\pi}{k^2}, t_{k+1}\right[$$

$$(k = 1, 2, \dots).$$

It is easy to verify that the function p is infinitely differentiable and nonincreasing in $\mathbf{R}_+$.

Let the sequence $(\tau_k)_{k=1}^{+\infty}$ satisfy the conditions (4.106), (4.117) and (4.118). If $[t_j, t_j + (j-1)\pi/j^2] \subset [\tau_{2k-1}, \tau_{2k}]$ and $t_{j-1} < \tau_{2k-1}$, then

$$\sqrt{|p(\tau_{2k-1})|}(\tau_{2k} - \tau_{2k-1}) \geq \sqrt{|p(t_{j-1})|}\frac{(j-1)\pi}{j^2} = \frac{(j-1)^2\pi}{j^2} > \frac{\pi}{4}.$$

This contradiction shows that neither interval $[t_j, t_j + (j-1)\pi/j^2]$ is contained within an interval $[\tau_{2k-1}, \tau_{2k}]$.

Now let $\tau_{2k+1} \in\,]t_j, t_{j+1}]$ and $\tau_{2k} \geq t_j - \pi/(j-1)^2$. Then

$$\int_{\tau_{2k}}^{\tau_{2k+1}} \sqrt{|p(t)|}\, dt \leq \pi\left(1 + \frac{1}{(j-1)^2} + \frac{1}{j^2}\right),$$

which contradicts the condition (4.117) for sufficiently large k. So if $\tau_{2k+1} \in\,]t_j, t_{j+1}]$, then $\tau_{2k} < t_j - \pi/(j-1)^2$ for sufficiently large k.

Hence, for sufficiently large k each interval $[\tau_{2k}, \tau_{2k+1}]$ contains at least one interval of the type $[t_j - \pi/(j-1)^2, t_j]$ and each interval $[\tau_{2k-1}, \tau_{2k}]$ intersects only one interval of that type. Consequently, for sufficiently large k,

$$\int_{\tau_{2k}}^{\tau_{2k+1}} d\ln|p(t)| > \int_{\tau_{2k+1}}^{\tau_{2k+2}} d\ln|p(t)|.$$

But since $\int_0^{+\infty} d\ln|p(t)| = +\infty$, the last inequality immediately implies (4.119). On the other hand,

$$\sum_{k=1}^{+\infty} \frac{1}{k} \int_{t_k}^{t_{k+1}} |p(t) + k^2| \, dt < \sum_{k=1}^{+\infty} \frac{2k+1}{k} \frac{\pi}{k^2} < +\infty.$$

Therefore, Theorem 4.10 yields $\dim Z^{(2)}(p) = 1$.

COROLLARY 4.4. Let (4.89) and (4.94) be fulfilled,

$$\lim_{t \to +\infty} \inf \frac{p_2(t)}{p_0(t)} > -1, \tag{4.122}$$

and let there exist a positive constant ϵ such that for any sequence $(\tau_k)_{k=1}^{+\infty}$ satisfying the conditions (4.106) and

$$\lim_{k \to +\infty} \sup \frac{\tau_{2k+2} - \tau_{2k+1}}{\tau_{2k+1} - \tau_{2k}} < \epsilon, \tag{4.123}$$

$$\lim_{k \to +\infty} \sqrt{|p_0(\tau_{2k+1})|}(\tau_{2k+1} - \tau_{2k}) \in]\pi - \epsilon, \pi[,$$

the relation (4.110) holds. Then $\dim Z^{(2)}(p) = 2$.

COROLLARY 4.5. Let (4.89), (4.94) and (4.122) be fulfilled and let there exist a positive nonincreasing function $\varphi \in C(\mathbf{R}_+)$ such that for sufficiently large t,

$$p_0'(t) \geq \varphi(t) p_0(t) \tag{4.124}$$

and

$$\int_0^{+\infty} \varphi(t) \, dt = +\infty. \tag{4.125}$$

Then $\dim Z^{(2)}(p) = 2$.

PROOF. If $\epsilon \in]0, 1[$ and the sequence $(\tau_k)_{k=1}^{+\infty}$ satisfies (4.106) and (4.123), then

$$\int_{\tau_{2k}}^{\tau_{2k+1}} \varphi(t) \, dt \geq \int_{\tau_{2k+1}}^{\tau_{2k+2}} \varphi(t) \, dt,$$

and so

$$\sum_{k=1}^{+\infty} \int_{\tau_{2k}}^{\tau_{2k+1}} \varphi(t) \, dt = +\infty.$$

Therefore, (4.124) implies (4.110), and by Corollary 4.4, $\dim Z^{(2)}(p) = 2$. $\square$

PROBLEM 4.9. Let $p \in \tilde{C}^1_{\text{loc}}(\mathbf{R}_+)$,

$$\lim_{t \to +\infty} p(t) = -\infty, \tag{4.126}$$

$$\lim_{t\to+\infty} p'(t)|p(t)|^{-3/2} = 0,$$

and

$$\lim_{t\to+\infty} \sup \frac{1}{\ln|p(t)|} \int_0^t |dp'(s)|p(s)|^{-3/2}| < 1.$$

Then $\dim Z^{(2)}(p) = 2$.

PROBLEM 4.10. If $p \in \tilde{C}^2_{\text{loc}}(\mathbf{R}_+)$, (4.126) holds and

$$\lim_{t\to+\infty} \sup \frac{1}{\sqrt{|p(t)|}} \int_0^t |d(|p(s)|^{-1/2})''| < 1,$$

then $\dim Z^{(2)}(p) = 2$.

PROBLEM 4.11. Let p be nonincreasing and let (4.126) hold. Then the equation (4.80) under the condition $\lim_{t\to+\infty} \sup |u(t)| > 0$ is solvable if and only if there exists a solution of this equation such that $\lim_{t\to+\infty} \sup \sqrt{|p(t)|}|u(t)| < +\infty$.

Notes. The question of the existence of nontrivial solutions of the equation (4.2) which vanish at infinity was studied by M. Biernacki [37] and M. Švec [343]. Theorems 4.1–4.6 are due to I.T. Kiguradze [193]. For Problems 4.1–4.3 see [193].

Theorem 4.7 improves a theorem of A. Kneser [209].

For some particular cases, Theorem 4.8 was proved by H. Milloux [271], G. Ascoli [5], M. Biernacki [36], R. Bellman [26], and L.I. Kamynin [167]. In the form stated here, this theorem is due to Z. Opial [298]. For Problem 4.8 see [298].

Theorem 4.9 with $p_1(t) \equiv p_2(t) \equiv 0$ was established by H. Milloux [271], and in the general case by G. Trevisan [361] and Z. Opial [297].

Examples of equations (4.80) having solutions which do not vanish at infinity while p monotonically tends to $-\infty$ as $t \to +\infty$ were constructed in [94, 117, 271, 369]. Theorem 4.10 was proved in [198].

Theorem 4.11, which generalizes a well-known result of G. Armellini, L. Tonelli and G. Sansone [4, 314, 317, 348], was established in [56]. Various particular cases of this theorem are contained in [36, 126, 242, 267, 268]. For Corollary 4.5 and Problems 4.9, 4.10 and 4.11 see [268], [174], [256, 340] and [9], respectively.

§5. Bounded and Unbounded Solutions

In this section we investigate for the differential equation

$$u^{(n)} = p(t)u, \tag{5.1}$$

with a locally integrable coefficient $p: \mathbf{R}_+ \to \mathbf{R}_+$, the dimension of the space of bounded solutions, study the behavior of such solutions at infinity and establish existence conditions for unbounded oscillatory solutions.

As well as in the previous section, n_0 denotes the integral part of $n/2$, $V^{(n,\sigma)}(p)$ is the set of solutions of the equation (5.1) satisfying the conditions (4.4), and $Z^{(n)}(p)$ stands for the set of solutions of this equation tending to zero as $t \to +\infty$. Furthermore, $B^{(n)}(p)$ is the set of all bounded solutions of (5.1).

5.1. Inequalities of Kolmogorov–Gorny type.

LEMMA 5.1. Let $-\infty < a < b < +\infty$, $u \in C^m([a,b])$, m being a nonnegative integer, and let the function $g \colon [a,b] \to \mathbf{R}_+$ be integrable and satisfy the condition

$$\delta = \min\left\{ \int_s^t g(\tau)\,d\tau \colon a \le s < t \le b,\, t = s + \frac{b-a}{2m+1} \right\} > 0.$$

Then

$$\min\{|u^{(m)}(t)| \colon a \le t \le b\}$$

$$\le (m+1)!\,(2m+1)^m \delta^{-1/2}(b-a)^{-m}\left(\int_a^b g(t)|u(t)|^2\,dt \right)^{1/2} \tag{5.2}$$

and

$$\min\{|u^{(m)}(t)| \colon a \le t \le b\} \le (m+1)!\,m^m(b-a)^{-m}\max\{|u(t)| \colon a \le t \le b\}. \tag{5.3}$$

PROOF. Set

$$a_k = a + \frac{k(b-a)}{2m+1} \quad (k=0,\dots,2m+1), \qquad r = \left(\int_a^b g(t)|u(t)|^2\,dt \right)^{1/2}$$

and choose the points $t_k \in [a_{2k}, a_{2k+1}]$ $(k=0,\dots,m)$ so that

$$|u(t_k)| = \min\{|u(t)| \colon a_{2k} \le t \le a_{2k+1}\} \quad (k=0,\dots,m).$$

Let

$$g_k(t) = \prod_{\substack{i=0 \\ i \ne k}}^m (t - t_i), \qquad v(t) = u(t) - \sum_{k=0}^m \frac{g_k(t)}{g_k(t_k)}u(t_k).$$

Then

$$|u(t_k)| \le \delta^{-1/2}r, \qquad |g_k(t_k)| \ge \left(\frac{b-a}{2m+1} \right)^m \quad (k=0,\dots,m)$$

and

$$v(t_k) = 0 \quad (k=0,\dots,m).$$

By the Rolle theorem there exists a point $s \in\,]a,b[$ for which $v^{(m)}(s) = 0$. Hence

$$u^{(m)}(s) = \sum_{k=0}^m \frac{m!}{g_k(t_k)}u(t_k)$$

and

$$|u^{(m)}(s)| \le (m+1)!\,(2m+1)^m \delta^{-1/2}r.$$

Therefore the inequality (5.2) is true.

The proof of (5.3) is quite similar, but in this case the points t_k should be defined as

$$t_k = a + \frac{k(b-a)}{m} \quad (k=0,\dots,m). \qquad \square$$

LEMMA 5.2. Let $-\infty < a < b < +\infty$, $u \in \tilde{C}^{m-1}([a,b])$ and $u^{(m)} \in L^{+\infty}([a,b])$, $m \geq 2$ being an integer. Then

$$\rho_i \leq \frac{m!\, m^m}{2}(b-a)^{-i}\rho_0$$

$$+2^{(i-1/m)(m-i)}(m!)^{(m-i)/m}m^{m-i}\rho_0^{(m-i)/m}\rho_m^{i/m} \qquad (i=1,\dots,m-1), \qquad (5.4)$$

where

$$\rho_i = \max\{|u^{(i)}(t)| : a \leq t \leq b\} \qquad (i=0,\dots,m-1),$$

$$\rho_m = \operatorname{ess\,sup}\{|u^{(m)}(t)| : a \leq t \leq b\}.$$

PROOF. In view of Lemma 5.1 there exist points $t_k \in [a,b]$ $(k=1,\dots,m-1)$ such that

$$|u^{(k)}(t_k)| \leq (k+1)!\, k^k(b-a)^{-k}\rho_0 \qquad (k=1,\dots,m-1).$$

Let $t_0 = a$ and let g_k $(k=1,\dots,m-1)$ be polynomials satisfying the conditions

$$g_k^{(i)}(t_i) = 0 \qquad (i=0,\dots,k-1), \qquad g_k^{(k)}(t) \equiv 1$$

for any k. Clearly,

$$|g_k^{(i)}(t)| \leq (b-a)^{k-i} \qquad \text{for } a \leq t \leq b \qquad (i=1,\dots,k-1).$$

Setting

$$w(t) = u(t) - \sum_{k=1}^{m-1} u^{(k)}(t_k)g_k(t),$$

$$r_k = \max\{|w^{(k)}(t)| : a \leq t \leq b\} \qquad (k=0,\dots,m-1),$$

$$r_m = \operatorname{ess\,sup}\{|w^{(m)}(t)| : a \leq t \leq b\},$$

we obtain

$$w^{(i)}(t_i) = 0 \qquad (i=1,\dots,m-1), \qquad (5.5)$$

$$r_0 \leq \rho_0 + \sum_{k=1}^{m-1}(k+1)!\, k^k\rho_0 \leq \left(1 + m!\,(m-1)^{m-2}\sum_{k=1}^{m-1}k\right)\rho_0 \leq \frac{m!\,m^m}{2}\rho_0 \qquad (5.6)$$

and

$$\rho_i \leq \sum_{k=1}^{m-1}(k+1)!\, k^k(b-a)^{-i}\rho_0 + r_i$$

$$\leq \frac{m!\,m^m}{2}(b-a)^{-i}\rho_0 + r_i \qquad (i=1,\dots,m-1), \qquad \rho_m = r_m. \qquad (5.7)$$

According to (5.5), for any $i \in \{1,\dots,m-1\}$ we can find points $t_{1i} \in [a,b[$ and $t_{2i} \in \,]t_{1i},b]$ such that

$$\max\{|w^{(i)}(t)| : t_{1i} \leq t \leq t_{2i}\} = r_i, \qquad \min\{|w^{(i)}(t)| : t_{1i} \leq t \leq t_{2i}\} = 0,$$

while the function $w^{(i)}$ does not change sign on $[t_{1i},t_{2i}]$. So,

$$r_i^2 \leq 2\int_{t_{1i}}^{t_{2i}} |w^{(i)}(t)w^{(i+1)}|\, dt \leq 2r\int_{t_{1i}}^{t_{2i}} |w^{(i)}(t)|\, dt$$

$$= 2r_{i+1}|w^{(i-1)}(t_{2i}) - w^{(i-1)}(t_{1i})| \leq 4r_{i+1}r_{i-1} \qquad (i=1,\dots,m-1).$$

By induction, we easily obtain

$$r_i \leq 2^i r_0^{1/(i+1)} r_{i+1}^{i/(i+1)} \qquad (i = 1, \ldots, m-1)$$

and

$$r_i \leq 2^{i(m-i)} r_0^{(m-i)/m} r_m^{i/m} \qquad (i = 1, \ldots, m-1).$$

In view of (5.6) and (5.7) these inequalities imply (5.4). $\square$

5.2. Lemmas on the solvability of related boundary value problems.

LEMMA 5.3. Let $n = 2n_0 + 1$, $a \in \mathbf{R}_+$, and

$$(-1)^{n_0+1} p(t) \geq 0 \qquad \text{for } t \geq a. \tag{5.8}$$

Then for any $c_i \in \mathbf{R}$ $(i = 0, \ldots, n_0)$ the equation (5.1) has at least one solution satisfying the boundary conditions

$$u^{(i)}(a) = c_i \qquad (i = 0, \ldots, n_0), \qquad \int_a^{+\infty} |p(t)| \, |u(t)|^2 \, dt < +\infty. \tag{5.9}$$

Moreover, if for a certain $\alpha \in \,]0, +\infty[$ the conditions

$$\lim_{t \to +\infty} \inf \int_t^{t+\alpha} |p(\tau)| \, d\tau > 0, \qquad \lim_{t \to +\infty} \sup \int_t^{t+\alpha} |p(\tau)| \, d\tau < +\infty \tag{5.10}$$

hold together with (5.8), then the problem (5.1),(5.9) has a unique solution u, and

$$\lim_{t \to +\infty} u^{(i)}(t) = 0 \qquad (i = 0, \ldots, n-1). \tag{5.11}$$

PROOF. According to (5.8), for any $b \in \,]a, +\infty[$ the equality

$$\sum_{i=0}^{n_0-1} (-1)^i (u^{(n-1-i)}(b) u^{(i)}(b) - u^{(n-1-i)}(a) u^{(i)}(a))$$

$$+ \frac{(-1)^{n_0}}{2} (|u^{(n_0)}(b)|^2 - |u^{(n_0)}(a)|^2) = \int_a^b p(t) |u(t)|^2 \, dt \tag{5.12}$$

yields

$$\int_a^b |p(t)| \, |u(t)|^2 \, dt \leq \sum_{i=0}^{n-1} |u^{(i)}(a)| \sum_{i=0}^{n_0} |u^{(i)}(a)| + \sum_{i=0}^{n-1} |u^{(i)}(b)| \sum_{i=0}^{n_0-1} |u^{(i)}(b)|.$$

So, if p is distinct from zero on a set of positive measure in $[a, +\infty[$, the problem (5.1), (5.9) has a solution, by Lemma 4.7. In case $p(t) \equiv 0$ the solvability is obvious.

Now we assume that the conditions (5.10) hold. Then for some $a_0 \in \,]a, +\infty[$ we have

$$\delta = \inf \left\{ \int_t^{t+\alpha} |p(\tau)| \, d\tau : \, t \geq a_0 \right\} > 0$$

and

$$\eta = \sup \left\{ \int_t^{t+(2m+1)\alpha} |p(\tau)| \, d\tau : \, t \geq a_0 \right\} < +\infty.$$

Let u be a solution of the problem (5.1), (5.9). Clearly,

$$\epsilon(t) = \left(\int_t^{t+(2m+1)\alpha} |p(\tau)| \, |u(\tau)|^2 \, d\tau \right)^{1/2} \to 0 \qquad \text{as } t \to +\infty$$

and

$$\int_t^{t+(2m+1)\alpha} |u^{(m)}(\tau)| \, d\tau \leq \eta^{1/2}\epsilon(t) \to 0 \qquad \text{as } t \to +\infty.$$

On the other hand, in view of Lemma 5.1,

$$|u^{(i)}(t)| \leq (i+1)! \, (2i+1)^i \delta^{-1/2} \epsilon(t) + \int_t^{t+(2m+1)\alpha} |u^{(i+1)}(\tau)| \, d\tau$$

$$\text{for } t \geq a_0 \qquad (i = 0, \ldots, n-1).$$

Thus, u satisfies (5.11).

It remains to show that the problem (5.1), (5.9) has at most one solution. Let u_1 and u_2 be solutions of this problem. Then, as already proved, $u = u_1 - u_2$ is a solution of (5.1) under the conditions

$$u^{(i)}(a) = 0 \qquad (i = 0, \ldots, n_0), \qquad \lim_{t \to +\infty} u^{(k)}(t) = 0 \qquad (k = 0, \ldots, n-1).$$

Therefore, by passing b to $+\infty$ in (5.12), we get

$$\int_a^{+\infty} |p(\tau)| \, |u(\tau)|^2 \, d\tau = 0,$$

i.e. $u(t) \equiv 0$. $\square$

LEMMA 5.4. Let $a \in \mathbf{R}_+$ and

$$p(t) \geq 0 \qquad \text{for } t \geq a. \tag{5.13}$$

Then for any $c_i \in \mathbf{R}$ $(i = 0, \ldots, n-2)$ the equation (5.1) has at least one solution satisfying the boundary conditions

$$u^{(i)}(a) = c_i \qquad (i = 0, \ldots, n-2), \qquad \lim_{t \to +\infty} \inf |u^{(n-1)}(t)| = 0. \tag{5.14}$$

PROOF. Choose $a_0 \in]a, a+1[$ so that

$$\int_a^{a_0} p(\tau) \, d\tau \leq \frac{1}{3}, \tag{5.15}$$

and set

$$r = 3(n-1)! \, (a_0 - a)^{1-n} \sum_{k=0}^{n-2} |c_k|.$$

Let $u(\cdot, \gamma)$ be the solution of (5.1) satisfying the initial conditions

$$u^{(i)}(a, \gamma) = c_i \qquad (i = 0, \ldots, n-2), \qquad u^{(n-1)}(a, \gamma) = \gamma, \tag{5.16}$$

and let

$$\rho(\gamma) = \max\{|u(t, \gamma)| : a \leq t \leq a_0\}.$$

In view of (5.15), the equality

$$u(t, \gamma) = \frac{\gamma(t-a)^{n-1}}{(n-1)!} + \sum_{k=0}^{n-2} \frac{c_k}{k!}(t-a)^k + \frac{1}{(n-1)!} \int_a^t (t-\tau)^{n-1} p(\tau) u(\tau, \gamma) \, d\tau$$

yields

$$\rho(\gamma) \leq \frac{(a_0-a)^{n-1}}{(n-1)!} \left(\gamma + \frac{r}{3}\right) + \frac{1}{3}\rho(\gamma)$$

and, thus,

$$\rho(\gamma) \leq \frac{3(a_0-a)^{n-1}}{2(n-1)!} \left(\gamma + \frac{r}{3}\right). \tag{5.17}$$

Suppose $\gamma > r$. Then by (5.15) and (5.17) we have

$$u^{(i)}(a_0, \gamma) = \frac{\gamma(a_0-a)^{n-i}}{(n-i)!} + \sum_{k=i}^{n-2} \frac{c_k}{k!}(a_0-a)^k$$

$$+ \frac{1}{(n-1)!} \int_a^{a_0} (a_0-\tau)^{n-1} p(\tau) u(\tau, \gamma) \, d\tau \geq \frac{\gamma(a_0-a)^{n-1}}{(n-1)!}$$

$$- \frac{r(a_0-a)^{n-1}}{3(n-1)!} - \frac{(a_0-a)^{n-1}}{2(n-1)!} \frac{1}{\left(\gamma + \frac{r}{3}\right)} > 0 \qquad (i = 0, \dots, n-1).$$

According to (5.13), this implies

$$u^{(i)}(t, \gamma) > 0 \qquad \text{for } \gamma > r, \, t \geq a_0 \qquad (i = 0, \dots, n-1).$$

Similarly,

$$u^{(i)}(t, \gamma) < 0 \qquad \text{for } \gamma < -r, \, t \geq a_0 \qquad (i = 0, \dots, n-1).$$

Since the function $u(a+m, \cdot)$, with m a positive integer, is continuous and

$$u(a+m, \gamma) < 0 \qquad \text{for } \gamma < -r, \qquad u(a+m, \gamma) > 0 \qquad \text{for } \gamma > r,$$

we can find a constant $\gamma_m \in [-r, r]$ such that

$$u(a+m, \gamma_m) = 0. \tag{5.18}$$

Let $(\gamma_{m_j})_{j=1}^{+\infty}$ be a convergent subsequence of $(\gamma_m)_{m=1}^{+\infty}$ and

$$\gamma_0 = \lim_{j \to +\infty} \gamma_{m_j}.$$

We will show that the function $u(t) = u(t, \gamma_0)$ is a solution of the problem (5.1), (5.14). In view of (5.16), for this it suffices to prove that

$$\liminf_{t \to +\infty} |u^{(n-1)}(t)| < +\infty.$$

Assume the contrary, i.e.

$$\lim_{t \to +\infty} |u^{(n-1)}(t)| = +\infty.$$

Then, for some $t_0 > a$,

$$u^{(i)}(t_0) u^{(n-1)}(t_0) > 0 \qquad (i = 0, \dots, n-1).$$

This, together with the continuity of $u^{(i)}(t_0, \cdot)$ $(i = 0, \dots, n-1)$, yields

$$u^{(i)}(t_0, \gamma_{m_j}) u^{(n-1)}(t_0, \gamma_{m_j}) > 0 \qquad (i = 0, \dots, n-1; \; j \geq j_0),$$

where j_0 is a sufficiently large positive integer. So, by (5.13),

$$u(t, \gamma_{m_j}) > 0 \qquad \text{for } t \geq t_0 \qquad (i = 0, \dots, n-1; \; j \geq j_0).$$

On the other hand, (5.18) implies

$$u(a + m_j, \gamma_{m_j}) = 0 \qquad (j = 1, 2, \dots).$$

Hence we arrive at a contradiction. $\square$

5.3. On the space $B^{(n)}(p)$.

THEOREM 5.1. If

$$\lim_{t \to +\infty} (-1)^{n-n_0-1} t^n p(t) = +\infty, \qquad \lim_{t \to +\infty} \sup \int_t^{t+1} |p(\tau)| \, d\tau < +\infty, \qquad (5.19)$$

then

$$B^{(n)}(p) = V^{(n,0)}(p), \qquad \dim B^{(n)}(p) = n_0. \qquad (5.20)$$

Moreover, any $u \in B^{(n)}(p)$ satisfies the conditions

$$\lim_{t \to +\infty} t^{i+1/2} u^{(i)}(t) = 0 \qquad (i = 0, \dots, n_0 - 1),$$

$$\lim_{t \to +\infty} t^{(n-1-i)(n_0-1)/(n-n_0)+1/2} u^{(i)}(t) = 0 \qquad (i = n_0, \dots, n-1). \qquad (5.21)$$

PROOF. By Theorem 4.4 and Lemma 4.5,

$$\dim V^{(n,2n_0)}(p) = n_0, \qquad V^{(n,2n_0)}(p) = V^{(n,0)}(p),$$

and for any $u \in V^{(n,2n_0)}(p)$,

$$\lim_{t \to +\infty} t^{i+1/2} u^{(i)}(t) = 0 \qquad (i = 0, \dots, n_0 - 1). \qquad (5.22)$$

Consequently, $V^{(n,2n_0)}(p) \subset B^{(n)}(p)$, and to prove the equality (5.20) it suffices to show that

$$B^{(n)}(p) \subset V^{(n,2n_0)}(p). \qquad (5.23)$$

In view of (5.19),

$$\eta = n! \, (n-1)^{n-1} + \sup \left\{ \int_t^{t+1} |p(\tau)| \, d\tau : t \geq 0 \right\} < +\infty \qquad (5.24)$$

and

$$(-1)^{n-n_0-1} p(t) > 0 \qquad \text{for } t \geq a, \qquad (5.25)$$

provided $a \in \,]0, +\infty[$ is sufficiently large.

Suppose that $u \in B^{(n)}(p)$ and set $c = \sup\{|u(t)|: t \geq a\}$. Then by Lemma 5.1,

$$\min\{|u^{(n-1)}(\tau)| : t \leq \tau \leq t+1\} \leq n! \, (n-1)^{n-1} c.$$

So, applying the condition (5.24) and Lemma 5.2, we obtain

$$|u^{(n-1)}(t)| \leq n! \, (n-1)^{n-1} c + \int_t^{t+1} |p(\tau) u(\tau)| \, d\tau \leq \eta c \qquad \text{for } t \geq a$$

and

$$|u^{(i)}(t)| \le c_0 \qquad \text{for } t \ge a \qquad (i = 0, \ldots, n-1), \tag{5.26}$$

where $c_0 = (n-1)! \, (n-1)^{n-1} c + 2^{n^2} (n-1)! \, (n-1)^{n-2} \eta c$.

If $n = 2n_0$, then (5.25) and (5.26) imply

$$\int_a^t |u^{(n_0)}(\tau)|^2 \, d\tau + \int_a^t |p(\tau)| \, |u(\tau)|^2 \, d\tau$$

$$= \sum_{i=0}^{n_0-1} (-1)^{n_0-1-i}(u^{(n-1-i)}(t)u^{(i)}(t) - u^{(n-1-i)}(a)u^{(i)}(a)) \le nc_0^2$$

for $t \ge a$,

$$\int_a^{+\infty} |u^{(n_0)}(\tau)|^2 \, d\tau < +\infty,$$

and

$$u \in V^{(n, 2n_0)}(p). \tag{5.27}$$

Therefore the inclusion (5.23) is true.

Let $n = 2n_0 + 1$. Via (5.25) and (5.26), setting

$$w(t) = \sum_{i=0}^{n_0-1} (-1)^{n_0+i} u^{(n-1-i)}(t)u^{(i)}(t) + \frac{1}{2}|u^{(n_0)}(t)|^2,$$

we obtain

$$w'(t) = |p(t)| \, |u(t)|^2 \ge 0 \qquad \text{for } t \ge a \tag{5.28}$$

and

$$\int_a^{+\infty} |p(\tau)| \, |u(\tau)|^2 \, d\tau \le nc_0^2. \tag{5.29}$$

According to (5.24) and (5.25), the equalities

$$\int_a^t |u^{(n_0+1)}(\tau)|^2 \, d\tau = \sum_{i=1}^{n_0} (-1)^{n_0-i}(u^{(n-i)}(t)u^{(i)}(t) - u^{(n-i)}(a)u^{(i)}(a))$$

$$+ (-1)^{n_0} \int_a^t p(\tau)u(\tau)u'(\tau) \, d\tau$$

and

$$\int_a^t |u^{(n_0)}(\tau)|^2 \, d\tau = u^{(n_0)}(t)u^{(n_0-1)}(t) - u^{(n_0)}(a)u^{(n_0-1)}(a)$$

$$- \int_a^t u^{(n_0+1)}(\tau)u^{(n_0-1)}(\tau) \, d\tau$$

imply

$$\int_a^t |u^{(n_0+1)}(\tau)|^2 \, d\tau \le nc_0^2 + c_0 \left(\int_a^t |p(\tau)| \, |u(\tau)|^2 \, d\tau \right)^{1/2} \left(\int_a^t |p(\tau)| \, d\tau \right)^{1/2}$$

$$\le nc_0^2 + n^{1/2} \eta^{1/2} c_0^2 (t+1)^{1/2} \le 2nc_0^2 \eta(t+1)^{1/2} \qquad \text{for } t \ge a$$

and

$$\int_a^t |u^{(n_0)}(\tau)|^2 \, d\tau \le 2c_0^2 + c_0 \int_a^t |u^{(n_0+1)}(\tau)|^2 \, d\tau$$

$$\le 2c_0^2 + c_0 t^{1/2} \left(\int_a^t |u^{(n_0+1)}(\tau)|^2 \, d\tau \right)^{1/2} \le 4nc_0^2(t+1)^{3/4} \qquad \text{for } t \ge a. \qquad (5.30)$$

On the other hand, by (5.28), either $w(t) \le 0$ for $t \ge a$ or there exists $a_0 \in [a, +\infty[$ such that $\delta = w(a_0) > 0$ and $w(t) \ge \delta$ for $t \ge a_0$. According to (5.26), in the former case

$$\frac{n_0+1}{2} \int_a^t |u^{(n_0)}(\tau)|^2 \, d\tau$$

$$\le \sum_{i=0}^{n_0-1} \sum_{j=i}^{n_0-1} \left[u^{(2n_0-1-j)}(a)u^{(j)}(a) - u^{(2n_0-1-j)}(t)u^{(j)}(t) \right] \le n^2 c_0^2 \qquad \text{for } t \ge a, \qquad (5.31)$$

while in the latter case

$$\frac{n_0+1}{2} \int_a^t |u^{(n_0)}(\tau)|^2 \, d\tau$$

$$\ge \delta(t-a_0) + \sum_{i=0}^{n_0-1} \sum_{j=i}^{n_0-1} \left[u^{(2n_0-1-j)}(a)u^{(j)}(a) - u^{(2n_0-1-j)}(t)u^{(j)}(t) \right]$$

$$\ge \delta(t-a_0) - n^2 c_0 \qquad \text{for } t \ge a_0.$$

But the last inequality contradicts the estimate (5.30). Hence (5.31) holds and since (5.26) and (5.31) immediately yield (5.27), the validity of the inclusion (5.23) is established for odd n as well.

It remains to show that every $u \in B^{(n)}(p)$ satisfies (5.21).

If $u \in B^{(n)}(p)$, then, as already proved, $u \in V^{(n,2n_0)}(p)$ and the conditions (5.22) are fulfilled. So

$$\epsilon(t) = \sup \left\{ \tau^{n_0-1/2} |u^{(n_0)}(\tau)| + \tau^{1/2} |u(\tau)| : \tau \ge t \right\} \to 0 \qquad \text{as } t \to +\infty.$$

Furthermore, by (5.24) and Lemma 5.1 we have

$$|u^{(n-1)}(t)| \le n! \, (n-1)^{n-1} \max\{|u(\tau)| : t \le \tau \le t+1\}$$

$$+ \int_t^{t+1} |p(\tau)u(\tau)| \, d\tau \le \eta \epsilon(t) t^{-1/2} \qquad \text{for } t > 0.$$

Since

$$\sup \left\{ |u^{(n_0-1)}(\tau)| : \tau \ge t \right\} \le t^{-n_0+1/2} \epsilon(t)$$

and

$$\sup\{ |u^{(n-1)}(\tau)| : \tau \ge t \} \le \eta t^{-1/2} \epsilon(t) \qquad \text{for } t > 0,$$

Lemma 5.2 implies

$$|u^{(i)}(t)| \le 2^{(n-n_0-1)^2} (n-n_0)! \, (n-n_0)^{n-n_0-1} \eta \epsilon(t) t^{-(n-1-i)(n_0-1)/(n-n_0)-1/2}$$

for $t > 0$.

So the conditions (5.21) are satisfied. $\square$

THEOREM 5.2. Suppose $n = 2n_0 + 1$, $(-1)^{n_0+1}p(t) \ge 0$ for t large, and there exists a positive constant α such that

$$\lim_{t \to +\infty} \inf \int_t^{t+\alpha} |p(\tau)| \, d\tau > 0, \qquad \lim_{t \to +\infty} \sup \int_t^{t+\alpha} |p(\tau)| \, d\tau < +\infty.$$

Then
$$B^{(n)}(p) = Z^{(n)}(p), \qquad \dim B^{(n)}(p) = n_0 + 1.$$

PROOF. Choose $a \in \mathbf{R}_+$ so that (5.8) hold. Let $\tilde{Z}^{(n)}(p)$ be the set of solutions of the equation (5.1) for which

$$\int_a^{+\infty} |p(t)| \, |u(t)|^2 \, dt < +\infty. \tag{5.32}$$

By Lemma 5.3,
$$\tilde{Z}^{(n)}(p) \subset Z^{(n)}(p), \qquad \dim \tilde{Z}^{(n)}(p) = n_0 + 1.$$

So it suffices to show that every $u \in B^{(n)}(p)$ satisfies (5.32).

Suppose $u \in B^{(n)}(p)$ and $c = \sup\{|u(t)|: t \geq a\}$. Then, according to Lemmas 5.1 and 5.2,

$$|u^{(n-1)}(t)| \leq n! \, (n-1)^{n-1} \alpha^{1-n} c + \int_t^{t+\alpha} |p(\tau) u(\tau)| \, d\tau \leq \eta c \qquad \text{for } t \geq a,$$

and the inequalities (5.26) are fulfilled with

$$\eta = n! \, (n-1)^{n-1} \alpha^{1-n} + \sup\left\{ \int_t^{t+\alpha} |p(\tau)| \, d\tau : t \geq a \right\} + 1,$$

$$c_0 = (n-1)! \, (n-1)^{n-1} (\alpha^{-1} + \alpha^{1-n}) c + 2^{n^2} (n-1)! \, (n-1)^{n-2} \eta c.$$

In view of (5.8) and (5.26) we have

$$\int_a^t |p(\tau)| \, |u(\tau)|^2 \, d\tau$$

$$= \sum_{i=0}^{n_0-1} (-1)^{n_0-i} \big(u^{(n-1-i)}(a) u^{(i)}(a) - u^{(n-1-i)}(t) u^{(i)}(t) \big)$$

$$+ \frac{1}{2} \big(|u^{(n_0)}(a)|^2 + |u^{(n_0)}(t)|^2 \big) \leq n c_0^2 \qquad \text{for } t \geq a.$$

Therefore (5.32) is valid. $\square$

To conclude this subsection we consider the third and fourth order equations

$$u''' = p(t) u \tag{5.33}$$

and

$$u^{(4)} = p(t) u. \tag{5.34}$$

THEOREM 5.3. If

$$\lim_{t \to +\infty} \inf p(t) > 0, \tag{5.35}$$

then
$$B^{(3)}(p) = Z^{(3)}(p), \qquad \dim B^{(3)}(p) = 2,$$

and the set of all oscillatory solutions of the equation (5.33) coincides with $B^{(3)}(p)$.

PROOF. Denote by $\tilde{B}^{(3)}(p)$ the set of solutions of the equation (5.33) for which

$$\lim_{t \to +\infty} \inf |u''(t)| = 0.$$

Lemma 5.4 yields dim $\tilde{B}^{(3)}(p) \geq 2$. On the other hand, in view of (5.35), for sufficiently large $a \in \mathbf{R}_+$ the solution u_0 of the equation (5.33) with the initial data

$$u_0^{(i)}(a) = 1 \qquad (i = 0, 1, 2)$$

satisfies the condition

$$\lim_{t \to +\infty} u_0''(t) = +\infty.$$

Hence, applying Corollary 1.7, we easily conclude that the set of oscillatory solutions of the equation (5.33) coincides with $\tilde{B}^{(3)}(p)$, and dim $\tilde{B}^{(3)}(p) = 2$.

Since $Z^{(3)}(p) \subset B^{(3)}(p) \subset \tilde{B}^{(3)}(p)$, to prove the theorem it suffices to show that every oscillatory solution of the equation (5.33) tends to zero as $t \to +\infty$.

According to (5.35), we can find $a \in \mathbf{R}_+$ for which

$$\delta = \operatorname{ess\,inf}\{p(t) : t \geq a\} > 0. \tag{5.36}$$

Let u be a nonzero oscillatory solution of the equation (5.33), and let $(t_k)_{k=1}^{+\infty}$ and $(\tau_k)_{k=1}^{+\infty}$ be the sequences of zeros of u and u'' in $[a, +\infty[$. By (5.36), setting

$$w(t) = \frac{1}{2}|u'(t)|^2 - u''(t)u(t),$$

we derive

$$w'(t) = -p(t)|u(t)|^2 \leq 0 \qquad \text{for } t \geq a, \qquad |u'(\tau_k)| \leq |u'(\tau_1)|$$

and

$$\int_a^{\tau_k} |u(\tau)|^2 \, d\tau \leq \frac{1}{\delta} \int_a^{\tau_k} p(\tau)|u(\tau)|^2 \, d\tau = w(a) - \frac{1}{2}|u'(\tau_k)|^2 \leq w(a)$$

$$(k = 1, 2, \dots).$$

Therefore,

$$\eta = \sup\{|u'(t)| : t \geq a\} < +\infty, \qquad \int_a^{+\infty} |u(\tau)|^2 \, d\tau < +\infty.$$

So the equality

$$|u(t)|^3 = \left| 3 \int_t^{t_k} |u(\tau)|^2 u'(\tau) \, d\tau \right|$$

implies

$$|u(t)|^3 \leq 3\eta \int_t^{+\infty} |u(\tau)|^2 \, d\tau \to 0 \qquad \text{as } t \to +\infty. \qquad \square$$

PROBLEM 5.1. Let $p(t) \geq 0$ for $t \geq a$. Then the condition

$$\int_a^{+\infty} t^4 p(t) \, dt = +\infty$$

is necessary and sufficient for the solution of the equation (5.33) under the boundary conditions

$$u(a) = c_0, \qquad u'(a) = c_1, \qquad \int_a^{+\infty} p(t)|u(t)|^2 \, dt < +\infty$$

to be unique.

THEOREM 5.4. If $p \in \tilde{C}_{\text{loc}}(\mathbf{R}_+)$ and for large t,

$$p(t) > 0, \qquad p'(t) \geq 0, \tag{5.37}$$

then $\dim B^{(4)}(p) = 3$ and every oscillatory solution of the equation (5.34) is bounded.
PROOF. Let $\tilde{B}^{(4)}(p)$ be the set of solutions of the equation (5.34) under the condition

$$\lim_{t \to +\infty} \inf |u'''(t)| = 0.$$

By Lemma 5.4, $\dim \tilde{B}^{(4)}(p) \geq 3$. On the other hand, (5.37) implies that the equation (5.34) has a solution u_0 for which

$$\lim_{t \to +\infty} u_0'''(t) = +\infty.$$

So, obviously, $\dim \tilde{B}^{(4)}(p) = 3$.

According to Corollary 1.7 and Definition 1.2, every $u \in \tilde{B}^{(4)}(p)$ either is oscillatory or monotonically tends to zero as $t \to +\infty$. Hence it remains to prove that every oscillatory solution of the equation (5.34) is bounded.

Assume, on the contrary, that the equation (5.34) has an unbounded oscillatory solution u. Then we can find an increasing sequence of positive constants $(t_k)_{k=1}^{+\infty}$ which converges to $+\infty$ and is such that

$$|u(t_k)| = \max\{|u(t)| : t_1 \leq t \leq t_k\}, \qquad u'(t_k) = 0 \qquad (k = 1, 2, \ldots), \tag{5.38}$$

and the inequalities (5.37) hold on $[t_1, +\infty[$. Multiplying both sides of (5.34) by $2u'$ and integrating from t_1 to t_k, we get

$$c_0 = |u''(t_k)|^2 + p(t_k)|u(t_k)|^2 - \int_{t_1}^{t_k} p'(\tau)|u(\tau)|^2 \, d\tau$$

with $c_0 = p(t_1)|u(t_1)|^2 + |u''(t_1)|^2$. But by (5.37) and (5.38),

$$\int_{t_1}^{t_k} p'(\tau)|u(\tau)|^2 \, d\tau \leq [p(t_k) - p(t_1)]|u(t_k)|^2 \qquad (k = 1, 2, \ldots).$$

So

$$p(t_1)|u(t_k)|^2 \leq c_0 \qquad (k = 1, 2, \ldots),$$

which contradicts our assumption on the unboudedness of u. $\square$

PROBLEM 5.2. Let $p(t) \leq 0$ for $t \geq 0$. Suppose that u is an oscillatory solution of the equation (5.34) and that $(t_k)_{k=1}^{+\infty}$ is the sequence of its zeros. Then $u \in V^{(4,4)}(p)$ if and only if

$$u'(t_k)u''(t_k)u'''(t_k) \neq 0, \qquad \text{sign } u'(t_k) = \text{sign } u'''(t_k) \neq \text{sign } u''(t_k)$$

$$(k = 1, 2, \ldots).$$

5.4. Existence theorems for unbounded oscillatory solutions.

THEOREM 5.5. If $n \geq 3$ and

$$\lim_{t \to +\infty} (-1)^{n-n_0-1} t^n p(t) = +\infty, \qquad \lim_{t \to +\infty} \sup \int_t^{t+1} |p(\tau)| \, d\tau < +\infty,$$

then the equation (5.1) has unbounded oscillatory solutions.

PROOF. By Theorem 5.1 the equalities (5.20) are fulfilled. So it is clear that the equation (5.1) has unbounded solutions.

If $n - n_0$ is even, then according to Corollary 1.7 the equation (5.1) has property A and any unbounded solution of it is oscillatory.

Consider the case $n - n_0$ odd. In view of Lemma 5.4, for any $k \in \{1, \ldots, n-1\}$ the equation (5.1) has a solution u_k satisfying the conditions

$$u_k^{(i-1)}(a) = \delta_{ik} \qquad (i = 1, \ldots, n-1), \qquad \lim_{t \to +\infty} \inf |u_k^{(n-1)}(t)| = 0,$$

where δ_{ik} is the Kronecker delta. Corollary 1.7 and Remark 1.3 imply that the solutions $u_1, \ldots, u_{n-1}$ oscillate. On the other hand, by (5.20) at least $n - 1 - n_0$ of them are unbounded. $\square$

THEOREM 5.6. Let $n = 2n_0 + 1 > 3$, $(-1)^{n_0+1} p(t) \geq 0$ for large t, and let there exist a positive constant α such that

$$\lim_{t \to +\infty} \inf \int_t^{t+\alpha} |p(\tau)| \, d\tau > 0, \qquad \lim_{t \to +\infty} \sup \int_t^{t+\alpha} |p(\tau)| \, d\tau < +\infty.$$

Then the equation (5.1) has unbounded oscillatory solutions.

This theorem can be proved in the same way as Theorem 5.5, provided Theorem 5.2 is applied instead of Theorem 5.1.

Note that Theorems 5.5 and 5.6 are worthy of interest only for odd $n - n_0$ and n_0, respectively, because the other cases are covered by more general propositions (see Corollaries 5.1 and 5.2 below).

In the sequel we will need some auxiliary statements.

LEMMA 5.5. Let $n \geq 3$, $q, g \in L_{\text{loc}}(\mathbf{R}_+)$,

$$0 \leq q(t) \leq -(n-2)p(t) \qquad \text{for } t \geq 0, \tag{5.39}$$

$$0 \leq g(t) \leq \frac{n(n-3)}{2(n-4)!} t^{n-4} \qquad \text{for } t \geq 0, \tag{5.40}$$

and let u_1 and u_2 be the solutions of the equation (5.1) under the initial conditions

$$u^{(i)}(0) = 0 \qquad (i = 0, \ldots, n-1; \, i \neq n-2), \qquad u^{(n-2)}(0) = 1 \tag{5.41}$$

and

$$u^{(i)}(0) = 0 \qquad (i = 0, \ldots, n-2), \qquad u^{(n-1)}(0) = 1, \tag{5.42}$$

respectively. Then

$$w(t) \geq v(t) \qquad \text{for } t \geq 0, \tag{5.43}$$

where $w = u_1 u_2' - u_1' u_2$, and v is the solution of the problem

$$v^{(n)} = q(t)v + g(t), \tag{5.44}$$

$$v^{(i)}(0) = 0 \quad (i = 0, \dots, n-2), \quad v^{(n-1)}(0) = 1. \tag{5.45}$$

PROOF. If $n = 3$, then, as is easily verified, $w = v$. Hence below we assume that $n \geq 4$. Set

$$w_{ij} = u_1^{(i)} u_2^{(j)} - u_1^{(j)} u_2^{(i)} \quad (i = 0, \dots, j; \; j = 1, \dots, n).$$

For $j \in \{1, \dots, n-1\}$ and $i \in \{0, \dots, j-1\}$ we have

$$w'_{ij} = w_{i+1j} + w_{ij+1}, \qquad w_{ii} = 0, \qquad w_{in} = -p(t)w_{0i}, \tag{5.46}$$

$$w_{ij}(0) = 0 \quad \text{if either } i \neq n-2 \text{ or } j \neq n-1, \qquad w_{n-2n-1}(0) = 1. \tag{5.47}$$

By (5.39), the equalities (5.46) and (5.47) yield

$$w_{ij}(t) > 0 \quad \text{for } t > 0 \quad (i = 0, \dots, j-1; \; j = 1, \dots, n-1). \tag{5.48}$$

First we will show that for any $i \in \{3, \dots, n-1\}$ the representation

$$w_{01}^{(i)} = w_{0i+1} + (i-1)w_{1i} + \frac{i(i-3)}{2}w_{2i-1} + f_i \tag{5.49$_i$}$$

holds on $\mathbf{R}_+$, where f_i is a function with continuous $(n-i+1)$-st derivative and $f_i^{(j)}(t) \geq 0$ for $t \in \mathbf{R}_+$ $(j = 0, \dots, n-i+1)$. Clearly, $w_{01}''' = w_{04} + 2w_{13}$, i.e. (5.49$_3$) is true. We claim that (5.49$_i$) implies (5.49$_{i+1}$). Indeed,

$$w_{01}^{(i+1)} = w'_{0i+1} + (i-1)w'_{1i} + \frac{i(i-3)}{2}w'_{2i-1} + f_i$$

$$= w_{0i+2} + iw_{1i+1} + \frac{(i+1)(i-2)}{2}w_{2i} + f_{i+1},$$

where

$$f_{i+1} = f'_i + \frac{i(i-3)}{2}w_{3i-1}.$$

Obviously, the $(n-i)$-th derivative of f_{i+1} is continuous and $f_{i+1}^{(j)}(t) \geq 0$ for $t \geq 0$ $(j = 0, \dots, n-i)$. Thus, (5.49$_i$) has been proved for any $i \in \{3, \dots, n-1\}$. According to (5.49$_{n-1}$),

$$w_{01}^{(n)}(t) \geq -(n-2)p(t)w_{01}(t) + \frac{n(n-3)}{2}w_{2n-1}(t). \tag{5.50}$$

Furthermore, in view of (5.46)–(5.48), $w'_{in-1}(t) \geq w_{i+1n-1}(t)$ $(i = 2, \dots, n-3)$ and $w_{n-2n-1}(t) \geq 1$ for $t \in \mathbf{R}_+$. Consequently,

$$w_{2n-1}(t) \geq \frac{t^{n-4}}{(n-4)!} \quad \text{for } t \geq 0.$$

Therefore, (5.39), (5.40) and (5.50) yield

$$w^{(n)}(t) \geq q(t)w(t) + g(t) \quad \text{for } t \geq 0,$$

which immediately implies (5.43). $\square$

The following two assertions are consequences of Lemma 5.2.

LEMMA 5.6. Let $u \in \tilde{C}^{m-1}_{\text{loc}}(\mathbf{R}_+)$ and

$$\lim_{t \to +\infty} \sup \left(\frac{|u(t)|}{\varphi(t)} + \frac{|u^{(n)}(t)|}{\psi(t)} \right) < +\infty, \tag{5.51}$$

where $\varphi, \psi \in C(\mathbf{R}_+)$ are positive functions such that either φ and ψ do not decrease and

$$\limsup_{t \to +\infty} \frac{\varphi(t)}{t^n \psi(t)} < +\infty,$$

or φ and ψ do not increase. Then

$$\limsup_{t \to +\infty} (\varphi(t))^{-(n-j)/n} (\psi(t))^{-j/n} |u^{(j)}(t)| < +\infty \qquad (j = 1, \dots, n). \qquad (5.52)$$

LEMMA 5.7. Let $u \in \tilde{C}_{loc}^{m-1}(\mathbf{R}_+)$ and

$$\limsup_{t \to +\infty} (t^{-\mu} |u(t)| + t^{-\mu-\sigma} |u^{(n)}(t)|) < +\infty,$$

with $\mu \in \mathbf{R}$ and $\sigma \geq -n$. Then

$$\limsup_{t \to +\infty} t^{-\mu-(j/n)\sigma} |u^{(j)}(t)| < +\infty \qquad (j = 1, \dots, n-1).$$

REMARK 5.1 If the condition (5.51) in Lemma 5.6 is replaced by either

$$\lim_{t \to +\infty} \frac{u(t)}{\varphi(t)} = 0, \qquad \limsup_{t \to +\infty} \frac{|u^{(n)}(t)|}{\psi(t)} < +\infty$$

or

$$\limsup_{t \to +\infty} \frac{|u(t)|}{\varphi(t)} < +\infty, \qquad \lim_{t \to +\infty} \frac{u^{(n)}(t)}{\psi(t)} = 0,$$

then instead of (5.52) we obtain

$$\lim_{t \to +\infty} (\varphi(t))^{-(n-j)/n} (\psi(t))^{-j/n} u^{(j)}(t) = 0 \qquad (j = 1, \dots, n).$$

LEMMA 5.8. Let $n \geq 3$ and let the conditions (5.39) and (5.40) hold. Furthermore, suppose that

$$\int_0^{+\infty} \alpha^2(t) v(t) \, dt = +\infty \qquad (5.53)$$

and

$$\liminf_{t \to +\infty} \frac{t^{n-1} \int_0^t |p(s)| \, ds}{\left(\int_0^t \alpha^2(s) v(s) \, ds \right)^n} = 0, \qquad (5.54)$$

where $\alpha \colon \mathbf{R}_+ \to \mathbf{R}_+$ is a continuous function and v is the solution of the problem (5.44), (5.45). Then the equation (5.1) has a solution u such that

$$\limsup_{t \to +\infty} \alpha(t) |u(t)| = +\infty. \qquad (5.55)$$

PROOF. Let u_1 and u_2 be the solutions of the problems (5.1), (5.41) and (5.1), (5.42), respectively. Set $w = u_1 u_2' - u_1' u_2$. Then in view of Lemma 5.5 the inequality (5.43) holds. Since $v(t) > 0$ and, thus, $w(t) > 0$ for $t > 0$, there exist functions $r, \varphi \in C^1(\mathbf{R}_+)$, defined by the equalities

$$u_1(t) = r(t) \cos \varphi(t), \qquad u_2(t) = r(t) \sin \varphi(t), \qquad \varphi(0) = 0.$$

It is easily verified that $\varphi'(t) > 0$ for $t \geq 0$ and, moreover,

$$\varphi'(t) = \frac{w(t)}{r^2(t)} \qquad \text{for } t > 0. \tag{5.56}$$

Assume

$$\lim_{t \to +\infty} \sup \alpha(t)|u_i(t)| < +\infty \qquad (i = 1, 2).$$

Then for a certain $c_0 > 0$ we have $r(t)\alpha(t) \leq c_0$ on $\mathbf{R}_+$. So (5.43) and (5.56) yield

$$\varphi'(t) \geq \frac{1}{c_0}\alpha^2(t)v(t) \qquad \text{for } t \geq 0.$$

According to (5.53) this implies $\varphi(t) \to +\infty$ as $t \to +\infty$. Hence u_1 and u_2 are oscillatory. Moreover, both u_1 and u_2 have only simple zeros in $]0, +\infty[$. If $(t_j)_{j=1}^{+\infty}$ is the increasing sequence of all zeros of u_1 in $]0, +\infty[$, then

$$\frac{\pi}{2} = \varphi(t_1) \geq \frac{1}{c_0} \int_0^{t_1} \alpha^2(s)v(s)\,ds,$$

$$\pi = \varphi(t_{j+1}) - \varphi(t_j) \geq \frac{1}{c_0} \int_{t_j}^{t_{j+1}} \alpha^2(s)v(s)\,ds \qquad (j = 1, 2, \ldots).$$

Therefore,

$$\pi\left(N_t + \frac{1}{2}\right) \geq \frac{1}{c_0} \int_0^t \alpha^2(s)v(s)\,ds \qquad \text{for } t > 0, \tag{5.57}$$

where N_t denotes the number of zeros of u_1 in the interval $]0, t]$.

On the other hand, if $0 = s_0 < s_1 < \ldots < s_k \leq t$ and u_1 has exactly n zeros in each interval $[s_{j-1}, s_j]$ $(j = 1, \ldots, k)$, then for a certain $c_1 > 0$ (see, e.g., Lemma 1.21),

$$(s_j - s_{j-1})^{n-1} \int_{s_{j-1}}^{s_j} |p(s)|\,ds > c_1 \qquad (j = 1, \ldots, k).$$

Since $(n-1)k + 1 \geq N_t$ for $t \geq s_2$, these inequalities yield

$$\int_0^t |p(s)|\,ds \geq c_1 k \left[\frac{1}{k}\sum_{j=1}^k (t_j - t_{j-1})\right]^{-n+1} \geq \frac{c_1}{t^{n-1}}\left(\frac{N_t - 1}{n - 1}\right)^n \qquad \text{for } t \geq s_2.$$

Consequently,

$$N_t \leq (n-1)\left(\frac{1}{c_1}t^{n-1}\int_0^t |p(s)|\,ds\right)^{1/n} + 1 \qquad \text{for } t \geq s_2. \tag{5.58}$$

But the inequalities (5.57) and (5.58) contradict the condition (5.54). $\square$

REMARK 5.2. If the condition (5.54) in Lemma 5.8 is replaced by

$$\lim_{t \to +\infty} \inf \frac{t^{n-1}\int_0^t |p(s)|\,ds}{\left(\int_0^t \alpha^2(s)v(s)\,ds\right)^n} < +\infty, \tag{5.59}$$

then instead of (5.55) we obtain

$$\lim_{t \to +\infty} \sup \alpha(t)|u(t)| > 0.$$

THEOREM 5.7. Let $n \geq 3$ and

$$-t^n p(t) \geq c \qquad \text{for } t \geq t_0,$$

c and t_0 being nonnegative constants. Then there exists a solution of the equation (5.1) under the condition

$$\lim_{t \to +\infty} \sup t^{j-(\mu+1)/2} |u^{(j)}(t)| > 0 \qquad (j = 1, \ldots, n-1), \tag{5.60}$$

where $\mu = \max\{\mu_0, 2n-4\}$ and $\mu_0 \geq n-1$ is such that

$$\mu_0(\mu_0 - 1) \ldots (\mu_0 - n + 1) = c(n-2).$$

If, in addition, for a certain $\sigma \geq -n$,

$$\lim_{t \to +\infty} \sup t^{-\sigma} |p(t)| < +\infty, \tag{5.61}$$

then the equation (5.1) has a solution which satisfies both the condition (5.60) and the condition

$$\lim_{t \to +\infty} \sup t^{-\mu/2 + \sigma/2n} |u(t)| > 0. \tag{5.62}$$

PROOF. Let u_1 and u_2 be the solutions of the problems (5.1), (5.41) and (5.1), (5.42), respectively. Set $w = u_1 u_2' - u_1' u_2$. We claim that either u_1 or u_2 satisfies (5.60). Indeed, otherwise

$$\lim_{t \to +\infty} t^{-(\mu-1)/2} u_i'(t) = \lim_{t \to +\infty} t^{-(\mu+1)/2} u_i(t) = 0 \qquad (i = 1, 2).$$

So

$$\lim_{t \to +\infty} t^{-\mu} w(t) = 0. \tag{5.63}$$

On the other hand, if v is the solution of the problem (5.44), (5.45) with $q(t) = c(n-2)t^{-n}$ for $t \geq t_0$ and $g(t) = (n(n-3)/2(n-4)!)t^{n-4}$ for $t \geq 0$, then we can easily conclude that for sufficiently small $\epsilon > 0$,

$$v(t) \geq \epsilon t^{\mu_0} + \frac{n(n-3)}{2(2n-4)!} t^{2n-4} \qquad \text{for } t \geq t_0,$$

i.e.

$$v(t) \geq \epsilon t^\mu \qquad \text{for } t \geq t_0. \tag{5.64}$$

But since $w(t) \geq v(t)$ for $t \geq 0$, (5.64) contradicts (5.63).

Now let (5.61) hold and

$$\lim_{t \to +\infty} t^{-\mu/2 + \sigma/2n} u_i(t) = 0 \qquad (i = 1, 2).$$

Then by (5.61) and Lemma 5.7,

$$\lim_{t \to +\infty} \sup t^{-\mu/2 - \sigma/2n} |u_i'(t)| < +\infty \qquad (i = 1, 2).$$

Consequently, the condition (5.63) is fulfilled and again we arrive at a contradiction.

Therefore, either u_1 or u_2 satisfies (5.62). It is clear that if u_1 satisfies only one of the conditions (5.60) and (5.62) and u_2 satisfies only the other, then $u = u_1 + u_2$ satisfies both conditions. $\square$

REMARK 5.3. Theorem 5.7 remains true if (5.61) is replaced by the condition

$$\lim_{t \to +\infty} \inf t^{-\sigma-1} \int_0^t |p(s)| \, ds < +\infty, \tag{5.65}$$

where $\sigma > -1$. In order to verify this it suffices to take into account Lemma 5.8, Remark 5.2 with $\alpha(t) = t^{-\mu/2+\sigma/2n}$ and the inequality (5.64).

REMARK 5.4. If (5.61) in Theorem 5.7 is replaced by either

$$\lim_{t \to +\infty} t^{-\sigma} p(t) = 0 \qquad \text{with } \sigma \geq -n$$

or

$$\lim_{t \to +\infty} \inf t^{-\sigma-1} \int_0^t |p(s)| \, ds = 0 \qquad \text{with } \sigma > -1,$$

then instead of (5.62) we obtain

$$\lim_{t \to +\infty} \sup t^{-\mu/2+\sigma/2n} |u(t)| = +\infty.$$

COROLLARY 5.1. If $n \geq 3$ and p is a nonpositive function, then there exists a solution of the equation (5.1) such that

$$\lim_{t \to +\infty} \sup t^{-n+j+3/2} |u^{(j)}(t)| > 0 \qquad (j = 1, \dots, n-1). \tag{5.66}$$

If, in addition, (5.61) is fulfilled for a certain $\sigma \geq -n$, then the equation (5.1) has a solution satisfying both the condition (5.66) and the condition

$$\lim_{t \to +\infty} \sup t^{-n+2+\sigma/2n} |u(t)| > 0.$$

COROLLARY 5.2. If

$$\lim_{t \to +\infty} t^n p(t) = -\infty, \tag{5.67}$$

then the equation (5.1) has a solution such that

$$\lim_{t \to +\infty} \sup t^{-\mu} |u^{(j)}(t)| = +\infty \tag{5.68}$$

for any $\mu > 0$ and $j \in \{1, \dots, n-1\}$. Moreover, if, together with (5.67), the condition (5.65) holds for a certain $\sigma > -1$, then there exists a solution of the equation (5.1) satisfying (5.68) for any $\mu > 0$ and $j \in \{0, \dots, n-1\}$.

THEOREM 5.8. Suppose $n \geq 3$, $q \in \tilde{C}^1_{\mathrm{loc}}(\mathbf{R}_+)$ satisfies (5.39) and

$$\int_0^{+\infty} |q''(t)| [q(t)]^{-1-1/n} \, dt < +\infty. \tag{5.69}$$

Then the equation (5.1) has a solution u such that

$$\lim_{t \to +\infty} \sup [q(t)]^{(n+1)/4n-j/n} \exp\left(-\frac{1}{2} \int_0^t [q(s)]^{1/n} \, ds\right) |u^{(j)}(t)| > 0$$

$$(j = 1, \dots, n-1). \tag{5.70}$$

If, in addition,

$$\lim_{t\to+\infty} \inf t^{n-1}[q(t)]^{\sigma(n+1)/2n} \exp\left(-\sigma\int_0^t [q(s)]^{1/n}\,ds\right)\int_0^t |p(s)|\,ds < +\infty \qquad (5.71)$$

for a certain $\sigma \in [0,n]$, then

$$\lim_{t\to+\infty} \sup [q(t)]^{(n+1)(n-\sigma)/4n^2} \exp\left(\frac{\sigma-n}{2n}\int_0^t [q(s)]^{1/n}\,ds\right)|u(t)| > 0. \qquad (5.72)$$

PROOF. First of all we note that if (5.69) holds, then

$$\int_0^{+\infty} [q(t)]^{1/n}\,dt = +\infty, \qquad \lim_{t\to+\infty} q'(t)[q(t)]^{-1-1/n} = 0, \qquad (5.73)$$

the function $[q(t)]^\alpha \exp(\int_0^t [q(s)]^{1/n}\,ds)$, $\alpha \in \mathbf{R}$ arbitrary, does not decrease for t large, and

$$\lim_{t\to+\infty} [q(t)]^\alpha \exp\left(\int_0^t [q(s)]^{1/n}\,ds\right) = +\infty. \qquad (5.74)$$

Let v be a solution of the problem (5.44), (5.45) with $g(t) \equiv 0$. By (5.69) and Corollary 6.5 (which is proved below),

$$v(t) = [q(t)]^{-(n-1)/2n} \exp\left(\int_0^t [q(s)]^{1/n}\,ds\right)(1 + o(1)). \qquad (5.75)$$

Assume that the first part of the theorem is not true and that u_1 and u_2 are the solutions of the equation (5.1) under the conditions (5.41) and (5.42), respectively. Then for a certain $j_i \in \{1,\ldots,n-1\}$,

$$\lim_{t\to+\infty} [q(t)]^{(n+1)/4n-j_i/n} \exp\left(-\frac{1}{2}\int_0^t [q(s)]^{1/n}\,ds\right)u_i^{(j_i)}(t) = 0 \qquad (i=1,2).$$

So, taking into account (5.73) and (5.74), we obtain

$$\lim_{t\to+\infty} [q(t)]^{(n+1)/4n-j/n} \exp\left(-\frac{1}{2}\int_0^t [q(s)]^{1/n}\,ds\right)u_i^{(j)}(t) = 0$$

$$(j=0,\ldots,j_i;\ i=1,2).$$

Hence if $w = u_1 u_2' - u_1' u_2$, then

$$\lim_{t\to+\infty} [q(t)]^{(n-1)/2n} \exp\left(-\int_0^t [q(s)]^{1/n}\,ds\right)w(t) = 0. \qquad (5.76)$$

But (5.75) and (5.76) contradict Lemma 5.5. So either u_1 or u_2 satisfies (5.70).

Now suppose the condition (5.71) holds. Then by (5.74) and (5.75) we get (5.53) and (5.59), where

$$\alpha(t) = [q(t)]^{(n+1)(n-\sigma)/4n^2} \exp\left(\frac{\sigma-n}{2n}\int_0^t [q(s)]^{1/n}\,ds\right).$$

Thus, according to Remark 5.2 and Lemma 5.8, either u_1 or u_2 satisfies (5.72), and it remains to apply the same argument which completed the proof of Theorem 5.7. $\square$

COROLLARY 5.3. Let $n \geq 3$ and

$$-(n-2)p(t) \geq ct^\alpha \qquad \text{for } t \geq t_0$$

with certain $c > 0$, $t_0 > 0$ and $\alpha > -n$. Then the equation (5.1) has a solution u such that

$$\lim_{t \to +\infty} \sup |u^{(j)}(t)| t^{\alpha(n+1)/4n - j\alpha/n} \exp\left(-\frac{\sqrt[n]{c}}{2} t^{(\alpha+n)/n}\right) > 0$$

$$(j = 1, \dots, n-1).$$

If, in addition,

$$\lim_{t \to +\infty} \inf t^{n-1+\alpha\sigma(n+1)/2n} \exp\left(-\sigma \sqrt[n]{c} t^{(\alpha+n)/n}\right) \int_0^t |p(s)| \, ds < +\infty,$$

where $\sigma \in [0, n]$, then

$$\lim_{t \to +\infty} \sup |u(t)| t^{\alpha(n+1)(n-\sigma)/4n^2} \exp\left(\frac{\sigma - n}{2n} \sqrt[n]{c} t^{(\alpha+n)/n}\right) > 0.$$

COROLLARY 5.4. Let $n \geq 3$ and

$$\lim_{t \to +\infty} t^{-\alpha} p(t) = -\infty,$$

where $\alpha > -n$. Then the equation (5.1) has a solution such that

$$\lim_{t \to +\infty} \sup |u^{(j)}(t)| \exp(-\mu t^{(\alpha+n)/n}) = +\infty \tag{5.77}$$

for any $\mu > 0$ and $j \in \{1, \dots, n-1\}$. If, in addition,

$$\lim_{t \to +\infty} \inf \exp(-\sigma t^{(\alpha+n)/n}) \int_0^t |p(s)| \, ds < +\infty$$

with a certain $\sigma > 0$, then there exists a solution of the equation (5.1) satisfying (5.77) for any $\mu > 0$ and $j \in \{0, \dots, n-1\}$.

REMARK 5.5. Corollary 1.5 implies that the solutions whose existence is guaranteed by Corollaries 5.2–5.4 are oscillatory.

We have proved above (Corollary 5.2) that if $n \geq 3$ and p is a nonpositive function, then the equation (5.1) possesses a solution with unbounded derivative. On the other hand, by Remark 5.4, the condition

$$\lim_{t \to +\infty} \inf t^{-2n(n-2)-1} \int_0^t |p(s)| \, ds = 0$$

together with the nonpositiveness of p is sufficient for the existence of an unbounded solution.

PROBLEM 5.3*. Let $n \geq 3$. Is the nonpositiveness of the function p sufficient by itself for the existence of an unbounded solution of the equation (5.1)?

PROBLEM 5.4*. Do the conditions

$$n \geq 5, \qquad \lim_{t \to +\infty} t^n p(t) = +\infty$$

guarantee the existence of at least one unbounded oscillatory solution of the equation (5.1)?

5.5. On second order equations. The following assertions concerning the unboundedness of solutions of the equation

$$u'' = p(t)u \tag{5.78}$$

with $p \in L_{\mathrm{loc}}(\mathbf{R}_+)$ can be proved similarly to the corresponding assertions in subsection 4.6.

THEOREM 5.9. Let

$$p(t) = -p_0(t) - p_1(t) - p_2(t), \qquad p_i \in \tilde{C}_{\mathrm{loc}}(\mathbf{R}_+) \qquad (i = 0, 1),$$

$$p_2 \in L_{\mathrm{loc}}(\mathbf{R}_+), \qquad p_0(t) > 0, \qquad p_0'(t) \le 0 \qquad \text{for } t \ge 0, \tag{5.79}$$

$$\int_0^{+\infty} \frac{|p_1'(t)|}{p_0(t)} \, dt < +\infty, \qquad \int_0^{+\infty} \frac{|p_2(t)|}{\sqrt{p_0(t)}} \, dt < +\infty$$

and

$$\lim_{t \to +\infty} \inf \frac{p_1(t)}{p_0(t)} > -1. \tag{5.80}$$

Then for any solution u of the equation (5.78) the limit

$$\lim_{t \to +\infty} \left(\frac{[u'(t)]^2}{p_0(t) + p_1(t)} + [u(t)]^2 \right) = \rho_0 > 0,$$

finite or infinite, exists.

THEOREM 5.10. Let the conditions (5.79) hold,

$$\lim_{t \to +\infty} p_0(t) = \lim_{t \to +\infty} p_1(t) = 0, \tag{5.81}$$

and let the equation (5.78) be oscillatory. Then this equation has a solution such that

$$\lim_{t \to +\infty} \left(\frac{[u'(t)]^2}{p_0(t) + p_1(t)} + [u(t)]^2 \right) = +\infty. \tag{5.82}$$

THEOREM 5.11. Let $(m_k)_{k=1}^{+\infty}$ be a sequence of positive integers and $(r_k)_{k=1}^{+\infty}$ a nonincreasing sequence of positive constants such that

$$r_k \to 0, \qquad t_k = \pi \sum_{i=1}^{k} \frac{m_i}{r_i} \to +\infty \qquad \text{as } t \to +\infty.$$

If, in addition,

$$\sum_{k=1}^{+\infty} \frac{1}{r_k} \int_{t_k}^{t_{k+1}} |p(t) + r_k^2| \, dt < +\infty,$$

then the equation (5.78) has a nonzero bounded solution.

This theorem implies that for all nonzero solutions of (5.78) to be unbounded, in addition to (5.79) and (5.81) we should impose some other restrictions on the function p. Such restrictions appear in the following assertion.

THEOREM 5.12. Let (5.79) and (5.81) hold,

$$-1 < \lim_{t \to +\infty} \inf \frac{p_2(t)}{p_0(t)} \le \lim_{t \to +\infty} \sup \frac{p_2(t)}{p_0(t)} < +\infty,$$

and let there exist a positive constant ϵ such that for any sequence $(\tau_k)_{k=1}^{+\infty}$ satisfying the conditions

$$0 \le \tau_k < \tau_{k+1} \qquad (k = 1, 2, \dots), \qquad \lim_{k \to +\infty} \tau_k = +\infty,$$

$$\lim_{k \to +\infty} \int_{\tau_{2k}}^{\tau_{2k+1}} \sqrt{p_0(t)}\, dt \in \,]\pi - \epsilon, \pi[,$$

$$\lim_{t \to +\infty} \inf \sqrt{p_0(\tau_{2k-1})(\tau_{2k} - \tau_{2k-1})} > 0,$$

$$\lim_{t \to +\infty} \sup \sqrt{p_0(\tau_{2k})(\tau_{2k} - \tau_{2k-1})} < \epsilon$$

we have

$$\sum_{k=1}^{+\infty} (\ln p_0(\tau_{2k}) - \ln p_0(\tau_{2k+1})) = +\infty.$$

Furthermore, suppose the equation (5.78) is oscillatory. Then every nonzero solution of it is subject to (5.82).

COROLLARY 5.5. Let the conditions (5.79) and (5.81) hold,

$$\lim_{t \to +\infty} \inf \frac{p_2(t)}{p_0(t)} > -1,$$

and let there exist a positive nonincreasing function $\varphi \in C(\mathbf{R}_+)$ such that

$$-p_0'(t) \ge p_0(t)\varphi(t) \qquad \text{for } t \ge 0, \qquad \int_0^{+\infty} \varphi(t)\, dt = +\infty.$$

Furthermore, suppose the equation (5.78) is oscillatory. Then every solution of it satisfies (5.82).

PROBLEM 5.5. If

$$\lim_{t \to +\infty} \inf \frac{1}{t} \int_0^t [p(s)]_- \, ds = 0,$$

then the equation (5.78) has an unbounded solution.

PROBLEM 5.6. (i) For any $\epsilon > 0$ there exist functions $p_i \in \tilde{C}_{\text{loc}}(\mathbf{R}_+)$ $(i = 0, 1)$ and $p_2 \in L_{\text{loc}}(\mathbf{R}_+)$ such that $p_0(t) > 0$, $p_0'(t) \ge 0$ for $t \ge 0$, $\lim_{t \to +\infty} p_0(t) = +\infty$,

$$\int_0^{+\infty} \frac{|p_1'(t)|\, dt}{[p_0(t)]^{1+\epsilon}} < +\infty, \qquad \int_0^{+\infty} \frac{|p_2(t)|\, dt}{[p_0(t)]^{1/2+\epsilon}} < +\infty,$$

and the equation (5.78) with $p(t) = -p_0(t) - p_1(t) - p_2(t)$ has an unbounded solution;

(ii) for any $\epsilon \in \,]0, 1[$ there exist functions $p_i \in \tilde{C}_{\text{loc}}(\mathbf{R}_+)$ $(i = 0, 1)$ and $p_2 \in L_{\text{loc}}(\mathbf{R}_+)$ such that $p_0(t) > 0$, $p_0'(t) \le 0$ for $t \ge 0$, $\lim_{t \to +\infty} p_0(t) = \lim_{t \to +\infty} p_1(t) = 0$,

$$\int_0^{+\infty} \frac{|p_1'(t)|\, dt}{[p_0(t)]^{1-\epsilon}} < +\infty, \qquad \int_0^{+\infty} \frac{|p_2(t)|\, dt}{[p_0(t)]^{1/2-\epsilon}} < +\infty,$$

and the equation (5.78) with $p(t) = -p_0(t) - p_1(t) - p_2(t)$ has a solution which vanishes at infinity.

PROBLEM 5.7. (i) There exists a function $p \in C(\mathbf{R}_+)$ such that $p(t) \to -1$ as $t \to +\infty$ and the equation (5.78) has an unbounded solution;

(ii) there exists a function $p \in C(\mathbf{R}_+)$ such that $p(t) \to -1$ as $t \to +\infty$ and the equation (5.78) has a solution which vanishes at infinity.

Notes. For Lemmas 5.1 and 5.2 see [125]. Lemmas 5.3 and 5.4 were established in [193] and [213], respectively.

The question of the existence of unbounded oscillatory solutions of third and fourth order differential equations was studied by V.A. Kondratyev [212] and M. Švec [343].

Theorems 5.1–5.6 were proved in [193].

Theorems 5.7 and 5.8 together with their corollaries are contained in [75, 77]. Theorems 5.9 and 5.10 sharpen some results of P. Hartman [126, 127]. Similar statements for two-dimensional differential systems were established by R.G. Koplatadze [215]. Theorem 5.11 was established in [198] and Theorem 5.12 in [56].

For Problems 5.1, 5.5, 5.6 and 5.7 see [193], [127], [298] and [27, 301], respectively.

§6. Asymptotic Formulas

6.1. Statement of the main theorem. In this section we derive asymptotic representations at infinity for solutions of the equation

$$u^{(n)} = \sum_{k=0}^{n-1} p_k(t) u^{(k)}. \tag{6.1}$$

First of all we give

DEFINITION 6.1. A system of locally integrable functions $\eta_k \colon [a, +\infty[\to \mathbf{R}$ $(k = 1, \dots, n)$ satisfies the *Levinson condition* if for any j and k either

$$\int_a^{+\infty} [\eta_k(\tau) - \eta_j(\tau)] \, d\tau = +\infty,$$

$$\inf \left\{ \int_s^t [\eta_k(\tau) - \eta_j(\tau)] \, d\tau : t \geq s \geq a \right\} > -\infty$$

or

$$\sup \left\{ \int_s^t [\eta_k(\tau) - \eta_j(\tau)] \, d\tau : t \geq s \geq a \right\} < +\infty.$$

The Levinson condition is satisfied if, e.g., each difference $\eta_k(t) - \eta_j(t)$ is of constant sign in a certain neighborhood of $+\infty$.

THEOREM 6.1. Let

$$p_k(t) = p_{0k}(t) + p_{1k}(t), \qquad p_{0k} \in \tilde{C}_{\mathrm{loc}}([a, +\infty[),$$

$$p_{1k} \in L_{\mathrm{loc}}([a, +\infty[) \qquad (k = 0, \dots, n-1), \tag{6.2}$$

and let there exist twice continuosly differentiable functions $\varphi, \psi \colon [a, +\infty[\to \,]0, +\infty[$ such that

(i) the functions $a_k(t) = \varphi^{-1}(t)\psi^{-1-k}(t)(\varphi(t)\psi^k(t))'$, $b_k(t) = \psi^{k-n}(t)p_{0k}(t)$ and $c_k(t) = \psi^{k-n+1}(t)p_{1k}(t)$ satisfy the conditions

$$\int_a^{+\infty} |a_k'(t)|\, dt < +\infty, \qquad \int_a^{+\infty} |b_k'(t)|\, dt < +\infty,$$

$$\int_a^{+\infty} |c_k(t)|\, dt < +\infty \qquad (k = 0,\dots,n-1);$$

(ii) the roots λ_{0k} $(k = 1,\dots,n)$ of the equation

$$\prod_{j=0}^{n-1}(\lambda + a_{0j}) = \sum_{k=1}^{n-1} b_{0k} \prod_{j=0}^{k-1}(\lambda + a_{0j}) + b_{00},$$

with $a_{0j} = \lim_{t\to+\infty} a_j(t)$, $b_{0j} = \lim_{t\to+\infty} b_j(t)$, are distinct;

(iii) the system of functions $\psi(t)\,\mathrm{Re}\,\lambda_k(t)(k = 1,\dots,n)$, where each $\lambda_k(t)$ is the root of the equation

$$\prod_{j=0}^{n-1}(\lambda + a_j(t)) = \sum_{k=1}^{n-1} b_k(t) \prod_{j=0}^{k-1}(\lambda + a_j(t)) + b_0(t)$$

tending to λ_{0k} as $t \to +\infty$, satisfies the Levinson condition.

Then the equation (6.1) possesses a fundamental system of solutions u_j $(j = 1,\dots,n)$ having the asymptotic representations

$$u_j^{(k-1)}(t) = \varphi(t)\psi^{k-1}(t) \exp\left(\int_a^t \lambda_j(\tau)\psi(\tau)\, d\tau\right) (g_{kj}^0 + o(1))$$

$$(j,k = 1,\dots,n) \tag{6.3}$$

with $g_{1j}^0 = 1$ and $g_{kj}^0 = \prod_{l=0}^{k-2}(\lambda_{0j} + a_{0l})$ $(k = 2,\dots,n)$.

6.2. Auxiliary assertions.

LEMMA 6.1. Let

$$\int_{x_0}^{+\infty} |h(x)|\, dx < +\infty, \tag{6.4}$$

and let the function $g\colon [x_0,+\infty[\times [x_0,+\infty[\to \mathbf{R}$ be locally integrable with respect to the second variable and

$$\lim_{x\to+\infty} g(x,s) = 0 \qquad \text{for } s \geq x_0, \qquad \gamma = \sup\{|g(x,s)| : x \geq s \geq x_0\} < +\infty. \tag{6.5}$$

Then

$$\lim_{x\to+\infty} \int_{x_0}^x g(x,s)h(s)\, ds = 0. \tag{6.6}$$

PROOF. By (6.4), for any $\epsilon > 0$ there exists $x_\epsilon > x_0$ such that

$$\gamma \int_{x_\epsilon}^{+\infty} |h(s)|\, ds < \epsilon.$$

Hence taking into account (6.4), we obtain

$$\left| \int_{x_0}^x g(x,s)h(s)\, ds \right| \leq \int_{x_0}^{x_\epsilon} |g(x,s)h(s)|\, ds + \gamma \int_{x_0}^{x_\epsilon} |h(s)|\, ds$$

$$< \int_{x_0}^{x_\epsilon} |g(x,s)h(s)|\, ds + \epsilon \qquad \text{for } x \geq x_\epsilon$$

and

$$\lim_{x \to +\infty} \sup \left| \int_{x_0}^{x} g(x,s)h(s)\, ds \right| \leq \epsilon.$$

Since ϵ is arbitrary, this implies (6.6). $\square$

Below $\mathbf{C}$ denotes the set of complex numbers and $(y_k)_{k=1}^{n}$ is an n-dimensional column vector with components $y_1, \ldots, y_n$.

LEMMA 6.2. Let $\omega_k, h_{kl} \colon \mathbf{R}_+ \to \mathbf{C}$ $(k, l = 1, \ldots, n)$ be locally integrable functions such that

$$\int_{0}^{+\infty} |h_{kl}(x)|\, dx < +\infty \qquad (k, l = 1, \ldots, n) \tag{6.7}$$

and the system of the functions $\operatorname{Re} \omega_k$ $(k = 1, \ldots, n)$ satisfies the Levinson condition. Then for any $j \in \{1, \ldots, n\}$ the system of differential equations

$$\frac{dy_k}{dx} = \omega_k(x)y_k + \sum_{l=1}^{n} h_{kl}(x)y_l \qquad (k = 1, \ldots, n) \tag{6.8}$$

possesses a solution $(y_{jk})_{k=1}^{n}$ having the asymptotic representation

$$y_{jk}(x) = \exp\left(\int_{0}^{x} \omega_j(\xi)\, d\xi\right)(\delta_{jk} + o(1)) \qquad (k = 1, \ldots, n), \tag{6.9}$$

where δ_{jk} is the Kronecker delta.

PROOF. Fix an arbitrary $j \in \{1, \ldots, n\}$ and set

$$g_k(x,s) = \exp\left(-\int_{s}^{x} [\omega_j(\xi) - \omega_k(\xi)]\, d\xi\right) \qquad (k = 1, \ldots, n). \tag{6.10}$$

By the Levinson condition, the set $\{1, \ldots, n\}$ can be decomposed into two subsets, N_{1n} and N_{2n}, so that

$$\lim_{x \to +\infty} |g_k(x,s)| = 0,$$

$$\eta_k = \sup\{|g_k(x,s)| : x \geq s \geq 0\} < +\infty \qquad \text{for } k \in N_{1n} \tag{6.11}$$

and

$$\eta_k = \sup\{|g_k(x,s)| : s \geq x \geq 0\} < +\infty \qquad \text{for } k \in N_{2n}. \tag{6.12}$$

According to (6.7) there exists $x_0 \in \mathbf{R}_+$ for which

$$\sum_{k=1}^{n} \eta_k \int_{x_0}^{+\infty} |h_{kl}(s)|\, ds < \frac{1}{2} \qquad (l = 1, \ldots, n). \tag{6.13}$$

Put

$$x_k = \begin{cases} x_0 & \text{for } k \in N_{1n} \\ +\infty & \text{for } k \in N_{2n}, \end{cases} \tag{6.14}$$

and denote by $C([x_0, +\infty[; \mathbf{C}^n)$ the Banach space of n-dimensional continuous vector functions $z = (z_k)_{k=1}^{n} \colon [x_0, +\infty[\to \mathbf{C}^n$ with the norm

$$\|z\| = \sup\left\{\sum_{k=1}^{n} |z_k(x)| : x \geq x_0\right\}$$

and by A the operator defined as

$$A(z) = (A_k(z))_{k=1}^n,$$

$$A_k(z)(x) = \delta_{jk} + \sum_{l=1}^n \int_{x_k}^x g_k(x,s) h_{kl}(s) z_l(s)\, ds. \tag{6.15}$$

In view of (6.11)–(6.15), A maps $C([x_0, +\infty[; \mathbf{C}^n)$ into itself and satisfies the condition

$$\|A(z) - A(\zeta)\| \le \frac{1}{2}\|z - \zeta\|.$$

According to the contractive mapping principle, the equation $z = A(z)$ has a unique solution $z = (z_k)_{k=1}^n$.

If $k \in N_{1n}$, then (6.7), (6.11), (6.14), (6.15), and Lemma 6.1 yield

$$|z_k(x) - \delta_{jk}| \le \|z\| \sum_{l=1}^n \int_{x_0}^x |g_k(x,s) h_{kl}(s)|\, ds \to 0 \qquad \text{as } x \to +\infty,$$

and if $k \in N_{2n}$, then (6.7), (6.12), (6.14), and (6.15) imply

$$|z_k(x) - \delta_{jk}| \le \|z\| \sum_{l=1}^n \eta_k \int_x^{+\infty} |h_{kl}(s)|\, ds \to 0 \qquad \text{as } x \to +\infty.$$

Therefore the functions

$$y_{jk}(x) = \exp\left(\int_0^x \omega_j(s)\, ds\right) z_k(x) \qquad (k = 1, \dots, n)$$

have the representations (6.9). On the other hand, it follows from (6.10) and (6.15) that the vector function $(y_{jk})_{k=1}^n$ is a solution of the differential system (6.8). $\square$

6.3. Proof of the main theorem. According to the condition (ii) there exists $t_0 \in [a, +\infty[$ such that

$$\lambda_k(t) \ne \lambda_j(t) \qquad \text{for } j \ne k,\, t \ge t_0; \tag{6.16}$$

moreover, every function λ_k is locally absolutely continuous on $[t_0, +\infty[$ and

$$\lambda_k'(t) = -\frac{f_t'(\lambda_k(t), t)}{f_\lambda'(\lambda_k(t), t)} = O\left(\sum_{l=0}^{n-1} [|a_l'(t)| + |b_l'(t)|]\right),$$

where $f(\lambda, t) = \prod_{j=0}^{n-1}(\lambda + a_j(t)) - \sum_{k=1}^{n-1} b_k(t) \prod_{j=0}^{k-1}(\lambda + a_j(t)) + b_0(t)$. By the condition (i) this yields

$$\int_{t_0}^{+\infty} |\lambda_k'(t)|\, dt < +\infty \qquad (k = 1, \dots, n). \tag{6.17}$$

We write the equation (6.1) as the system

$$\frac{d}{dt}(u^{(k-1)})_{k=1}^n = [P_0(t) + P_1(t)](u^{(k-1)})_{k=1}^n, \tag{6.18}$$

with

$$P_0(t) = \begin{pmatrix} 0 & 1 & 0 & \ldots & 0 \\ 0 & 0 & 1 & \ldots & 0 \\ \cdot & \cdot & \cdot & \ldots & \cdot \\ 0 & 0 & 0 & \ldots & 1 \\ p_{00}(t) & p_{01}(t) & p_{02}(t) & \ldots & p_{0n-1}(t) \end{pmatrix}$$

and

$$P_1(t) = \begin{pmatrix} 0 & 0 & \ldots & 0 \\ \cdot & \cdot & \ldots & \cdot \\ 0 & 0 & \ldots & 0 \\ p_{10}(t) & p_{11}(t) & \ldots & p_{1n-1}(t) \end{pmatrix},$$

and consider the matrices

$$F(t) = \begin{pmatrix} \varphi(t) & 0 & \ldots & 0 \\ 0 & \varphi(t)\psi(t) & \ldots & 0 \\ \cdot & \cdot & \ldots & \cdot \\ 0 & 0 & \ldots & \varphi(t)\psi^{n-1}(t) \end{pmatrix}$$

and

$$G(t) = \begin{pmatrix} g_{11}(t) & \ldots & g_{1n}(t) \\ \cdot & \ldots & \cdot \\ g_{n1}(t) & \ldots & g_{nn}(t) \end{pmatrix},$$

where $g_{1j}(t) = 1$, $g_{kj}(t) = \prod_{l=0}^{k-2}(\lambda_j(t) + a_l(t))$ $(k = 2, \ldots, n; j = 1, \ldots, n)$. In view of the conditions (i), (ii), (6.16), and (6.17),

$$\int_{t_0}^{+\infty} |g'_{kj}(t)|\, dt < +\infty \qquad (k, j = 1, \ldots, n), \tag{6.19}$$

$$\det G(t) = \prod_{k=1}^{n-1} \prod_{j=k+1}^{n} [\lambda_j(t) - \lambda_k(t)] \neq 0 \qquad \text{for } t \geq t_0,$$

$$\lim_{t \to +\infty} \det G(t) \neq 0,$$

and the entries of the matrices G and G^{-1} are bounded on $[t_0, +\infty[$.

Clearly,

$$\psi^{-1}(t)G^{-1}(t)[F^{-1}(t)P_0(t)F(t) - F^{-1}(t)F'(t)]G(t)$$
$$= \begin{pmatrix} \lambda_1(t) & 0 & \ldots & 0 \\ \cdot & \cdot & \ldots & \cdot \\ 0 & 0 & \ldots & \lambda_n(t) \end{pmatrix}.$$

So the transformation

$$x = \int_{t_0}^{t} \psi(\tau)\, d\tau, \qquad (u^{(k-1)}(t))_{k=1}^{n} = F(t)G(t)(y_k(x))_{k=1}^{n} \tag{6.20}$$

turns the system (6.18) into (6.8) with

$$\omega_k(x) = \lambda_k(t) \qquad (k = 1, \ldots, n),$$

$$\begin{pmatrix} h_{11}(x) & \cdots & h_{1n}(x) \\ \cdot & \cdots & \cdot \\ h_{n1}(x) & \cdots & h_{nn}(x) \end{pmatrix} = \psi^{-1}(t)G^{-1}(t)[Q(t) - G'(t)], \tag{6.21}$$

and

$$Q(t) = F^{-1}(t)P_1(t)F(t)G(t) = \begin{pmatrix} 0 & \cdots & 0 \\ \cdot & \cdots & \cdot \\ 0 & \cdots & 0 \\ q_1(t) & \cdots & q_n(t) \end{pmatrix},$$

$$q_j(t) = \sum_{k=0}^{n-1} c_k(t)g_{k+1j}(t) \qquad (j = 1, \dots, n). \tag{6.22}$$

We claim that

$$x = \int_{t_0}^{t} \psi(\tau)\, d\tau \to +\infty \qquad \text{as } t \to +\infty. \tag{6.23}$$

Indeed, otherwise, taking into account the boundedness of a_k $(k = 0, 1)$, we would have

$$\int_{t_0}^{+\infty} |[\ln \varphi(t)\psi^k(t)]'|\, dt = \int_{t_0}^{+\infty} \psi(t)|a_k(t)|\, dt < +\infty \qquad (k = 0, 1),$$

giving

$$0 < \delta_k = \lim_{t \to +\infty} \varphi(t)\psi^k(t) < +\infty \qquad (k = 0, 1), \qquad \lim_{t \to +\infty} \psi(t) = \frac{\delta_1}{\delta_0} > 0,$$

which contradicts the integrability of ψ on $[t_0, +\infty[$. So (6.23) is true.

By the conditions (iii) and (6.23), the equalities

$$\int_0^x \operatorname{Re}\omega_k(\xi)\, d\xi = \int_{t_0}^t \psi(\tau) \operatorname{Re}\lambda_k(\tau)\, d\tau \qquad (k = 1, \dots, n)$$

imply that the system of functions $\operatorname{Re}\omega_k$ $(k = 1, \dots, n)$ satisfies the Levinson condition.

Since the matrices G and G^{-1} are bounded, (6.21) and (6.22) yield

$$|h_{kl}(x)| \le \gamma \psi^{-1}(t) \left[\sum_{j=0}^{n-1} |c_j(t)| + \sum_{j,\nu=1}^{n} |g'_{j\nu}(t)| \right] \qquad (k, l = 1, \dots, n),$$

where γ is a sufficiently large positive constant. Hence the conditions (i), (6.19) and (6.23) imply

$$\int_0^{+\infty} |h_{kl}(x)|\, dx \le \gamma \left[\sum_{j=0}^{n-1} \int_{t_0}^{+\infty} |c_j(t)|\, dt + \sum_{j,\nu=1}^{n} \int_{t_0}^{+\infty} |g'_{j\nu}(t)|\, dt \right] < +\infty$$

$$(k, l = 1, \dots, n).$$

Therefore, all the hypotheses of Lemma 6.2 are satisfied. So, for any $j \in \{1, \dots, n\}$ the system (6.8) possesses a solution $(y_{jk})_{k=1}^{n}$ having the asymptotic representation (6.9).

It follows from (6.9) and (6.20) that each solution $(y_{jk})_{k=1}^{n}$ of the system (6.8) corresponds to a solution u_j of the equation (6.1) whose behavior at $+\infty$ is described by the asymptotic formulas (6.3). $\quad\square$

6.4. Equations with almost-constant coefficients. In case $\varphi(t) \equiv \psi(t) \equiv 1$, Theorem 6.1 reduces to the following assertion.

THEOREM 6.2. Let, together with (6.2), the conditions

$$\int_a^{+\infty} |p'_{0k}(t)|\, dt < +\infty, \qquad \int_a^{+\infty} |p_{1k}(t)|\, dt < +\infty \qquad (k = 0, \ldots, n-1) \qquad (6.24)$$

hold. Suppose, in addition, that the roots λ_{0j} $(j = 1, \ldots, n)$ of the equation

$$\lambda^n = \sum_{k=0}^{n-1} q_k \lambda^k, \qquad (6.25)$$

$q_k = \lim_{t \to +\infty} p_{0k}(t)$, are distinct and that the system of functions $\operatorname{Re}\lambda_j(t)$ $(j = 1, \ldots, n)$, where each $\lambda_j(t)$ is the root of the equation

$$\lambda^n = \sum_{k=0}^{n-1} p_{0k}(t)\lambda^k \qquad (6.26)$$

tending to λ_{0j} as $t \to +\infty$, satisfies the Levinson condition. Then the equation (6.1) possesses a fundamental system of solutions u_j $(j = 1, \ldots, n)$ having the asymptotic representations

$$u_j^{(k-1)}(t) = \exp\left(\int_a^t \lambda_j(\tau)\, d\tau\right)\left(\lambda_{0j}^{k-1} + o(1)\right) \qquad (j, k = 1, \ldots, n). \qquad (6.27)$$

COROLLARY 6.1. Let the conditions (6.2) and (6.24) hold and let the roots λ_{0j} $(j = 1, \ldots, n)$ of the equation (6.25) be distinct. Suppose, in addition, that the real parts of any two roots of (6.25) which are neither equal nor conjugate differ.[6] Then the equation (6.1) possesses a fundamental system of solutions u_j $(j = 1, \ldots, n)$ having the asymptotic representations (6.27), where each $\lambda_j(t)$ is the root of the equation (6.26) tending to λ_{0j} as $t \to +\infty$.

To justify this assertion it suffices to note that the system of functions $\operatorname{Re}\lambda_j(t)$ $(j = 1, \ldots, n)$ satisfies the Levinson condition, because for any j and k the difference $\operatorname{Re}\lambda_j(t) - \operatorname{Re}\lambda_k(t)$ either vanishes identically or does not change sign in some neighborhood of $+\infty$.

COROLLARY 6.2. Let

$$p_k(t) = p_{0k} + p_{1k}(t), \qquad p_{0k} = \text{const},$$

$$\int_a^{+\infty} |p_{1k}(t)|\, dt < +\infty \qquad (k = 0, \ldots, n-1),$$

and let the roots λ_{0j} $(j = 1, \ldots, n)$ of the equation

$$\lambda^n = \sum_{k=0}^{n-1} p_{0k}\lambda^k \qquad (6.28)$$

---

[6] In case of $n = 2$ this condition is automatically fulfilled.

be distinct. Then the equation (6.1) possesses a fundamental system of solutions u_j $(j = 1, \ldots, n)$ having the asymptotic representations

$$u_j^{(k-1)}(t) = \exp(\lambda_{0j}t)(\lambda_{0j}^{k-1} + o(1)) \qquad (j, k = 1, \ldots, n).$$

PROBLEM 6.1. Let $p_k \in L_{\text{loc}}([a, +\infty[)$ $(k = 0, \ldots, n-1)$ and let the finite limits

$$p_{0k} = \lim_{t \to +\infty} p_k(t) \qquad (k = 0, \ldots, n-1)$$

exist. Suppose, in addition, that the roots λ_{0j} $(j = 1, \ldots, n)$ of the equation (6.28) satisfy the condition $\text{Re}\,\lambda_{0j} \neq \text{Re}\,\lambda_{0k}$ for $j \neq k$. Then the equation (6.1) possesses a fundamental system of solutions u_j $(j = 1, \ldots, n)$ having the asymptotic representations

$$u_j^{(k-1)}(t) = (\lambda_{0j}^{k-1} + o(1)) \exp(\lambda_{0j}t + o(t)) \qquad (j, k = 1, \ldots, n).$$

6.5. Equations with asymptotically small coefficients. For $a > 0$, $\varphi(t) \equiv 1$ and $\psi(t) \equiv 1/t$, Theorem 6.1 implies

THEOREM 6.3. Let the conditions (6.2) hold,

$$\int_a^{+\infty} |(t^{n-k}p_{0k}(t))'|\, dt < +\infty, \qquad \int_a^{+\infty} t^{n-k-1}|p_{1k}(t)|\, dt < +\infty$$

$$(k = 0, \ldots, n-1), \tag{6.29}$$

and let the roots λ_{0j} $(j = 1, \ldots, n)$ of the equation

$$\prod_{j=0}^{n-1}(\lambda - j) = \sum_{k=1}^{n-1} q_k \prod_{j=0}^{k-1}(\lambda - j) + q_0, \tag{6.30}$$

with $q_k = \lim_{t \to +\infty} t^{n-k}p_{0k}(t)$, be such that for any $j \neq k$ either $\text{Re}\,\lambda_{0j} \neq \text{Re}\,\lambda_{0k}$ or λ_{0j} and λ_{0k} are complex conjugate. Then the equation (6.1) possesses a fundamental system of solutions u_j $(j = 1, \ldots, n)$ having the asymptotic representations

$$u_j^{(k-1)}(t) = t^{1-k} \exp\left(\int_a^t \frac{\lambda_j(\tau)}{\tau}\, d\tau\right)(l_{kj} + o(1)) \qquad (j, k = 1, \ldots, n), \tag{6.31}$$

where $l_{1j} = 1$, $l_{kj} = \prod_{\nu=0}^{k-2}(\lambda_{0j} - \nu)$ $(k = 2, \ldots, n)$ and each $\lambda_j(t)$ is the root of the equation

$$\prod_{j=0}^{n-1}(\lambda - j) = \sum_{k=1}^{n-1} t^{n-k}p_{0k}(t) \prod_{j=0}^{k-1}(\lambda - j) + t^n p_{00}(t) \tag{6.32}$$

tending to λ_{0j} as $t \to +\infty$.

COROLLARY 6.3. Let the conditions (6.2), (6.29) and

$$\lim_{t \to +\infty} t^{n-k}p_{0k}(t) = 0 \qquad (k = 0, \ldots, n-1) \tag{6.33}$$

hold. Then the equation (6.1) possesses a fundamental system of solutions u_j $(j = 1, \ldots, n)$ having the asymptotic representations

$$u_j^{(k-1)}(t) = t^{j-k+\epsilon_j(t)}(l_{kj} + o(1)) \qquad (j, k = 1, \ldots, n),$$

where $l_{1j} = 1$, $l_{kj} = \prod_{\nu=0}^{k-2}(j - \nu)$ $(k = 2, \ldots, n)$ and

$$\epsilon_j(t) = O\left(\frac{1}{\ln t}\sum_{k=0}^{n-1}\int_a^t \tau^{n-k-1}|p_{0k}(\tau)|\,d\tau\right) \to 0$$

as $t \to +\infty$ $(j = 1, \ldots, n)$.

To verify this statement it suffices to note that by (6.33), $\lambda_{0j} = j - 1$ and $\lambda_j(t) = j - 1 + O(\sum_{k=0}^{n-1} t^{n-k}|p_{0k}(t)|)$ $(j = 1, \ldots, n)$ are the roots of the equations (6.30) and (6.32), respectively.

COROLLARY 6.4. Let $p_k \in L_{\mathrm{loc}}([a, +\infty[)$ $(k = 0, \ldots, n - 1)$. Then the condition

$$\int_a^{+\infty} t^{n-k-1}|p_k(t)|\,dt < +\infty \qquad (k = 0, \ldots, n - 1) \tag{6.34}$$

is sufficient, and if, in addition, p_k $(k = 0, \ldots, n - 1)$ are functions of the same constant sign, then this condition is also necessary, for the equation (6.1) to possess a fundamental system of solutions u_j $(j = 1, \ldots, n)$ having the asymptotic representations

$$u_j^{(k-1)}(t) = t^{j-k}(l_{kj} + o(1)) \qquad (j, k = 1, \ldots, n), \tag{6.35}$$

where $l_{1j} = 1$, $l_{kj} = \prod_{\nu=0}^{k-2}(j - \nu)$ $(k = 2, \ldots, n)$.

PROOF. If the conditions (6.34) hold, then the hypotheses of Theorem 6.3 with $p_{0k}(t) \equiv 0$, $q_k = 0$ and $\lambda_j(t) \equiv \lambda_{0j} = j - 1$ $(j = 1, \ldots, n)$ are satisfied. So (6.31) becomes (6.35).

Now suppose p_k $(k = 0, \ldots, n - 1)$ are functions of the same constant sign and the equation (6.1) has a fundamental system of solutions satisfying (6.35). Choose $t_0 \in [a, +\infty[$ so that

$$\frac{1}{2}t^{n-1-k} < u_n^{(k-1)}(t) \qquad (k = 1, \ldots, n),$$

$$|u_n^{(n-1)}(t) - u_n^{(n-1)}(t_0)| < 1 \qquad \text{for } t \geq t_0.$$

Then

$$\sum_{k=0}^{n-1}\int_{t_0}^t \tau^{n-1-k}|p_k(\tau)|\,d\tau \leq 2\left|\sum_{k=0}^{n-1}\int_{t_0}^t p_k(\tau)u_n^{(k-1)}(\tau)\,d\tau\right|$$

$$= 2|u_n^{(n-1)}(t) - u_n^{(n-1)}(t_0)| < 2 \qquad \text{for } t \geq t_0.$$

Consequently, the conditions (6.34) are fulfilled. $\square$

6.6. Equations asymptotically equivalent to two-term ones.

THEOREM 6.4. Let the conditions (6.2) hold and let there exist a continuously differentiable function $p: [a, +\infty[\to \mathbf{R}$ such that $p(t) \neq 0$ for $t \geq a$,

$$\int_a^{+\infty} |d(p'(t)|p(t)|^{-(n+1)/n})|\,dt < +\infty, \qquad \int_a^{+\infty} |(p_{0k}(t)|p(t)|^{(k-n)/n})'|\,dt < +\infty \tag{6.36}$$

$$\int_a^{+\infty} |p_{1k}(t)|\,|p(t)|^{(k+1-n)/n}\,dt < +\infty \qquad (k = 0, \ldots, n - 1),$$

and

$$\lim_{t\to+\infty} p'(t)|p(t)|^{-(n+1)/n} = 0, \qquad \lim_{t\to+\infty} \frac{p_{00}(t)}{p(t)} = 1,$$

(6.37)

$$\lim_{t\to+\infty} p_{0k}(t)|p(t)|^{(k-n)/n} = 0 \qquad (k = 1,\dots,n-1).$$

Then the equation (6.1) possesses a fundamental system of solutions u_j $(j = 1,\dots,n)$ having the asymptotic representations

$$u_j^{(k-1)}(t) = |p(t)|^{(2k-n-1)/2n} \exp\left(\int_a^t \lambda_j(\tau)|p(\tau)|^{1/n}\,d\tau\right) (\lambda_{0j}^{k-1} + o(1))$$

$$(j,k = 1,\dots,n),$$

(6.38)

where λ_{0j} $(j = 1,\dots,n)$ are the roots of the equation

$$\lambda^n = \sigma,$$

(6.39)

$\sigma = \operatorname{sign} p(a)$ and each $\lambda_j(t)$ is the root the equation

$$\prod_{j=0}^{n-1}\left(\lambda + \frac{2j-n+1}{2n}\sigma p'(t)|p(t)|^{-(n+1)/n}\right)$$

$$= \sum_{k=1}^{n-1} p_{0k}(t)|p(t)|^{(k-n)/n} \prod_{j=0}^{k-1}\left(\lambda + \frac{2j-n+1}{2n}\sigma p'(t)|p(t)|^{-(n+1)/n}\right) + \frac{p_{00}(t)}{p(t)}\sigma \quad (6.40)$$

tending to λ_{0j} as $t \to +\infty$.

PROOF. Set

$$q(t) = \sigma p'(t)|p(t)|^{-(n+1)/n}$$

and

$$\omega_k = 4^{-k} \min\left\{(1 + |p(\tau)|^{2/n})^{-1} : a \le t \le a+k\right\} \qquad (k = 1,2,\dots),$$

and for any positive integer k choose a positive integer m_k so that

$$\max\left\{|q(t) - q(\tau)| : a+k-1 \le \tau \le t \le a+k,\ t-\tau \le \frac{1}{m_k}\right\} < \omega_k.$$

Let $t_{kj} = a + k - 1 + j/m_k$,

$$\tilde{q}(t) = \frac{1}{2}[q(t_{kj+1}) + q(t_{kj})] + \frac{1}{2}[q(t_{kj+1}) - q(t_{kj})]\sin\left(m_k\left((t - t_{kj})\pi - \frac{\pi}{2}\right)\right)$$

for $t_{kj} \le t \le t_{kj+1}$ $(j = 0,\dots,m_k;\ k = 1,2,\dots)$,

$$\epsilon(t) = q(t) - \tilde{q}(t).$$

It is easy to verify that the continuously differentiable function $\tilde{q}\colon \mathbf{R}_+ \to \mathbf{R}$ is monotone on each interval $[t_{kj}, t_{kj+1}]$,

$$\tilde{q}(t_{kj}) = q(t_{kj}) \qquad (j = 0,\dots,m_k;\ k = 1,2,\dots),$$

and

$$\int_t^{+\infty} |\epsilon(\tau)|\,d\tau \le 2^{-t}(1 + |p(t)|^{2/n})^{-1},$$

$$|\epsilon(t)| \leq 4^{-t}(1 + |p(t)|^{2/n})^{-1} \qquad \text{for } t \geq a. \tag{6.41}$$

Hence, (6.36) and (6.37) yield

$$\lim_{t \to +\infty} \tilde{q}(t) = 0, \qquad \int_a^{+\infty} |\tilde{q}'(t)| \, dt < +\infty. \tag{6.42}$$

The continuous differentiability of p and $\tilde{q}$ together with the estimates (6.41) imply that the function

$$\tilde{p}(t) = p(t) \left[1 - \frac{1}{n} |p(t)|^{1/n} \int_t^{+\infty} \epsilon(\tau) \, d\tau \right]^{-n}$$

is twice continuously differentiable,

$$\sigma \tilde{p}'(t) |\tilde{p}(t)|^{-(n+1)/n} = \tilde{q}(t),$$

$$\left(1 - \frac{1}{n} 2^{-t} \right)^n < \frac{p(t)}{\tilde{p}(t)} = 1 + O\left[2^{-t}(1 + |p(t)|^{1/n})^{-1} \right], \tag{6.43}$$

and

$$\left(\frac{p(t)}{\tilde{p}(t)} \right)' = \left(\frac{p(t)}{\tilde{p}(t)} \right)^{(n-1)/n} \left[\epsilon(t)|p(t)|^{1/n} - \frac{1}{n}|p(t)|^{2/n}q(t) \int_t^{+\infty} \epsilon(\tau) \, d\tau \right] = O(2^{-t}). \tag{6.44}$$

Let

$$\varphi(t) = |\tilde{p}(t)|^{-(n-1)/2n}, \qquad \psi(t) = |\tilde{p}(t)|^{1/n},$$

and let a_k, b_k and c_k be the functions appearing in Theorem 5.1, i.e.

$$a_k(t) = \frac{2k - n + 1}{2n} \tilde{q}(t), \qquad b_k(t) = |\tilde{p}(t)|^{(k-n)/n} p_{0k}(t),$$

$$c_k(t) = |\tilde{p}(t)|^{(k-n+1)/n} p_{1k}(t) \qquad (k = 0, \dots, n - 1).$$

By (6.36), (6.37) and (6.42)–(6.44),

$$\lim_{t \to +\infty} a_k(t) = 0 \qquad (k = 0, \dots, n-1), \qquad \lim_{t \to +\infty} b_0(t) = \sigma, \tag{6.45}$$

$$\lim_{t \to +\infty} b_k(t) = 0 \qquad (k = 1, \dots, n-1),$$

$$b_k'(t) = (p_{0k}(t)|p(t)|^{(k-n)/n})' \left(\frac{p(t)}{\tilde{p}(t)} \right)^{1-k/n} - \frac{k}{n} \left(\frac{\tilde{p}(t)}{p(t)} \right)^{k/n} \left(\frac{p(t)}{\tilde{p}(t)} \right)'$$

$$= O\left(|(p_{0k}(t)|p(t)|^{(k-n)/n})'| + 2^{-t} \right),$$

$$c_k(t) = O(|p(t)|^{(k-n+1)/n}|p_{1k}(t)|) \qquad (k = 0, \dots, n-1),$$

and the hypothesis (i) of Theorem 6.1 is satisfied. In view of (6.45), the equation appearing in the hypothesis (ii) of the same theorem coincides with (6.39). So this hypothesis is fulfilled as well.

Suppose that $\tilde{\lambda}_j(t)$ $(j = 1, \dots, n)$ are the roots of the equation

$$\prod_{j=0}^{n-1} \left(\lambda + \frac{2j - n + 1}{2n} \tilde{q}(t) \right)$$

$$= \sum_{k=1}^{n-1} p_{0k}(t)|\tilde{p}(t)|^{(k-n)/n} \prod_{j=0}^{k-1} \left(\lambda + \frac{2j-n+1}{2n}\tilde{q}(t)\right) + \frac{p_{00}(t)}{\tilde{p}(t)}\sigma \qquad (6.46)$$

tending, respectively, to λ_{0j} $(j = 1, \dots, n)$ as $t \to +\infty$. Then each difference $\operatorname{Re} \tilde{\lambda}_j(t) - \operatorname{Re} \tilde{\lambda}_k(t)$ either does not change sign in some neighborhood of $+\infty$ or vanishes identically. Thus, the system of functions $\psi(t) \operatorname{Re} \tilde{\lambda}_k(t)$ $(k = 1, \dots, n)$ satisfies the Levinson condition.

By Theorem 6.1 the equation (6.1) has a fundamental system of solutions $\tilde{u}_j$ $(j = 1, \dots, n)$ such that

$$\tilde{u}_j^{(k-1)}(t) = |\tilde{p}(t)|^{(2k-n-1)/2n} \exp\left(\int_a^t \tilde{\lambda}_j(\tau)|\tilde{p}(\tau)|^{1/n}\tau\right)(\lambda_{0j}^{k-1} + o(1))$$

$$(j, k = 1, \dots, n). \qquad (6.47)$$

According to (6.41) and (6.43),

$$\tilde{q}(t) = \sigma p'(t)|p(t)|^{-(n+1)/n} + O\left[2^{-t}(1 + |p(t)|^{1/n})^{-1}\right],$$

$$p_{0k}(t)|p(t)|^{(k-n)/n} = p_{0k}(t)|\tilde{p}(t)|^{(k-n)/n} + O\left[2^{-t}(1 + |p(t)|^{1/n})^{-1}\right]$$

$$(k = 0, \dots, n-1).$$

Therefore, (6.40) and (6.46) yield

$$\tilde{\lambda}_j(t) = \lambda_j(t) + O\left[2^{-t}(1 + |p(t)|^{1/n})^{-1}\right] \qquad (j = 1, \dots, n)$$

and

$$\delta_j(t) = \lambda_j(t)|p(t)|^{1/n} - \tilde{\lambda}_j(t)|\tilde{p}(t)|^{1/n} = O(2^{-t}) \qquad (j = 1, \dots, n).$$

In view of (6.47) it is clear that

$$u_j(t) = \exp\left(\int_a^{+\infty} \delta_j(\tau)\, d\tau\right)\tilde{u}_j(t) \qquad (j = 1, \dots, n) \qquad (6.48)$$

is a fundamental system of solutions of the equation (6.1) having the asymptotic representations (6.38). $\square$

REMARK 6.1. Under the hypotheses of Theorem 6.4 the equation (6.1) is asymptotically equivalent to the two-term equation

$$u^{(n)} = p(t)u,$$

because the solutions of these equations have the same asymptotic representations.

COROLLARY 6.5. Let there exist a continuously differentiable function $p: [a, +\infty[\to \mathbf{R}$ such that $p(t) \neq 0$ for $t \geq a$,

$$\int_a^{+\infty} \left|d(p'(t)|p(t)|^{-(n+1)/n})\right| < +\infty,$$

$$\int_a^{+\infty} |p'(t)|^2|p(t)|^{-(2n+1)/n}\, dt < +\infty \qquad (6.49)$$

and

$$\int_a^{+\infty} |p_k(t)| \, |p(t)|^{(k+1-n)/n} \, dt < +\infty \qquad (k = 1, \dots, n-1),$$

$$(6.50)$$

$$\int_a^{+\infty} |p_0(t) - p(t)| \, |p(t)|^{(1-n)/n} \, dt < +\infty.$$

Then the equation (6.1) possesses a fundamental system of solutions u_j $(j = 1, \dots, n)$ having the asymptotic representations

$$u_j^{(k-1)}(t) = |p(t)|^{(2k-n-1)/2n} \exp\left(\lambda_{0j} \int_a^t |p(\tau)|^{1/n} \, d\tau\right) (\lambda_{0j}^{k-1} + o(1))$$

$$(j, k = 1, \dots, n), \tag{6.51}$$

where λ_{0j} $(j = 1, \dots, n)$ are the roots of the equation $\lambda^n = \operatorname{sign} p(a)$.

PROOF. First of all we show that

$$\lim_{t \to +\infty} p'(t) |p(t)|^{-(n+1)/n} = 0. \tag{6.52}$$

Indeed, according to the first condition in (6.49) the finite limit

$$\lim_{t \to +\infty} p'(t) |p(t)|^{-(n+1)/n} = \eta$$

exists. If $\eta \neq 0$, then the last equality implies

$$|p(t)|^{1/n} = t^{-1} \left(\frac{n}{|\eta|} + o(1)\right)$$

and

$$|p'(t)|^2 |p(t)|^{-(2n+1)/n} = t^{-1}(n|\eta| + o(1)).$$

But this contradicts the second condition in (6.49). So (6.52) is true.

By (6.49), (6.50) and (6.52), the conditions (6.36) and (6.37) hold with $p_{00}(t) = p(t)$, $p_{10}(t) = p_0(t) - p(t)$, $p_{0k}(t) = 0$, $p_{1k}(t) = p_k(t)$ $(k = 1, \dots, n-1)$. Hence, in view of Theorem 6.4 the equation (6.1) possesses a fundamental system of solutions $\tilde{u}_j$ $(j = 1, \dots, n)$ having the asymptotic representations (6.47), where each $\tilde{\lambda}_j(t)$ is the root of the equation

$$\prod_{j=0}^{n-1} \left(\lambda + \frac{2j - n + 1}{2n} p'(t) |p(t)|^{-(n+1)/n}\right) = \sigma \tag{6.53}$$

tending to λ_{0j} as $t \to +\infty$ and $\sigma = \operatorname{sign} p(a)$.

Since in (6.53) the coefficient of λ^{n-1} is equal to zero, we get

$$\tilde{\lambda}_j(t) = \lambda_{0j} + O\left(|p'(t)|^2 |p(t)|^{-(2n+2)/n}\right) \qquad (j = 1, \dots, n).$$

According to (6.49) this yields

$$\int_a^{+\infty} |\delta_j(t)| \, dt < +\infty \qquad (j = 1, \dots, n),$$

where
$$\delta_j(t) = [\lambda_{0j} - \lambda_j(t)]|p(t)|^{1/n} \qquad (j = 1, \ldots, n).$$

It follows from (6.47) that the solutions (6.48) of the equation (6.1) have the asymptotic representations (6.51). □

COROLLARY 6.6. Let there exist a constant $\alpha \in [0, (n+1)/n[$ and a continuously differentiable function $p: [a, +\infty[\to \mathbf{R}$ such that
$$\inf\{|p(t)| : t \geq a\} > 0, \qquad \int_a^{+\infty} |d(p'(t)|p(t)|^{-\alpha})| < +\infty, \qquad (6.54)$$

and let the conditions (6.50) hold. Then the equation (6.1) possesses a fundamental system of solutions u_j $(j = 1, \ldots, n)$ having the asymptotic representations (6.51), where λ_{0j} $(j = 1, \ldots, n)$ are the roots of the equation $\lambda^n = \operatorname{sign} p(a)$.

PROOF. By (6.54),
$$\delta = \inf\{|p(t)| : t \geq a\} > 0,$$
$$\eta = \sup\{|p'(t)| \, |p(t)|^{-(n+1)/n} : t \geq a\} < +\infty. \qquad (6.55)$$

Set $\beta = -\alpha + (n+1)/n$ and $\sigma = \operatorname{sign} p(a)$. Then applying (6.55) to the equality
$$\int_a^t |p'(\tau)|^2 |p(\tau)|^{-(2n+1)/n} \, d\tau = -\frac{\sigma}{\beta} \int_a^t p'(\tau)|p(\tau)|^{-\alpha} \, d|p(\tau)|^{-\beta}$$

$$= \frac{\sigma}{\beta}[p'(a)|p(a)|^{-(n+1)/n} - p'(t)|p(t)|^{-(n+1)/n}]$$

$$+ \frac{\sigma}{\beta} \int_a^t |p(\tau)|^{-\beta} \, d(p'(\tau)|p(\tau)|^{-\alpha}),$$

we obtain
$$\int_a^t |p'(\tau)|^2 |p(\tau)|^{-(2n+1)/n} \, d\tau \leq \frac{\eta}{\beta} + \frac{\delta^{-\beta}}{\beta} \int_a^t |d(p'(\tau)|p(\tau)|^{-\alpha})| \qquad \text{for } t \geq a.$$

Besides, as is easily verified,
$$\int_a^{+\infty} |d(p'(t)|p(t)|^{-(n+1)/n})| \leq \delta^{-\beta} \int_a^t |d(p'(\tau)|p(\tau)|^{-\alpha})|$$

$$+ \beta \int_a^t |p'(\tau)|^2 |p(\tau)|^{-(2n+1)/n} \, d\tau \qquad \text{for } t \geq a.$$

In view of (6.54) the last two inequalities imply that the function p satisfies the conditions (6.49). Thus the hypotheses of Corollary 6.5 are fulfilled. □

REMARK 6.2. It is readily shown that Theorem 6.4 and Corollaries 6.5 and 6.6 remain valid if instead of continuous differentiability of p we require p to be piecewise continuously differentiable.

COROLLARY 6.7. Let there exist a constant $\beta \in [0, 1/n[$ and a continuous function $p: [a, +\infty[\to \mathbf{R}$ such that
$$\inf\{|p(t)| : t \in \mathbf{R}_+\} > 0, \qquad |p(t)|^{-\beta} \text{ is downward convex} \qquad (6.56)$$

and the conditions (6.50) hold. Then the equation (6.1) possesses a fundamental system of solutions u_j $(j = 1, \ldots, n)$ having the asymptotic representations (6.51), where λ_{0j} $(j = 1, \ldots, n)$ are the roots of the equation $\lambda^n = \operatorname{sign} p(a)$.

In order to justify this assertion it suffices to note that the continuity and convexity of $|p|^{-\beta}$ imply the piecewise continuous differentiability of $p^{*)}$[7] and that the conditions (6.56) yield (6.54).

PROBLEM 6.2. Suppose $p, q \in L_{\text{loc}}([a, +\infty[)$ and either

$$\int_a^{+\infty} |p(t) - q(t)| v_k^2(t)\, dt < +\infty \qquad (k = 1, 2),$$

where v_1 and v_2 are linearly independent solutions of the equation

$$v'' = q(t)v, \tag{6.57}$$

or the equation (6.57) is nonoscillatory and

$$\int_a^{+\infty} |p(t) - q(t)| v_1(t) v_2(t)\, dt < +\infty,$$

where v_1 and v_2 are linearly independent solutions of (6.57) such that

$$\int_a^{+\infty} v_1^{-2}(t)\, dt < +\infty, \qquad \int_a^{+\infty} v_2^{-2}(t)\, dt = +\infty.$$

Then for every nontrivial solution of the equation

$$u'' = p(t)u \tag{6.58}$$

the asymptotic formulas

$$u^{(k)}(t) = c_1(1 + o(1))v_1^{(k)}(t) + c_2(1 + o(1))v_2^{(k)}(t) \qquad (k = 0, 1) \tag{6.59}$$

hold, with constants c_1 and c_2 at least one of which is not zero.

PROBLEM 6.3. Let $p \in L_{\text{loc}}([a, +\infty[)$ and let there exist a continuously differentiable function $h \colon [a, +\infty[\ \to\ ]0, +\infty[$ such that

$$\int_a^{+\infty} |p(t)h(t) - h''(t)h(t) + h^{-2}(t)|\, dt < +\infty$$

$$\left(\int_a^{+\infty} |p(t)h(t) - h''(t)h(t) - h^{-2}(t)|\, dt < +\infty \right).$$

Then for every nontrivial solution of the equation (6.58) the asymptotic formulas (6.59) hold, with

$$v_k(t) = h(t) \sin\left(\frac{k\pi}{2} - \int_a^t h^{-2}(\tau)\, d\tau \right)$$

$$\left(v_k(t) = h(t) \exp\left((-1)^k \int_a^t h^{-2}(\tau)\, d\tau \right) \right) \qquad (k = 1, 2)$$

and constants c_1 and c_2 at least one of which is not zero.

---

[7] See [235], p. 15.

Notes. Lemma 6.2 is due to N. Levinson [89, 260]. Theorems 6.1–6.4, which were proved in [177], generalize some results of M. Matell [266]. Similar assertions are contained in [111] and [311].

For second order equations a proposition analogous to Corollary 6.1 was established by J. Mařik and M. Ráb [264].

Corollary 6.2 was first derived by M. Bôcher [41] for $n = 2$, and by M. Matell [266] for arbitrary n.

For Problem 6.1 see the paper of O. Perron [300] and also the monograph of R. Bellman [27].

Corollary 6.4 was proved by M. Bôcher [41] in case $n = 2$, and by O. Haupt [134], J.E. Wilkins [368] and I.M. Sobol [339] in the general case. W.F. Trench [359, 360] obtained results which are somewhat more general.

Statements concerning the equation (6.58) and similar to Theorem 6.4 and Corollaries 6.5 and 6.7 can be found in the works of F. Hartman and A. Wintner [132], A. Wintner [371], M. Zlamal [377] and V. Doležal [96].

For Problems 6.2 and 6.3 see the papers of M. Ráb [304, 306] and the book of F. Hartman [127].

CHAPTER II

QUASILINEAR DIFFERENTIAL EQUATIONS

§7. Statement of the Problem. Auxiliary Assertions

In this chapter the question of the asymptotic equivalence of solutions of the differential equations

$$u^{(n)} = \sum_{k=0}^{n-1} p_k(t) u^{(k)} + q(t, u, u', \dots, u^{(n-1)}) \tag{7.1}$$

and

$$v^{(n)} = \sum_{k=0}^{n-1} p_k(t) v^{(k)}, \tag{7.2}$$

where $p_k \in L_{\mathrm{loc}}([a, +\infty[)$ $(k = 0, \dots, n-1)$ and $q \in K_{\mathrm{loc}}([a, +\infty[\times \mathbf{R}^n)$, is studied.

Let $h = (h_k)_{k=1}^n$ be a vector function with continuous components $h_k \colon [a, +\infty[\to]0, +\infty[$ $(k = 1, \dots, n)$. For any $n-1$ times continuously differentiable function $v \colon [a, +\infty[\to \mathbf{R}$ and $k \in \{1, \dots, n\}$, we set

$$\sigma_h(v)(t) = \sum_{j=1}^n \frac{|v^{(j-1)}(t)|}{h_j(t)}, \qquad \sigma_{k,h}(v)(t) = h_k(t) \sigma_h(v)(t). \tag{7.3}$$

DEFINITION 7.1. A fundamental system of solutions v_k $(k = 1, \dots, n)$ of the equation (7.2) has the *Levinson property with respect to the vector function* h if

$$\prod_{k=1}^n \sigma_{k,h}(v_k)(t) = O\left(\exp\left[\int_a^t p_{n-1}(\tau)\, d\tau\right]\right) \tag{7.4}$$

and for any $k, l \in \{1, \dots, n\}$ either

$$\lim_{t \to +\infty} \gamma_{kl}(t, a) = 0, \qquad \sup\{\gamma_{kl}(t, \tau) : t \geq \tau \geq a\} < +\infty \tag{7.5}$$

or

$$\sup\{\gamma_{kl}(t, \tau) : \tau \geq t \geq a\} < +\infty, \tag{7.6}$$

where

$$\gamma_{kl}(t, \tau) = \frac{\sigma_h(v_k)(t)}{\sigma_h(v_k)(\tau)} \left[\frac{\sigma_h(v_l)(t)}{\sigma_h(v_l)(\tau)}\right]^{-1}. \tag{7.7}$$

DEFINITION 7.2. The equation (7.2) has the *Levinson property with respect to the vector function h* if it possesses a fundamental system of solutions (which are complex-valued, in general) having this property.

REMARK 7.1. Taking into account the results of §6, we can easily conclude that the equation (7.2) has the Levinson property with respect to the vector function h if, e.g.: (i) $h_k(t) = \psi^{k-1}(t)$ $(k = 1, \ldots, n)$ and the hypotheses of Theorem 6.1 are satisfied; (ii) $h_k(t) = 1$ $(k = 1, \ldots, n)$ and the hypotheses of Theorem 6.2 are satisfied; (iii) $h_k(t) = t^{1-k}$ $(k = 1, \ldots, n)$ and the hypotheses of Theorem 6.3 are satisfied; or (iv) $h_k(t) = |p(t)|^{(2k-n-1)/(2n)}$ $(k = 1, \ldots, n)$ and the hypotheses of Theorem 6.4 are satisfied.

DEFINITION 7.3. A nonzero solution v_0 of the equation (7.2) is said to be *h-maximal* (*h-minimal*) if every nonzero solution v of this equation satisfies the estimate

$$\sigma_h(v)(t) = O(\sigma_h(v_0)(t)) \qquad (\sigma_h(v_0)(t) = O(\sigma_h(v)(t)))\,.$$

DEFINITION 7.4. A solution u of the equation (7.1), defined in some neighborhood of $+\infty$, is said to be an $\mathcal{L}_h$ *type solution* if there exists a nonzero solution v of the equation (7.2) such that

$$u^{(j-1)}(t) = v^{(j-1)}(t) + o(\sigma_{j,h}(v)(t)) \qquad (j = 1, \ldots, n). \tag{7.8}$$

DEFINITION 7.5. The equation (7.1) is said to be an $\mathcal{L}_h^0$ *type equation* (an $\mathcal{L}_h^\infty$ *type equation*) if there exists a positive constant δ such that every solution u of this equation satisfying the condition

$$0 < \sum_{j=1}^{n} |u^{(j-1)}(a)| < \delta \qquad \left(\sum_{j=1}^{n} |u^{(j-1)}(a)| > \delta\right)$$

is of $\mathcal{L}_h$ type.

DEFINITION 7.6. The equation (7.1) is said to be an $\mathcal{L}_h$ *type equation* if every nonzero solution of it is an $\mathcal{L}_h$ type solution.

DEFINITION 7.7. The family of $\mathcal{L}_h$ type solutions of the equation (7.1) is said to be *stable* if for any solution u from this family and any point t_0 belonging to the interval of definition of u there exists $\delta > 0$ such that every solution $\overline{u}$ of the equation (7.1) satisfying the condition

$$\sum_{j=1}^{n} |u^{(j-1)}(t_0) - \overline{u}^{(j-1)}(t_0)| < \delta$$

is of $\mathcal{L}_h$ type.

In §8 and §9, assuming that (7.2) has the Levinson property with respect to a vector function h, we derive conditions under which: (a) (7.1) has at least one $\mathcal{L}_h$

type solution; (b) the family of $\mathcal{L}_h$ type solutions of the equation (7.1) is stable; (c) (7.1) is either an $\mathcal{L}_h^0$, $\mathcal{L}_h^\infty$ or $\mathcal{L}_h$ type equation.

Below we give some assertions which we need in the future.

7.1. Linear equations having the Levinson property. Let v_k $(k = 1, \dots, n)$ be a fundamental system of solutions of the equation (7.2). We call it *h-ordered* if for any $k \in \{1, \dots, n-1\}$,

$$\sigma_h(v_k)(t) = O(\sigma_h(v_{k+1})(t)).$$

LEMMA 7.1. *Let v_k $(k = 1, \dots, n)$ be a fundamental system of solutions of the equation (7.2) such that for any $k, l \in \{1, \dots, n\}$ either (7.5) or (7.6) holds. Then this system can be renumbered so that it becomes h-ordered.*

PROOF. We have

$$\lim_{t \to +\infty} \frac{\sigma_h(v_k)(t)}{\sigma_h(v_l)(t)} = 0$$

for (7.5) and

$$\sigma_h(v_l)(t) = O(\sigma_h(v_k)(t))$$

for (7.6). Clearly, this implies the existence of distinct numbers $m_k \in \{1, \dots, n\}$ $(k = 1, \dots, n)$ for which the system $\tilde{v}_k(t) = v_{m_k}(t)$ $(k = 1, \dots, n)$ is h-ordered. $\square$

LEMMA 7.2. *Let a fundamental system of solutions v_k $(k = 1, \dots, n)$ of the equation (7.2) have the Levinson property with respect to a vector function h and let v_0 be a solution of (7.2) such that*

$$\sigma_h(v_0)(t) = O(\sigma_h(v_l)(t)) \tag{7.9}$$

for a certain $l \in \{1, \dots, n\}$. If the system of functions $\tilde{v}_k$ $(k = 1, \dots, n)$, where

$$\tilde{v}_k(t) = \begin{cases} v_k(t) & \text{for } k \neq l \\ v_0(t) & \text{for } k = l, \end{cases} \tag{7.10}$$

is linearly independent, then

$$\sigma_h(v_l)(t) = O(\sigma_h(v_0)(t)), \tag{7.11}$$

and $\tilde{v}_k$ $(k = 1, \dots, n)$ also has the Levinson property with respect to the vector function h.

PROOF. First we will show that the estimate (7.11) is true.

Assume the contrary. Then there exists a sequence $t_k \in [a, +\infty[$ $(k = 1, 2, \dots)$ converging to $+\infty$ and satisfying the condition

$$\lim_{k \to +\infty} \frac{\sigma_h(v_0)(t_k)}{\sigma_h(v_l)(t_k)} = 0. \tag{7.12}$$

By the Liouville formula

$$\begin{vmatrix} \dfrac{\tilde{v}_1(t_k)}{h_1(t_k)\sigma_h(v_1)(t_k)} & \cdots & \dfrac{\tilde{v}_n(t_k)}{h_1(t_k)\sigma_h(v_n)(t_k)} \\ \cdots & \cdots & \cdots \\ \dfrac{\tilde{v}_1^{(n-1)}(t_k)}{h_n(t_k)\sigma_h(v_1)(t_k)} & \cdots & \dfrac{\tilde{v}_n^{(n-1)}(t_k)}{h_n(t_k)\sigma_h(v_n)(t_k)} \end{vmatrix}$$

$$= c_0 \left[\prod_{j=1}^{n} \sigma_{j,h}(v_j)(t_k) \right]^{-1} \exp \left[\int_a^{t_k} p_{n-1}(\tau) \, d\tau \right],$$

where c_0 is a nonzero constant. According to (7.10) and (7.12), the determinant at the left-hand side of the last equality tends to zero as $k \to +\infty$, which is impossible, because by (7.4),

$$\liminf_{k \to +\infty} \left(\left[\prod_{j=1}^{n} \sigma_{j,h}(v_j)(t_k) \right]^{-1} \exp \left[\int_a^{t_k} p_{n-1}(\tau) \, d\tau \right] \right) > 0.$$

This contradiction proves (7.11).

Taking into account (7.9)–(7.11), it is easy to conclude that $\tilde{v}_k$ $(k = 1, \dots, n)$ has the Levinson property with respect to the vector function h. $\square$

LEMMA 7.3. Let the equation (7.2) have the Levinson property with respect to a vector function h. Then for any nontrivial real-valued solution v_0 of this equation there exists a fundamental system of real-valued solutions v_k $(k = 1, \dots, n)$, $v_1(t) \equiv v_0(t)$, having the Levinson property with respect to the vector function h.

PROOF. In view of Lemma 7.1, the equation (7.2) possesses an h-ordered fundamental system of solutions w_k $(k = 1, \dots, n)$ having the Levinson property with respect to the vector function h. As $v_0(t) \not\equiv 0$, there exist $l \in \{1, \dots, n\}$ and complex constants c_k $(k = 1, \dots, l)$ such that

$$c_l \neq 0, \qquad v_0(t) = \sum_{k=1}^{l} c_k w_k(t).$$

Since w_k $(k = 1, \dots, n)$ is h-ordered, this yields

$$\sigma_h(v_0)(t) = O(\sigma_h(w_l)(t)).$$

Furthermore, it is obvious that the system of functions $\tilde{w}_k$ $(k = 1, \dots, n)$,

$$\tilde{w}_k(t) = \begin{cases} v_0(t) & \text{for } k = 1 \\ w_{k-1}(t) & \text{for } 2 \leq k \leq l \\ w_k(t) & \text{for } k \geq l+1, \end{cases}$$

is linearly independent. Thus, by Lemma 7.2 this system has the Levinson property with respect to the vector function h.

For any $k \in \{2, \dots, n-1\}$ the function $\tilde{w}_k$ can be represented in the form

$$\tilde{w}_k(t) = v_{1k}(t) + i v_{2k}(t),$$

where v_{1k} and v_{2k} are real-valued solutions of the equation (7.2) and i is the imaginary unit. Let $v_1(t) \equiv v_0(t)$. Denote by v_k $(k = 2, \dots, n-1)$ either v_{1k} or v_{2k}, so that the system $v_1, \dots, v_k, \tilde{w}_{k+1}, \dots, \tilde{w}_n$ be linearly independent, and by v_n either v_{1n} or v_{2n}, so that $v_1, \dots, v_n$ be linearly independent. Lemma 7.2 clearly implies that v_k $(k = 1, \dots, n)$ is the fundamental system of solutions of the equation (7.2) we search for. $\square$

LEMMA 7.4. Let v_k $(k = 1, \ldots, n)$ be a fundamental system of solutions of the equation (7.2) having the Levinson property with respect to a vector function h, and let

$$v_0(t) = \sum_{k=1}^{n} c_k v_k(t),$$

where c_k $(k = 1, \ldots, n)$ are complex constants. Then

$$\sum_{k=1}^{n} |c_k| \sigma_h(v_k)(t) = O(\sigma_h(v_0)(t)).$$

PROOF. By Lemma 7.1 we may assume that the system v_k $(k = 1, \ldots, n)$ is h-ordered.

Note that only the case when $\sum_{k=1}^{n} |c_k| \neq 0$ is worthy of interest. Let l be the maximal index k for which $c_k \neq 0$. Clearly,

$$v_0(t) = \sum_{k=1}^{l} c_k v_k(t), \qquad \sigma_h(v_0)(t) = O(\sigma_h(v_l)(t))$$

and the system $v_1, \ldots, v_{l-1}, v_0, v_{l+1}, \ldots, v_n$ is linearly independent. According to Lemma 7.2 this implies the equality (7.9). So,

$$\sum_{k=1}^{n} |c_k| \sigma_h(v_k)(t) = \sum_{k=1}^{l} |c_k| \sigma_h(v_k)(t) = O(\sigma_h(v_l)(t)) = O(\sigma_h(v_0)(t)). \qquad \square$$

LEMMA 7.5. Let v_k $(k = 1, \ldots, n)$ be a fundamental system of solutions of the equation (7.2) having the Levinson property with respect to a vector function h, and let

$$u_k \colon [a, +\infty[\to \mathbf{R} \qquad (k = 1, \ldots, n) \tag{7.13}$$

be a system of $n - 1$ times differentiable functions such that

$$u_k^{(j-1)}(t) = v_k^{(j-1)}(t) + o(\sigma_{j,h}(v_k)(t)) \qquad (j, k = 1, \ldots, n). \tag{7.14}$$

Then the system (7.13) is linearly independent.

PROOF. By (7.14) we can find a sequence $t_m \in [a, +\infty[$ $(m = 1, 2, \ldots)$ such that $t_m \to +\infty$ as $m \to +\infty$ and the finite limits

$$\lim_{m \to +\infty} \frac{u_k^{(j-1)}(t_m)}{\sigma_{j,h}(v_k)(t_m)} = \lim_{m \to +\infty} \frac{v_k^{(j-1)}(t_m)}{\sigma_{j,h}(v_k)(t_m)} \qquad (j, k = 1, \ldots, n) \tag{7.15}$$

exist.

Let W be the Wronskian of the system (7.13). According to the Liouville formula and the conditions (7.4) and (7.15),

$$W(t_m) = \begin{vmatrix} v_1(t_m) & \cdots & v_n(t_m) \\ \cdots & \cdots & \cdots \\ v_1^{(n-1)}(t_m) & \cdots & v_n^{(n-1)}(t_m) \end{vmatrix}$$

$$+ \prod_{k=1}^{n} \sigma_{k,h}(v_k)(t_m) \left[\left| \begin{array}{ccc} \dfrac{u_1(t_m)}{\sigma_{1,h}(v_1)(t_m)} & \cdots & \dfrac{u_n(t_m)}{\sigma_{1,h}(v_n)(t_m)} \\ \cdots & \cdots & \cdots \\ \dfrac{u_1^{(n-1)}(t_m)}{\sigma_{n,h}(v_1)(t_m)} & \cdots & \dfrac{u_n^{(n-1)}(t_m)}{\sigma_{n,h}(v_n)(t_m)} \end{array} \right| \right.$$

$$\left. - \left| \begin{array}{ccc} \dfrac{v_1(t_m)}{\sigma_{1,h}(v_1)(t_m)} & \cdots & \dfrac{v_n(t_m)}{\sigma_{1,h}(v_n)(t_m)} \\ \cdots & \cdots & \cdots \\ \dfrac{v_1^{(n-1)}(t_m)}{\sigma_{n,h}(v_1)(t_m)} & \cdots & \dfrac{v_n^{(n-1)}(t_m)}{\sigma_{n,h}(v_n)(t_m)} \end{array} \right| \right] = c_0 \exp\left[\int_a^{t_m} p_{n-1}(\tau)\, d\tau\right] + o(1) \prod_{k=1}^{n} \sigma_{k,h}(v_k)(t_m)$$

$$= [c_0 + o(1)] \exp\left[\int_a^{t_m} p_{n-1}(\tau)\, d\tau\right],$$

where $c_0 \neq 0$. Hence $W(t_m) \neq 0$ for large m and so the system (7.13) is linearly independent. $\square$

7.2. On a system of nonlinear integral equations. Let m be a positive integer, $t_0 \in \mathbf{R}$, $r \in]0, +\infty[$,

$$D_1 = \{(\tau, t) \in \mathbf{R}^2 : t_0 \leq \tau \leq t\}, \qquad D_2 = \{(\tau, t) \in \mathbf{R}^2 : t_0 \leq t \leq \tau\},$$

$$I^n = \{(y_j)_{j=1}^{n} \in \mathbf{R}^n : |y_1| \leq r, \ldots, |y_n| \leq r\},$$

and let g_{1k} and g_{2k} ($k = 1, \ldots, n$) be real-valued functions defined on the sets $D_1 \times I^n$ and $D_2 \times I^n$, respectively. Suppose, in addition, that for any $t \in]t_0, +\infty[$ and $(y_j)_{j=1}^{n} \in I^n$ the functions $g_{1k}(\cdot, t, y_1, \ldots, y_n): [t_0, t] \to \mathbf{R}$ and $g_{2k}(\cdot, t, y_1, \ldots, y_n): [t, +\infty[\to \mathbf{R}$ are integrable, while for any $\tau \in]t_0, +\infty[$ the functions $g_{1k}(\tau, \cdot, \ldots, \cdot): [\tau, +\infty[\times I^n \to \mathbf{R}$ and $g_{2k}(\tau, \cdot, \ldots, \cdot): [t_0, \tau] \times I^n \to \mathbf{R}$ are continuous. Consider the system of integral equations

$$y_k(t) = \int_{t_0}^{t} g_{1k}(\tau, t, y_1(\tau), \ldots, y_n(\tau))\, d\tau$$

$$+ \int_{t}^{+\infty} g_{2k}(\tau, t, y_1(\tau), \ldots, y_n(\tau))\, d\tau \qquad (k = 1, \ldots, n), \tag{7.16}$$

whose solution we search for in the set of continuous functions $(y_k)_{k=1}^{n}: [t_0, +\infty[\to I^n$.

LEMMA 7.6. Let

$$|g_{1k}(\tau, t, y_1, \ldots, y_n)| \leq \eta(t, \tau) g(\tau) \qquad (k = 1, \ldots, n) \tag{7.17}$$

on the set $D_1 \times I^n$ and

$$|g_{2k}(\tau, t, y_1, \ldots, y_n)| \leq g(\tau) \qquad (k = 1, \ldots, n) \tag{7.18}$$

on the set $D_2 \times I^n$, where $\eta: D_1 \to [0, 1]$ is continuous and

$$\lim_{t \to +\infty} \eta(t, \tau) = 0 \qquad \text{for } \tau \geq t_0, \tag{7.19}$$

while $g: [t_0, +\infty[\to \mathbf{R}_+$ is integrable and

$$\int_{t_0}^{+\infty} g(\tau)\, d\tau \leq r. \tag{7.20}$$

Then the system (7.16) has at least one solution $(y_k)_{k=1}^n$, and

$$\lim_{t \to +\infty} y_k(t) = 0 \qquad (k = 1, \ldots, n). \tag{7.21}$$

PROOF. Set

$$\epsilon(t) = \int_{t_0}^t \eta(t, \tau) g(\tau) \, d\tau + \int_t^{+\infty} g(\tau) \, d\tau.$$

By Lemma 6.1 the conditions (7.19) and (7.20) yield

$$\lim_{t \to +\infty} \epsilon(t) = 0. \tag{7.22}$$

For any $\delta > 0$ and $b > t_0$ we define the functions

$$\omega_{1b}(\tau, \delta) = \max \left\{ \sum_{k=1}^n |g_{1k}(\tau, t_1, y_1, \ldots, y_n) - g_{2k}(\tau, t_2, y_1, \ldots, y_n)| : \right.$$

$$\left. \tau \le t_1 \le t_2 \le b, \ t_2 - t_1 \le \delta, \ (y_j)_{j=1}^n \in I^n \right\} \qquad \text{for } t_0 \le \tau \le b,$$

$$\omega_{2b}(\tau, \delta) = \max \left\{ \sum_{k=1}^n |g_{2k}(\tau, t_1, y_1, \ldots, y_n) - g_{2k}(\tau, t_2, y_1, \ldots, y_n)| : \right.$$

$$\left. t_0 \le t_1 \le t_2 \le \min\{\tau, b\}, \ t_2 - t_1 \le \delta, \ (y_j)_{j=1}^n \in I^n \right\} \qquad \text{for } t_0 \le \tau < +\infty,$$

and

$$\omega_b(\delta) = \int_{t_0}^b \omega_{1b}(\tau, \delta) \, d\tau + \int_{t_0}^{+\infty} \omega_{2b}(\tau, \delta) \, d\tau + \sup \left\{ \int_t^{t+\delta} g(\tau) \, d\tau : t \ge t_0 \right\}.$$

According to the restrictions imposed on g_{1k} and g_{2k} $(k = 1, \ldots, n)$,

$$\lim_{\delta \to 0} \omega_b(\delta) = 0. \tag{7.23}$$

Let $C([t_0, +\infty[; \mathbf{R}^n)$ be the Banach space of continuous bounded vector functions $y = (y_k)_{k=1}^n \colon [t_0, +\infty[\to \mathbf{R}^n$ with the norm

$$\|y\| = \sup \left\{ \sum_{k=1}^n |y_k(t)| : t \ge t_0 \right\},$$

and let S be the set of $(y_k)_{k=1}^n \in C([t_0, +\infty[; \mathbf{R}^n)$ satisfying the inequalities

$$|y_k(t)| \le r \qquad \text{for } t \ge t_0 \qquad (k = 1, \ldots, n).$$

Consider the operator $A(y) = (A_k(y))_{k=1}^n$, where

$$A_k(y)(t) = \int_{t_0}^t g_{1k}(\tau, t, y_1(\tau), \ldots, y_n(\tau)) \, d\tau$$

$$+ \int_t^{+\infty} g_{2k}(\tau, t, y_1(\tau), \ldots, y_n(\tau)) \, d\tau.$$

By (7.17) and (7.18), for any $y \in S$ and $b \in \,]t_0, +\infty[$ we obtain

$$|A_k(y)(t)| \le \epsilon(t) \le r \qquad \text{for } t \ge t_0 \qquad (k = 1, \ldots, n), \tag{7.24}$$

$$|A_k(y)(t_1) - A_k(y)(t_2)| \leq \omega_b(t_2 - t_1) \qquad \text{for } t_0 \leq t_1 \leq t_2 \leq b. \tag{7.25}$$

In view of the Arzelá–Ascoli lemma the conditions (7.22)–(7.25) imply that A maps S into a compact subset of S. Besides, as we can easily verify, A is continuous. So, according to the Schauder principle [168] the system (7.16) has a solution $(y_k)_{k=1}^n \in S$. On the other hand, (7.24) yields

$$|y_k(t)| \leq \epsilon(t) \qquad \text{for } t \geq t_0 \qquad (k = 1, \ldots, n).$$

Consequently, the inequalities (7.21) are true. $\square$

§8. The Family of $\mathcal{L}_h$ Type Solutions of the Equation (7.1)

Throughout this section, $W(v_1, \ldots, v_n)(\cdot)$ denotes the Wronskian of a system of functions $v_1, \ldots, v_n$, while $W_j(v_1, \ldots, v_n)(\cdot)$ denotes the cofactor of the element of this Wronskian at the intersection of the n-th row and j-th column.

We call $c\colon [a, +\infty[\times [a, +\infty[\to \mathbf{R}$ the *Cauchy function* of the equation (7.2) if for any $\tau \in [a, +\infty[$ the function $v(\cdot) = c(\cdot, \tau)$ is the solution of this equation satisfying the initial conditions

$$v(\tau) = \ldots = v^{(n-2)}(\tau) = 0, \qquad v^{(n-1)}(\tau) = 1.$$

If $v_1, \ldots, v_n$ is a fundamental system of solutions of the equation (7.2), then c can be represented in the form

$$c(t, \tau) = \sum_{j=1}^n \frac{W_j(v_1, \ldots, v_n)(\tau)}{W(v_1, \ldots, v_n)(\tau)} v_j(t). \tag{8.1}$$

8.1. Necessary and sufficient conditions for the existence of $\mathcal{L}_h$ type solutions

THEOREM 8.1. Let the equation (7.2) have the Levinson property with respect to a vector function h and let there exist a nontrivial real-valued solution v_0 of this equation, a positive constant r and a measurable function $q^*\colon [a, +\infty[\to \mathbf{R}_+$ satisfying the conditions

$$|q(t, x_1, \ldots, x_n)| \leq q^*(t)$$
$$\text{for } t \geq a, \ |x_k - v_0^{(k-1)}(t)| \leq r\sigma_{k,h}(v_0)(t) \ (k = 1, \ldots, n) \tag{8.2}$$

and

$$\int_a^{+\infty} \frac{q^*(t)}{\sigma_{n,h}(v_0)(t)} \, dt < +\infty. \tag{8.3}$$

Then the equation (7.1) possesses a solution u such that

$$u^{(k-1)}(t) = v_0^{(k-1)}(t) + o(\sigma_{k,h}(v_0)(t)) \qquad (k = 1, \ldots, n). \tag{8.4}$$

PROOF. By Lemma 7.3, the equation (7.2) possesses a fundamental system of real-valued solutions v_k $(k = 1, \ldots, n)$, $v_1(t) \equiv v_0(t)$, having the Levinson property with respect to the vector function h. Let

$$\gamma_j(t, \tau) = \frac{\sigma_h(v_j)(t)}{\sigma_h(v_j)(\tau)} \left[\frac{\sigma_h(v_0)(t)}{\sigma_h(v_0)(\tau)} \right]^{-1}.$$

Then the set $\{1, \ldots, n\}$ can be represented as the union of two nonintersecting sets N_{1n} and N_{2n} so that

$$\lim_{t \to +\infty} \gamma_j(t, \tau) = 0, \qquad \gamma_j^* = \sup\{\gamma_j(t, \tau) : t \geq \tau \geq a\} < +\infty \qquad \text{for } j \in N_{1n} \quad (8.5)$$

and

$$\gamma_j^* = \sup\{\gamma_j(t, \tau) : \tau \geq t \geq a\} < +\infty \qquad \text{for } j \in N_{2n}. \tag{8.6}$$

According to the Liouville formula and the notation (7.3),

$$W(v_1, \ldots, v_n)(\tau) = W(v_1, \ldots, v_n)(a) \exp\left[\int_a^\tau p_{n-1}(s)\, ds\right]$$

and

$$|W_j(v_1, \ldots, v_n)(\tau)| \leq \frac{(n-1)!}{h_n(\tau)\sigma_h(v_j)(\tau)} \prod_{k=1}^n \sigma_{k,h}(v_k)(\tau) \qquad (j = 1, \ldots, n).$$

In view of (7.4), (8.5) and (8.6) this implies the estimates

$$\left| \frac{W_j(v_1, \ldots, v_n)(\tau)}{W(v_1, \ldots, v_n)(\tau)} v_j^{(k-1)}(t) \right| \leq \rho\gamma_j(t, \tau) \frac{\sigma_{k,h}(v_0)(t)}{\sigma_{n,h}(v_0)(\tau)}$$

$$\leq \rho\gamma_j^* \frac{\sigma_{k,h}(v_0)(t)}{\sigma_{n,h}(v_0)(\tau)} \qquad \text{for } a \leq \tau \leq t \qquad (j \in N_{1n};\ k = 1, \ldots, n) \tag{8.7}$$

and

$$\left| \frac{W_j(v_1, \ldots, v_n)(\tau)}{W(v_1, \ldots, v_n)(\tau)} v_j^{(k-1)}(t) \right| \leq \rho\gamma_j^* \frac{\sigma_{k,h}(v_0)(t)}{\sigma_{n,h}(v_0)(t)}$$

$$\text{for } a \leq t \leq \tau \qquad (j \in N_{2n};\ k = 1, \ldots, n), \tag{8.8}$$

where ρ is a positive constant.

By (8.3), for sufficiently large t_0 the function

$$g(\tau) = \rho \sum_{j=1}^n \gamma_j^* \frac{q^*(\tau)}{\sigma_{n,h}(v_0)(\tau)} \tag{8.9}$$

satisfies the inequality (7.20).

Consider the integro-differential equation[1]

$$u(t) = v_0(t)$$

$$+ \sum_{j \in N_{1n}} \int_{t_0}^t \frac{W_j(v_1, \ldots, v_n)(\tau)}{W(v_1, \ldots, v_n)(\tau)} v_j(t) q(\tau, u(\tau), \ldots, u^{(n-1)}(\tau))\, d\tau$$

$$- \sum_{j \in N_{2n}} \int_t^{+\infty} \frac{W_j(v_1, \ldots, v_n)(\tau)}{W(v_1, \ldots, v_n)(\tau)} v_j(t) q(\tau, u(\tau), \ldots, u^{(n-1)}(\tau))\, d\tau. \tag{8.10}$$

Note that if (8.10) has a solution satisfying (8.4), then it is the solution of the equation (7.1) we search for.

---

[1] If $N_{1n} = \emptyset$, then the sum $\sum_{j \in N_{1n}}$ should be omitted.

Via the transformation

$$y_k(t) = \frac{u^{(k-1)}(t) - v_0^{(k-1)}(t)}{\sigma_{k,h}(v_0)(t)} \qquad (k = 1, \ldots, n), \tag{8.11}$$

(8.10) can be reduced to the system of integral equations (7.16) with

$$g_{1k}(\tau, t, y_1, \ldots, y_n) = \sum_{j \in N_{1n}} \frac{W_j(v_1, \ldots, v_n)(\tau)}{W(v_1, \ldots, v_n)(\tau)} \frac{v_j^{(k-1)}(t)}{\sigma_{k,h}(v_0)(t)}$$

$$\times q(\tau, v_0(\tau) + \sigma_{1,h}(v_0)(\tau)y_1, \ldots, v_0^{(n-1)}(\tau) + \sigma_{n,h}(v_0)(\tau)y_n) \quad \text{if } N_{1n} \neq \emptyset,$$

$$g_{1k}(\tau, t, y_1, \ldots, y_n) = 0 \quad \text{if } N_{1n} = \emptyset$$

and

$$g_{2k}(\tau, t, y_1, \ldots, y_n) = - \sum_{j \in N_{2n}} \frac{W_j(v_1, \ldots, v_n)(\tau)}{W(v_1, \ldots, v_n)(\tau)} \frac{v_j^{(k-1)}(t)}{\sigma_{k,h}(v_0)(t)}$$

$$\times q(\tau, v_0(\tau) + \sigma_{1,h}(v_0)(\tau)y_1, \ldots, v_0^{(n-1)}(\tau) + \sigma_{n,h}(v_0)(\tau)y_n).$$

The inequalities (8.2), (8.7) and (8.8) yield (7.17) and (7.18), where

$$\eta(t, \tau) = \left(\sum_{j \in N_{1n}} \gamma_j^* \right)^{-1} \sum_{j \in N_{1n}} \gamma_j(t, \tau) \quad \text{if } N_{1n} \neq \emptyset,$$

$$\eta(t, \tau) = 0 \quad \text{if } N_{1n} = \emptyset$$

and g is the function defined by (8.9). In view of (8.5),

$$\eta(t, \tau) \leq 1 \quad \text{for } t \geq \tau \geq t_0, \qquad \lim_{t \to +\infty} \eta(t, \tau) = 0.$$

Hence the hypotheses of Lemma 7.6 are fulfilled. Therefore the system (7.16) under the conditions (7.21) possesses a solution $(y_k)_{k=1}^n$. According to (8.11), the function $u(t) = v_0(t) + y_1(t)\sigma_{1,h}(v_0)(t)$ is a solution of the equation (8.10) having the asymptotic representation (8.4). □

By Remark 7.1 this theorem implies

COROLLARY 8.1. Suppose one of the following possibilities occurs: (i) $h_k(t) = \psi^{k-1}(t)$ $(k = 1, \ldots, n)$ and the hypotheses of Theorem 6.1 are fulfilled; (ii) $h_k(t) = 1$ $(k = 1, \ldots, n)$ and the hypotheses of Theorem 6.2 are fulfilled; (iii) $h_k(t) = t^{1-k}$ $(k = 1, \ldots, n)$ and the hypotheses of Theorem 6.3 are fulfilled; or (iv) $h_k(t) = |p(t)|^{(2k-n-1)/(2n)}$ $(k = 1, \ldots, n)$ and the hypotheses of Theorem 6.4 are fulfilled. Suppose, in addition, that there exist a nontrivial real-valued solution v_0 of the equation (7.2), a positive constant r and a measurable function q^*: $[a, +\infty[\to \mathbf{R}_+$ for which the conditions (8.2) and (8.3) hold. Then the equation (7.1) possesses a solution u having the asymptotic representation (8.4).

In case $p_k(t) \equiv 0$ $(k = 0, \ldots, n-1)$, the last assertion implies

COROLLARY 8.2. Let there exist $l \in \{1, \ldots, n\}$, $r \in {]0, +\infty[}$, $\alpha \neq 0$, and a measurable function q^*: $[a, +\infty[\to \mathbf{R}_+$ such that

$$|q(t, x_1, \ldots, x_n)| \leq q^*(t) \quad \text{for } t \geq a,$$

$$\left| x_k - \alpha \frac{(l-1)!}{(l-k)!} t^{l-k} \right| \leq r t^{l-k} \qquad (k=1,\dots,l),$$

$$|x_k| \leq r t^{l-k} \qquad (k=l+1,\dots,n)$$

and

$$\int_a^{+\infty} t^{n-l} q^*(t)\, dt < +\infty.$$

Then the equation

$$u^{(n)} = q(t,u,u',\dots,u^{(n-1)}) \tag{8.12}$$

possesses a solution u having the asymptotic representation

$$u^{(k-1)}(t) = \alpha \frac{(l-1)!}{(l-k)!} t^{l-k} + o(t^{l-k}) \qquad (k=1,\dots,l),$$

$$u^{(k-1)}(t) = o(t^{l-k}) \qquad (k=l+1,\dots,n).$$

PROBLEM 8.1. Let the equation (7.2) have a fundamental system of real-valued solutions v_k $(k=1,\dots,n)$ such that $v_1(t) \neq 0$ for $t \geq a$ and $|v_k(t)/v_1(t)|$ $(k=1,\dots,n)$ are monotone. Suppose, in addition, $q \in K_{loc}([a,+\infty[\times \mathbf{R})$ and

$$|q(t,x)| \leq q^*(t) \qquad \text{for } t \geq a,\ |x - v_1(t)| \leq r|v_1(t)|,$$

where $r \in]0,+\infty[$, the function $q^*: [a,+\infty[\to \mathbf{R}_+$ is measurable and

$$\int_a^{+\infty} \exp\left(-\int_a^t p_{n-1}(\tau)\, d\tau \right) \left(\sum_{j=1}^n \left| W_j(v_1,\dots,v_n)(t) \frac{v_j(t)}{v_1(t)} \right| \right) q^*(t)\, dt < +\infty.$$

Then the equation

$$u^{(n)} = \sum_{k=0}^{n-1} p_k(t) u^{(k)} + q(t,u) \tag{8.13}$$

possesses a solution u having the asymptotic representation

$$u(t) = v_1(t)(1 + o(1)).$$

THEOREM 8.2. Suppose the equation (7.2) has the Levinson property with respect to a vector function h, its Cauchy function c satisfies the inequality

$$c(t,\tau) \geq 0 \qquad \text{for } t \geq \tau \geq a, \tag{8.14}$$

and there exists an h-maximal real-valued solution v_0 of this equation such that

$$v_0^{(k-1)}(t) \geq \delta \sigma_{k,h}(v_0)(t) \qquad \text{for } t \geq a \qquad (k=1,\dots,n) \tag{8.15}$$

and

$$\lim_{t \to +\infty} \inf \frac{c(t,\tau)}{v_0(t)} \geq \frac{\delta}{v_0^{(n-1)}(\tau)} \qquad \text{for } \tau \geq a, \tag{8.16}$$

where δ is a positive constant. Moreover, let on each of the sets $[a, +\infty[\times \mathbf{R}_-^n$ and $[a, +\infty[\times \mathbf{R}_+^n$ the function q be either nonpositive or nonnegative, and

$$|q(t, x_1, \dots, x_n)| \leq |q(t, y_1, \dots, y_n)|$$

$$\text{for } t \geq a, \, x_k x_1 > 0, \, y_k x_1 > 0, \, |x_k| \leq |y_k| \qquad (k = 1, \dots, n). \tag{8.17}$$

Then the condition

$$\int_a^{+\infty} \frac{1}{v_0^{(n-1)}(t)} |q(t, \alpha v_0(t), \dots, \alpha v_0^{(n-1)}(t))| \, dt < +\infty \qquad \text{for } \alpha \in \mathbf{R} \tag{8.18}$$

is necessary and sufficient for the equation (7.1) to possess, for any real $\alpha_0 \neq 0$, a solution u having the asymptotic representation

$$u^{(k-1)}(t) = v_0^{(k-1)}(t)(\alpha_0 + o(1)) \qquad (k = 1, \dots, n). \tag{8.19}$$

PROOF. Sufficiency. Let $\alpha_0 \neq 0$ be a real constant,

$$r = \frac{1}{2}|\alpha_0|\delta, \qquad \alpha = (1 + \delta)\alpha_0,$$

and

$$q^*(t) = q(t, \alpha v_0(t), \dots, \alpha v_0^{(n-1)}(t)).$$

Then by (8.15) and (8.17),

$$|q(t, x_1, \dots, x_n)| \leq q^*(t)$$

for $t \geq a$, $|x_k - \alpha_0 v_0^{(k-1)}(t)| \leq r\sigma_{k,h}(v_0)(t)$ $(k = 1, \dots, n)$.

On the other hand, it follows from (8.15) and (8.18) that q^* satisfies the condition (8.3). So, in view of Theorem 8.1 and the inequalities (8.15), the equation (7.1) possesses a solution u having the asymptotic representation (8.19).

Necessity. Let α be a real constant,

$$\alpha_0 = \begin{cases} 1 & \text{if } \alpha = 0 \\ 2\alpha & \text{if } \alpha \neq 0, \end{cases}$$

and let u be a solution of the equation (7.1) having the asymptotic representation (8.19). Without loss of generality we may assume that

$$2|\alpha_0|v_0^{(k-1)}(t) \geq u^{(k-1)}(t) \operatorname{sign} \alpha_0 > |\alpha|v_0^{(k-1)}(t)$$

$$\text{for } t \geq a \qquad (k = 1, \dots, n). \tag{8.20}$$

By the Cauchy formula,

$$u(t) = \sum_{k=1}^n \alpha_k v_k(t) + \int_a^t c(t, \tau) q(\tau, u(\tau), \dots, u^{(n-1)}(\tau)) \, d\tau, \tag{8.21}$$

where v_k $(k = 1, \dots, n)$ is a fundamental system of solutions of the equation (7.2) and α_k $(k = 1, \dots, n)$ are constants. Since the solution v_0 is h-maximal and q does not change sign on the sets $[a, +\infty[\times \mathbf{R}^n_-$ and $[a, +\infty[\times \mathbf{R}^n_+$, applying (8.14), (8.15), (8.17), and (8.20) to (8.21), we obtain

$$\int_a^t \frac{c(t, \tau)}{u(t)} |q(\tau, \alpha v_0(\tau), \dots, \alpha v_0^{(n-1)}(\tau))| \, d\tau \le \rho \qquad \text{for } t \ge a,$$

where $\rho = \text{const} > 0$. This inequality together with (8.16) yields (8.18). $\square$

COROLLARY 8.3. Suppose $p_k \in \tilde{C}_{\mathrm{loc}}([a, +\infty[)$ $(k = 0, \dots, n-1)$,

$$p_k(t) \ge 0 \qquad \text{for } t \ge a \qquad (k = 0, \dots, n-2), \tag{8.22}$$

and there exist twice continuously differentiable functions $\varphi, \psi: [a, +\infty[\to \,]0, +\infty[$ such that

(a) $a_k(t) = (\varphi(t)\psi^k(t))'/\varphi(t)\psi^{k+1}(t)$ and $b_k(t) = \psi^{k-n}p_k(t)$ satisfy the conditions

$$\int_a^{+\infty} |a'_k(t)| \, dt < +\infty, \qquad \int_a^{+\infty} |b'_k(t)| \, dt < +\infty \qquad (k = 0, \dots, n-1); \tag{8.23}$$

(b) the roots λ_{0k} $(k = 1, \dots, n)$ of the equation

$$\prod_{j=0}^{n-1} (\lambda + a_{0j}) = \sum_{k=1}^{n-1} b_{0k} \prod_{j=0}^{k-1} (\lambda + a_{0j}) + b_{00}, \tag{8.24}$$

with $a_{0j} = \lim_{t \to +\infty} a_j(t)$ and $b_{0j} = \lim_{t \to +\infty} b_j(t)$, are distinct, $\lambda_{0n} \in \mathbf{R}$, and

$$\lambda_{0n} > \max\{\mathrm{Re}\,\lambda_{01}, \dots, \mathrm{Re}\,\lambda_{0n-1}, -a_{00}, \dots, a_{0n-1}\}; \tag{8.25}$$

(c) the system of functions $\psi(t) \,\mathrm{Re}\, \lambda_k(t)$ $(k = 1, \dots, n)$, where each $\lambda_k(t)$ is the root of the equation

$$\prod_{j=0}^{n-1} (\lambda + a_j(t)) = \sum_{k=1}^{n-1} b_k(t) \prod_{j=0}^{k-1} (\lambda + a_j(t)) + b_0(t) \tag{8.26}$$

tending to λ_{0k} as $t \to +\infty$, satisfies the Levinson condition.

Moreover, let on each of the sets $[a, +\infty[\times \mathbf{R}^n_-$ and $[a, +\infty[\times \mathbf{R}^n_+$ the function q be either nonpositive or nonnegative and let (8.17) hold. Then the condition

$$\int_a^{+\infty} \frac{1}{v_n(t)} |q(t, \alpha v_1(t), \dots, \alpha v_n(t))| \, dt < +\infty \qquad \text{for } \alpha \in \mathbf{R}, \tag{8.27}$$

where

$$v_k(t) = \varphi(t)\psi^{k-1}(t) \exp\left(\int_a^t \lambda_n(\tau)\psi(\tau) \, d\tau \right) \qquad (k = 1, \dots, n), \tag{8.28}$$

is necessary and sufficient for the equation (7.1) to possess, for any real $\alpha_0 \ne 0$, a solution u having the asymptotic representation

$$u(t) = v_1(t)(\alpha_0 + o(1)),$$

$$\tag{8.29}$$

$$u^{(k-1)}(t) = v_k(t) \prod_{l=0}^{k-2} (\lambda_{0n} + a_{0l})(\alpha_0 + o(1)) \qquad (k = 2, \dots, n).$$

PROOF. First of all we note that the inequalities (8.22) guarantee the validity of (8.14).

By Theorem 6.1 the equation (7.2) has a fundamental system of solutions v_j $(j = 1, \ldots, n)$ such that

$$v_j(t) = \varphi(t) \exp \left(\int_a^t \lambda_j(\tau)\psi(\tau) \, d\tau \right) (1 + o(1)),$$

(8.30)

$$v_j^{(k-1)}(t) = \varphi(t)\psi^{k-1}(t) \exp \left(\int_a^t \lambda_j(\tau)\psi(\tau) \, d\tau \right) \left(\prod_{l=0}^{k-2} (\lambda_{0j} + a_{0l}) + o(1) \right)$$

$$(j = 1, \ldots, n; \; k = 2, \ldots, n).$$

In view of the equalities (8.1), (8.28), (8.30) and inequalities (8.14), (8.25),

$$\lim_{t \to +\infty} \frac{v_n(t)}{v_1(t)} = 1,$$

(8.31)

$$\lim_{t \to +\infty} \frac{v_n^{(k-1)}(t)}{v_k(t)} = \prod_{l=0}^{k-2} (\lambda_{0n} + a_{0l}) > 0 \qquad (k = 2, \ldots, n),$$

$$\lim_{t \to +\infty} \frac{c(t, \tau)}{v_n(t)} = \frac{W_n(v_1, \ldots, v_n)(\tau)}{W(v_1, \ldots, v_n)(\tau)} \geq 0 \qquad \text{for } \tau \geq 0,$$

(8.32)

and

$$\frac{W_n(v_1, \ldots, v_n)(\tau)}{W(v_1, \ldots, v_n)(\tau)} = \varphi^{n-1}(\tau)\psi^{(n-1)(n-2)/2}(\tau)$$

$$\times \exp \left[\int_a^\tau \left(\psi(s) \sum_{j=1}^{n-1} \lambda_j(s) - p_{n-1}(s) \right) ds \right] (c_0 + o(1)),$$

where

$$c_0 = \frac{1}{W(v_1, \ldots, v_n)(a)} \begin{vmatrix} 1 & \cdots & 1 \\ \lambda_{01} + \alpha_{00} & \cdots & \lambda_{0n-1} + \alpha_{00} \\ \cdots & \cdots & \cdots \\ \prod_{l=0}^{n-2}(\lambda_{01} + \alpha_{0l}) & \cdots & \prod_{l=0}^{n-2}(\lambda_{0n-1} + \alpha_{0l}) \end{vmatrix} \neq 0$$

On the other hand, by the Viète theorem,

$$\sum_{j=1}^{n} \lambda_n(t) = \sum_{j=0}^{n-1} \frac{(\varphi(t)\psi^j(t))'}{\varphi(t)\psi^{j+1}(t)} + \psi^{-1}(t)p_{n-1}(t).$$

Thus

$$\frac{W_n(v_1, \ldots, v_n)(\tau)}{W(v_1, \ldots, v_n)(\tau)} = \nu_n(t)(c_1 + o(1)),$$

(8.33)

where

$$c_1 = \varphi^n(a)\psi^{(n(n-1)/2}(a)c_0 \neq 0.$$

Let $h(t) = (\psi^{k-1}(t))_{k=1}^n$. According to (8.30)–(8.33), without loss of generality we may assume that v_n is an h-maximal real-valued solution and that the inequalities (8.15), (8.16) and

$$\frac{1}{\delta}\nu_k(t) \geq v_0^{(k-1)}(t) \geq \delta\nu_k(t) \qquad \text{for } t \geq a \qquad (k = 1, \dots, n), \tag{8.34}$$

where $v_0(t) \equiv v_n(t)$ and δ is a positive constant, hold.

In view of Theorem 8.2 the condition (8.18) is necessary and sufficient for the equation (7.1) to have, for any $\alpha_0 \neq 0$, a solution u satisfying (8.29). But by (8.17) and (8.34) the conditions (8.18) and (8.27) are equivalent. $\square$

According to Theorems 6.2–6.4, Corollary 8.3 implies the following assertions.

COROLLARY 8.4. Suppose $p_k(t) \geq 0$ for $t \geq a$ $(k = 0, \dots, n-2)$,

$$p_k \in \tilde{C}_{\mathrm{loc}}([a, +\infty[), \qquad \int_a^{+\infty} |p_k'(t)|\, dt < +\infty \qquad (k = 0, \dots, n-1),$$

the roots λ_{0j} $(j = 1, \dots, n)$ of the equation

$$\lambda^n = \sum_{k=0}^{n-1} p_{0k}\lambda^k,$$

$p_{0k} = \lim_{t \to +\infty} p_k(t)$, are all distinct,

$$\lambda_{0n} \in \mathbf{R}, \qquad \lambda_{0n} > \mathrm{Re}\,\lambda_{0j} \qquad (j = 1, \dots, n-1),$$

and the system of functions $\mathrm{Re}\,\lambda_j(t)$ $(j = 1, \dots, n)$, where $\lambda_j(t)$ is the root of the equation

$$\lambda^n = \sum_{k=0}^{n-1} p_k(t)\lambda^k,$$

tending to λ_{0j} as $t \to +\infty$, satisfies the Levinson condition. Moreover, let on each of the sets $[a, +\infty[\times \mathbf{R}_-^n$ and $[a, +\infty[\times \mathbf{R}_+^n$ the function q be either nonpositive or nonnegative and let (8.17) hold. Then the condition

$$\int_a^{+\infty} \frac{1}{\nu(t)} |q(t, \alpha\nu(t), \dots, \alpha\nu(t))|\, dt < +\infty \qquad \text{for } \alpha \in \mathbf{R},$$

with

$$\nu(t) = \exp\left(\int_a^t \lambda_n(\tau)\, d\tau\right),$$

is necessary and sufficient for the equation (7.1) to possess, for any real $\alpha_0 \neq 0$, a solution u having the asymptotic representation

$$u^{(k-1)}(t) = \exp\left(\int_a^t \lambda_n(\tau)d\tau\right)(\alpha_0\lambda_{0n}^{k-1} + o(1)) \qquad (k = 1, \dots, n).$$

COROLLARY 8.5. Suppose $p_k(t) \geq 0$ for $t \geq a$ $(k = 0, \dots, n-2)$, $p_k \in \tilde{C}_{\mathrm{loc}}([a, +\infty[)$ $(k = 0, \dots, n-1)$,

$$\int_a^{+\infty} |(t^{n-k}p_k(t))'|\, dt < +\infty, \qquad \lim_{t \to +\infty} t^{n-k}p_k(t) = 0 \qquad (k = 0, \dots, n-1),$$

and $\lambda_n(t)$ is the root of the equation

$$\prod_{j=0}^{n-1}(\lambda - j) = \sum_{k=1}^{n-1} t^{n-k} p_k(t) \prod_{j=0}^{k-1}(\lambda - j) + t^n p_0(t)$$

tending to $n - 1$ as $t \to +\infty$. Moreover, let on each of the sets $[a, +\infty[\times \mathbf{R}_-^n$ and $[a, +\infty[\times \mathbf{R}_+^n$ the function q be either nonpositive or nonnegative and let (8.17) hold. Then the condition (8.27), with

$$\nu_k(t) = t^{1-k} \exp\left(\int_a^t \frac{\lambda_n(\tau)}{\tau} d\tau\right) \qquad (k = 1, \ldots, n),$$

is necessary and sufficient for the equation (7.1) to possess, for any real $\alpha_0 \neq 0$, a solution u having the asymptotic representation

$$u^{(k-1)}(t) = \frac{(n-1)!}{(n-k)!} t^{1-k} \exp\left(\int_a^t \frac{\lambda_n(\tau)}{\tau} d\tau\right)(\alpha_0 + o(1)) \qquad (k = 1, \ldots, n).$$

COROLLARY 8.6. Let on each of the sets $[a, +\infty[\times \mathbf{R}_-^n$ and $[a, +\infty[\times \mathbf{R}_+^n$ the function q be either nonpositive or nonnegative and let (8.17) hold. Then the condition

$$\int_a^{+\infty} |q(t, \alpha t^{n-1}, \alpha t^{n-2}, \ldots, \alpha)| \, dt < +\infty \qquad \text{for } \alpha \in \mathbf{R}$$

is necessary and sufficient for the equation (8.12) to possess, for any real $\alpha_0 \neq 0$, a solution u having the asymptotic representation

$$u^{(k-1)}(t) = \frac{(n-1)!}{(n-k)!} t^{n-k}(\alpha_0 + o(1)) \qquad (k = 1, \ldots, n).$$

COROLLARY 8.7. Suppose the function p_0 is continuously differentiable, $p_k \in \tilde{C}_{\text{loc}}([a, +\infty[)$ $(k = 1, \ldots, n-1)$, $p_0(t) > 0$, $p_k(t) \geq 0$ for $t \geq a$ $(k = 1, \ldots, n-2)$,

$$\int_a^{+\infty} |d(p_0'(t) p_0^{-(n+1)/n}(t))| < +\infty,$$

$$\int_a^{+\infty} |p_k(t) p_0^{(k-n)/n}(t))'| \, dt < +\infty \qquad (k = 1, \ldots, n-1),$$

$$\lim_{t \to +\infty} p_0'(t) p_0^{-(n+1)/n}(t) = \lim_{t \to +\infty} p_k(t) p_0^{(k-n)/n}(t) = 0 \qquad (k = 1, \ldots, n-1),$$

and $\lambda_n(t)$ is the root of the equation

$$\prod_{j=0}^{n-1}\left(\lambda + \frac{2j - n + 1}{2n} p_0'(t) p_0^{-(n+1)/n}(t)\right)$$

$$= \sum_{k=1}^{n-1} p_k(t) p_0^{(k-n)/n}(t) \prod_{j=0}^{k-1}\left(\lambda + \frac{2j - n + 1}{2n} p_0'(t) p_0^{-(n+1)/n}(t)\right) + 1$$

tending to 1 as $t \to +\infty$. Moreover, let on each of the sets $[a, +\infty[\times \mathbf{R}^n_-$ and $[a, +\infty[\times \mathbf{R}^n_+$ the function q be either nonpositive or nonnegative and let (8.17) hold. Then the condition (8.27), with

$$\nu_k(t) = p_0^{(2k-n-1)/2n}(t)\exp\left(\int_a^t \lambda_n(\tau)p_0^{1/n}(\tau)\,d\tau\right) \qquad (k = 1, \ldots, n),$$

is necessary and sufficient for the equation (7.1) to possess, for any real $\alpha_0 \neq 0$, a solution u having the asymptotic representation

$$u^{(k-1)}(t) = p_0^{(2k-n-1)/2n}(t)\exp\left(\int_a^t \lambda_n(\tau)p_0^{1/n}(\tau)\,d\tau\right)(\alpha_0 + o(1))$$

$$(k = 1, \ldots, n).$$

COROLLARY 8.8. Suppose $p: [a, +\infty[\to]0, +\infty[$ is continuously differentiable and

$$\int_a^{+\infty} \left|d(p'(t)p^{-(n+1)/n}(t))\right| < +\infty, \qquad \int_a^{+\infty} |p'(t)|^2 p^{-(2n+1)/n}(t)\,dt < +\infty.$$

Moreover, let on each of the sets $[a, +\infty[\times \mathbf{R}^n_-$ and $[a, +\infty[\times \mathbf{R}^n_+$ the function q be either nonpositive or nonnegative and let (8.17) hold. Then the condition (8.27), with

$$\nu_k(t) = p^{(2k-n-1)/2n}(t)\exp\left(\int_a^t p^{1/n}(\tau)\,d\tau\right) \qquad (k = 1, \ldots, n),$$

is necessary and sufficient for the equation

$$u^{(n)} = p(t)u + q(t, u, u', \ldots, u^{(n-1)}) \tag{8.35}$$

to possess, for any real $\alpha_0 \neq 0$, a solution u having the asymptotic representation

$$u^{(k-1)}(t) = p^{(2k-n-1)/2n}(t)\exp\left(\int_a^t p^{1/n}(\tau)\,d\tau\right)(\alpha_0 + o(1)) \qquad (k = 1, \ldots, n).$$

PROBLEM 8.2. Let on each of the sets $[a, +\infty[\times \mathbf{R}^l_-$ and $[a, +\infty[\times \mathbf{R}^l_+$, $l \in \{1, \ldots, n\}$, the function $q \in K_{\mathrm{loc}}([a, +\infty[\times \mathbf{R}^l)$ be either nonpositive or nonnegative, and

$$|q(t, x_1, \ldots, x_l)| \leq |q(t, y_1, \ldots, y_l)|$$

for $t \geq a$, $x_k x_1 > 0$, $y_k x_1 > 0$, $|x_k| \leq |y_k|$ $(k = 1, \ldots, l)$.
    Then the condition

$$\int_a^{+\infty} t^{n-l}|q(t, \alpha t^{l-1}, \alpha t^{l-2}, \ldots, \alpha)|\,dt < +\infty \qquad \text{for } \alpha \in \mathbf{R}$$

is necessary and sufficient for the equation

$$u^{(n)} = q(t, u, u', \ldots, u^{(l-1)})$$

to possess, for any real $\alpha_0 \neq 0$, a solution u having the asymptotic representation

$$u^{(k-1)}(t) = \frac{(l-1)!}{(l-k)!}t^{l-k}(\alpha_0 + o(1)) \qquad (k = 1, \ldots, l).$$

8.2. Asymptotic representations for solutions of a related differential inequality. For the sake of future discussion we consider, together with the equations (7.1) and (7.2), the differential inequality

$$\left| u^{(n)}(t) - \sum_{k=0}^{n-1} p_k(t) u^{(k)}(t) \right| \leq \sum_{k=0}^{n-1} g_k(t) |u^{(k)}(t)| \tag{8.36}$$

where $g_k \in L_{\text{loc}}([a, +\infty[)$ $(k = 0, \dots, n-1)$ are nonnegative functions.

A function $u \in \tilde{C}_{\text{loc}}^{n-1}([t_0, +\infty[)$, where $t_0 \in [a, +\infty[$, is said to be a *solution* of the differential inequality (8.36) if it satisfies this inequality almost everywhere on $[t_0, +\infty[$. Furthermore, u is said to be an $\mathcal{L}_h$ *type solution* if there exists a nonzero solution v of the equation (7.2) such that the asymptotic formulas (7.8) hold.

LEMMA 8.1. Let the equation (7.2) have the Levinson property with respect to a vector function h and

$$\int_a^{+\infty} \frac{h_{k+1}(t)}{h_n(t)} g_k(t) \, dt < +\infty \qquad (k = 0, \dots, n-1). \tag{8.37}$$

Then every nonzero solution of the differential inequality (8.36) is of $\mathcal{L}_h$ type.

PROOF. Suppose u is a nonzero solution of the inequality (8.36) defined on a certain interval $[t_0, +\infty[\subset [a, +\infty[$. Set

$$\rho(t) = \sum_{k=0}^{n-1} g_k(t) |u^{(k)}(t)|,$$

$$\delta(t) = \begin{cases} 0 & \text{for } \rho(t) = 0 \\ \frac{1}{\rho(t)} \left[u^{(n)}(t) - \sum_{k=0}^{n-1} p_k(t) u^{(k-1)}(t) \right] & \text{for } \rho(t) \neq 0, \end{cases}$$

and

$$q_k(t) = \delta(t) g_k(t) \operatorname{sign} u^{(k)}(t) \qquad (k = 0, \dots, n-1).$$

In view of (8.36) and (8.37),

$$|q_k(t)| \leq g_k(t) \qquad \text{for } t \geq t_0 \qquad (k = 0, \dots, n-1)$$

and

$$\int_{t_0}^{+\infty} \frac{h_{k+1}(t)}{h_n(t)} |q_k(t)| \, dt < +\infty \qquad (k = 0, \dots, n-1). \tag{8.38}$$

On the other hand, it is obvious that u is a solution of the differential equation

$$u^{(n)}(t) = \sum_{k=0}^{n-1} [p_k(t) + q_k(t)] \, u^{(k)}. \tag{8.39}$$

Let v_k $(k = 1, \dots, n)$ be a fundamental system of real-valued solutions of the equation (7.2) having the Levinson property with respect to the vector function h. By Theorem 8.1 and Lemma 7.5, the inequalities (8.38) guarantee the existence of a fundamental system of solutions u_k $(k = 1, \ldots, n)$ of the equation (8.39) which satisfy (7.14). Choose constants c_k $(k = 1, \ldots, n)$ so that the equality

$$u(t) = \sum_{k=1}^{n} c_k u_k(t)$$

be fulfilled. Then Lemma 7.4 yields (7.8), where $v = \sum_{k=1}^{n} c_k v_k$ is a solution of the equation (7.2). $\square$

8.3. Stability of the family of $\mathcal{L}_h$ type solutions. Let $\mu > 1$, $\gamma > 1 + \mu(n-1)$, and $h(t) = (t^{1-k})_{k=1}^{n}$. Then by Corollary 8.2, for any $\alpha \in \mathbf{R}$ and $j \in \{1, \ldots, n\}$ the equation

$$u^{(n)} = t^{-\gamma} |u|^{\mu} \operatorname{sign} u \tag{8.40}$$

possesses a solution u having the asymptotic representation

$$u^{(k-1)}(t) = \alpha t^{k-j}(l_{kj} + o(1)) \qquad (k = 1, \ldots, n),$$

with $l_{1j} = 1$, $l_{kj} = j(j-1)\ldots(j-k+2)$ $(k = 2, \ldots, n)$. Moreover, these and only these solutions form the family of $\mathcal{L}_h$ type solutions. On the other hand, (8.40) has the solution

$$\overline{u}(t) = \left[\frac{\gamma - n}{\mu - 1} \left(\frac{\gamma - n}{\mu - 1} - 1 \right) \cdots \left(\frac{\gamma - n}{\mu - 1} - n + 1 \right) \right]^{1/(\mu-1)} t^{(\gamma-n)/(\mu-1)},$$

which does not belong to this family. Therefore, under the hypotheses of Theorem 8.1 the equation (7.1) may have both $\mathcal{L}_h$ type solutions and solutions which are not of $\mathcal{L}_h$ type. This naturally raises the problem of the stability of the above-mentioned family.

THEOREM 8.3. Let the equation (7.2) have the Levinson property with respect to a vector function h and let on $[a, +\infty[\times \mathbf{R}$ the function q satisfy the inequality

$$|q(t, \sigma_{1,h}(v_0)(t)x_1, \ldots, \sigma_{n,h}(v_0)(t)x_n)| \leq g\left(t, \sum_{k=1}^{n} |x_k|\right) \sum_{k=1}^{n} |x_k|, \tag{8.41}$$

where v_0 is an h-maximal real-valued solution of the equation (7.2), $g \in K_{\mathrm{loc}}([a, +\infty[\times \mathbf{R}_+)$ does not decrease with respect to the second variable and

$$\int_{a}^{+\infty} \frac{g(t, \rho)}{\sigma_{n,h}(v_0)(t)} \, dt < +\infty \qquad \text{for } \rho > 0. \tag{8.42}$$

Suppose, in addition, that the Cauchy problem for the equation (7.1) is uniquely solvable with any initial data. Then the family of $\mathcal{L}_h$ type solutions of this equation is stable.

PROOF. By Lemma 7.3 there exists a fundamental system of real-valued solutions v_k $(k = 1, \ldots, n)$, $v_1(t) \equiv v_0(t)$, of the equation (7.2) having the Levinson property with respect to the vector function h. Since v_0 is h-maximal,

$$\sup\left\{ \frac{\sigma_h(v_j)(t)}{\sigma_h(v_j)(\tau)} \left[\frac{\sigma_h(v_0)(t)}{\sigma_h(v_0)(\tau)} \right]^{-1} : t \geq \tau \geq a \right\} < +\infty \qquad (j = 1, \ldots, n). \tag{8.43}$$

For any $j, l \in \{1, \ldots, n\}$ and $t \in [a, +\infty[$ we denote by $W_{jl}(v_1, \ldots, v_n)(\cdot)$ the cofactor of the element of the Wronskian $W(v_1, \ldots, v_n)(\cdot)$ at the intersection of the j-th row and l-th column. Set

$$c_j(t, \tau) = \sum_{l=1}^{n} \frac{W_{jl}(v_1, \ldots, v_n)(\tau)}{W(v_1, \ldots, v_n)(\tau)} v_l(t) \qquad (j = 1, \ldots, n). \tag{8.44}$$

Clearly, c_n is the Cauchy function of the equation (7.2). Furthermore, for any $j \in \{1, \ldots, n\}$ and $\tau \in [a, +\infty[$ the function $v(\cdot) = c_j(\cdot, \tau)$ is the solution of the equation (7.2) under the initial conditions

$$v^{(j-1)}(\tau) = 1, \qquad v^{(k-1)}(\tau) = 0 \qquad (k \neq j; \; k = 1, \ldots, n).$$

From (7.4), (8.43) and (8.44) we derive the estimates

$$\left| \frac{\partial^{k-1} c_j(t, \tau)}{\partial t^{k-1}} \right| \leq \rho_0 \frac{\sigma_{k,h}(v_0)(t)}{\sigma_{j,h}(v_0)(\tau)} \qquad \text{for } t \geq \tau \geq a \qquad (j, k = 1, \ldots, n), \tag{8.45}$$

with $\rho_0 = \text{const} \geq 1$.

Theorem 8.1 implies that the family of $\mathcal{L}_h$ type solutions of the equation (7.1) is nonempty. Let u_0 be a solution from this family defined on a certain $[t_0, +\infty[\subset [a, +\infty[$. By the h-maximality of v_0 and the condition (8.42), we can find constants $\rho_1 > 0$ and $t_1 \in]t_0, +\infty[$ so that

$$\frac{\sigma_h(u_0)(t)}{\sigma_h(v_0)(t)} < \rho_1 \qquad \text{for } t \geq t_0 \tag{8.46}$$

and

$$n \rho_0 \rho_2 \int_{t_1}^{+\infty} \frac{g(t, \rho_2)}{\sigma_{n,h}(v_0)(t)} \, dt < 1, \tag{8.47}$$

where $\rho_2 = n \rho_0 \rho_1 + 1$.

Since the solutions of the equation (7.1) continuously depend on the initial data, by (8.46) there exists a positive constant δ such that every solution u of this equation for which

$$\sum_{k=1}^{n} |u^{(k-1)}(t_0) - u_0^{(k-1)}(t_0)| < \delta$$

satisfies the estimate

$$0 < \frac{\sigma_h(u)(t)}{\sigma_h(v_0)(t)} < \rho_1 \qquad \text{for } t_0 \leq t \leq t_1. \tag{8.48}$$

We claim that u is defined throughout $[t_0, +\infty[$ and

$$\frac{\sigma_h(u)(t)}{\sigma_h(v_0)(t)} < \rho_2 \qquad \text{for } t \geq t_0. \tag{8.49}$$

Assume the contrary. Then, in view of (8.48), we can find $t_2 \in]t_1, +\infty[$ such that

$$\frac{\sigma_h(u)(t)}{\sigma_h(v_0)(t)} < \rho_2 \qquad \text{for } t_1 \leq t < t_2, \qquad \frac{\sigma_h(u)(t_2)}{\sigma_h(v_0)(t_2)} = \rho_2. \tag{8.50}$$

From the representation

$$u(t) = \sum_{j=1}^{n} c_j(t, t_1) u^{(j-1)}(t_1)$$

$$+ \int_{t_1}^{t} c_n(t, \tau) q(\tau, u(\tau), \dots, u^{(n-1)}(\tau)) \, d\tau \qquad \text{for } t_1 \le t \le t_2,$$

taking into account (8.41), (8.45) and (8.48), we get

$$\frac{\sigma_h(u)(t)}{\sigma_h(v_0)(t)} \le n \rho_0 \rho_1$$

$$+ n\rho_0 \int_{t_1}^{t} \frac{1}{\sigma_{n,h}(v_0)(\tau)} g\left(\tau, \frac{\sigma_h(u)(\tau)}{\sigma_h(v_0)(\tau)}\right) \frac{\sigma_h(u)(\tau)}{\sigma_h(v_0)(\tau)} \, d\tau \qquad \text{for } t_1 \le t \le l_2.$$

Hence, (8.47) and (8.50) yield

$$\rho_2 \le n\rho_0\rho_1 + n\rho_0\rho_2 \int_{t_1}^{t_2} \frac{g(\tau, \rho_2)}{\sigma_{n,h}(v_0)(\tau)} \, d\tau < \rho_2.$$

This contradiction proves the validity of (8.49).

By (8.41) and (8.49), u is a nonzero solution of the differential inequality (8.36) with

$$g_k(t) = \frac{g(t, \rho_2)}{h_{k+1}(t)\sigma_h(v_0)(t)} \qquad (k = 0, \dots, n-1).$$

On the other hand, (8.42) implies that the functions g_k $(k = 0, \dots, n-1)$ satisfy the conditions (8.37). So, according to Lemma 8.1, u is an $\mathcal{L}_h$ type solution. Consequently, the family of such solutions is stable. $\square$

Theorems 6.1–6.4 and 8.3 and Remark 7.1 imply the following assertions.

COROLLARY 8.9. Let $p_k \in \tilde{C}_{\mathrm{loc}}([a, +\infty[)$ $(k = 0, \dots, n-1)$ and let there exist twice continuously differentiable functions $\varphi, \psi : [a, +\infty[\to]0, +\infty[$ such that:

(a) $a_k(t) = (\varphi(t)\psi^k(t))'/\varphi(t)\psi^{k+1}(t)$ and $b_k(t) = \psi^{k-n}p_k(t)$ are subject to the conditions (8.23);

(b) the equation (8.24) has distinct roots λ_{0k} $(k = 1, \dots, n)$;

(c) the system of functions $\psi(t) \operatorname{Re} \lambda_k(t)$ $(k = 1, \dots, n)$, where each $\lambda_k(t)$ is the root of the equation (8.26) tending to λ_{0k} as $t \to +\infty$, satisfies the Levinson condition and the inequality

$$\sup\left\{\operatorname{Re} \int_{a}^{t} [\lambda_k(\tau) - \lambda_n(\tau)]\psi(\tau) \, d\tau : t \ge a\right\} < +\infty \qquad (k = 1, \dots, n-1).$$

$$(8.51)$$

Suppose, in addition, that the Cauchy problem for the equation (7.1) is uniquely solvable with any initial data and

$$|q(t, \nu_1(t)x_1, \dots, \nu_n(t)x_n)| \le g\left(t, \sum_{k=1}^{n} |x_k|\right) \sum_{k=1}^{n} |x_k| \qquad (8.52)$$

on the set $[a, +\infty[\times \mathbf{R}^n$, where

$$\nu_k(t) = \varphi(t)\psi^{k-1}(t) \exp\left(\operatorname{Re} \int_{a}^{t} \lambda_n(\tau)\psi(\tau) \, d\tau\right) \qquad (k = 1, \dots, n), \qquad (8.53)$$

the function $g \in K_{\text{loc}}([a, +\infty[\times \mathbf{R}_+)$ does not decrease with respect to the second variable and

$$\int_a^{+\infty} \frac{g(t, \rho)}{\nu_n(t)} \, dt < +\infty \qquad \text{for } \rho > 0. \tag{8.54}$$

Then the family of solutions of the equation (7.1) having the asymptotic representations[2]

$$u^{(k-1)}(t) = \varphi(t)\psi^{k-1}(t) \sum_{j=1}^n \alpha_j \exp\left(\int_a^t \lambda_j(\tau)\psi(\tau) \, d\tau\right) (l_{kj} + o(1))$$

$$(k = 1, \dots, n), \tag{8.55}$$

with $l_{1j} = 1$, $l_{kj} = (\lambda_{00} + a_{00}) \dots (\lambda_{0k-2} + a_{0k-2})$ $(k = 2, \dots, n)$, is stable.

COROLLARY 8.10. Suppose $p_k \in \tilde{C}_{\text{loc}}([a, +\infty[)$,

$$\int_a^{+\infty} |p_k'(t)| \, dt < +\infty, \qquad p_{0k} = \lim_{t \to +\infty} p_k(t) \qquad (k = 0, \dots, n-1),$$

the roots λ_{0k} $(k = 1, \dots, n)$ of the equation

$$\lambda^n = \sum_{k=0}^{n-1} p_{0k} \lambda^k$$

are distinct, and the system of functions $\operatorname{Re} \lambda_k(t)$ $(k = 1, \dots, n)$, where each $\lambda_k(t)$ is the root of the equation

$$\lambda^n = \sum_{k=0}^{n-1} p_k(t) \lambda^k$$

tending to λ_{0k} as $t \to +\infty$, satisfies the Levinson condition and the inequality

$$\sup\left\{\operatorname{Re} \int_a^t [\lambda_1(\tau) - \lambda_k(\tau)] \, d\tau : t \ge a\right\} < +\infty \qquad (k = 1, \dots, n).$$

Moreover, let the Cauchy problem for the equation (7.1) be uniquely solvable with any initial data and

$$|q(t, \nu(t)x_1, \dots, \nu(t)x_n)| \le g\left(t, \sum_{k=1}^n |x_k|\right) \sum_{k=1}^n |x_k|$$

on the set $[a, +\infty[\times \mathbf{R}^n$, where

$$\nu(t) = \exp\left(\operatorname{Re} \int_a^t \lambda_n(\tau) \, d\tau\right), \tag{8.56}$$

the function $g \in K_{\text{loc}}([a, +\infty[\times \mathbf{R}_+)$ does not decrease with respect to the second variable and

$$\int_a^{+\infty} \frac{g(t, \rho)}{\nu(t)} \, dt < +\infty \qquad \text{for } \rho > 0.$$

---

[2] In asymptotic formulas of the type (8.55), $\alpha_1, \dots, \alpha_n$ always denote arbitrary, in general complex, constants which are not all zero.

Then the family of solutions of the equation (7.1) having the asymptotic representations

$$u^{(k-1)}(t) = \sum_{j=1}^{n} \alpha_j \exp\left(\int_a^t \lambda_j(\tau)\,d\tau\right)(\lambda_{0j}^{k-1} + o(1)) \qquad (k = 1, \ldots, n) \qquad (8.57)$$

is stable.

COROLLARY 8.11. Suppose $p_k \in \tilde{C}_{\mathrm{loc}}([a, +\infty[)$,

$$\int_a^{+\infty} |(t^{n-k}p_k(t))'|\,dt < +\infty, \qquad \lim_{t \to +\infty} t^{n-k}p_k(t) = 0 \qquad (k = 0, \ldots, n-1),$$

and each $\lambda_k(t)$ $(k = 1, \ldots, n)$ is the root of the equation

$$\prod_{j=0}^{n-1}(\lambda - j) = \sum_{k=1}^{n-1} t^{n-k}p_k(t)\prod_{j=0}^{k-1}(\lambda - j) + t^n p_0(t)$$

tending to $k-1$ as $t \to +\infty$. Moreover, let the Cauchy problem for the equation (7.1) be uniquely solvable with any initial data, and let on $[a, +\infty[\times \mathbf{R}^n$ the inequality (8.52) hold, where

$$\nu_k(t) = t^{1-k}\exp\left(\int_a^t \frac{\lambda_n(\tau)}{\tau}\,d\tau\right) \qquad (k = 1, \ldots, n) \qquad (8.58)$$

and the function $g \in K_{\mathrm{loc}}([a, +\infty[\times \mathbf{R}_+)$ does not decrease with respect to the second variable and satisfies (8.54). Then the family of solutions of the equation (7.1) having the asymptotic representations

$$u^{(k-1)}(t) = \sum_{j=1}^{n} \alpha_j t^{1-k}\exp\left(\int_a^t \frac{\lambda_j(\tau)}{\tau}\,d\tau\right)(l_{kj} + o(1)) \qquad (k = 1, \ldots, n), \qquad (8.59)$$

with $l_{1j} = 1$, $l_{kj} = j(j-1)\ldots(j-k+2)$ $(k = 2, \ldots, n)$, is stable.

COROLLARY 8.12. Let the Cauchy problem for the equation (8.12) be uniquely solvable with any initial data and

$$|q(t, t^{n-1}x_1, t^{n-2}x_2, \ldots, x_n)| \le g\left(t, \sum_{k=1}^{n}|x_k|\right)\sum_{k=1}^{n}|x_k|$$

on $[a, +\infty[\times \mathbf{R}^n$, where the function $g \in K_{\mathrm{loc}}([a, +\infty[\times \mathbf{R}_+)$ does not decrease with respect to the second variable and

$$\int_a^{+\infty} g(t, \rho)\,dt < +\infty \qquad \text{for } \rho > 0.$$

Then the family of solutions of the equation (8.12) having the asymptotic representations

$$u^{(k-1)}(t) = \sum_{j=1}^{n} \alpha_j t^{j-k}(l_{kj} + o(1)) \qquad (k = 1, \ldots, n), \qquad (8.60)$$

with $l_{1j} = 1$, $l_{kj} = j(j-1)\ldots(j-k+2)$ $(k = 2, \ldots, n)$, is stable.

COROLLARY 8.13. Suppose the function p_0 is continuously differentiable and everywhere nonzero, $p_k \in \tilde{C}_{\text{loc}}([a, +\infty[)$ $(k = 1, \dots, n-1)$,

$$\int_a^{+\infty} \left| d(p_0'(t)|p_0(t)|^{-(n+1)/n}) \right| < +\infty,$$

$$\int_a^{+\infty} \left| (p_k(t)|p_0(t)|^{(k-n)/n})' \right| dt < +\infty \qquad (k = 1, \dots, n-1),$$

$$\lim_{t \to +\infty} p_0'(t)|p_0(t)|^{-(n+1)/n} = \lim_{t \to +\infty} p_k(t)|p_0(t)|^{(k-n)/n} = 0 \qquad (k = 1, \dots, n-1).$$

λ_{0k} $(k = 1, \dots, n)$, $\operatorname{Re}\lambda_{0n} = \max\{\operatorname{Re}\lambda_{01}, \dots, \operatorname{Re}\lambda_{0n}\}$, are the roots of the equation $\lambda^n = \gamma$ with $\gamma = \operatorname{sign} p_0(a)$ and each $\lambda_k(t)$, $k \in \{1, \dots, n\}$, is the root of the equation

$$\prod_{j=0}^{n-1} \left(\lambda + \frac{2j-n+1}{2n} \gamma p_0'(t)|p_0(t)|^{-(n+1)/n} \right)$$

$$= \sum_{k=1}^{n-1} p_k(t)|p_0(t)|^{(k-n)/n} \prod_{j=0}^{k-1} \left(\lambda + \frac{2j-n+1}{2n} \gamma p_0'(t)|p_0(t)|^{-(n+1)/n} \right) + \gamma$$

tending to λ_{0k} as $t \to +\infty$. Moreover, let the Cauchy problem for the equation (7.1) be uniquely solvable with any initial data, and let the inequality (8.52) hold on $[a, +\infty[\times \mathbf{R}^n$, where

$$\nu_k(t) = |p_0(t)|^{(2k-n-1)/2n} \exp\left(\operatorname{Re} \int_a^t \lambda_n(\tau)|p_0(\tau)|^{1/n} d\tau \right) \qquad (k = 1, \dots, n) \quad (8.61)$$

and the function $g \in K_{\text{loc}}([a, +\infty[\times \mathbf{R}_+)$ does not decrease with respect to the second variable and satisfies (8.54). Then the family of solutions of the equation (7.1) having the asymptotic representations

$$u^{(k-1)}(t) = \sum_{j=1}^n \alpha_j |p_0(t)|^{(2k-n-1)/2n} \exp\left(\int_a^t \lambda_j(\tau)|p_0(\tau)|^{1/n} d\tau \right) (\lambda_{0j}^{k-1} + o(1))$$

$$(k = 1, \dots, n) \tag{8.62}$$

is stable.

COROLLARY 8.14. Suppose $p \colon [a, +\infty[\to \mathbf{R}$ is continuously differentiable, $p(t) \neq 0$ for $t \geq a$,

$$\int_a^{+\infty} \left| d(p'(t)|p(t)|^{-(n+1)/n}) \right| < +\infty, \qquad \int_a^{+\infty} |p'(t)|^2 |p(t)|^{-(2n+1)/n} dt < +\infty$$

and λ_{0k} $(k = 1, \dots, n)$, $\operatorname{Re}\lambda_{0n} = \max\{\operatorname{Re}\lambda_{01}, \dots, \operatorname{Re}\lambda_{0n}\}$, are the roots of the equation $\lambda^n = \operatorname{sign} p(a)$. Moreover, let the Cauchy problem for the equation (8.35) be uniquely solvable with any initial data, and let the inequality (8.52) hold on $[a, +\infty[\times \mathbf{R}^n$, where

$$\nu_k(t) = |p(t)|^{(2k-n-1)/2n} \exp\left(\operatorname{Re}\lambda_{0n} \int_a^t |p(\tau)|^{1/n} d\tau \right) \qquad (k = 1, \dots, n) \tag{8.63}$$

and the function $g \in K_{\text{loc}}([a, +\infty[\times \mathbf{R}_+)$ does not decrease with respect to the second variable and satisfies (8.54). Then the family of solutions of the equation (8.35) having the asymptotic representations

$$u^{(k-1)}(t) = \sum_{j=1}^{n} \alpha_j |p(t)|^{(2k-n-1)/2n} \exp\left(\lambda_{0j} \int_a^t |p(\tau)|^{1/n} d\tau\right) (\lambda_{0j}^{k-1} + o(1))$$

$$(k = 1, \ldots, n) \tag{8.64}$$

is stable.

PROBLEM 8.3. Let the equation (7.2) have a fundamental system of solutions $v_k: [a, +\infty[\to]0, +\infty[$ $(k = 1, \ldots, n)$ such that the functions v_k/v_{k+1} $(k = 1, \ldots, n-1)$ do not decrease. Moreover, suppose $q \in K_{\text{loc}}([a, +\infty[\times \mathbf{R})$, the Cauchy problem for the equation (8.13) is uniquely solvable with any initial data, and

$$|q(t, x)| \le g(t, |x|)|x|$$

on $[a, +\infty[\times \mathbf{R}$, where the function $g \in K_{\text{loc}}([a, +\infty[\times \mathbf{R}_+)$ does not decrease with respect to the second variable,

$$\int_a^{+\infty} \sigma(t) g(t, \rho v_n(t)) \, dt < +\infty$$

and

$$\sigma(t) = \sum_{j=1}^{n} |W_j(v_1, \ldots, v_n)(t) v_j(t)| \exp\left(-\int_a^t p_{n-1}(\tau) \, d\tau\right). \tag{8.65}$$

Then the family of solutions of the equation (8.13) having the asymptotic representations

$$u(t) = \sum_{j=1}^{n} \alpha_j v_j(t)(1 + o(1)) \tag{8.66}$$

is stable.

Notes. Theorems 8.1–8.3 are due to I.T. Kiguradze and generalize some results in [176, 350]. Questions similar to those considered in this section were studied in [48, 50] for second order differential equations, in [51, 124, 255, 349, 351–353] for differential systems and in [39] for differential equations in a Banach space.

§9. $\mathcal{L}_h^0$, $\mathcal{L}_h^\infty$ and $\mathcal{L}_h$ Type Equations

9.1. Lemmas on integral inequalities. This subsection contains lemmas involving estimates of continuous functions satisfying one of the following integral inequalities:

$$x(t) \le c_0 + \left[\int_{t_0}^t \omega(\tau, x(\tau)) \, d\tau\right] \operatorname{sign}(t - t_0) \quad \text{for } t \in I, \tag{9.1}$$

$$x(t) \le c_0 + \left[\int_{t_0}^t g(\tau) \omega_0(x(\tau)) \, d\tau\right] \operatorname{sign}(t - t_0) \quad \text{for } t \in I, \tag{9.2}$$

and

$$x(t) \le c_0 + \left[\int_{t_0}^t g(\tau) x(\tau) \, d\tau\right] \operatorname{sign}(t - t_0) \quad \text{for } t \in I. \tag{9.3}$$

Everywhere we assume that I is an interval, $t_0 \in I$, $c_0 \in \mathbf{R}$, $\omega \in K_{\text{loc}}(I \times \mathbf{R})$, $g \in L_{\text{loc}}(I)$, and $\omega_0 \colon \mathbf{R}_+ \to \mathbf{R}_+$ is a continuous function.

LEMMA 9.1. Let ω be nondecreasing with respect to the second variable and let the Cauchy problem

$$\frac{dy}{dt} = \omega(t, y)\operatorname{sign}(t - t_0), \qquad y(t_0) = c_0 \tag{9.4}$$

have an upper solution y defined on the whole of I. Then every continuous function $x \colon I \to \mathbf{R}$ satisfying the integral inequality (9.1) is subject to the estimate

$$x(t) \le y(t) \qquad \text{for } t \in I. \tag{9.5}$$

PROOF. Set

$$z(t) = c_0 + \left[\int_{t_0}^{t} \omega(\tau, x(\tau))\, d\tau \right]\operatorname{sign}(t - t_0)$$

and

$$\bar{\omega}(t, s) = \begin{cases} \omega(t, z(t)) & \text{for } s \le z(t) \\ \omega(t, s) & \text{for } s > z(t). \end{cases} \tag{9.6}$$

Then, by (9.1),

$$x(t) \le z(t) \qquad \text{for } t \in I \tag{9.7}$$

and

$$z'(t)\operatorname{sign}(t - t_0) = \omega(t, x(t)) \le \omega(t, z(t)) \qquad \text{for } t \in I.$$

Since the upper solution of the problem (9.4) is defined on the whole of I, the problem

$$\frac{dv}{dt} = \bar{\omega}(t, v)\operatorname{sign}(t - t_0), \qquad v(t_0) = c_0$$

also has a solution v defined on I. Besides,

$$z(t_0) = v(t_0), \qquad [z'(t) - v'(t)]\operatorname{sign}(t - t_0) \le 0 \qquad \text{for } t \in I.$$

Hence

$$z(t) \le v(t) \qquad \text{for } t \in I. \tag{9.8}$$

From (9.6) and (9.8) we easily conclude that v is also a solution of the problem (9.4). So, $v(t) \le y(t)$ for $t \in I$. According to (9.7) and (9.8) this yields the estimate (9.5). $\square$

The following assertions are immediate consequences of the previous lemma.

LEMMA 9.2. Let $c_0 \ge 0$, $g(t) \ge 0$ for $t \in I$, $\omega_0(s) > 0$ for $s \ge c_0$, and

$$\int_{c_0}^{+\infty} \frac{ds}{\omega_0(s)} = +\infty.$$

Then every continuous function $x \colon I \to \mathbf{R}_+$ satisfying the integral inequality (9.2) is subject to the estimate

$$x(t) \le \Omega^{-1}\left(\left| \int_{t_0}^{t} g(\tau)\, d\tau \right| \right) \qquad \text{for } t \in I,$$

where the function Ω^{-1} is the inverse of

$$\Omega(s) = \int_{c_0}^s \frac{d\tau}{\omega_0(\tau)}.$$

LEMMA 9.3. Let $c_0 \geq 0$ and $g(t) \geq 0$ for $t \in I$. Then every continuous function $x: I \to \mathbf{R}_+$ satisfying the integral inequality (9.3) is subject to the estimate

$$x(t) \leq c_0 \exp\left(\left|\int_{t_0}^t g(\tau)\,d\tau\right|\right) \qquad \text{for } t \in I.$$

9.2. $\mathcal{L}_h^0$ type equations.

THEOREM 9.1. Let the equation (7.2) have the Levinson property with respect to a vector function h and

$$|q(t, \sigma_{1,h}(v_0)(t)x_1, \ldots, \sigma_{n,h}(v_0)(t)x_n)| \leq g(t) \sum_{k=1}^n |x_k| \qquad \text{for } t \geq a, \sum_{k=1}^n |x_k| \leq r, \quad (9.9)$$

where $r > 0$, v_0 is an h-maximal real-valued solution of the equation (7.2), $g \in L_{\text{loc}}([a, +\infty[)$, and

$$\int_a^{+\infty} \frac{g(t)}{\sigma_{n,h}(v_0)(t)}\,dt < +\infty. \qquad (9.10)$$

Then (7.1) is an $\mathcal{L}_h^0$ type equation.

PROOF. Let c_j $(j = 1, \ldots, n)$ be the functions introduced in the proof of Theorem 8.3. Then, as shown above, there exists a constant $\rho_0 \geq 1$ for which the inequalities (8.45) hold.

Set

$$\delta = \frac{r\sigma_h(v_0)(a)}{n\rho_0\sigma_h(v_0)(1)} \exp\left(-\int_a^{+\infty} g(\tau)\,d\tau\right)$$

and consider a solution u of the equation (7.1) under the condition

$$0 < \sum_{k=1}^n |u^{(k-1)}(a)| < \delta.$$

Clearly,

$$\frac{\sigma_h(u)(a)}{\sigma_h(v_0)(a)} \leq \frac{\sigma_h(1)(a)}{\sigma_h(v_0)(a)} \sum_{k=1}^n |u^{(k-1)}(a)| < \frac{\sigma_h(1)(a)}{\sigma_h(v_0)(a)}\delta, \qquad (9.11)$$

and so

$$\frac{\sigma_h(u)(a)}{\sigma_h(v_0)(a)} < r.$$

On the other hand, by (9.9),

$$\sum_{k=1}^n |u^{(k-1)}(t)| > 0 \qquad \text{for } t \geq a.$$

We have to prove that u is an $\mathcal{L}_h$ type solution. First of all we show that

$$\frac{\sigma_h(u)(t)}{\sigma_h(v_0)(t)} < r \qquad \text{for } t \geq a. \qquad (9.12)$$

Assume the contrary. Then for a certain $b \in]a, +\infty[$,

$$\frac{\sigma_h(u)(t)}{\sigma_h(v_0)(t)} < r \qquad \text{for } a \le t < b \tag{9.13}$$

and

$$\frac{\sigma_h(u)(b)}{\sigma_h(v_0)(b)} = r. \tag{9.14}$$

In view of (8.45), (9.9) and (9.13), the equality

$$u(t) = \sum_{k=1}^{n} c_k(t,a) u^{(k-1)}(a)$$

$$+ \int_a^t c_n(t,\tau) q(\tau, u(\tau), \dots, u^{(n-1)}(\tau)) \, d\tau \qquad \text{for } a \le t \le b$$

yields

$$\frac{\sigma_h(u)(t)}{\sigma_h(v_0)(t)} \le \frac{n\rho_0 \sigma_h(u)(a)}{\sigma_h(v_0)(a)} + n\rho_0 \int_a^t \frac{g(\tau)}{\sigma_{n,h}(v_0)(\tau)} \frac{\sigma_h(u)(\tau)}{\sigma_h(v_0)(\tau)} \, d\tau \qquad \text{for } a \le t \le b.$$

Thus, Lemma 9.3 and the condition (9.11) imply

$$\frac{\sigma_h(u)(b)}{\sigma_h(v_0)(b)} \le \frac{n\rho_0 \sigma_h(u)(a)}{\sigma_h(v_0)(a)} \exp\left(n\rho_0 \int_a^b \frac{g(\tau)}{\sigma_{n,h}(v_0)(\tau)} \, d\tau \right) < r,$$

which contradicts (9.14). So the estimate (9.12) is true.

According to (9.9), (9.10) and (9.12), u satisfies the differential inequality (8.36), with

$$g_k(t) = \frac{g(t)}{h_{k+1}(t) \sigma_h(v_0)(t)} \qquad (k = 0, \dots, n-1),$$

and the conditions (8.37) hold. Therefore, by Lemma 8.1, u is an $\mathcal{L}_h$ type solution. $\square$

In view of Theorems 6.1–6.4 and Remark 7.1, Theorem 9.1 implies the following assertions.

COROLLARY 9.1. Let the functions p_k $(k = 1, \dots, n)$ satisfy the hypotheses of Corollary 8.9 and

$$|q(t, \nu_1(t)x_1, \dots, \nu_n(t)x_n)| \le g(t) \sum_{k=1}^{n} |x_k| \qquad \text{for } t \ge a, \ \sum_{k=1}^{n} |x_k| \le r, \tag{9.15}$$

where $r > 0, \nu_k$ $(k = 1, \dots, n)$ are the functions defined by the equalities (8.53), $g \in L_{\text{loc}}([a, +\infty[)$, and

$$\int_a^{+\infty} \frac{g(t)}{\nu_n(t)} \, dt < +\infty. \tag{9.16}$$

Then every nonzero solution u of the equation (7.1) with sufficiently small initial data is of the form (8.55).

COROLLARY 9.2. Let the functions p_k $(k = 1, \dots, n)$ satisfy the hypotheses of Corollary 8.10 (Corollary 8.11) and let the inequality (9.15) be fulfilled, where $r > 0$, $\nu_1(t) = \dots = \nu_n(t) = \exp\left(\int_a^t \lambda_n(\tau) \, d\tau \right)$ $(\nu_k$ $(k = 1, \dots, n)$ are defined by (8.58)),

$g \in L_{\mathrm{loc}}([a, +\infty[)$ and (9.16) holds. Then every nonzero solution u of the equation (7.1) with sufficiently small initial data is of the form (8.57) (of the form (8.59)).

COROLLARY 9.3. Let

$$|q(t, t^{n-1}x_1, t^{n-2}x_2, \dots, x_n)| \leq g(t) \sum_{k=1}^{n} |x_k| \quad \text{for } t \geq a, \ \sum_{k=1}^{n} |x_k| \leq r,$$

where $r > 0$, $g \in L_{\mathrm{loc}}([a, +\infty[)$ and

$$\int_a^{+\infty} g(t)\, dt < +\infty.$$

Then every nonzero solution u of the equation (8.12) with sufficiently small initial data is of the form (8.60).

COROLLARY 9.4. Let the functions p_k ($k = 1, \dots, n$) (the function p) satisfy the hypotheses of Corollary 8.13 (Corollary 8.14), and let the inequality (9.15) be fulfilled, where $r > 0$, ν_k ($k = 1, \dots, n$) are defined by (8.61) (by (8.63)), $g \in L_{\mathrm{loc}}([a, +\infty[)$, and (9.16) holds. Then every nonzero solution u of the equation (7.1) (the equation (8.35)) with sufficiently small initial data is of the form (8.62) (of the form (8.64)).

PROBLEM 9.1 Let the equation (7.2) have a fundamental system of solutions v_k: $[a, +\infty[\to]0, +\infty[$ ($k = 1, \dots, n$) such that the functions v_k/v_{k+1} ($k = 1, \dots, n-1$) do not decrease, and let $q \in K_{\mathrm{loc}}([a, +\infty[\times \mathbf{R})$ satisfy the inequality

$$|q(t, x)| \leq g(t)|x| \quad \text{for } t \geq a, \ |x| \leq rv_n(t),$$

where $g \in L_{\mathrm{loc}}([a, +\infty[)$,

$$\int_a^{+\infty} \sigma(t)g(t)\, dt < +\infty$$

and σ is defined by (8.65). Then every nonzero solution u of the equation (8.13) with sufficiently small initial data is of the form (8.66).

9.3. $\mathcal{L}_h^\infty$ type equations.

THEOREM 9.2. Let the equation (7.2) have the Levinson property with respect to a vector function h and

$$|q(t, \sigma_{1,h}(v_0)(t)x_1, \dots, \sigma_{n,h}(v_0)(t)x_n)| \leq g(t) \sum_{k=1}^{n} |x_k| \quad \text{for } t \geq a, \ \sum_{k=1}^{n} |x_k| \geq r,$$
$$(9.17)$$

where v_0 is an h-minimal solution of the equation (7.2), $g \in L_{\mathrm{loc}}([a, +\infty[)$ and the condition (9.10) holds. Then (7.1) is an $\mathcal{L}_h^\infty$ type equation.

PROOF. Suppose c_j ($j = 1, \dots, n$) are the functions introduced in the proof of Theorem 8.3. Since v_0 is h-minimal, we can easily establish the existence of a constant $\rho_0 > 1$ such that

$$\left| \frac{\partial^{k-1} c_j(t, \tau)}{\partial t^{k-1}} \right| \leq \rho_0 \frac{\sigma_{k,h}(v_0)(t)}{\sigma_{j,h}(v_0)(\tau)} \quad \text{for } \tau \geq t \geq a \quad (j, k = 1, \dots, n). \qquad (9.18)$$

Set

$$r_1 = n\rho_0 r \exp\left(\int_a^{+\infty} \frac{g(\tau)}{\sigma_{n,h}(v_0)(\tau)} \, d\tau\right), \qquad \delta = r_1 \sigma_h(v_0)(a) \sum_{k=1}^{n} h_k(a),$$

and consider a solution u of the equation (7.1) satisfying the condition

$$\sum_{k=1}^{n} |u^{(k-1)}(a)| > \delta. \tag{9.19}$$

Then

$$\frac{\sigma_h(u)(a)}{\sigma_h(v_0)(a)} \geq \frac{\sum_{k=1}^{n} |u^{(k-1)}(a)|}{\sigma_h(v_0)(a) \sum_{k=1}^{n} h_k(a)} > r_1. \tag{9.20}$$

By (9.17), u is defined on the whole of $[a, +\infty[$. We have to show that it is an $\mathcal{L}_h$ type solution.

First of all we prove that

$$\frac{\sigma_h(u)(t)}{\sigma_h(v_0)(t)} > r \qquad \text{for } t \geq a. \tag{9.21}$$

Assume the contrary. Then for a certain $b \in \,]a, +\infty[$,

$$\frac{\sigma_h(u)(t)}{\sigma_h(v_0)(t)} > r \qquad \text{for } a \leq t < b, \tag{9.22}$$

and the equality (9.14) holds.

From the equality

$$u(t) = \sum_{k=1}^{n} c_k(t, b) u^{(k-1)}(b) - \int_t^b c_n(t, \tau) q(\tau, u(\tau), \dots, u^{(n-1)}(\tau)) \, d\tau \qquad \text{for } a \leq t \leq b,$$

applying (9.17), (9.18) and (9.22), we get

$$\frac{\sigma_h(u)(t)}{\sigma_h(v_0)(t)} \leq \frac{n\rho_0 \sigma_h(u)(b)}{\sigma_h(v_0)(b)} + n\rho_0 \int_t^b \frac{g(\tau)}{\sigma_{n,h}(v_0)(\tau)} \frac{\sigma_h(u)(\tau)}{\sigma_h(v_0)(\tau)} \, d\tau \qquad \text{for } a \leq t \leq b.$$

By Lemma 9.3 and the condition (9.14), this yields

$$\frac{\sigma_h(u)(a)}{\sigma_h(v_0)(a)} \leq n\rho_0 r \exp\left(n\rho_0 \int_a^b \frac{g(\tau)}{\sigma_{n,h}(v_0)(\tau)} \, d\tau\right) \leq r_1,$$

which contradicts (9.20). Hence (9.21) is proved.

According to (9.10), (9.17) and (9.21), u satisfies the differential inequality (8.36), with

$$g_k(t) = \frac{g(t)}{h_{k+1}(t) \sigma_h(v_0)(t)} \qquad (k = 0, \dots, n-1),$$

and the conditions (8.37) hold. Therefore, by Lemma 8.1, u is an $\mathcal{L}_h$ type solution. $\square$

COROLLARY 9.5. Let the functions p_k $(k = 1, \dots, n)$ satisfy the hypotheses of Corollary 8.9,

$$\sup\left\{ \operatorname{Re} \int_a^t [\lambda_1(\tau) - \lambda_k(\tau)] \psi(\tau) \, d\tau : t \geq a \right\} < +\infty \qquad (k = 1, \dots, n),$$

and

$$|q(t, \mu_1(t)x_1, \ldots, \mu_n(t)x_n)| \leq g(t) \sum_{k=1}^{n} |x_k| \quad \text{for } t \geq a, \ \sum_{k=1}^{n} |x_k| \geq r, \qquad (9.23)$$

where $r > 0$,

$$\mu_k(t) = \varphi(t)\psi^{k-1}(t) \exp\left(\text{Re} \int_a^t \lambda_1(\tau)\,d\tau\right) \qquad (k = 1, \ldots, n),$$

$g \in L_{\text{loc}}([a, +\infty[)$, and

$$\int_a^{+\infty} \frac{g(t)}{\mu_n(t)}\,dt < +\infty. \qquad (9.24)$$

Then there exists $\delta > 0$ such that every nonzero solution u of the equation (7.1) under the condition (9.19) is of the form (8.55).

COROLLARY 9.6. Let the functions p_k $(k = 1, \ldots, n)$ satisfy the hypotheses of Corollary 8.10 (Corollary 8.11),

$$\sup\left\{\text{Re} \int_a^t [\lambda_1(\tau) - \lambda_k(\tau)]\,d\tau : t \geq a\right\} < +\infty \qquad (k = 1, \ldots, n),$$

and let the inequality (9.23) be fulfilled, where $r > 0$,

$$\mu_k(t) = \exp\left(\int_a^t \lambda_1(\tau)\,d\tau\right) \quad \left(\mu_k(t) = t^{1-k}\exp\left(\int_a^t \frac{\lambda_1(\tau)}{\tau}\,d\tau\right)\right)$$

$$(k = 1, \ldots, n),$$

$g \in L_{\text{loc}}([a, +\infty[)$, and (9.24) holds. Then there exists $\delta > 0$ such that every nonzero solution u of the equation (7.1) under the condition (9.19) is of the form (8.57) (of the form (8.59)).

COROLLARY 9.7. Let

$$|q(t, x_1, t^{-1}x_2, \ldots, t^{1-n}x_n)| \leq g(t) \sum_{k=1}^{n} |x_k| \quad \text{for } t \geq a, \ \sum_{k=1}^{n} |x_k| \geq r,$$

where $r > 0$, $g \in L_{\text{loc}}([a, +\infty[)$, and

$$\int_a^{+\infty} t^{n-1}g(t)\,dt < +\infty.$$

Then there exists $\delta > 0$ such that every nonzero solution u of the equation (8.12) under the condition (9.19) is of the form (8.60).

COROLLARY 9.8. Let the functions p_k $(k = 1, \ldots, n)$ (the function p) satisfy the hypotheses of Corollary 8.13 (Corollary 8.14), $\text{Re}\,\lambda_{01} = \min\{\text{Re}\,\lambda_{01}, \ldots, \text{Re}\,\lambda_{0n}\}$, and let the inequality (9.23) be fulfilled, where $r > 0$,

$$\mu_k(t) = |p_0(t)|^{(2k-n-1)/2n} \exp\left(\text{Re} \int_a^t \lambda_1(\tau)|p_0(\tau)|^{1/n}\,d\tau\right) \qquad (k = 1, \ldots, n)$$

$$\left(\mu_k(t) = |p(t)|^{(2k-n-1)/2n} \exp\left(\text{Re}\,\lambda_{01} \int_a^t |p(\tau)|^{1/n}\,d\tau\right) \qquad (k = 1, \ldots, n)\right),$$

$g \in L_{loc}([a, +\infty[)$, and (9.24) holds. Then there exists $\delta > 0$ such that every nonzero solution u of the equation (7.1) (the equation (8.34)) under the condition (9.19) is of the form (8.62) (of the form (8.64)).

PROBLEM 9.2. Let the equation (7.2) have a fundamental system of solutions $v_k: [a, +\infty[\to]0, +\infty[$ $(k = 1, \dots, n)$ such that the functions v_k/v_{k+1} $(k = 1, \dots, n-1)$ do not decrease and let $q \in K_{loc}([a, +\infty[\times \mathbf{R})$ satisfy the inequality

$$|q(t, x)| \leq g(t)|x| \qquad \text{for } t \geq a, \ |x| \geq rv_1(t),$$

where $g \in L_{loc}([a, +\infty[)$,

$$\int_a^{+\infty} \sigma(t)|g(t)| \, dt < +\infty,$$

and σ is defined by (8.65). Then there exists $\delta > 0$ such that every nonzero solution u of the equation (8.13) under the condition (9.19) is of the form (8.66).

9.4. $\mathcal{L}_h$ type equations.

THEOREM 9.3. Let the equation (7.2) have the Levinson property with respect to a vector function h and

$$|q(t, \sigma_{1,h}(v_0)(t)x_1, \dots, \sigma_{n,h}(v_0)(t)x_n)| \leq g(t)\omega\left(\sum_{k=1}^n |x_k|\right) \qquad (9.25)$$

on the set $[a, +\infty[\times \mathbf{R}^n$, where v_0 is an h-maximal real-valued solution of the equation (7.2), $g \in L_{loc}([a, +\infty[)$, the condition (9.10) holds, and $\omega: \mathbf{R}_+ \to \mathbf{R}_+$ is a continuous nondecreasing function such that

$$\limsup_{x \to 0} \frac{\omega(x)}{x} < +\infty, \qquad \omega(x) > 0, \qquad \int_x^{+\infty} \frac{ds}{\omega(s)} = +\infty \qquad \text{for } x > 0. \quad (9.26)$$

Then (7.1) is an $\mathcal{L}_h$ type equation.

PROOF. Suppose c_j $(j = 1, \dots, n)$ are the functions introduced in the proof of Theorem 8.3 and $\rho_0 > 1$ is a constant for which the inequalities (8.45) are fulfilled.

Consider a nonzero solution u of the equation (7.1) defined on an interval $[t_0, t^*[$. Then by (8.45) the equality

$$u(t) = \sum_{j=1}^n c_j(t, t_0)u^{(j-1)}(t_0) + \int_{t_0}^t c_n(t, \tau)q(\tau, u(\tau), \dots, u^{(n-1)}(\tau)) \, d\tau \qquad \text{for } t_0 \leq t \leq t^*$$

yields

$$\frac{\sigma(u)(t)}{\sigma(v_0)(t)} \leq r_0 + n\rho_0 \int_{t_0}^t \frac{g(\tau)}{\sigma_{n,h}(v_0)(\tau)} \omega\left(\frac{\sigma(u)(\tau)}{\sigma(v_0)(\tau)}\right) d\tau \qquad \text{for } t_0 \leq t \leq t^*,$$

where

$$r_0 = n\rho_0 \frac{\sigma(u)(t_0)}{\sigma(v_0)(t_0)}.$$

Hence, according to Lemma 9.2,

$$\frac{\sigma(u)(t)}{\sigma(v_0)(t)} \leq \Omega^{-1}\left(n\rho_0 \int_{t_0}^t \frac{g(\tau)}{\sigma_{n,h}(v_0)(\tau)} d\tau\right) \qquad \text{for } t_0 \leq t \leq t^*,$$

where the function Ω^{-1} is the inverse of

$$\Omega(x) = \int_{r_0}^{x} \frac{ds}{\omega(s)}.$$

Since u is maximally continued to the right, the last estimate implies that $t^* = +\infty$. On the other hand, in view of (9.10) we have

$$\frac{\sigma(u)(t)}{\sigma(v_0)(t)} \le r_1 \qquad \text{for } t \ge t_0, \tag{9.27}$$

with

$$r_1 = \Omega^{-1}\left(n\rho_0 \int_{t_0}^{+\infty} \frac{g(\tau)}{\sigma_{n,h}(v_0)(\tau)} d\tau\right).$$

The first condition in (9.26) yields

$$r_2 = \sup\left\{\frac{\omega(x)}{x} : 0 < x < r_1\right\} < +\infty.$$

So, applying (9.10), (9.25) and (9.27), we conclude that u is a solution of the differential inequality (8.36), where

$$g_k(t) = \frac{r_2 g(t)}{h_{k+1}(t)\sigma_h(v_0)(t)} \qquad (k = 0,\dots,n-1)$$

satisfy the conditions (8.37). By Lemma 8.1, u is an $\mathcal{L}_h$ type solution. $\square$

COROLLARY 9.9. Let the functions p_k ($k = 1,\dots,n$) satisfy the hypotheses of Corollary 8.9 and

$$|q(t, \nu_1(t)x_1, \dots, \nu_n(t)x_n)| \le g(t)\omega\left(\sum_{k=1}^{n} |x_k|\right) \tag{9.28}$$

on the set $[a, +\infty[\times \mathbf{R}^n$, where ν_k ($k = 1,\dots,n$) are defined by (8.53), $g \in L_{\text{loc}}([a, +\infty[)$, $\omega \colon \mathbf{R}_+ \to \mathbf{R}_+$ is a continuous nondecreasing function, and the conditions (9.16) and (9.26) hold. Then every nonzero solution u of the equation (7.1) is of the form (8.55).

COROLLARY 9.10. Let the functions p_k ($k = 1,\dots,n$) satisfy the hypotheses of Corollary 8.10 (Corollary 8.11), and let on the set $[a, +\infty[\times \mathbf{R}^n$ the inequality (9.28) be fulfilled, where $\nu_1(t) = \dots = \nu_n(t) = \exp\left(\int_a^t \lambda_n(\tau) d\tau\right)$ (ν_k ($k = 1,\dots,n$) are defined by (8.58)), $g \in L_{\text{loc}}([a, +\infty[)$, $\omega \colon \mathbf{R}_+ \to \mathbf{R}_+$ is a continuous nondecreasing function, and the conditions (9.16) and (9.26) hold. Then every nonzero solution u of the equation (7.1) is of the form (8.57) (of the form (8.59)).

COROLLARY 9.11. Let

$$|q(t, t^{n-1}x_1, t^{n-2}x_2, \dots, x_n)| \le g(t)\omega\left(\sum_{k=1}^{n} |x_k|\right)$$

on the set $[a, +\infty[\times \mathbf{R}^n$, where $g \in L_{\text{loc}}([a, +\infty[)$,

$$\int_a^{+\infty} g(t)\, dt < +\infty,$$

and $\omega: \mathbf{R}_+ \to \mathbf{R}_+$ is a continuous nondecreasing function satisfying (9.26). Then every nonzero solution u of the equation (8.12) is of the form (8.60).

COROLLARY 9.12. Let the functions p_k $(k = 1, \ldots, n)$ (the function p) satisfy the hypotheses of Corollary 8.13 (Corollary 8.14), and let on the set $[a, +\infty[\times \mathbf{R}^n$ the inequality (9.28) be fulfilled, where ν_k $(k = 1, \ldots, n)$ are defined by (8.61) (by (8.63)), $g \in L_{\text{loc}}([a, +\infty[)$, $\omega: \mathbf{R}_+ \to \mathbf{R}_+$ is a continuous nondecreasing function, and the conditions (9.16) and (9.26) hold. Then every nonzero solution u of the equation (7.1) (the equation (8.35)) is of the form (8.62) (of the form (8.64)).

PROBLEM 9.3. Let the equation (7.2) have a fundamental system of solutions $v_k: [a, +\infty[\to]0, +\infty[$ $(k = 1, \ldots, n)$ such that the functions v_k/v_{k+1} $(k = 1, \ldots, n-1)$ do not decrease and let $q \in K_{\text{loc}}([a, +\infty[\times \mathbf{R})$ satisfy the inequality

$$|q(t, v_n(t)x)| \le g(t)\omega(|x|)$$

on the set $[a, +\infty[\times \mathbf{R}^n$, where $g \in L_{\text{loc}}([a, +\infty[)$,

$$\int_a^{+\infty} \frac{\sigma(t)}{v_n(t)} g(t)\, dt < +\infty,$$

σ is defined by (8.65), $\omega: \mathbf{R}_+ \to \mathbf{R}_+$ is a continuous nondecreasing function, and the conditions (9.26) hold. Then every nonzero solution u of the equation (8.13) is of the form (8.66).

PROBLEM 9.4. Let $p \in L_{\text{loc}}([1, +\infty[)$ and

$$\int_1^{+\infty} |p(t)| t^{n-1} \ln t\, dt < +\infty.$$

Then every nonzero solution u of the equation

$$u^{(n)} = p(t) u \ln(1 + |u|)$$

is of the form (8.60).

Notes. Lemma 9.1 goes back to G. Peano, O. Perron (see [127]) and S.A. Chaplygin [87]. Lemma 9.2 is due to I. Bihari [38] and Lemma 9.3 to T. Gronwall and R. Bellman [27].

Theorems 9.1 and 9.3 are due to I.T. Kiguradze, Theorem 9.2 to T.A. Chanturia, while Corollaries 9.3, 9.4, 9.11, and 9.12 are due to I.A. Toroshelidze [350].

For Problem 9.4 see [354].

Some questions similar to those considered in this section were studied in [51, 55, 122, 349, 352, 353] for differential systems and in [116, 362] for functional-differential equations.

CHAPTER III

GENERAL NONLINEAR DIFFERENTIAL EQUATIONS

§10. Theorems on the Classification of Equations with Respect to Their Oscillation Properties

In this section we study the differential equation

$$u^{(n)} = f(t, u, \dots, u^{(n-1)}), \tag{10.1}$$

with $a > 0$, $f \in K_{\text{loc}}([a, +\infty[\times \mathbf{R}^n)$, and either

$$f(t, x_1, \dots, x_n)x_1 \leq 0 \tag{10.2}$$

or

$$f(t, x_1, \dots, x_n)x_1 \geq 0 \tag{10.3}$$

on the set $[a, +\infty[\times \mathbf{R}^n$.

DEFINITION 10.1. A solution u of the equation (10.1) is said to be *proper* if it is defined in a neighborhood of $+\infty$ and satisfies the condition

$$\sup\{|u(s)| : s \geq t\} > 0$$

for any t in this neighborhood.

DEFINITION 10.2. A proper solution of the equation (10.1) is said to be *oscillatory* (*nonoscillatory*) if it has (does not have) a sequence of zeros converging to $+\infty$.

DEFINITION 10.3. The equation (10.1) has *property A* if any proper solution of this equation in case n even is oscillatory and in case n odd either is oscillatory or satisfies the condition

$$|u^{(i)}(t)| \downarrow 0 \quad \text{as } t \uparrow +\infty \quad (i = 0, \dots, n-1). \tag{10.4}$$

DEFINITION 10.4. The equation (10.1) has *property B* if any proper solution of this equation in case n even either is oscillatory, satisfies the condition (10.4), or satisfies the condition

$$|u^{(i)}(t)| \uparrow +\infty \quad \text{as } t \uparrow +\infty \quad (i = 0, \dots, n-1) \tag{10.5}$$

and in case n odd either is oscillatory or satisfies the condition (10.5).

DEFINITION 10.5. Let $k \in \{1, \dots, n-1\}$. The equation (10.1) has *property A_k* (*property B_k*) if any proper solution of this equation either is oscillatory or satisfies the condition

$$|u^{(i)}(t)| \downarrow 0 \qquad \text{as } t \uparrow +\infty \qquad (i = k, \dots, n-1) \tag{10.6}$$

(either is oscillatory, satisfies the condition (10.5), or satisfies the condition (10.6)).

DEFINITION 10.6. Let $k \in \{1, \dots, n-1\}$. The equation (10.1) has *property A_k^** (*property B_k^**) if any proper solution of this equation in case n even either is oscillatory or satisfies the condition

$$\lim_{t \to +\infty} |u^{(n-k)}(t)| > 0 \tag{10.7}$$

(either is oscillatory, satisfies the condition (10.4), or satisfies the condition (10.7)) and in case n odd either is oscillatory, satisfies the condition (10.4), or satisfies the condition (10.7) (either is oscillatory or satisfies the condition (10.7)).

Below we establish necessary and sufficient conditions for the equation (10.1) to have properties A, B, A_k, B_k, A_k^*, or B_k^*.

10.1. Comparison theorem.

Together with (10.1) we consider the differential equation

$$u^{(n)} = g(t, u, \dots, u^{(n-1)}), \tag{10.8}$$

where $g \in K_{\text{loc}}([a, +\infty[\times \mathbf{R}^n)$.

THEOREM 10.1. Let the equation (10.8) have property A (property B) and

$$(-1)^m f(t, x_1, \dots, x_n) \operatorname{sign} x_1 \geq (-1)^m g(t, y_1, \dots, y_n) \operatorname{sign} y_1 \geq 0$$

for $t \geq a$, $x_k y_k \geq 0$, $|x_k| \geq |y_k|$ $\quad (k = 1, \dots, n-1)$, $\quad x_n = y_n$, $\tag{10.9}$

where $m = 1$ ($m = 2$). Then the equation (10.1) also has property A (property B).

Before starting the proof of this theorem, we derive two auxiliary assertions. Note that below $C^{n-1}([\alpha, \beta])$, $-\infty < \alpha < \beta < +\infty$, is the Banach space of $n-1$ times continuously differentiable functions $u: [\alpha, \beta] \to \mathbf{R}$ with the norm

$$\|u\| = \max\{|u^{(k)}(t)| : \alpha \leq t \leq \beta, \, k = 0, \dots, n-1\}.$$

LEMMA 10.1. Let $\tilde{g} \in K([\alpha, \beta] \times \mathbf{R}^n)$,

$$|\tilde{g}(t, x_1, \dots, x_n)| \leq g^*(t) \tag{10.10}$$

on $[\alpha, \beta] \times \mathbf{R}^n$, where $g^* \in L([\alpha, \beta])$, and let $\varphi_i : C^{n-1}([\alpha, \beta]) \to \mathbf{R}$ $(i = 1, \dots, n)$ be continuous linear functionals such that the boundary value problem

$$v^{(n)} = 0, \qquad \varphi_i(v) = 0 \qquad (i = 1, \dots, n) \tag{10.11}$$

does not have a nonzero solution. Then for any $\gamma_i \in \mathbf{R}$ $(i = 1, \dots, n)$ the equation

$$u^{(n)} = \tilde{g}(t, u, \dots, u^{(n-1)}) \tag{10.12}$$

has at least one solution satisfying the boundary conditions

$$\varphi_i(u) = \gamma_i \qquad (i = 1, \dots, n). \tag{10.13}$$

PROOF. Set

$$v_j(t) = t^{j-1}, \qquad \varphi_{ij} = \varphi_i(v_j) \qquad (i,j = 1,\dots,n).$$

Then

$$\Delta = \begin{vmatrix} \phi_{11} & \cdots & \phi_{1n} \\ \cdot & \cdots & \cdot \\ \phi_{n1} & \cdots & \phi_{nn} \end{vmatrix} \neq 0,$$

because the problem (10.1) does not have a nonzero solution.

On the space $C^{n-1}([\alpha,\beta])$ we define the operators

$$h_0(u)(t) = \frac{1}{(n-1)!} \int_\alpha^t (t-\tau)^{n-1} \tilde{g}(\tau, u(\tau), \dots, u^{(n-1)}(\tau))\, d\tau$$

and

$$h(u)(t) = \sum_{j,k=1}^n \frac{\Delta_{jk}}{\Delta} [\alpha_j - \varphi_j(h_0(u))] t^{k-1} + h_0(u)(t), \qquad (10.14)$$

where Δ_{jk} is the cofactor of the element of the determinant Δ at the intersection of the j-th row and k-th column.

Taking into account (10.10), we easily conclude that the operator h is completely continuous and, for a sufficiently large $r > 0$, maps the ball $\{u \in C^{n-1}([\alpha,\beta]) : \|u\| \le r\}$ into itself. So, by the Schauder principle [168] there exists $u \in C^{n-1}([\alpha,\beta])$ such that $u(t) = h(u)(t)$ for $\alpha \le t \le \beta$. According to (10.14), u is a solution of the problem (10.12), (10.13). $\square$

LEMMA 10.2. Let $t_0 \in\,]a, +\infty[$ and let for every positive integer k the equation (10.8) have a solution u_k whose domain includes the interval $[t_0, t_0 + k]$. Moreover, suppose that whatever $t^* \in\,]t_0, +\infty[$,

$$\sum_{i=0}^{n-1} |u_k^{(i)}(t)| \le r(t^*) \qquad \text{for } t \in [t_0, t^*] \cap [t_0, t_0 + k] \qquad (k = 1, 2, \dots), \qquad (10.15)$$

with a certain constant $r(t^*)$. Then $(u_k)_{k=1}^{+\infty}$ contains a subsequence $(u_{k_j})_{j=1}^{+\infty}$ such that $(u_{k_j}^{(i)})_{j=1}^{+\infty}$ $(i = 0, \dots, n-1)$ converge uniformly on each finite interval in $[t_0, +\infty[$, and

$$u(t) = \lim_{j\to+\infty} u_{k_j}(t) \qquad \text{for } t \ge t_0 \qquad (10.16)$$

is a solution of the equation (10.8).

PROOF. Since $g \in K_{\mathrm{loc}}([a, +\infty[\times \mathbf{R}^n)$, the condition (10.5) implies that the sequences $(u_k^{(i)})_{k=1}^{+\infty}$ $(i = 0, \dots, n-1)$ are uniformly bounded and equicontinuous on each finite interval in $[t_0, +\infty[$. Hence by the Arzelá–Ascoli lemma there exists a sequence of positive integers $(k_j)_{j=1}^{+\infty}$ for which $(u_{k_j}^{(i)})_{j=1}^{+\infty}$ $(i = 0, \dots, n-1)$ converge uniformly on each of these intervals. Clearly, the function u defined by the equality (10.16) is a solution of the equation (10.8). $\square$

PROOF OF THEOREM 10.1. Assume, on the contrary, that the equation (10.1) does not have property A (property B). Then, in view of Lemma 1.1 and the inequality

(10.9), there exists a solution u_0 of the equation (10.1) defined on a certain interval $[t_0, +\infty[\subset [a, +\infty[$ and satisfying the conditions

$$u_0^{(i)}(t)u_0(t) > 0 \qquad (i = 0, \ldots, l-1),$$

$$(-1)^{i-l}u_0^{(i)}(t)u_0(t) \geq 0 \qquad (i = l, \ldots, n) \qquad \text{for } t \geq t_0, \tag{10.17}$$

$$r_0 = \inf\{|u_0(t)| : t \geq t_0\} > 0, \qquad r = \sup\{|u_0^{(n-1)}(t)| : t \geq t_0\} < +\infty, \tag{10.18}$$

where $l \in \{0, \ldots, n\}$ and the number $n + l + m$ is even. For the sake of being specific, suppose

$$u_0(t) > 0 \qquad \text{for } t \geq t_0. \tag{10.19}$$

Set

$$\chi_i(t, x) = \begin{cases} 0 & \text{for } x < 0 \\ x & \text{for } 0 \leq x \leq u_0^{(i-1)}(t) \\ u_0^{(i-1)}(t) & \text{for } x > u_0^{(i-1)}(t) \end{cases}$$

in case $i \in \{1, \ldots, l\}$, and

$$\chi_i(t, x) = \begin{cases} 0 & \text{for } (-1)^{i-1-l}x < 0 \\ x & \text{for } 0 \leq (-1)^{i-1-l}x \leq (-1)^{i-1-l}u_0^{(i-1)}(t) \\ u_0^{(i-1)}(t) & \text{for } (-1)^{i-1-l}x > (-1)^{i-1-l}u_0^{(i-1)}(t) \end{cases}$$

in case $i \in \{l+1, \ldots, n\}$. Consider the function

$$\tilde{g}(t, x_1, \ldots, x_n) = g(t, \chi_1(t, x_1), \ldots, \chi_n(t, x_n)). \tag{10.20}$$

According to (10.9) and (10.19), the inequalities (10.10),

$$(-1)^m \tilde{g}(t, x_1, \ldots, x_n) \geq 0 \tag{10.21}$$

and

$$(-1)^m \tilde{g}(t, x_1, \ldots, x_n) \leq (-1)^m u_0^{(n)}(t) \qquad \text{for } (-1)^{l-n+1}x_n > (-1)^{l-n+1}u_0^{(n-1)}(t) \tag{10.22}$$

hold on the set $[t_0, +\infty[\times \mathbf{R}^n$.

By Lemma 10.1, for any positive integer k the equation (10.12) has a solution u_k satisfying the boundary conditions[1]

$$u_k^{(i)}(t_0) = u_0^{(i)}(t_0) \qquad (i = 0, \ldots, l-1),$$
$$\tag{10.23}$$
$$u_k^{(i)}(t_0 + k) = u_0^{(i)}(t_0 + k) \qquad (i = l, \ldots, n-1)^{*)}.$$

Since $n + l + m$ is even, from (10.17), (10.19) and (10.21) we obtain

$$u_k^{(i)}(t) > 0 \qquad (i = 0, \ldots, l-1),$$

$$(-1)^{i-l}u_k^{(i)}(t) \geq 0 \qquad (i = l, \ldots, n) \qquad \text{for } t_0 \leq t \leq t_0 + k, \tag{10.24}$$

and

$$u_k(t) \geq \inf\{u_0(t) : t_0 \leq t \leq t_0 + k\}. \tag{10.25}$$

---

[1]If $l = 0$ ($l = n$) the conditions at t_0 (at $t_0 + k$) should be omitted.

We claim that

$$|u_k^{(n-1)}(t)| \le |u_0^{(n-1)}(t)| \qquad \text{for } t_0 \le t \le t_0 + k. \tag{10.26}$$

Indeed, by (10.23) and (10.24) there would otherwise exist points $t_{1k} \in [t_0, t_0 + k[$ and $t_{2k} \in]t_{1k}, t_0 + k]$ such that if $l \le n - 1(l = n)$, then

$$(-1)^{l-n+1}[u_k^{(n-1)}(t) - u_0^{(n-1)}(t)] > 0 \qquad \text{for } t_{1k} \le t < t_{2k},$$

$$u_k^{(n-1)}(t_{2k}) - u_0^{(n-1)}(t_{2k}) = 0$$

$$\left(u_k^{(n-1)}(t) - u_0^{(n-1)}(t) > 0 \qquad \text{for } t_{1k} < t \le t_{2k},\right.$$

$$\left. u_k^{(n-1)}(t_{1k}) - u_0^{(n-1)}(t_{1k}) = 0\right).$$

But this is impossible, because in view of (10.22) we would have

$$(-1)^{l-n+1}[u_k^{(n)}(t) - u_0^{(n)}(t)] \ge 0 \qquad \left(u_k^{(n)}(t) - u_0^{(n)}(t) \le 0\right) \qquad \text{for } t_{1k} \le t \le t_{2k}.$$

Hence (10.26) is true.

According to (10.23), (10.24) and (10.26),

$$0 < u_k^{(i)}(t) < u_0^{(i)}(t) \qquad (i = 0, \dots, l - 1),$$

$$0 \le (-1)^{i-l} u_k^{(i)}(t) \le (-1)^{i-l} u_0^{(i)}(t) \qquad (i = l, \dots, n - 1)$$

$$\text{for } t_0 \le t \le t_0 + k. \tag{10.27}$$

Lemma 10.2 implies that $(u_k)_{k=1}^{+\infty}$ contains a subsequence $(u_{k_j})_{j=1}^{+\infty}$ such that $(u_{k_j}^{(i)})_{j=1}^{+\infty}$ $(i = 0, \dots, n - 1)$ converge uniformly on each finite interval in $[t_0, +\infty[$, and the function u defined by (10.16) is a solution of the equation (10.8). On the other hand, (10.18), (10.25) and (10.27) yield

$$u(t) \ge r_0, \qquad |u^{(n-1)}(t)| \le r \qquad \text{for } t \ge t_0.$$

Since the equation (10.8) has property A (property B), we arrive at a contradiction. $\square$

PROBLEM 10.1. Let the equation (10.8) have property A (property B) and

$$(-1)^m g(t, x_1, \dots, x_n) \operatorname{sign} x_1 \ge (-1)^m g(t, y_1, \dots, y_n) \operatorname{sign} y_1 \ge 0$$

for $t \ge a$, $x_k y_k \ge 0$, $|x_k| \ge |y_k|$ $\quad (k = 1, \dots, n - 1)$, $\qquad x_n = y_n$,

where $m = 1$ $(m = 2)$. Then the differential inequality

$$(-1)^m[u^{(n)} - g(t, u(t), \dots, u^{(n-1)}(t))] \operatorname{sign} u(t) \ge 0$$

also has property A (property B).

10.2. Equations having property A or B.

THEOREM 10.2. Let the condition (10.2) (the condition (10.3)) hold, and let for any sufficiently large positive constant c the inequality

$$|f(t, x_1, \ldots, x_n)| \geq \varphi_c(t, x_1) \tag{10.28}$$

be fulfilled on the set

$$\left\{ (t, x_1, \ldots, x_n) : t \geq a, \; \frac{1}{c} \leq |x_1| \leq ct^{n-1}, \right.$$

$$\left. |x_i| \leq c|x_1|^{(n-i)/(n-1)} \quad (i = 2, \ldots, n) \right\}, \tag{10.29}$$

where $\varphi_c \in K_{\text{loc}}([a, +\infty[\times \mathbf{R}),$

$$\varphi_c(t, x) \geq \varphi_c(t, y) \geq 0 \quad \text{for } xy \geq 0, \; |x| \geq |y|. \tag{10.30}$$

Suppose, in addition, that for each $m \in \{1, 2\}$ either the equation

$$\frac{dx}{dt} = \frac{t^{n-1}}{(n-1)!} \varphi_c(t, (-1)^m x) \tag{10.31}$$

or the equation

$$\frac{dx}{dt} = -\varphi_c\left(t, (-1)^m \frac{t^{n-1}}{(n-1)!} x\right) \tag{10.32}$$

does not possess a positive proper solution. Then the equation (10.1) has property A (property B).

In order to prove this theorem we need three lemmas; these we prove first.

LEMMA 10.3. Suppose $\psi \in K_{\text{loc}}([a, +\infty[\times \mathbf{R}),$

$$\psi(t, x) \geq \psi(t, y) \quad \text{for } t \geq a, \; x \geq y \geq 0, \tag{10.33}$$

and the differential equation

$$\frac{dx}{dt} = \psi(t, x) \tag{10.34}$$

does not have a positive proper solution. Let, in addition, $c_0 \in]0, +\infty[$, $t_0 \in [a, +\infty[$, and let $h : [t_0, +\infty[\to]0, +\infty[$ be a continuous nondecreasing function. Then there is no continuous function $y : [t_0, +\infty[\to]0, +\infty[$ satisfying the integral inequality

$$y(t) \geq c_0 + \int_{t_0}^{t} \frac{\psi(\tau, h(\tau)y(\tau))}{h(\tau)} \, d\tau \quad \text{for } t \geq t_0. \tag{10.35}$$

PROOF. Assume that there exists a continuous function $y : [t_0, +\infty[\to]0, +\infty[$ satisfying (10.35).

Let x be the solution of the equation (10.34) under the initial condition $x(t_0) = c_0 h(t_0)/2$. By the hypotheses of the lemma this solution is defined on a finite interval $[t_0, t^*[$, and

$$\lim_{t \to t^*} x(t) = +\infty.$$

Thus, there exists $t_1 \in]t_0, t^*[$ such that

$$0 < x(t) < h(t)y(t) \quad \text{for } t_0 \leq t < t_1, \quad x(t_1) = h(t_1)y(t_1).$$

So, taking into account the inequalities (10.33) and (10.35), we obtain a contradiction:

$$y(t_1) = \frac{x(t_1)}{h(t_1)} = \frac{c_0 h(t_0)}{2h(t_1)} + \frac{1}{h(t_1)} \int_{t_0}^{t_1} \psi(\tau, x(\tau))\, d\tau$$

$$\leq \frac{c_0}{2} + \int_{t_0}^{t_1} \frac{\psi(\tau, h(\tau)y(\tau))}{h(\tau)}\, d\tau < y(t_1). \qquad \square$$

LEMMA 10.4. Suppose the condition (10.33) with $\psi \in K_{\text{loc}}([a, +\infty[\times \mathbf{R})$ holds and the differential equation

$$\frac{dx}{dt} = -\psi(t, x) \tag{10.36}$$

does not have a positive proper solution. Moreover, let $h: [t_0, +\infty[\rightarrow]0, +\infty[$, $t_0 \in [a, +\infty[$, be a continuous nondecreasing function. Then there is no continuous function $y: [t_0, +\infty[\rightarrow]0, +\infty[$ satisfying the integral inequality

$$y(t) \geq \int_t^{+\infty} h(\tau)\psi\left(\tau, \frac{y(\tau)}{h(\tau)}\right) d\tau \quad \text{for } t \geq t_0. \tag{10.37}$$

PROOF. Assume that there exists a continuous function $y: [t_0, +\infty[\rightarrow]0, +\infty[$ satisfying (10.37). Then by (10.33) either

$$\int_t^{+\infty} h(\tau)\psi\left(\tau, \frac{y(\tau)}{h(\tau)}\right) d\tau > 0 \quad \text{for } t \geq t_0 \tag{10.38}$$

or we can find $t^* \in [t_0, +\infty[$ such that

$$\psi(t, x) = 0 \quad \text{for } t \geq t^*, 0 \leq x \leq \frac{y(t)}{h(t)}. \tag{10.39}$$

First suppose (10.38) holds. Consider the solution x of the equation (10.36) under the initial condition $h(t_0)x(t_0) = y(t_0)$. Since (10.36) does not have a positive proper solution, there exist points $t_1 \in [t_0, +\infty[$ and $t_2 \in]t_1, +\infty[$ such that $h(t_1)x(t_1) = y(t_1) = 0$,

$$0 < h(t)x(t) < y(t) \quad \text{for } t_1 < t < t_2, x(t_2) = 0. \tag{10.40}$$

Then from (10.33), (10.37) and (10.38) we derive

$$y(t_1) = h(t_1)x(t_1) = h(t_1)\int_{t_1}^{t_2} \psi(\tau, x(\tau))\, d\tau \leq \int_{t_1}^{t_2} h(\tau)\psi\left(\tau, \frac{y(\tau)}{h(\tau)}\right) d\tau < y(t_1),$$

which is a contradiction.

It remains to consider the case when (10.39) holds. Let x be a solution of the equation (10.36) such that $x(t^*) > 0$. Then for certain $t_1 \in [t^*, +\infty[$ and $t_2 \in]t_1, +\infty[$ the conditions (10.40) are satisfied. But this is impossible, because by (10.39), $x'(t) = 0$ for $t_1 \leq t \leq t_2$. $\square$

LEMMA 10.5. Suppose the condition (10.33) with $\psi \in K_{loc}([a, +\infty[\times \mathbf{R})$ holds and the equation (10.34) (the equation (10.36)) does not have a positive proper solution. Then for any continuous nondecreasing function $h: [a, +\infty[\rightarrow]0, +\infty[$,

$$\int_a^{+\infty} \frac{\psi(\tau, h(\tau))}{h(\tau)} \, d\tau = +\infty \qquad \left(\int_a^{+\infty} h(\tau)\psi\left(\tau, \frac{1}{h(\tau)}\right) d\tau = +\infty \right). \qquad (10.41)$$

PROOF. If (10.41) is violated, then $y(t) \equiv 1$ would satisfy the inequality (10.35) (the inequality (10.37)) with $c_0 = 1/2$ and a sufficiently large t_0. But this contradicts Lemma 10.3 (Lemma 10.4). $\square$

PROOF OF THEOREM 10.2. First of all we note that by Lemma 10.5,

$$\int_a^{+\infty} t^{n-k}\varphi_c(t, (-1)^m\delta t^{k-1}) \, dt = +\infty \qquad (k = 1, \ldots, n) \qquad (10.42)$$

for any $c > 0$, $\delta > 0$ and $m \in \{1, 2\}$.

Let u be a proper nonoscillatory solution of the equation (10.1) and let $t_0 \in [a, +\infty[$ and $m \in \{1, 2\}$ be such that $(-1)^m u(t) > 0$ for $t > t_0$. The condition (10.2) (the condition (10.3)) implies

$$p(t) \equiv \frac{u^{(n)}(t)}{u(t)} \leq 0 \qquad (p(t) \geq 0) \qquad \text{for } t \geq t_0.$$

Furthermore, (10.28) and (10.42) yield

$$\text{meas}\{s \geq t : p(s) \neq 0\} > 0 \qquad \text{for } t \geq t_0.$$

Hence, in view of Lemma 1.1 there exist $t_1 \in [t_0, +\infty[$ and $l \in \{0, \ldots, n\}$ such that $l + n$ is odd (even) and

$$(-1)^m u^{(i)}(t) > 0 \qquad (i = 0, \ldots, l-1),$$

$$(-1)^{m+i+l}u^{(i)}(t) > 0 \qquad (i = l, \ldots, n-1), \qquad (-1)^{m+n+l}u^{(n)}(t) \geq 0$$

on $[t_1, +\infty[$.

We will show that $l \in \{0, n\}$. Assume the contrary. Then $l \in \{1, \ldots, n-1\}$ and $|u(t)| \geq \delta t^{l-1}$ for $t \geq t_1$, where

$$\delta = \frac{|u^{(l-1)}(t_1)|}{(l-1)!} > 0.$$

On the other hand, by Lemma 5.2 we can choose a positive constant c such that for $t \geq t_1$ the point $(t, u(t), \ldots, u^{(n-1)}(t))$ belongs to the set (10.29). Hence (10.28), (10.30) and (10.42) imply

$$\int_{t_1}^{+\infty} t^{n-l}|u^{(n)}(t)| \, dt \geq \int_{t_1}^{+\infty} t^{n-l}\varphi_c(t, (-1)^m\delta t^{l-1}) \, dt = +\infty \qquad (10.44)$$

and

$$|u^{(n)}(t)| \geq \varphi_c(t, (-1)^m|u(t)|) \qquad \text{for } t \geq t_1. \qquad (10.45)$$

From (10.43), (10.44) and the equality

$$\sum_{i=l}^{n} \frac{(-1)^{m+l+i}}{(i-l)!} t^{i-l} u^{(i-1)}(t) = \sum_{i=l}^{n} \frac{(-1)^{m+l+i}}{(i-l)!} t_1^{i-l} u^{(i-1)}(t_1)$$

$$+ \frac{(-1)^{m+l+n}}{(n-l)!} \int_{t_1}^{t} \tau^{n-l} u^{(n)}(\tau) \, d\tau$$

we derive

$$|u^{(l-1)}(t)| \geq 1 + \frac{1}{(n-l)!} \int_{t_2}^{t} \tau^{n-l} |u^{(n)}(\tau)| \, d\tau \qquad \text{for } t \geq t_2, \qquad (10.46)$$

where t_2 is a point in $[t_1, +\infty[$. According to Lemma 1.3, t_2 can be taken so large that

$$|u^{(l-1)}(t)| \geq \frac{t}{(n-l)!} \int_{t}^{+\infty} \tau^{n-l-1} |u^{(n)}(\tau)| \, d\tau \qquad \text{for } t \geq t_2 \qquad (10.47)$$

and

$$|u(t)| \geq \frac{t^{l-1}}{l!} |u^{(l-1)}(t)| \qquad \text{for } t \geq t_2. \qquad (10.48)$$

In view of (10.30), the relations (10.45)–(10.48) yield

$$y(t) \geq 1 + \frac{1}{(n-1)!} \int_{t_2}^{t} \frac{\tau^{n-1}}{g(\tau)} \varphi_c(\tau, (-1)^m g(\tau) y(\tau)) \, d\tau \qquad \text{for } t \geq t_2$$

and

$$z(t) \geq \frac{1}{(n-1)!} \int_{t}^{+\infty} h(\tau) \varphi_c \left(\tau, (-1)^m \frac{\tau^{n-1} z(\tau)}{h(\tau)} \right) d\tau \qquad \text{for } t \geq t_2,$$

where $y(t) = |u^{(l-1)}(t)|$, $g(t) = t^{l-1}/l!$, $z(t) = |u^{(l-1)}(t)|/t$, and $h(t) = l! t^{n-1-l}$. But by Lemmas 10.3 and 10.4 these inequalities cannot hold simultaneously, because either the equation (10.31) or the equation (10.32) does not have a positive proper solution. Therefore, $l \in \{0, n\}$.

Let $l = 0$. Then $|u^{(i)}(t)| \downarrow 0$ as $t \uparrow +\infty$ $(i = 1, \ldots, n - 1)$. We claim that $|u(t)| \downarrow 0$ as $t \uparrow +\infty$. Indeed, otherwise we can find $c > 0$ so that for $t \geq t_1$ the point $(t, u(t), \ldots, u^{(n-1)}(t))$ belongs to the set (10.29). Consequently, applying (10.28), (10.30) and (10.44), we would have

$$\int_{t_1}^{+\infty} t^{n-1} |u^{(n)}(t)| \, dt \geq \int_{t_1}^{+\infty} t^{n-1} \varphi_c \left(t, \frac{(-1)^m}{c} \right) dt = +\infty,$$

which contradicts Lemma 1.2.

It remains to show that $|u^{(n-1)}(t)| \uparrow +\infty$ as $t \uparrow +\infty$ whenever $l = n$. Assume the contrary. Then there exists $c > 0$ such that for $t \geq t_1$ the point $(t, u(t), \ldots, u^{(n-1)}(t))$ belongs to the set (10.29). Moreover,

$$(-1)^m u(t) > \delta t^{n-1} \qquad \text{for } t \geq t_1,$$

where $\delta > 0$. Thus, (10.28), (10.30) and (10.44) imply a contradiction:

$$c \geq |u^{(n-1)}(t)| \geq \int_{t_1}^{t} \varphi_c(\tau, (-1)^m \delta \tau^{m-1}) \, d\tau \to +\infty \qquad \text{as } t \to +\infty. \qquad \square$$

COROLLARY 10.1. Let the condition (10.2) (the condition (10.3)) hold, and let for any sufficiently large positive constant c the inequality

$$|f(t, x_1, \ldots, x_n)| \geq p_c(t)\omega_c(|x_1|)$$

be fulfilled on the set (10.29), where $\omega_c \colon]0, +\infty[\ \to\ ]0, +\infty[$ is a continuous nondecreasing function satisfying the condition

$$\int_1^{+\infty} \frac{ds}{\omega_c(s)} < +\infty, \tag{10.49}$$

$p_c \in L_{\mathrm{loc}}([a, +\infty[)$ is nonnegative, and

$$\int_a^{+\infty} t^{n-1} p_c(t)\, dt = +\infty. \tag{10.50}$$

Then the equation (10.1) has property A (property B).

PROOF. By Theorem 10.2 it suffices to show that, whenever $c > 0$ is large enough, the equation

$$\frac{dx}{dt} = \frac{t^{n-1}}{(n-1)!} p_c(t)\omega_c(|x|)$$

does not have a positive proper solution. Assume, on the contrary, that this equation has a solution $x \colon [t_0, +\infty[\ \to\ ]0, +\infty[$. Then, in view of (10.49),

$$\int_{t_0}^t \tau^{n-1} p_c(\tau)\, d\tau = (n-1)! \int_{x(t_0)}^{x(t)} \frac{ds}{\omega_c(s)} \leq (n-1)! \int_{x(t_0)}^{+\infty} \frac{ds}{\omega_c(s)} < +\infty$$

on $[t_0, +\infty[$, which contradicts (10.50). $\square$

COROLLARY 10.2. Let the condition (10.2) (the condition (10.3)) hold, and let for any sufficiently large positive constant c the inequality

$$|f(t, x_1, \ldots, x_n)| \geq p_c(t)\omega_c\left(\frac{|x_1|}{t^{n-1}}\right)$$

be fulfilled on the set (10.29), where $\omega_c \colon]0, +\infty[\ \to\ ]0, +\infty[$ is a continuous nondecreasing function satisfying the condition

$$\int_0^1 \frac{ds}{\omega_c(s)} < +\infty, \tag{10.51}$$

$p_c \in L_{\mathrm{loc}}([a, +\infty[)$ is nonnegative, and

$$\int_a^{+\infty} p_c(t)\, dt = +\infty. \tag{10.52}$$

Then the equation (10.1) has property A (property B).

PROOF. By Theorem 10.2 it suffices to show that, whenever $c > 0$ is large enough, the equation

$$\frac{dx}{dt} = -\frac{1}{(n-1)!} p_c(t)\omega_c(|x|)$$

does not have a positive proper solution. Assume, on the contrary, that this equation has a solution $x: [t_0, +\infty[\to]0, +\infty[$. Then, according to (10.51),

$$\int_{t_0}^{t} p_c(\tau) \, d\tau = (n-1)! \int_{x(t)}^{x(t_0)} \frac{ds}{w_c(s)} \le (n-1)! \int_{0}^{x(t_0)} \frac{ds}{w_c(s)} < +\infty$$

on $[t_0, +\infty[$, which contradicts the condition (10.52). $\square$

Corollaries 8.2, 10.1 and 10.2 imply the following assertions.

THEOREM 10.3. Let the condition (10.2) (the condition (10.3)) hold, and let for any sufficiently large positive constant c the inequality

$$p(t)w_c(|x_1|) \le |f(t, x_1, \dots, x_n)| \le p(t)w_c^*(|x_1|)$$

be fulfilled on the set (10.29), where $p \in L_{\text{loc}}([a, +\infty[)$, $w_c^*: \mathbf{R} \to]0, +\infty[$ is a continuous function, and $w_c:]0, +\infty[\to]0, +\infty[$ is a continuous nondecreasing function satisfying (10.49). Then the equation (10.1) has property A (property B) if and only if

$$\int_{a}^{+\infty} t^{n-1} p(t) \, dt = +\infty.$$

THEOREM 10.4. Let the condition (10.2) (the condition (10.3)) hold, and let for any sufficiently large positive constant c the inequality

$$p(t)w_c\left(\frac{|x_1|}{t^{n-1}}\right) \le |f(t, x_1, \dots, x_n)| \le p(t)w_c^*\left(\frac{|x_1|}{t^{n-1}}\right)$$

be fullfilled on the set (10.29), where $p \in L_{\text{loc}}([a, +\infty[)$, $w_c^*: \mathbf{R} \to]0, +\infty[$ is a continuous function, and $w_c:]0, +\infty[\to]0, +\infty[$ is a continuous nondecreasing function satisfying (10.51). Then the equation (10.1) has property A (property B) if and only if

$$\int_{a}^{+\infty} p(t) \, dt = +\infty.$$

As an example we consider the equation

$$u^{(n)} = g(t)\frac{|u|^{\lambda}}{(1+|u|)^{\mu}} \operatorname{sign} u \tag{10.53}$$

with $g \in L_{\text{loc}}([a, +\infty[)$, $\lambda \in]0, +\infty[$ and $\mu \in \mathbf{R}$.

Theorems 10.3 and 10.4 yield

COROLLARY 10.3. Let $\lambda \ne \mu + 1$ and $g(t) \le 0$ ($g(t) \ge 0$) for $t \ge a$. Then the equation (10.53) has property A (property B) if and only if

$$\int_{a}^{+\infty} t^{\nu} |g(t)| \, dt = +\infty,$$

where $\nu = n - 1$ for $\lambda > \mu + 1$ and $\nu = (n-1)(\lambda - \mu)$ for $\lambda < \mu + 1$.

The following assertion can be readily proved by means of Theorem 10.1.

THEOREM 10.5. Let the condition (10.2) (the condition (10.3)) hold and let for any sufficiently large positive constant c the inequality

$$|f(t, x_1, \ldots, x_n)| \geq p_c(t)|x_1| \tag{10.54}$$

be fulfilled on the set (10.29), where $p_c \in L_{\text{loc}}([a, +\infty[)$ is nonnegative and the equation $u^{(n)} = -p_c(t)u$ (the equation $u^{(n)} = p_c(t)u$) has property A (property B). Then the equation (10.1) also has property A (property B).

Using Theorems 1.6, 1.7 and 1.7', from Theorem 10.5 we can obtain

COROLLARY 10.4. Let the condition (10.2) (the condition (10.3)) hold and let for any sufficiently large positive constant c the inequality (10.54) be fulfilled on the set (10.29), where $p_c \in L_{\text{loc}}([a, +\infty[)$ is nonnegative and

$$\lim_{t \to +\infty} \sup t \int_t^{+\infty} s^{n-2} p_c(s)\, ds = +\infty.$$

Then the equation (10.1) has property A (property B).

COROLLARY 10.5. Let $\lambda = \mu + 1$, $g(t) \leq 0$ ($g(t) \geq 0$) for $t \geq a$ and

$$\lim_{t \to +\infty} \sup t \int_t^{+\infty} s^{n-2}|g(s)|\, ds = +\infty.$$

Then the equation (10.53) has property A (property B).

Theorems 10.6 and 10.7, given below, also relate to the case when the function f is subject to an estimate of the type (10.28). But, in contrast with Theorems 10.1–10.5, the restriction (10.30) does not appear in these theorems.

THEOREM 10.6. Let the condition (10.2) (the condition (10.3)) hold and let for any sufficiently large positive constant c the inequality (10.28) be fulfilled on the set (10.29), where the function $\varphi_c \colon [a, +\infty[\times \mathbf{R} \to \mathbf{R}_+$ is continuous, does not decrease with respect to the first variable, $\varphi_c(t, x) > 0$ for $t \geq a$, $x \neq 0$, and

$$\int_a^{+\infty} \varphi_c(t, (-1)^m ct^{n-1})\, dt = +\infty \qquad (m = 1, 2). \tag{10.55}$$

Then the equation (10.1) has property A (property B).
PROOF. First we note that under the hypotheses of Theorem 10.6,

$$f(t, x_1, \ldots, x_n)x_1 < 0 \qquad (f(t, x_1, \ldots, x_n)x_1 > 0) \qquad \text{for } x_1 \neq 0. \tag{10.56}$$

Assume the theorem to be false. Then by the inequality (10.56) and Lemma 1.1, the equation (10.1) has a proper nonoscillatory solution $u \colon [t_0, +\infty[\to \mathbf{R}$ satisfying the conditions

$$\delta_0 = \inf\{|u(t)| : t \geq t_0\} > 0, \qquad \sup\{|u^{(n-1)}(t)| : t \geq t_0\} < +\infty. \tag{10.57}$$

Moreover, there exist $m \in \{1, 2\}$ and $l \in \{0, \ldots, n\}$ such that $l + n$ is odd (even) and the inequalities (10.43) hold on $[t_0, +\infty[$.
According to (10.43), (10.57) and Lemma 5.2,

$$\int_{t_0}^{+\infty} |u^{(n)}(t)|\, dt < +\infty,$$

and for a certain $c > 1$ the point $(t, u(t), \ldots, u^{(n-1)}(t))$, $t \geq t_0$, belongs to the set (10.29). Hence (10.28) yields

$$r = \int_{t_0}^{+\infty} \varphi_c(t, u(t))\, dt < +\infty. \qquad (10.58)$$

Assuming that $\delta_1 = \sup\{|u(t)|: t \geq t_0\} < +\infty$ and applying (10.57) and the monotonicity of φ_c with respect to the first variable, we obtain

$$\varphi_c(t, u(t)) \geq \eta \qquad \text{for } t \geq t_0,$$

where $\eta = \min\{\varphi_c(a, x): \delta_0 \leq |x| \leq \delta_1\} > 0$. But this contradicts the condition (10.58). So $l \in \{1, \ldots, n\}$ and $(-1)^m u(t) \uparrow +\infty$ as $t \uparrow +\infty$.

Set

$$v(t) = \left[\frac{(-1)^m u(t)}{c}\right]^{1/(n-1)}$$

and suppose t_0 to be so large that $v(t_0) > a$. For $t \geq t_0$ the point $(t, u(t), \ldots, u^{(n-1)}(t))$ belongs to the set (10.29). Therefore,

$$v(t) \leq t \qquad \text{for } t \geq t_0 \qquad (10.59)$$

and

$$0 < v'(t) = \frac{1}{n-1} c^{-1/(n-1)} |u'(t)|\, |u(t)|^{-(n-2)/(n-1)} \leq c \qquad \text{for } t \geq t_0. \qquad (10.60)$$

Since φ_c does not decrease with respect to the first variable, (10.58) and (10.59) imply

$$\int_{t_0}^{+\infty} \varphi_c(v(t), u(t))\, dt \leq r.$$

By (10.60) and the change of variable $x = v(t)$ in the integral at the left-hand side of the last inequality, we obtain

$$\int_{v(t_0)}^{+\infty} \varphi_c(x, (-1)^m c x^{n-1})\, dx \leq rc,$$

which contradicts (10.55). $\square$

In the sequel we will need

LEMMA 10.6. *Let $k \in \{0, \ldots, n-1\}$, $\delta > 0$, $\alpha, \alpha_0 \in \mathbf{R}$, $|\alpha_0| > |\alpha|$, $\alpha_0 \alpha > 0$, $v_0(t) = \alpha_0 t^{n-k-1}$, and*

$$|f(t, x_1, \ldots, x_n)| \leq \varphi(t, x_1)$$

for $t \geq a$, $|x_i - v_0^{(i-1)}(t)| \leq \delta t^{n-k-i}$ $(i = 1, \ldots, n)$, (10.61) where $\varphi: [a, +\infty[\times \mathbf{R} \to \mathbf{R}_+$ is continuous, $t^k \varphi(t, x)$ is nondecreasing in t, and

$$\int_a^{+\infty} t^k \varphi(t, \alpha t^{n-k-1})\, dt < +\infty. \qquad (10.62)$$

Then the equation (10.1) possesses a solution u having the asymptotic representation

$$u^{(i-1)}(t) = v_0^{(i-1)}(t) + o(t^{n-k-i}) \qquad (i = 1, \ldots, n). \qquad (10.63)$$

PROOF. Without loss of generality we may assume that $\delta < |\alpha_0 - \alpha|$.

Set

$$\eta = \left(\frac{|\alpha_0| + \delta}{|\alpha|}\right)^{(n-2-k)/(n-1-k)}.$$

By (10.62),

$$\epsilon(t) = \int_t^{+\infty} s^k \varphi(s, \alpha s^{n-1-k})\, ds \to 0 \qquad \text{as } t \to +\infty. \qquad (10.64)$$

Choose $t_0 > a$ so that $\epsilon(t_0) < \delta$. Clearly,

$$\epsilon_0(t) = \frac{1}{t_0} \int_{t_0}^t \epsilon(s)\, ds \to 0 \qquad \text{as } t \to +\infty \qquad (10.65)$$

and

$$\epsilon_0(t) < \epsilon(t_0) < \delta \qquad \text{for } t \geq t_0. \qquad (10.66)$$

We denote by Φ the Banach space of $n - 1$ times differentiable functions u: $[t_0, +\infty[\to \mathbf{R}$ satisfying the conditions

$$\sup\{t^{i+k-n}|u^{(i-1)}(t)| : t \geq t_0\} < +\infty \qquad (i = 1, \ldots, n),$$

with the norm

$$\|u\| = \sup\left\{\sum_{i=0}^{n-1} t^{i+k-n}|u^{(i-1)}(t)| : t \geq t_0\right\}.$$

Let

$$\varphi^*(t) = \max\{\varphi(t, x) : |x - v_0(t)| \leq \delta t^{n-k-1}\},$$

and let Φ_0 be the set of all $u \in \Phi$ such that

$$|u^{(i-1)}(t) - v_0^{(i-1)}(t)| \leq t^{n-k-i}\epsilon_0(t) \qquad (i = 1, \ldots, n - k),$$

$$|u^{(i-1)}(t)| \leq t^{n-k-i}\epsilon(t) \qquad (i = n - k + 1, \ldots, n) \qquad \text{for } t \geq t_0, \qquad (10.67)$$

and

$$|u^{(n-1)}(t) - u^{(n-1)}(s)| \leq \int_s^t \omega^*(\tau)\, d\tau \qquad \text{for } t_0 \leq s \leq t < +\infty.$$

The Arzelá–Ascoli lemma, together with the conditions (10.64), (10.65) and (10.67), implies that Φ_0 is a convex closed compact set. On the other hand, according to (10.66) and (10.67) any $u \in \Phi_0$ satisfies the inequalities

$$(|\alpha_0| + \delta)t^{n-1-k}u(t)\operatorname{sign}\alpha > (|\alpha_0| - \delta)t^{n-1-k} > |\alpha|t^{n-1-k},$$

$$u'(t)\operatorname{sign}\alpha > (n - 1 - k)|\alpha|t^{n-2-k} \qquad \text{for } t \geq t_0. \qquad (10.68)$$

We define the operator h on Φ_0 as

$$h(u)(t) = v_0(t) - \frac{1}{(n - k - 2)!\, k!}$$

$$\times \int_{t_0}^t (t - \tau)^{n-k-2}\, d\tau \int_\tau^{+\infty} (\tau - s)^k f(s, u(s), \ldots, u^{(n-1)}(s))\, ds. \qquad (10.69)$$

We will show that this operator maps Φ_0 into itself.

Suppose $u \in \Phi_0$. Put

$$\xi(t) = \left[\frac{u(t)}{\alpha}\right]^{1/(n-1-k)}.$$

Then (10.68) yields

$$\xi(t) > t, \qquad \xi'(t) > \frac{1}{\eta} \qquad \text{for } t \geq t_0.$$

Since f satisfies (10.61) and $t^k \varphi(t, x)$ is nondecreasing in t, we obtain

$$\int_t^{+\infty} s^k |f(s, u(s), \ldots, u^{(n-1)}(s))| \, ds \leq \int_t^{+\infty} s^k \varphi(s, u(s)) \, ds$$

$$\leq \eta \int_t^{+\infty} \xi^k(s) \varphi(\xi(s), u(s)) \, ds = \eta \int_{\xi(t)}^{+\infty} x^k \varphi(x, \alpha x^{n-1-k}) \, dx \leq \epsilon(t) \quad \text{for } t \geq t_0.$$

Hence by (10.61), (10.67) and (10.69), $h(u) \in \Phi_0$, i.e. h maps Φ_0 into itself. Moreover, h is continuous. Consequently, by the Schauder principle [168] there exists $u \in \Phi_0$ such that $u(t) = h(u)(t)$ for $t \geq t_0$. It follows from (10.67) and (10.69) that u is a solution of the equation (10.1) having the asymptotic representation (10.63). $\square$

THEOREM 10.7. Let the condition (10.2) (the condition (10.3)) hold, $c_0 > a^{(1-n)/2}$, and let for any $c > c_0$ the inequality

$$\varphi_c(t, x_1) \leq |f(t, x_1, \ldots, x_n)| \leq r_c \varphi_c(t, x_1) \tag{10.70}$$

be fulfilled on the set (10.29), where r_c is a positive constant, the continuous function $\varphi_c \colon [a, +\infty[\times \mathbf{R} \to \mathbf{R}_+$ does not decrease with respect to the first variable, and $\varphi_c(t, x) > 0$ for $t \geq a$, $x \neq 0$. Then the equation (10.1) has property A (property B) if and only if

$$\int_a^{+\infty} \varphi_c(t, (-1)^m ct^{n-1}) \, dt = +\infty \qquad \text{for } c > c_0 \qquad (m = 1, 2). \tag{10.71}$$

PROOF. If (10.71) is satisfied, then by Theorem 10.6 the equation (10.1) has property A (property B).

Suppose (10.71) is violated, i.e.

$$\int_a^{+\infty} \varphi_{c_1}(t, \alpha t^{n-1}) \, dt < +\infty \tag{10.72}$$

for $\alpha = (-1)^m c_1$ and certain $m \in \{1, 2\}$ and $c_1 > c_0$.

Let $\alpha_0 = (-1)^m (c_1 + 2)$, $v_0(t) = \alpha_0 t^{n-1}$ and $c = n! |\alpha_0|$. Then the set

$$\{(t, x_1, \ldots, x_n) : t \geq a, |x - v_0^{(i-1)}(t)| \leq t^{n-i} \ (i = 1, \ldots, n)\} \tag{10.73}$$

is contained in the set (10.29). Thus, the inequality

$$|f(t, x_1, \ldots, x_n)| \leq \varphi(t, x_1),$$

with $\varphi(t, x) = r_c \varphi_c(t, x)$, holds on the set (10.73). Moreover, (10.72) yields

$$\int_a^{+\infty} \varphi(t, \alpha t^{n-1}) \, dt \leq r_c r_{c_1} \int_a^{+\infty} \varphi_{c_1}(t, \alpha t^{n-1}) \, dt < +\infty.$$

So, by Lemma 10.6, the equation (10.1) possesses a solution u having the asymptotic representation $u(t) = (\alpha_0 + o(1)) t^{n-1}$. Therefore, this equation has property A (property B). $\square$

COROLLARY 10.6. Let the function $f: [a, +\infty[\times \mathbf{R} \to \mathbf{R}$ be continuous,

$$f(t, x)\operatorname{sign} x < 0 \qquad (f(t, x)\operatorname{sign} x > 0) \qquad \text{for } t \geq a,\ x \neq 0,$$

and let $|f(t, x)|$ be nondecreasing in t. Then the equation

$$u^{(n)} = f(t, u) \tag{10.74}$$

has property A (property B) if and only if for any $\alpha \neq 0$,

$$\int_a^{+\infty} |f(t, \alpha t^{n-1})|\, dt = +\infty.$$

10.3. Equations having property A_k, B_k, A_k^*, or B_k^*. For any $c \in\ ]0, +\infty[$ and $k \in \{1, \dots, n-1\}$ we denote by $D_{nk}(a, c)$ the set of points $(t, x_1, \dots, x_n) \in [a, +\infty[\times \mathbf{R}^n$ whose coordinates satisfy the inequalities

$$\frac{(n-1-k)!}{2(n-i)!} \leq t^{i-k-1}\frac{x_i}{x_{k+1}} \leq \frac{2}{(k-i+1)!} \qquad (i = 1, \dots, k),$$

$$\frac{1}{c} \leq x_{k+1} \leq ct^{n-1-k}, \qquad |x_i| \leq c|x_{k+1}|^{(n-i)/(n-k-1)} \qquad (i = k+2, \dots, n).$$

THEOREM 10.8. Let the condition (10.2) (the condition (10.3)) be fulfilled and $k \in \{1, \dots, n-1\}$. Moreover, suppose that for any sufficiently large positive constant c the inequality

$$|f(t, x_1, \dots, x_n)| \geq \varphi_c(t, x_{k+1}) \tag{10.75}$$

holds on the set $D_{nk}(a, c)$, where $\varphi_c \in K_{\mathrm{loc}}([a, +\infty[\times \mathbf{R})$ satisfies (10.30), and for neither $m \in \{1, 2\}$ the equation

$$\frac{dx}{dt} = \frac{t^{n-1-k}}{(n-1-k)!}\varphi_c(t, (-1)^m x) \tag{10.76}$$

possesses a positive proper solution. Then the equation (10.1) has property A_k (property B_k).

PROOF. Assume the theorem to be false. Then, by the condition (10.2) (the condition (10.3)), the equation (10.1) has a solution $u: [t_0, +\infty[\to \mathbf{R}$ satisfying the inequalities

$$\inf\{|u^{(k)}(t)| : t \geq t_0\} > 0, \qquad \sup\{|u^{(n-1)}(t)| : t \geq t_0\} < +\infty,$$

and

$$u^{(i)}(t)u(t) > 0 \qquad (i = 0, \dots, l-1),$$
$$(-1)^{l+i}u^{(i)}(t)u(t) \geq 0 \qquad (i = l, \dots, n) \qquad \text{for } t \geq t_0,$$

where $l \in \{k, \dots, n\}$ and $l+n$ is odd (even). This immediately implies the existence of $t_1 \in [t_0, +\infty[$ such that

$$\frac{(n-1-k)!}{2(n-i)!} \leq t^{i-k-1}\frac{u^{(i-1)}(t)}{u^{(k)}(t)} \leq \frac{2}{(k-i+1)!} \qquad \text{for } t \geq t_1 \qquad (i = 1, \dots, k). \tag{10.77}$$

Clearly, the function $w(t) = u^{(k)}(t)$ is a solution of the differential equation

$$w^{(n-k)} = g(t, w, \dots, w^{(n-1-k)}) \tag{10.78}$$

with

$$g(t, x_1, \dots, x_{n-k}) = f\left(t, \frac{u(t)}{u^{(k)}(t)} x_1, \dots, \frac{u^{(k-1)}(t)}{u^{(k)}(t)} x_1, x_1, \dots, x_{n-k}\right)$$

and

$$\inf\{|w(t)| : t \geq t_1\} > 0, \qquad \sup\{|w^{(n-k-1)}(t)| : t \geq t_1\} < +\infty. \tag{10.79}$$

According to the condition (10.2) (the condition (10.3)),

$$g(t, x_1, \dots, x_{n-k}) x_1 \leq 0 \qquad (g(t, x_1, \dots, x_{n-k}) x_1 \geq 0).$$

On the other hand, by (10.75) and (10.77), for any sufficiently large positive constant c the inequality

$$|g(t, x_1, \dots, x_{n-k})| \geq \varphi_c(t, x_1)$$

holds on the set

$$\left\{ (t, x_1, \dots, x_{n-k}) : t \geq t_1, \frac{1}{c} \leq |x_1| \leq c t^{n-1-k}, \right.$$

$$\left. |x_i| \leq c|x_1|^{(n-k-i)/(n-k-1)} \quad (i = 2, \dots, n-k) \right\}.$$

Since, moreover, for neither $m \in \{1, 2\}$ the equation (10.76) possesses a positive proper solution, Theorem 10.2 implies that the equation (10.78) has property A (property B). Therefore, this equation cannot possess a solution satisfying (10.79), i.e. we have arrived at a contradiction. $\square$

REMARK 10.1. It is easily verified that if $k \in \{1, \dots, n-1\}$, $n+k$ is even (odd), the inequality (10.2) (the inequality (10.3)) holds and the equation (10.1) has property A_k (property B_k), then every proper nonoscillatory solution u of this equation satisfies the condition

$$\lim_{t \to +\infty} |u^{(k-1)}(t)| < +\infty \tag{10.80}$$

(either the condition (10.5) or the condition (10.80)).

COROLLARY 10.7. Let the condition (10.2) (the condition (10.3)) be fullfilled and $k \in \{1, \dots, n-1\}$. Moreover, suppose that for any sufficiently large positive constant c the inequality

$$|f(t, x_1, \dots, x_n)| \geq p_c(t)\omega_c(|x_{k+1}|)$$

holds on the set $D_{nk}(a, c)$, where $\omega_c: \,]0, +\infty[\, \to \,]0, +\infty[$ is a continuous nondecreasing function satisfying (10.49), $p_c \in L_{\mathrm{loc}}([a, +\infty[)$ is nonnegative and

$$\int_a^{+\infty} t^{n-1-k} p_c(t) \, dt = +\infty.$$

Then the equation (10.1) has property A_k (property B_k).

Corollaries 8.2 and 10.7 imply

THEOREM 10.9. Let the condition (10.2) (the condition (10.3)) be fulfilled and $k \in \{1, \dots, n-1\}$. Moreover, suppose that for any sufficiently large positive constant c the inequality

$$p(t)\omega_c(|x_{k+1}|) \leq |f(t, x_1, \dots, x_n)| \leq p(t)\omega_c^*(x_{k+1})$$

holds on the set $D_{nk}(a,c)$, where $p \in L_{\mathrm{loc}}([a, +\infty[)$, $\omega_c^*\colon \mathbf{R} \to \,]0, +\infty[$ is a continuous function, and $\omega_c\colon \,]0, +\infty[\,\to\,]0, +\infty[$ is a continuous nondecreasing function satisfying (10.49). Then the equation (10.1) has property A_k (property B_k) if and only if

$$\int_a^{+\infty} t^{n-1-k} p(t)\, dt = +\infty.$$

COROLLARY 10.8. Let $\lambda > \mu + 1$, $k \in \{1, \dots, n-1\}$, and $g(t) \leq 0$ ($g(t) \geq 0$) for $t \geq a$. Then the equation (10.53) has property A_k (property B_k) if and only if

$$\int_a^{+\infty} t^{n-1+(\lambda-\mu-1)k} |g(t)|\, dt = +\infty.$$

The proof of the following proposition is similar to that of Theorem 10.2.

THEOREM 10.10. Let $n \geq 3$ and let the condition (10.2) (the condition (10.3)) be fulfilled. Moreover, suppose that $k \in \{1, \dots, n-1\}$ is even (odd) and that for any sufficiently large positive constant c the inequality (10.28) holds on the set (10.29), where $\varphi_c \in K_{\mathrm{loc}}([a, +\infty[\times\mathbf{R})$ satisfies (10.30), and for neither $m \in \{1, 2\}$ the equation

$$\frac{dx}{dt} = -\frac{t^k}{(n-1)!\,k!}\varphi_c(t, (-1)^m t^{n-1-k} x)$$

possesses a positive proper solution. Then the equation (10.1) has property A_k^* (property B_k^*).

COROLLARY 10.9. Let $n \geq 3$ and let the condition (10.2) (the condition (10.3)) be fulfilled. Moreover, suppose that $k \in \{1, \dots, n-1\}$ is even (odd) and that for any sufficiently large positive constant c the inequality

$$|f(t, x_1, \dots, x_n)| \geq p_c(t)\omega_c\left(\frac{|x_1|}{t^{n-1-k}}\right)$$

holds on the set (10.29), where $\omega_c\colon \,]0, +\infty[\,\to\,]0, +\infty[$ is a continuous nondecreasing function satisfying (10.51), $p_c \in L_{\mathrm{loc}}([a, +\infty[)$ is nonnegative and

$$\int_a^{+\infty} t^k p_c(t)\, dt = +\infty.$$

Then the equation (10.1) has property A_k^* (property B_k^*).

From Corollaries 8.2 and 10.9 we can easily derive

THEOREM 10.11. Let $n \geq 3$ and let the condition (10.2) (the condition (10.3)) be fulfilled. Moreover, suppose that $k \in \{1, \dots, n-1\}$ is even (odd) and that for any sufficiently large positive constant c the inequality

$$p(t)\omega_c\left(\frac{|x_1|}{t^{n-1-k}}\right) \leq |f(t, x_1, \dots, x_n)| \leq p(t)\omega_c^*\left(\frac{x_1}{t^{n-1-k}}\right)$$

holds on the set (10.29), where $p \in L_{\text{loc}}([a, +\infty[)$, $\omega_c^* : \mathbf{R} \to]0, +\infty[$ is a continuous function, and $\omega_c :]0, +\infty[\to]0, +\infty[$ is a continuous nondecreasing function satisfying (10.51). Then the equation (10.1) has property A_k^* (property B_k^*) if and only if

$$\int_a^{+\infty} t^k p(t) \, dt = +\infty.$$

COROLLARY 10.10. Let $n \geq 3$, $\lambda < \mu + 1$, $g(t) \leq 0$ ($g(t) \geq 0$) for $t \geq a$, and let $k \in \{1, \ldots, n-1\}$ be even (odd). Then the equation (10.53) has property A_k^* (property B_k^*) if and only if

$$\int_a^{+\infty} t^{k+(\lambda-\mu)(n-1-k)} |g(t)| \, dt = +\infty.$$

Similarly to Theorem 10.6 we can prove

THEOREM 10.12. Let $n \geq 3$ and let the condition (10.2) (the condition (10.3)) be fulfilled. Moreover, suppose that $k \in \{1, \ldots, n-1\}$ is even (odd) and that for any sufficiently large positive constant c the inequality (10.28) holds on the set (10.29), where $\varphi_c : [a, +\infty[\times \mathbf{R} \to \mathbf{R}_+$ is continuous, $t^k \varphi_c(t, x)$ is nondecreasing in t, $\varphi_c(t, x) > 0$ for $t \geq a, x \neq 0$, and

$$\int_a^{+\infty} t^k \varphi_c(t, (-1)^m c t^{n-1-k}) \, dt = +\infty \qquad (m = 1, 2).$$

Then the equation (10.1) has property A_k^* (property B_k^*).

The following proposition is a consequence of Lemma 10.6 and Theorem 10.12.

THEOREM 10.13. Let $n \geq 3$ and let the condition (10.2) (the condition (10.3)) be fulfilled. Moreover, suppose that $k \in \{1, \ldots, n-1\}$ is even (odd), $c_0 > a^{(1-n)/2}$, and for any $c > c_0$ the inequality (10.70) holds on the set (10.29), where r_c is a positive constant, the function $\varphi_c : [a, +\infty[\times \mathbf{R} \to \mathbf{R}_+$ is continuous, $t^k \varphi_c(t, x)$ is nondecreasing in t, and $\varphi_c(t, x) > 0$ for $t \geq a, x \neq 0$. Then the equation (10.1) has property A_k^* (property B_k^*) if and only if

$$\int_a^{+\infty} t^k \varphi_c(t, (-1)^m c t^{n-1-k}) \, dt = +\infty \qquad \text{for } c > c_0 \qquad (m = 1, 2).$$

COROLLARY 10.11. Let $n \geq 3$ and let the continuous function $f : [a, +\infty[\times \mathbf{R} \to \mathbf{R}$ satisfy the inequality

$$f(t, x) \operatorname{sign} x < 0 \qquad (f(t, x) \operatorname{sign} x > 0) \qquad \text{for } t \geq a, \ x \neq 0.$$

Moreover, suppose that $k \in \{1, \ldots, n-1\}$ is even (odd), and that $t^k |f(t, x)|$ is nondecreasing in t. Then the equation (10.74) has property A_k^* (property B_k^*) if and only if for any $\alpha \neq 0$

$$\int_a^{+\infty} t^k |f(t, \alpha t^{n-1-k})| \, dt = +\infty.$$

Notes. Theorem 10.1 somewhat modifies a result in [60]. In one particular case, for equations of the type (10.74), it was proved by A.G. Kartsatos [169]. Analogues of this theorem for differential systems and functional-differential equations are contained in [70, 230, 303, 330]. For second order nonlinear differential equations and for two-dimensional differential systems, K. Schrader [318] and Ch.A. Skhalyakho [330] have established comparison theorems that are more general.

Theorems 10.2–10.4 and 10.8–10.11 are due to I.T. Kiguradze [172, 180, 185–187], while Theorems 10.6, 10.7, 10.12 and 10.13 are due to D.V. Izyumova and I.T. Kiguradze [155, 190]. Besides, the following works concerning the oscillation of solutions of nonlinear differential equations should be mentioned: [28, 29, 31, 33, 45, 47, 64, 92, 137, 138, 145, 146, 148, 150, 157, 161–166, 189, 290, 291, 313, 323, 326].

Oscillation properties of differential systems were investigated in [18, 42–44, 76, 156, 272–279, 328–338].

From the large amount of studies dealing with the oscillation theory of functional-differential equations we indicate the monographs [97, 230, 285, 324] and the papers [10, 60, 66, 79, 83, 98, 119, 155, 158, 169, 216–228, 244, 253, 254, 265, 282, 286, 287, 289, 319–321, 325, 327, 355–358, 366, 367].

§11. Singular Solutions

Consider the differential equation

$$u^{(n)} = f(t, u, \ldots, u^{(n-1)}), \tag{11.1}$$

where $f \in K_{\mathrm{loc}}([a, +\infty[\times \mathbf{R}^n)$, $a \in]0, +\infty[$, and

$$f(t, 0, \ldots, 0) = 0 \qquad \text{for } t \geq a. \tag{11.2}$$

DEFINITION 11.1. A solution u of the equation (11.1) defined on an interval $[t_*, +\infty[\subset [a, +\infty[$ is said to be *first kind singular* if there exists $t^* \in]t_*, +\infty[$ such that

$$\max\{|u(\tau)| : t \leq \tau \leq t^*\} > 0 \qquad \text{for } t_* < t < t^*, \qquad u(t) = 0 \qquad \text{for } t \geq t^*.$$

The interval $[t_*, t^*[$ is said to be the *carrier* of the solution u.

DEFINITION 11.2. A first kind singular solution of the equation (11.1) with the carrier $[t_*, t^*[$ is said to be *oscillatory* (*nonoscillatory*) if it has (does not have) a sequence of zeros converging to t^*.

DEFINITION 11.3. A solution u of the equation (11.1) defined on a finite interval $[t_*, t^*[\subset [a, +\infty[$ is said to be *second kind singular* if

$$\lim_{t \to t^*} \sup |u^{(n-1)}(t)| = +\infty,$$

and is said to be *oscillatory* (*nonoscillatory*) if it has (does not have) a sequence of zeros converging to t^*.

11.1. Nonoscillatory first kind singular solutions.

THEOREM 11.1. Let $\lambda \in]0,1[$, $m \in \{0,1\}$, $k \in \{1,\ldots,n\}$, $t^* \in]a,+\infty[$, and let on the set $\{(t,x_1,\ldots,x_n): a < t < t^*, (-1)^{m+i-1}x_i \geq 0 \ (i=1,\ldots,n)\}$ the inequality

$$(-1)^{m+n} f(t,x_1,\ldots,x_n) \geq g(t)|x_k|^\lambda \tag{11.3}$$

hold, where $g \in L([a,t^*])$,

$$g(t) \geq 0, \qquad \operatorname{meas}\{\tau \in [t,t^*] : g(\tau) > 0\} > 0 \qquad \text{for } a < t < t^*. \tag{11.4}$$

Then there exist $t_* \in]a,t^*[$ and a first kind singular solution u of the equation (11.1) with carrier $[t_*,t^*[$ such that

$$(-1)^{m+i} u^{(i)}(t) > 0 \qquad \text{for } t_* \leq t < t^* \qquad (i=0,\ldots,n-1). \tag{11.5}$$

In order to prove this theorem we need

LEMMA 11.1. Let $u \in \tilde{C}^{n-1}([t_*,t^*])$ and

$$(-1)^i u^{(i)}(t)u(t) > 0 \qquad (i=0,\ldots,n-1),$$

$$(-1)^n u^{(n)}(t)u(t) \geq 0 \qquad \text{for } t_* \leq t < t^*. \tag{11.6}$$

Then, whatever $\lambda \in]0,1[$, $k \in \{1,\ldots,n\}$ and

$$\alpha \in \left[\frac{2}{(\lambda+1)(n-k)+2}, \frac{1}{\lambda(n-k)+1} \right[,$$

we have

$$\prod_{i=k-1}^{n-1} |u^{(i)}(t)|^{2\beta/(n-k+1)(n+k-2)}$$

$$\geq \gamma \int_t^{t^*} |u^{(n)}(\tau)|^\alpha |u^{(k-1)}(\tau)|^{-\lambda\alpha} |u^{(n-1)}(\tau)|^{(\beta+1-(n-k\lambda+\lambda)\alpha)/(n-1)} \, d\tau$$

$$\text{for } t_* \leq t < t^*, \tag{11.7}$$

where $\beta = (n+k-2)[1-(1+\lambda n - \lambda k)\alpha]/(n-k)$ and

$$\gamma = \frac{2\beta\alpha^{-\alpha}}{(n-k+1)(n+k-2)}$$

$$\times \prod_{i=1}^{n-k} \left[\lambda\alpha + \frac{2i\beta}{(n-k+1)(n+k-2)} \right]^{-\lambda\alpha - 2i\beta/(n-k+1)(n+k-2)}.$$

PROOF. Set

$$x(t) = \prod_{i=1}^{n-k+1} |u^{(i+k-2)}(t)|, \qquad \epsilon = \frac{2\beta}{(n-k+1)(n+k-2)},$$

$$\alpha_i = \lambda\alpha + i\epsilon \qquad (i=1,\ldots,n-k), \qquad \alpha_{n-k+1} = \alpha.$$

Note that $x(t) > 0$ for $t_* \leq t < t^*$, $\epsilon > 0$, and

$$\alpha_i > 0 \qquad (i=1,\ldots,n-k+1), \qquad \sum_{i=1}^{n-k+1} \alpha_i = 1. \tag{11.8}$$

From (11.6) and (11.8) we obtain

$$x'(t) = -x(t) \sum_{j=1}^{n-k+1} \frac{|u^{(j+k-1)}(t)|}{|u^{(j+k-2)}(t)|}$$

and[2]

$$\sum_{j=1}^{n-k+1} \frac{|u^{(j+k-1)}(t)|}{|u^{(j+k-2)}(t)|} \geq \prod_{j=1}^{n-k+1} \left(\frac{|u^{(j+k-1)}(t)|}{\alpha_j |u^{(j+k-2)}(t)|} \right)^{\alpha_j} = \frac{\gamma}{\epsilon} v(t)[x(t)]^{-\epsilon},$$

where

$$v(t) = |u^{(n)}(t)|^\alpha |u^{(k-1)}(t)|^{-\lambda\alpha} |u^{(n-1)}(t)|^{(\beta+1-(n-k\lambda+\lambda)\alpha)/(n-1)}.$$

Hence

$$-[x(t)]^{1-\epsilon} x'(t) \geq \frac{\gamma}{\epsilon} v(t) \qquad \text{for } t_* < t < t^*.$$

By integrating this inequality from t to t^*, we obtain (11.7). $\square$

PROOF OF THEOREM 11.1. Let

$$\alpha \in \left[\frac{2}{(\lambda+1)(n-k)+2}, \frac{1}{\lambda(n-k)+1} \right[, \qquad \beta = 1 - [1 + \lambda(n-k)]\alpha,$$

$$\delta(t) = \left(\gamma \int_t^{t^*} [g(\tau)]^\alpha \, d\tau \right)^{(n-k+1)(n+k-2)/2\beta},$$

$$\tilde{f}(t, x_1, \dots, x_n)$$

$$= \begin{cases} f(t, x_1, \dots, x_n) & \text{for } (-1)^m x_k \geq \delta(t) \\ f(t, x_1, \dots, x_{k-1}, (-1)^m \delta(t), x_{k+1}, \dots, x_n) & \text{for } (-1)^m x_k < \delta(t), \end{cases} \tag{11.9}$$

$$f^*(t) = \max\{|f(t, x_1, \dots, x_n)| : |x_1| \leq 1, \dots, |x_n| \leq 1\},$$

and let $t_* \in [a, t^*[$ be chosen so that

$$\delta(t_*) < 1, \qquad t^* - t_* < 1, \qquad \int_{t_*}^{t^*} f^*(t) \, dt < 1.$$

Then the Cauchy problem

$$u^{(n)} = \tilde{f}(t, u, \dots, u^{(n-1)}), \qquad u^{(i)}(t^*) = 0 \qquad (i = 0, \dots, n-1)$$

has a solution u on $[t_*, t^*]$, and

$$|u^{(i)}(t)| < 1 \qquad \text{for } t_* \leq t \leq t^* \qquad (i = 0, \dots, n-1). \tag{11.10}$$

On the other hand, according to (11.3), (11.4) and (11.9),

$$(-1)^{m+i} u^{(i)}(t) = \frac{(-1)^{m+n}}{(n-1-i)!} \int_t^{t^*} (t-\tau)^{n-i-1} u^{(n)}(\tau) \, d\tau$$

$$\geq \frac{1}{(n-1-i)!} \int_t^{t^*} (t-\tau)^{n-i-1} g(\tau)[\delta(\tau)]^\lambda \, d\tau > 0$$

$$\text{for } t_* \leq t < t^* \qquad (i = 0, \dots, n-1)$$

---

[2]See [125], p. 38.

and
$$(-1)^{n+m}u^{(n)}(t) \geq g(t)[u(t)]^\lambda \qquad \text{for } t_* \leq t < t^*. \tag{11.11}$$

By Lemma 11.1 the estimate (11.7) holds, and
$$\beta + 1 - (n - k\lambda + \lambda)\alpha = \frac{n-1}{n-k}[2 - ((n-k)(\lambda+1)+2)\alpha] \leq 0.$$

From (11.7), taking into account (11.5), (11.10) and (11.11), we get
$$(-1)^m u^{(k-1)}(t) \geq \delta(t) \qquad \text{for } t_* \leq t \leq t^*. \tag{11.12}$$

Extend u to the whole of $[t_*, +\infty[$, letting it be identically zero on $[t^*, +\infty[$. The relations (11.2), (11.9) and (11.12) imply that u is a solution of (11.1). $\square$

PROBLEM 11.1. Let $m \in \{0,1\}$, $k \in \{1, \ldots, n\}$, $t^* \in \,]a, +\infty[$, and
$$(-1)^{m+n}f(t, x_1, \ldots, x_n) \geq \varphi(|x_k|)$$
for $a < t < t^*$, $(-1)^{m+i-1}x_i \geq 0$ $(i = 1, \ldots, n)$, where $\varphi: \mathbf{R}_+ \to \mathbf{R}_+$ is continuous, $\varphi(x) > 0$ for $x > 0$, and the function
$$\varphi_0(x) = \left[\frac{1}{(n-k-1)!}\int_0^x (x-s)^{n-k-1}\varphi(s)\,ds\right]^{1/(n-k+1)}$$
satisfies the condition
$$\psi(x) = \int_0^x \frac{ds}{\varphi_0(s)} < +\infty \qquad \text{for } x > 0.$$

Then there exist $t_* \in \,]a, t^*[$ and a first kind singular solution u of the equation (11.1) with carrier $[t_*, t^*[$ such that the inequalities (11.5) hold and
$$\psi(|u^{(n-k)}(t)|) \geq t^* - t \qquad \text{for } t_* \leq t \leq t^*.$$

THEOREM 11.2. Let $\lambda \in \,]0,1[$, $m \in \{0,1\}$, $k \in \{1, \ldots, n\}$,
$$\text{meas}\{s \in [t, +\infty[\, : \, |f(s, x, 0, \ldots, 0)| > 0\} > 0 \qquad \text{for } t > a, \ (-1)^m x > 0, \tag{11.13}$$
and let on the set $\{(t, x_1, \ldots, x_n): t \geq a, (-1)^{m+i-1}x_i \geq 0 \ (i = 1, \ldots, n)\}$ the condition (11.3) with a nonnegative $g \in L_{\text{loc}}([a, +\infty[)$ hold. Suppose, in addition, that for a certain $\alpha \in [2/((\lambda+1)(n-k)+2), 1/(\lambda(n-k)+1)[$,
$$\lim_{t \to +\infty} \sup t^\beta \int_t^{+\infty} \tau^{(n-k\lambda+\lambda)\alpha-\beta-1}[g(\tau)]^\alpha \, d\tau > 0, \tag{11.14}$$
where $\beta = (n+k-2)[1-(1+\lambda n - \lambda k)\alpha]/(n-k)$.[3] Then every nonzero solution u of the equation (11.1) defined on $[t_*, +\infty[\, \subset [a, +\infty[$ and satisfying the inequalities
$$(-1)^{m+i}u^{(i)}(t) \geq 0 \qquad \text{for } t \geq t_* \qquad (i = 0, \ldots, n-1) \tag{11.15}$$
is first kind singular.

PROOF. Assume, on the contrary, that the equation (11.1) has a proper solution u satisfying (11.15). Then $(-1)^m u(t) > 0$ for $t \geq t_*$. If $u^{(i)}(t_0) = 0$ for certain

---

[3] For $k = 1$ the condition (11.13) follows from (11.3) and (11.14).

$t_0 \in \,]t_*, +\infty[$ and $i \in \{1, \dots, n-1\}$, then by (11.15), $u^{(j)}(t) = 0$ $(j = 1, \dots, n-1)$ and $u(t) = u(t_0) \neq 0$ for $t \geq t_0$. But this is impossible, because in view of (11.13),

$$|u^{(n-1)}(t)| = \int_{t_0}^{t} |f(\tau, u(t_0), 0, \dots, 0)| \, d\tau > 0$$

for sufficiently large t. So

$$(-1)^{m+i} u^{(i)}(t) > 0 \qquad \text{for } t \geq t_* \qquad (i = 0, \dots, n-1).$$

On the other hand, (11.3) yields

$$(-1)^{n+m} u^{(n)}(t) \geq g(t)|u^{(k-1)}(t)|^\lambda \qquad \text{for } t \geq t_*. \tag{11.16}$$

According to Lemma 11.1 the estimate (11.7) holds for any $t^* \in [t_*, +\infty[$. Hence by (11.16),

$$\epsilon(t) \geq t^\beta \int_t^{+\infty} \tau^{(n-k\lambda+\lambda)\alpha - \beta - 1} [g(\tau)]^\alpha \, d\tau \qquad \text{for } t \geq t^*, \tag{11.17}$$

where

$$\epsilon(t) = \frac{1}{\gamma} [\epsilon_{n-1}(t)]^{((n-k\lambda+\lambda)\alpha - \beta - 1)/(n-1)} \prod_{i=k-1}^{n-1} [\epsilon_i(t)]^{2\beta/(n-k+1)(n+k-2)}$$

and

$$\epsilon_i(t) = \sup\{s^i |u^{(i)}(s)| : \ s \geq t\} \qquad (i = 0, \dots, n-1).$$

But $\epsilon_i(t) \to 0$ as $t \to +\infty$ $(i = 1, \dots, n-1)$ and so $\epsilon_i(t) \to 0$ as $t \to +\infty$. On the other hand, (11.14) and (11.17) imply

$$\lim_{t \to +\infty} \sup \epsilon(t) > 0.$$

This is a contradiction. $\square$

COROLLARY 11.1. Let $\lambda \in \,]0, 1[$, $m \in \{0, 1\}$, $k \in \{1, \dots, n\}$, and let the condition (11.13) hold. Suppose, in addition, that

$$(-1)^{m+n} f(t, x_1, \dots, x_n) \geq \delta t^{(k-1)\lambda - n} |x_k|^\lambda$$

on the set $\{(t, x_1, \dots, x_n): \ t \geq a, \ (-1)^{m+i-1} x_i \geq 0 \ (i = 1, \dots, n)\}$, where δ is a positive constant. Then every nonzero solution u of the equation (11.1) defined on $[t_*, +\infty[\subset [a, +\infty[$ and satisfying the inequalities (11.15) is first kind singular.

11.2. Nonoscillatory second kind singular solutions.

THEOREM 11.3. Let $\lambda \in \,]1, +\infty[$, $m \in \{0, 1\}$, $k \in \{1, \dots, n\}$, $\delta \in \,]0, +\infty[$, $t^* \in \,]a, +\infty[$, and

$$(-1)^m f(t, x_1, \dots, x_n) \geq g(t)|x_k|^\lambda \tag{11.18}$$

on the set $\{(t, x_1, \dots, x_n): \ a < t < t^*, \ (-1)^m x_i \geq \delta \ (i = 1, \dots, n)\}$, where the function $g \in L([a, t^*])$ is subject to (11.4). Then for any $t_* \in \,]a, t^*[$ there exists $\eta(t_*) \in \,]\delta, +\infty[$ such that every solution u of the equation (11.1) satisfying the conditions

$$(-1)^m u^{(i)}(t_*) \geq \delta \quad (i = 0, \dots, n-2), \qquad (-1)^m u^{(n-1)}(t_*) \geq \eta(t_*) \tag{11.19}$$

is second kind singular.

The proof of this theorem is based on

LEMMA 11.2. Let $u \in \tilde{C}_{\text{loc}}^{n-1}([t_*, t^*[)$ and

$$u^{(i)}(t)u(t) > 0 \qquad (i = 0, \ldots, n-1),$$

$$u^{(n)}(t)u(t) \geq 0 \qquad \text{for } t_* < t < t^*. \tag{11.20}$$

Then, whatever $k \in \{1, \ldots, n\}$ and $\alpha \in]0, 1/(n-k+1)[$, we have

$$\prod_{i=k-1}^{n-1} |u^{(i)}(t)|^{-\epsilon} \geq \gamma \int_t^{t^*} |u^{(n)}(\tau)|^\alpha |u^{(k-1)}(\tau)|^{((n-k+2)\alpha-2)/(n-k)} \, d\tau$$

$$\text{for } t_* < t < t^*, \tag{11.21}$$

where

$$\epsilon = \frac{2[1 - (n-k+1)\alpha]}{(n-k+1)(n-k)}, \qquad \gamma = \epsilon \prod_{i=1}^{n-k+1} [\alpha + (n-k+1-i)\epsilon]^{-\alpha-(n-k+1-i)\epsilon}.$$

PROOF. Set

$$x(t) = \prod_{i=1}^{n-k+1} |u^{(i+k-2)}(t)|, \qquad \alpha_j = \alpha + (n-k+1-j)\epsilon \qquad (j = 1, \ldots, n-k+1).$$

From (11.8) and (11.20) we derive

$$[x(t)]^{-1-\epsilon} x'(t) = [x(t)]^{-\epsilon} \sum_{j=1}^{n-k+1} \frac{|u^{(j+k-1)}(t)|}{|u^{(j+k-2)}(t)|}$$

$$\geq [x(t)]^{-\epsilon} \prod_{j=1}^{n-k+1} \left(\frac{|u^{(j+k-1)}(t)|}{\alpha_j |u^{(j+k-2)}(t)|} \right)^{\alpha_j}$$

$$= \frac{\gamma}{\epsilon} |u^{(n)}(t)|^\alpha |u^{(k-1)}(t)|^{((n-k+2)\alpha-2)/(n-k)} \qquad \text{for } t_* < t < t^*.$$

Integrating this inequality from t to t^*, we get (11.21). $\square$

PROOF OF THEOREM 11.3. Let $t_* \in [a, t^*[$,

$$\alpha \in \left[\frac{2}{n-k+2+(n-k)\lambda}, \frac{1}{n-k+1} \right[, \qquad \mu = \frac{[n-k+2+(n-k)\lambda]\alpha - 2}{n-k},$$

and let ϵ and γ be constants appearing in Lemma 11.2. By (11.4) we can choose a number $\eta(t_*)$ so that $\eta(t_*) > \delta$ and

$$[\eta(t_*)]^{-\epsilon} < \delta^{\epsilon(n-k)+\mu} \gamma \int_{t_*}^{t^*} [g(\tau)]^\alpha \, d\tau. \tag{11.22}$$

Consider a solution u of the equation (11.1) under the conditions (11.19). We will show that u is second kind singular. Assume the contrary. Then the domain of u contains $[t_*, t^*]$. On the other hand, (11.18) and (11.19) yield

$$(-1)^m u^{(i)}(t) \geq \delta \qquad (i = 0, \ldots, n-2), \qquad (-1)^m u^{(n-1)}(t) \geq \eta(t_*) > \delta,$$

$$(-1)^m u^{(n)}(t) \geq g(t)|u^{(k-1)}(t)|^\lambda \qquad \text{for } t_* < t < t^*. \tag{11.23}$$

According to Lemma 11.2 the estimate (11.21) is valid. Hence, taking into account (11.22) and (11.23), we obtain a contradiction:

$$[\eta(t_*)]^{-\epsilon} \geq \delta^{\epsilon(n-k)}\gamma \int_{t_*}^{t^*} [g(\tau)]^\alpha |u^{(k-1)}(\tau)|^\mu \, d\tau$$

$$\geq \delta^{\epsilon(n-k)+\mu}\gamma \int_{t_*}^{t^*} [g(\tau)]^\alpha \, d\tau > [\eta(t_*)]^{-\epsilon}. \qquad \square$$

PROBLEM 11.2. Let $m\{0,1\}$, $k \in \{1,\dots,n\}$, $\delta \in \,]0,+\infty[$, $t^* \in \,]a,+\infty[$, and

$$(-1)^m f(t,x_1,\dots,x_n) \geq \varphi(|x_k|) \qquad \text{for } a < t < t^*, \ (-1)^m x_i \geq \delta \quad (i=1,\dots,n),$$

where $\varphi \colon \mathbf{R}_+ \to \mathbf{R}_+$ is continuous, $\varphi(x) > 0$ for $x > 0$, and

$$\int_x^{+\infty} \frac{ds}{\varphi_0(s)} < +\infty \qquad \text{for } x > 0,$$

with

$$\varphi_0(x) = \left[\frac{1}{(n-k-1)!} \int_0^x (x-s)^{n-k-1}\varphi(s)\,ds \right]^{1/(n-k+1)}.$$

Then for any $t_* \in \,]a,t^*[$ there exists $\eta(t_*) \in \,]\delta,+\infty[$ such that every solution u of the equation (11.1) satisfying the inequalities (11.19) is second kind singular.

THEOREM 11.4. Let $\lambda \in \,]1,+\infty[$, $m \in \{0,1\}$, $k \in \{1,\dots,n\}$, $\delta \in \,]0,+\infty[$, and let on the set $\{(t,x_1,\dots,x_n)\colon t > a, \ (-1)^m x_i \geq (\delta/(n-i)!)t^{n-i} \ (i=1,\dots,n)\}$ the condition (11.18) hold, where $g \in L_{\mathrm{loc}}([a,+\infty[)$ is nonnegative. Suppose, in addition, that for a certain $\alpha \in [2/(n-k+2+(n-k)\lambda), 1/(n-k+1)[$,

$$\lim_{t\to+\infty} \sup t^{1-(n-k+1)\alpha} \int_t^{+\infty} \tau^{[\lambda(n-k)+n-k+2]\alpha-2}[g(\tau)]^\alpha \, d\tau > 0. \qquad (11.24)$$

Then every nonzero solution u of the equation (11.1) satisfying the inequalities

$$(-1)^m u^{(i-1)}(t_*) \geq \frac{\delta}{(n-i)!} t_*^{n-i} \qquad (i=1,\dots,n-1)$$

$$\tag{11.25}$$

$$(-1)^m u^{(n-1)}(t_*) > \delta$$

for a certain $t_* \in [a,+\infty[$ is second kind singular.

PROOF. First of all we note that (11.24) yields

$$\int_a^{+\infty} \tau^{\lambda(n-k)} g(\tau) \, d\tau = +\infty. \qquad (11.26)$$

Indeed, otherwise by the Hölder inequality we would obtain

$$t^{1-(n-k+1)\alpha} \int_t^{+\infty} \tau^{[\lambda(n-k)+n-k+2]\alpha-2}[g(\tau)]^\alpha \, d\tau$$

$$\leq t^{1-(n-k+1)\alpha} \left(\int_t^{+\infty} \tau^{((n-k+2)\alpha-2)/(1-\alpha)} \, d\tau \right)^{1-\alpha} \left(\int_t^{+\infty} \tau^{\lambda(n-k)} g(\tau) \, d\tau \right)^\alpha$$

$$\leq \left[\frac{1-\alpha}{1-(n-k+1)\alpha}\right]^{1-\alpha} \left(\int_t^{+\infty} \tau^{\lambda(n-k)} g(\tau)\, d\tau\right)^\alpha \to 0 \qquad \text{as } t \to +\infty,$$

which is impossible in view of (11.24).

Now we assume that the theorem is false, i.e. the equation (11.1) has a proper solution u satisfying (11.25) for a certain $t_* \in [a, +\infty[$. Then (11.18) and (11.26) imply

$$(-1)^m u^{(i-1)}(t) \geq \frac{\delta}{(n-i)!} t^{n-i} \qquad \text{for } t \geq t_* \qquad (i = 1, \dots, n), \tag{11.27}$$

$$(-1)^m u^{(n)}(t) \geq g(t) |u^{(k-1)}(t)|^\lambda \qquad \text{for } t \geq t_* \tag{11.28}$$

and

$$(-1)^m u^{(n-1)}(t) \geq \left[\frac{\delta}{(n-k)!}\right]^\lambda \int_{t_*}^t \tau^{\lambda(n-k)} g(\tau)\, d\tau \to +\infty \qquad \text{as } t \to +\infty.$$

Hence

$$v_i(t) = t^{i-n} |u^{(i-1)}(t)| \to +\infty \qquad \text{as } t \to +\infty \qquad (i = 1, \dots, n). \tag{11.29}$$

According to Lemma 11.2 the estimate (11.21) is true for any $t^* \in \,]t_*, +\infty[$. So, taking into account (11.27)–(11.29), we get

$$\eta(t) \geq t^{1-(n-k+1)\alpha} \int_t^{+\infty} \tau^{[\lambda(n-k)+n-k+2]\alpha-2} [g(\tau)]^\alpha\, d\tau \qquad \text{for } t > t_*,$$

with

$$\eta(t) = \frac{1}{\gamma} \delta^{(2-[\lambda(n-k)+n-k+2]\alpha)/(n-k)} \prod_{i=k}^n [v_i(t)]^{-\epsilon} \to 0 \qquad \text{as } t \to +\infty.$$

But this contradicts (11.24). $\square$

COROLLARY 11.2. Let $\lambda \in \,]1, +\infty[$, $m \in \{0, 1\}$, $k \in \{1, \dots, n\}$, $\delta \in \,]0, +\infty[$, and

$$(-1)^m f(t, x_1, \dots, x_n) \geq r t^{-\lambda(n-k)-1} |x_k|^\lambda$$

on the set $\{(t, x_1, \dots, x_n): t > a, (-1)^m x_i \geq (\delta/(n-i)!) t^{n-i}\ (i = 1, \dots, n)\}$, where r is a positive constant. Then every nonzero solution u of the equation (11.1) satisfying the inequalities (11.25) for a certain $t_* \in [a, +\infty[$ is second kind singular.

11.3. Nonexistence theorem for singular solutions.

THEOREM 11.5. Let the inequality

$$|f(t, x_1, \dots, x_n)| \leq h(t) \omega \left(\sum_{k=1}^n |x_k|\right) \tag{11.30}$$

hold on the set $[a, +\infty[\times \mathbf{R}^n$, where $h \in L_{\text{loc}}([a, +\infty[)$, the continuous function $\omega: \mathbf{R}_+ \to \mathbf{R}_+$ is nondecreasing, and $\omega(x) > 0$ for $x > 0$. If, in addition,

$$\int_0^1 \frac{dx}{\omega(x)} = +\infty, \tag{11.31}$$

then the equation (11.1) does not have a first kind singular solution, and if

$$\int_1^{+\infty} \frac{dx}{\omega(x)} = +\infty, \tag{11.32}$$

then this equation does not have a second kind singular solution.

PROOF. First we consider the case when the condition (11.31) is fulfilled. Assume that the equation (11.1) has a first kind singular solution with carrier $[t_*, t^*[\subset [a, +\infty[$. Then, by (11.30),

$$\rho(t) \leq \eta_0 \int_t^{t^*} h(\tau)\omega(\rho(\tau))\, d\tau \qquad \text{for } t_* \leq t \leq t^*,$$

where

$$\rho(t) = \sum_{i=1}^n |u^{(i-1)}(t)|, \qquad \eta_0 = \sum_{i=1}^n \frac{(t^* - t_*)^{n-i}}{(n-i)!}. \tag{11.33}$$

Hence, according to Lemma 9.2,

$$\int_0^{\delta} \frac{dx}{\omega(x)} \leq \eta_0 \int_{t_*}^{t^*} h(\tau)\, d\tau < +\infty,$$

where $\delta = \rho(t^*) > 0$. But this contradicts the condition (11.31). So we have proved that if (11.31) is valid, then the equation (11.1) does not have a first kind singular solution.

It remains to show that if (11.32) holds, then the equation (11.1) does not have a second kind singular solution. Let u be such a solution on the interval $[t_*, t^*[$. In view of (11.30),

$$\rho(t) \leq \eta_1 + \eta_0 \int_{t_*}^t h(\tau)\omega(\rho(\tau))\, d\tau \qquad \text{for } t_* \leq t \leq t^*, \tag{11.34}$$

where

$$\eta_1 = \sum_{k=1}^n \sum_{j=k}^n \frac{|u^{(j-1)}(t_*)|}{(j-k)!}(t^* - t_*)^{j-k} > 0,$$

and η_0 and ρ are the constant and function given by the equalities (11.33). Moreover,

$$\lim_{t \to t^*-} \rho(t) = +\infty. \tag{11.35}$$

From (11.34), applying Lemma 9.2 and the condition (11.35), we obtain

$$\int_{\eta_1}^{+\infty} \frac{dx}{\omega(x)} \leq \eta_0 \int_{t_*}^{t^*} h(\tau)\, d\tau < +\infty,$$

which contradicts (11.32). $\square$

Notes. Lemmas 11.1 and 11.2 are due to N.A. Izobov [141, 143] and Theorems 11.1 and 11.3 to I.T. Kiguradze [188]. Theorem 11.2 generalizes Theorem 12.3 in [188] and Theorem 1 in [143]. Theorem 11.4 generalizes Theorem 12.5 in [188] and Theorem 1 in [141]. Theorem 11.5 somewhat modifies well-known results of A. Wintner and W. Osgood [127]. For Problems 11.1 and 11.2 see [199, 245].

Criteria for the existence of singular solutions of nonlinear differential systems, as well as asymptotic estimates of them are contained in the monograph by N.P. Erugin [107] and the papers by T.A. Chanturia [58, 59].

§12. Fast Growing Solutions

Consider the differential equation

$$u^{(n)} = f(t, u, \dots, u^{(n-1)}), \tag{12.1}$$

where $f \in K_{loc}([a, +\infty[\times \mathbf{R}^n)$ and $a \in]0, +\infty[$.

DEFINITION 12.1. A proper solution u of the equation (12.1) is said to be *fast growing* if

$$\lim_{t \to +\infty} |u^{(n-1)}(t)| = +\infty$$

and *slowly growing* if

$$\lim_{t \to +\infty} \sup |u^{(n-1)}(t)| < +\infty.$$

In general, the question of whether (12.1) has fast growing solutions is worthy of interest when the growth exponent of the function f in some phase variable exceeds 1. This is the case treated in the present section.

12.1. Existence theorem.

THEOREM 12.1. Let

$$(-1)^m f(t, x_1, \dots, x_n) \geq g(t) |x_k|^\lambda$$

$$\text{for } t \geq a, \ (-1)^m x_i \geq \frac{\delta}{(n-i)!} t^{n-i} \qquad (i = 1, \dots, n), \tag{12.2}$$

and

$$\int_a^{+\infty} f_\delta(t; r)\, dt < +\infty \qquad \text{for } r > \delta, \tag{12.3}$$

where $m \in \{0, 1\}$, $\lambda \in]1, +\infty[$, $k \in \{1, \dots, n\}$, $\delta \in]0, +\infty[$, $g \in L_{loc}([a, +\infty[)$ is a nonnegative function which is distinct from zero on a set of positive measure in any neighborhood of $+\infty$, and

$$f_\delta(t; r) = \max \left\{ |f(t, x_1, \dots, x_n)| : \right.$$

$$\left. \frac{\delta}{(n-i)!} \leq (-1)^m t^{i-n} x_i \leq \frac{r}{(n-i)!} \quad (i = 1, \dots, n) \right\}. \tag{12.4}$$

Moreover, suppose the equation (12.1) under the initial conditions $u^{(i-1)}(t_0) = c_i$ $(i = 1, \dots, n)$ is uniquely solvable whenever $t_0 \geq a$ and $(-1)^m c_i > (\delta/(n-i)!)t_0^{n-i}$

$(i = 1, \ldots, n)$. Then (12.1) has an $(n-1)$-parameter family of fast growing solutions satisfying the condition

$$(-1)^m u^{(n-1)}(t) \to +\infty \qquad \text{as } t \to +\infty. \tag{12.5}$$

To prove this assertion we need

LEMMA 12.1. Let $m \in \{0, 1\}$, $\delta \in {]}0, +\infty{[}$, $r_0 \in {]}\delta, +\infty{[}$, $t_0 \in {]}a, +\infty{[}$,

$$(-1)^m f(t, x_1, \ldots, x_n) \geq 0$$

$$\text{for } t \geq a, \ (-1)^m x_i \geq \frac{\delta}{(n-i)!} t^{n-i} \qquad (i = 1, \ldots, n), \tag{12.6}$$

and

$$\int_{t_0}^{+\infty} f_\delta(t; 2r_0) \, dt \leq r_0, \tag{12.7}$$

where f_δ is the function defined by (12.4). Then every solution u of the equation (12.1) satisfying the conditions

$$\frac{\delta}{(n-i)!} t_0^{n-i} < (-1)^m u^{(i-1)}(t_0) \leq \frac{r_0}{(n-i)!} t_0^{n-i} \qquad (i = 1, \ldots, n) \tag{12.8}$$

is slowly growing.

PROOF. Let u be a solution of the equation (12.1) satisfying (12.8). We claim that it is defined on the whole of $[t_0, +\infty{[}$ and that

$$\frac{\delta}{(n-i)!} t^{n-i} < (-1)^m u^{(i-1)}(t) < \frac{2r_0}{(n-i)!} t^{n-i} \qquad (i = 1, \ldots, n) \tag{12.9}$$

on this half-line. Assume the contrary. Then by (12.6) there exist $i_0 \in \{1, \ldots, n\}$ and $t_1 \in {]}t_0, +\infty{[}$ such that (12.9) holds on $[t_0, t_1{[}$ and

$$|u^{(i_0-1)}(t_1)| = \frac{2r_0}{(n-i_0)!} t_1^{n-i_0}.$$

On the other hand, the equality

$$u^{(i_0-1)}(t_1) = \sum_{i=i_0}^{n} \frac{(t_1 - t_0)^{i-i_0}}{(i - i_0)!} u^{(i-1)}(t_0)$$

$$+ \frac{1}{(n - i_0)!} \int_{t_0}^{t_1} (t_1 - \tau)^{n-i_0} f(\tau, u(\tau), \ldots, u^{(n-1)}(\tau)) \, d\tau,$$

together with (12.4), (12.7) and (12.8), implies

$$|u^{(i_0-1)}(t_1)| \leq \frac{1}{(n - i_0)!} \left[r_0 + \int_{t_0}^{t_1} f_\delta(\tau; 2r_0) \, d\tau \right] t_1^{n-i_0} < \frac{2r_0}{(n - i_0)!} t_1^{n-i_0},$$

which is a contradiction. $\square$

PROOF OF THEOREM 12.1. According to Theorem 11.3 there exists a function $\eta \colon [a, +\infty{[} \to {]}\delta, +\infty{[}$ such that any solution u of the equation (12.1) satisfying the conditions

$$(-1)^m u^{(i-1)}(t_*) > \frac{\delta}{(n-i)!} t_*^{n-i} \qquad (i = 1, \ldots, n-1),$$

$$(-1)^m u^{(n-1)}(t_*) > \eta(t_*)$$

for a certain $t_* \in [a, +\infty[$ is second kind singular.

Let $r_0 \in]\delta, +\infty[$. By (12.3), for a sufficiently large $t_0 \in [a, +\infty[$ the inequality (12.7) is fulfilled. We will show that if

$$\frac{\delta}{(n-i)!} t_0^{n-i} < (-1)^m c_i \le \frac{r_0}{(n-i)!} t_0^{n-i} \qquad (i = 1, \ldots, n-1),$$

then the equation (12.1) has at least one fast growing solution u which satisfies (12.5) and the equalities

$$u^{(i-1)}(t_0) = c_i \qquad (i = 1, \ldots, n-1). \tag{12.10}$$

For each $c \in]\delta, +\infty[$ we denote by u_c the solution of (12.1) under the conditions (12.10) and

$$u^{(n-1)}(t_0) = (-1)^m c. \tag{12.11}$$

If $c > \eta(t_0)$, then u_c is a second kind singular solution and if $c \le r_0$, then, in view of Lemma 12.1, it is slowly growing. Let c^* be the least upper bound of the set of those $c \in]\delta, +\infty[$ for which u_c is slowly growing. By (12.2),

$$(-1)^m u_{c^*}^{(i-1)}(t) > \frac{\delta}{(n-i)!} t^{n-i} \qquad \text{for } t \ge t_0 \qquad (i = 1, \ldots, n). \tag{12.12}$$

First we assume u_{c^*} to be slowly growing. Then

$$\frac{\delta}{(n-i)!} t^{n-i} < (-1)^m u_{c^*}^{(i-1)}(t) < \frac{2r_1}{(n-i)!} t^{n-i} \qquad \text{for } t \ge t_0 \qquad (i = 1, \ldots, n)$$

with a certain $r_1 \in]\delta, +\infty[$. Choose $t_1 \in]t_0, +\infty[$ so that

$$\int_{t_1}^{+\infty} f_\delta(t; 2r_1) \, dt < r_1.$$

The unique solvability of the problem (12.1), (12.10), (12.11) for any $c \in]\delta, +\infty[$ implies that if $c_0 > c^*$ and $c_0 - c^*$ is sufficiently small, then the domain of u_c contains $[t_0, t_1]$ and

$$\frac{\delta}{(n-i)!} t_1^{n-i} < (-1)^m u_{c_0}^{(i-1)}(t_1) < \frac{r_1}{(n-i)!} t_1^{n-i} \qquad (i = 1, \ldots, n).$$

According to Lemma 12.1, u_{c_0} is a slowly growing solution, which contradicts the definition of c^*. So u_{c^*} cannot be slowly growing.

Let u_{c^*} be a second kind singular solution on $[t_0, t^*[$. Then (12.2) and (12.12) yield

$$(-1)^m u_{c^*}^{(n-1)}(t) \to +\infty \qquad \text{as } t \to t^*.$$

Thus, for a certain $t_* \in]t_0, t^*[$ we have

$$(-1)^m u_{c^*}^{(n-1)}(t_*) > \eta(t^*). \tag{12.13}$$

By (12.11), (12.12) and (12.13) we can find $\epsilon > 0$ so that for any $c \in]c^* - \epsilon, c^*[$ the domain of u_c contains $[t_0, t_*]$ and

$$(-1)^m u_c^{(i-1)}(t_*) > \frac{\delta}{(n-i)!} t_*^{n-i} \qquad (i = 1, \dots, n-1),$$

$$(-1)^m u_c^{(n-1)}(t_*) > \eta(t^*).$$

Hence, taking into account (12.2) and the definition of the function η, we conclude that u_c is a second kind singular solution. But this contradicts the choice of c^*. Therefore, u_{c^*} cannot be second kind singular.

Since u_{c^*} is neither slowly growing nor second kind singular, by (12.2) and (12.12) it is a fast growing solution satisfying the condition (12.5). $\square$

PROBLEM 12.1. Let the condition (12.3) hold and

$$(-1)^m f(t, x_1, \dots, x_n) \geq \varphi(t, |x_k|, \dots, |x_n|)$$

$$\text{for } t \geq a, \ (-1)^m x_i \geq \frac{\delta}{(n-i)!} t^{n-i} \quad (i = 1, \dots, n),$$

where $m \in \{0, 1\}$, $k \in \{1, \dots, n\}$, $\delta \in]0, +\infty[$, $\varphi \colon [a, +\infty[\times (]0, +\infty[)^{n-k+1} \rightarrow]0, +\infty[$ is a continuous function nonincreasing in the first variable and nondecreasing in the other variables, and

$$\int_x^{+\infty} \frac{dy}{\psi(t, y)} < +\infty \qquad \text{for } t > a, \ x > 0,$$

with

$$\psi(t, x) = \left[\int_0^x (x - y)^{n-k-2} \varphi(t, y, t^{-1} y, \dots, (n-k)! t^{k-n} y) \, dy \right]^{1/(n-k+1)}$$

for $k < n$ and $\psi(t, x) = \varphi(t, x)$ for kn. Moreover, suppose that the equation (12.1) under the initial conditions $u^{(i-1)}(t_0) = c_i$ $(i = 1, \dots, n)$ is uniquely solvable whenever $t_0 \geq a$ and $(-1)^m c_i > (\delta/(n-i)!) t_0^{n-i}$ $(i = 1, \dots, n)$. Then (12.1) has an $(n-1)$-parameter family of fast growing solutions satisfying the condition (12.5).

12.2. Asymptotic estimates.

THEOREM 12.2. Let the inequality (12.2) hold, where $m \in \{0, 1\}$, $\lambda \in]1, +\infty[$, $k \in \{1, \dots, n\}$, $\delta \in]0, +\infty[$, and $g \in L_{\text{loc}}([a, +\infty[)$ is a nonnegative function which is distinct from zero on a set of positive measure in any neighborhood of $+\infty$. Suppose, in addition, that the equation (12.1) has at least one fast growing solution satisfying the condition (12.5). Then for any $\alpha \in [2/(n-k+2+(n-k)\lambda), 1/(n-k+1)[$,

$$\lim_{t \to +\infty} t^{1-(n-k+1)\alpha} \int_t^{+\infty} \tau^{[\lambda(n-k)+n-k+2]\alpha-2} [g(\tau)]^\alpha \, d\tau = 0, \qquad (12.14)$$

and every solution of this type satisfies the estimate

$$|u(t)| \leq \gamma_0 t^{k-1} \left(t^{(n-k+1)\alpha-1} \int_t^{+\infty} [g(\tau)]^\alpha \, d\tau \right)^{-1/(\lambda-1)\alpha} \qquad (12.15)$$

in a certain neighborhood of $+\infty$, where γ_0 is a positive constant depending on n, k, λ, and α only.

PROOF. Let $\alpha \in [2/(n-k+2+(n-k)\lambda), 1/(n-k+1)[$. According to Theorem 11.4, the existence of a fast growing solution of the equation (12.1) satisfying the condition (12.5) guarantees the validity of the equality (12.14).

Taking into account (12.2), we easily conclude that if u satisfies (12.5) and $t_0 \in \,]a, +\infty[$ is sufficiently large, then

$$(-1)^m u^{(i-1)}(t) > \frac{\delta}{(n-i)!} t^{n-i} \qquad (i = 1, \dots, n),$$

$$(-1)^m u^{(n)}(t) \geq g(t)|u^{(k-1)}(t)|^\lambda \qquad \text{for } t \geq t_0 \qquad (12.16)$$

and

$$|u^{(i)}(t)| \geq (i-j)!\, t^{j-i} |u^{(j)}(t)|$$
$$\text{for } t \geq t_0 \qquad (j = 1, \dots, n-1;\ i = j, \dots, n-1). \qquad (12.17)$$

By (12.16) and Lemma 11.2 we obtain

$$\prod_{i=k-1}^{n-1} |u^{(i)}(t)|^{-\epsilon}$$

$$\geq \gamma |u^{(k-1)}(t)|^{([n-k+2+(n-k)\lambda]\alpha-2)/(n-k)} \int_t^{+\infty} [g(\tau)]^\alpha \, d\tau \qquad \text{for } t \geq t_0,$$

where

$$\epsilon = \frac{2[1-(n-k+1)\alpha]}{(n-k+1)(n-k)}, \qquad \gamma = \epsilon \prod_{i=1}^{n-k+1} [\alpha+(n-k+1-i)\epsilon]^{-\alpha-(n-k+1-i)\epsilon}.$$

In view of (12.17), this yields

$$|u^{(k-1)}(t)| \leq \gamma_1 \left(t^{(n-k+1)\alpha-1} \int_t^{+\infty} [g(\tau)]^\alpha \, d\tau \right)^{-1/(\lambda-1)\alpha} \qquad \text{for } t \geq t_0,$$

with

$$\gamma_1 = \left(\gamma \prod_{i=k-1}^{n-1} [(i-k+1)!]^\epsilon \right)^{-1/(\lambda-1)\alpha}.$$

Applying (12.17) once again, from the last inequality we derive the estimate (12.15), where $\gamma_0 = \gamma_1/(k-1)!$. $\square$

PROBLEM 12.2. Let the inequality (12.2) hold, where $m \in \{0,1\}$, $\lambda \in \,]1, +\infty[$, $k \in \{1, \dots, n\}$, $\delta \in \,]0, +\infty[$, and $g: [a, +\infty[\,\to\,]0, +\infty[$ is a continuous nonincreasing function. Suppose, in addition, that the equation (12.1) has at least one fast growing solution satisfying the condition (12.5). Then

$$\lim_{t \to +\infty} t^{-1+(1+(n-k)\lambda)/(1+n-k)} \int_t^{+\infty} [g(\tau)]^{1/(n-k+1)} \, d\tau = 0$$

and every solution of this type satisfies the estimate

$$|u(t)| \leq \gamma^* t^{k-1} \left(\int_t^{+\infty} [g(\tau)]^{1/(n-k+1)} \, d\tau \right)^{-(n-k+1)/(\lambda-1)}$$

in a certain neighborhood of $+\infty$, where γ^* is a positive constant depending on n, k and λ only.

THEOREM 12.3. Let

$$0 \leq (-1)^m f(t, x_1, \ldots, x_n) \leq \sum_{k=1}^{n} h_k(t)|x_k|^{\lambda}$$

$$\text{for } t \geq a, \ (-1)^m x_i \geq \frac{\delta}{(n-i)!} t^{n-i} \qquad (i = 1, \ldots, n), \qquad (12.18)$$

where $m \in \{0, 1\}$, $\lambda \in]1, +\infty[$, $\delta \in]0, +\infty[$, the sum of the nonnegative functions $h_k \in L_{\text{loc}}([a, +\infty[)$ $(k = 1, \ldots, n)$ is distinct from zero on a set of positive measure in any neighborhood of $+\infty$, and

$$\int_a^{+\infty} t^{(n-k)\lambda} h_k(t)\, dt < +\infty \qquad (k = 1, \ldots, n).$$

Then any solution u of the equation (12.1) satisfying the condition (12.5) is subject to the estimate

$$|u^{(n-1)}(t)| \geq \left((\lambda - 1) \sum_{k=1}^{n} [(n-k)!]^{-\lambda} \int_t^{+\infty} \tau^{(n-k)\lambda} h_k(\tau)\, d\tau \right)^{-1/(\lambda-1)} \qquad (12.19)$$

in a certain neighborhood of $+\infty$.

PROOF. By (12.5) and (12.18), for a sufficiently large $t_0 \in]a, +\infty[$ the relations (12.17) hold and

$$\frac{|u^{(n-1)}(t)|'}{|u^{(n-1)}(t)|^{\lambda}} \leq \sum_{k=1}^{n} [(n-k)!]^{-\lambda} t^{(n-k)\lambda} h_k(t) \qquad \text{for } t \geq t_0.$$

Integrating this inequalities from t to $+\infty$, we get (12.19).

Notes. Theorems 12.1–12.3 are due to I.T. Kiguradze and G.G. Kvinikadze [199, 250]. The proof of Theorem 12.2 follows the paper of N.A. Izobov [141].

An analogue of Theorem 12.1 for the equation $(p(t)u')' = f(t, u, u')$ was established by H. Usami [363].

For Problems 12.1 and 12.2 see [199].

§13. Kneser Solutions

Consider the differential equation

$$u^{(n)} = f(t, u, \ldots, u^{(n-1)}), \qquad (13.1)$$

where $f \in K_{\text{loc}}([a, +\infty[\times \mathbf{R}^n)$ and $a \in]0, +\infty[$.

DEFINITION 13.1. A solution u of the equation (13.1) defined on an interval $[t_0, +\infty[\subset [a, +\infty[$ is said to be a *Kneser solution* if

$$(-1)^i u^{(i)}(t)u(t) \geq 0 \qquad \text{for } t \geq t_0 \qquad (i = 0, \ldots, n-1).$$

Below we establish existence criteria for Kneser solutions of (13.1) and study the asymptotics of these solutions.

13.1. Existence theorem.

THEOREM 13.1. Let
$$f(t, 0, \dots, 0) = 0 \qquad \text{for } t \geq a, \tag{13.2}$$
$m \in \{0, 1\}$, $r \in]0, +\infty[$ and
$$(-1)^{m+n} f(t, x_1, \dots, x_n) \geq 0$$
$$\text{for } t \geq a, \ 0 \leq (-1)^{m+i-1} x_i \leq rt^{1-i} \quad (i = 1, \dots, n). \tag{13.3}$$
Then the equation (13.1) has a continuum of Kneser solutions satisfying the conditions
$$(-1)^{m+i} u^{(i)}(t) \geq 0 \qquad \text{for } t \geq a \qquad (i = 0, \dots, n-1). \tag{13.4}$$

PROOF. Set
$$f^*(t) = \max\{|f(t, x_1, \dots, x_n)| : 0 \leq (-1)^{m+i-1} x_i \leq rt^{1-i} \quad (i = 1, \dots, n)\}$$
and choose $\delta \in]0, 1/(a+1)[$ so small that
$$2(a+1)^{n-1} \int_a^{a+\delta} f^*(\tau) \, d\tau \leq r. \tag{13.5}$$
To prove the theorem it suffices to show that if
$$r_0 = \left(\frac{\delta}{a+\delta}\right)^{n-1} \frac{r}{2n!},$$
then for any $c_0 \in [0, r_0]$ the equation (13.1) has at least one solution $u: [a, +\infty[$ $\to \mathbf{R}$ satisfying the inequalities (13.4) together with the condition
$$u(a) = (-1)^m c_0. \tag{13.6}$$

Consider the functions
$$\chi_i(t, x) = \begin{cases} 0 & \text{for } (-1)^{m+i-1} x < 0 \\ x & \text{for } 0 \leq (-1)^{m+i-1} x \leq rt^{1-i} \\ rt^{1-i} & \text{for } (-1)^{m+i-1} x > rt^{1-i} \end{cases}$$
and
$$\tilde{f}(t, x_1, \dots, x_n) = f(t, \chi_1(t, x_1), \dots, \chi_n(t, x_n)).$$
Clearly,
$$\tilde{f}(t, x_1, \dots, x_n) = f(t, x_1, \dots, x_n)$$
$$\text{for } t \geq a, \ 0 \leq (-1)^{m+i-1} x_i \leq rt^{1-i} \quad (i = 1, \dots, n) \tag{13.7}$$
and
$$|\tilde{f}(t, x_1, \dots, x_n)| \leq f^*(t) \qquad \text{for } t \geq a, \ (x_1, \dots, x_n) \in \mathbf{R}^n. \tag{13.8}$$
On the other hand, (13.2) and (13.3) yield
$$\tilde{f}(t, x_1, \dots, x_n) = 0 \qquad \text{for } t \geq a, \ (-1)^{m+i-1} x_i \leq 0 \qquad (i = 1, \dots, n) \tag{13.9}$$
and
$$(-1)^{m+n} \tilde{f}(t, x_1, \dots, x_n) \geq 0 \qquad \text{for } t \geq a, \ (-1)^{m+i-1} x_i \geq 0 \quad (i = 1, \dots, n). \tag{13.10}$$

By Lemma 10.1, for any positive integer k the differential equation

$$u^{(n)} = \tilde{f}(t, u, \dots, u^{(n-1)})$$

has a solution u_k satisfying the boundary conditions

$$u_k(a) = (-1)^m c_0, \qquad u_k^{(i-1)}(a+k) = 0 \qquad (i = 1, \dots, n-1). \tag{13.11}$$

Assuming that $(-1)^{m+n-1} u_k^{(n-1)}(a+k) < 0$, from (13.9) we obtain

$$u_k(t) = \frac{1}{(n-1)!} u_k^{(n-1)}(a+k)(t-a-k)^{n-1} \qquad \text{for } a \le t \le a+k,$$

$$(-1)^m u_k(a) < 0.$$

But this is impossible, because $(-1)^m u_k(a) = c_0 \ge 0$. So

$$(-1)^{m+n-1} u_k^{(n-1)}(a+k) \ge 0. \tag{13.12}$$

According to (13.10)–(13.12),

$$(-1)^{m+i} u_k^{(i)}(t) \ge 0 \qquad \text{for } a \le t \le a+k \qquad (i = 0, \dots, n), \tag{13.13}$$

$$\sum_{i=0}^{n-1} \frac{|u_k^{(i)}(t)|}{i!}(t-a)^i \le |u(a)| \le r_0 \qquad \text{for } a \le t \le a+k, \tag{13.14}$$

and

$$|u_k^{(i)}(a+\delta)| \le i! r_0 \delta^{-i} \qquad (i = 0, \dots, n-1). \tag{13.15}$$

Applying (13.5), (13.7) and (13.15), we get

$$|u_k^{(i)}(t)| \le \sum_{j=i}^{n-1} \frac{|u_k^{(j)}(a+\delta)|}{(j-i)!}(a+\delta-t)^{j-i} + \frac{1}{(n-i-1)!}\int_t^{a+\delta}(\tau-t)^{n-i+1} f^*(\tau) d\tau$$

$$\le n! r_0 \delta^{-i} + \int_a^{a+\delta} f^*(\tau)\, d\tau \le \left[n! r_0 \left(\frac{a+\delta}{\delta}\right)^i + \frac{(a+\delta)^i}{2(a+1)^{n-1}} r \right] t^{-i} \le rt^{-i}$$

$$\text{for } a \le t \le a+\delta \qquad (i = 0, \dots, n-1).$$

On the other hand, (13.14) implies

$$|u_k^{(i)}(t)| \le rt^{-i} \qquad \text{for } a+\delta \le t \le a+k \qquad (i = 0, \dots, n-1).$$

Therefore,

$$|u_k^{(i)}(t)| \le rt^{-i} \qquad \text{for } a \le t \le a+k \qquad (i = 0, \dots, n-1). \tag{13.16}$$

It follows from (13.7), (13.13) and (13.16) that for any positive integer k the function u_k is a solution of the equation (13.1) on $[a, a+k]$. By Lemma 10.2, $(u_k)_{k=1}^{+\infty}$ contains a subsequence $(u_{k_l})_{l=1}^{+\infty}$ such that $(u_{k_l}^{(i)})_{l=1}^{+\infty}$ $(i = 0, \dots, n-1)$ converge uniformly on every finite subinterval of $[a, +\infty[$, and

$$u(t) = \lim_{l \to +\infty} u_{k_l}(t) \qquad \text{for } t \ge a$$

is a solution of the equation (13.1). In view of (13.11) and (13.13), u satisfies the conditions (13.4) and (13.6). $\square$

PROBLEM 13.1. Let (13.2) hold, $m \in \{0,1\}$, $r \in\ ]0,+\infty[$, and

$$(-1)^{m+n}f(t,x_1,\dots,x_n) \geq -g_0(t)x_n^2 - g_1(t)|x_n|$$

for $t \geq a$, $0 \leq (-1)^{m+i-1}x_i \leq rt^{1-i}$ $(i=1,\dots,n-1)$, $x_n \in \mathbf{R}$,

where $g_0\colon [a,+\infty[\ \to \mathbf{R}_+$ is continuous and $g_1\colon [a,+\infty[\ \to \mathbf{R}_+$ is locally integrable. Then the equation (13.1) has a continuum of Kneser solutions satisfying (13.4).

13.2. Kneser solutions vanishing at infinity. We say that a Kneser solution *vanishes at infinity* if it tends to zero as $t \to +\infty$.

THEOREM 13.2. Let

$$(-1)^{m+n}f(t,x_1,\dots,x_n) \geq g(t,|x_1|) \qquad \text{for } t \geq a, (-1)^{m+i-1}x_i \geq 0 \quad (i=1,\dots,n),$$
$$(13.17)$$

where $m \in \{0,1\}$, the nonnegative function $g \in K_{\text{loc}}([a,+\infty[\ \times \mathbf{R}_+)$ is nondecreasing in the second variable, and

$$\int_a^{+\infty} t^{n-1}g(t,x)\,dt = +\infty \qquad \text{for } x > 0. \tag{13.18}$$

Then every Kneser solution of the equation (13.1) satisfying, for sufficiently large t, the inequality

$$(-1)^m u(t) \geq 0 \tag{13.19}$$

vanishes at infinity.

PROOF. Let $u\colon [t_0,+\infty[\ \to \mathbf{R}$ be a Kneser solution of the equation (13.1) satisfying (13.19). Then by (13.17), for any $t \in [t_0,+\infty[$ we have

$$|u(t_0)| = \sum_{i=0}^{n-1} \frac{|u^{(i)}(t)|}{i!}t^i + \frac{1}{(n-1)!}\int_{t_0}^t (\tau-t_0)^{n-1}|u^{(n)}(\tau)|\,d\tau$$

$$\geq \frac{1}{(n-1)!}\int_{t_0}^t (\tau-t_0)^{n-1}g(\tau,|u(\tau)|)\,d\tau \geq$$

$$\geq \frac{1}{(n-1)!}\int_{t_0}^t (\tau-t_0)^{n-1}g(\tau,|u(t)|)\,d\tau.$$

Hence, according to (13.18), $u(t) \to 0$ as $t \to +\infty$. □

REMARK 13.1. If

$$(-1)^{m+n}f(t,x_1,\dots,x_n) \geq 0 \qquad \text{for } t \geq a,\ (-1)^{m+i-1}x_i \geq 0 \quad (i=1,\dots,n),$$
$$(13.20)$$

then every Kneser solution of the equation (13.1) satisfying, for sufficiently large t, the inequality (13.19) is subject to the conditions

$$\lim_{t\to+\infty} t^i u^{(i)}(t) = 0 \qquad (i=1,\dots,n-1). \tag{13.21}$$

Indeed, (13.19) and (13.20) yield

$$|u(s)| = |u(t)| + \sum_{i=1}^{n-1} \frac{|u^{(i)}(t)|}{i!}t^i + \frac{1}{(n-1)!}\int_s^t (\tau-s)^{n-1}|u^{(n)}(\tau)|\,d\tau$$

for sufficiently large $s \in]a, +\infty[$ and $t \in]s, +\infty[$. From this equality, by passing first t and then s to $+\infty$, we obtain (13.21).

Theorem 13.2 and Corollary 8.2 imply

THEOREM 13.2'. Let

$$g(t, |x_1|) \leq (-1)^{m+n} f(t, x_1, \dots, x_n) \leq \rho g(t, |x_1|)$$

$$\text{for } t \geq a, \ (-1)^{m+i-1} x_i \geq 0 \quad (i = 1, \dots, n),$$

where $m \in \{0, 1\}$, $\rho \in]1, +\infty[$, and the nonnegative function $g \in K_{\text{loc}}([a, +\infty[\times \mathbf{R}_+)$ is nondecreasing in the second variable. Then every Kneser solution of the equation (13.1) satisfying, for sufficiently large t, the inequality (13.19) vanishes at infinity if and only if the condition (13.20) holds.

PROBLEM 13.2. Let

$$|f(t, x_1, \dots, x_n)| \leq \sum_{i=1}^{n} h_i(t) |x_i| \qquad \text{for } t \geq a, \ (-1)^{m+i-1} x_i \geq 0 \quad (i = 1, \dots, n),$$

where $m \in \{0, 1\}$, $h_i \in L_{\text{loc}}([a, +\infty[)$, and

$$\int_a^{+\infty} t^{n-i} h_i(t) \, dt < +\infty \qquad (i = 1, \dots, n).$$

Then every Kneser solution of the equation (13.1) satisfying, for sufficiently large t, the inequality (13.19) tends to a nonzero limit as $t \to +\infty$.

THEOREM 13.3. Let $k \in \{1, \dots, n\}$, $m \in \{0, 1\}$, $r \in]0, +\infty[$, and

$$g(t) |x_k|^\lambda \leq (-1)^{m+n} f(t, x_1, \dots, x_n) \leq h(t)$$

$$\tag{13.22}$$

$$\text{for } t \geq a, \ 0 \leq (-1)^{m+i-1} x_i \leq r t^{1-i} \quad (i = 1, \dots, n),$$

where $\lambda \in]0, 1[$, $g \in L_{\text{loc}}([a, +\infty[)$ is a nonnegative function which does not vanish on a set of positive measure in any neighborhood of $+\infty$, $h \in L_{\text{loc}}([a, +\infty[)$, and

$$\int_a^{+\infty} t^{n-1} h(t) \, dt < +\infty. \tag{13.23}$$

Then the equation (13.1) possesses at least one Kneser solution vanishing at infinity and satisfying the inequalities

$$(-1)^{m+i} u^{(i)}(t) > 0 \qquad (i = 0, \dots, n-1) \tag{13.24}$$

for sufficiently large t.

PROOF. According to the restrictions imposed on h there exists $t_0 \in]a, +\infty[$ such that

$$0 < \int_t^{+\infty} \tau^{n-1} h(\tau) \, d\tau < r \qquad \text{for } t \geq t_0.$$

Hence,

$$0 < (-1)^i w^{(i)}(t) < r t^{-i} \qquad \text{for } t \geq t_0 \quad (i = 0, \dots, n-1) \tag{13.25}$$

and
$$\lim_{t \to +\infty} w(t) = 0, \tag{13.26}$$

where
$$w(t) = \frac{1}{(n-1)!} \int_t^{+\infty} (\tau - t)^{n-1} h(\tau) \, d\tau.$$

By Theorem 11.1 and the inequalities (13.22) and (13.25), for any positive integer p the differential equation

$$\frac{dv}{dt} = (-1)^{m+n} g(t) [v^{(k-1)}]^\lambda$$

has a solution $v_p \colon [t_0, t_0 + p] \to \mathbf{R}$ satisfying the conditions

$$0 < (-1)^{m+i} v_p^{(i)}(t) < (-1)^i w^{(i)}(t) < rt^{-i} \qquad \text{for } t_0 \le t < t_0 + p,$$

$$v_p^{(i)}(t_0 + p) = 0 \qquad (i = 0, \dots, n-1). \tag{13.27}$$

Moreover, it is readily verified that

$$0 < (-1)^{m+i} v_p^{(i)}(t) < (-1)^{m+i} v_{p+1}^{(i)}(t)$$

$$\text{for } t_0 \le t < t_0 + p \qquad (i = 0, \dots, n-1; \ p = 1, 2, \dots). \tag{13.28}$$

Consider the Cauchy problem

$$u^{(i)}(t_0 + p) = \frac{(-1)^m}{2} w^{(i)}(t_0 + p) \qquad (i = 0, \dots, n-1)$$

for the equation (13.1). In view of (13.22) and (13.27), this problem has a solution $u_p \colon [t_0, t_0 + p] \to \mathbf{R}$ such that

$$(-1)^{m+i} v_p^{(i)}(t) < (-1)^{m+i} u_p^{(i)}(t) < (-1)^i w^{(i)}(t)$$

$$\text{for } t_0 \le t < t_0 + p \qquad (i = 0, \dots, n-1). \tag{13.29}$$

By Lemma 10.2, the inequalities (13.28) and (13.29) imply the existence of a solution $u \colon [t_0, +\infty[\to \mathbf{R}$ of the equation (13.1) satisfying the conditions (13.24) together with the estimate $|u(t)| \le w(t)$ for $t \ge t_0$. So from (13.26) we conclude that $u(t) \to 0$ as $t \to +\infty$. $\square$

PROBLEM 13.3. Let $k \in \{1, \dots, n\}$, $m \in \{0, 1\}$, $r \in \,]0, +\infty[$, and

$$g(t, |x_k|) \le (-1)^{m+n} f(t, x_1, \dots, x_n) \le h(t)$$

$$\text{for } t \ge a, \ 0 \le (-1)^{m+i-1} x_i \le rt^{1-i} \quad (i = 1, \dots, n),$$

where $g \in K_{\text{loc}}([a, +\infty[\times \mathbf{R}_+)$ is nonincreasing in the first variable, $g(t, x) > 0$ for $t \ge a$, $x > 0$,

$$\int_0^x \frac{ds}{g_0(t, s)} < +\infty \qquad \text{for } t \ge a, \ x > 0,$$

with
$$g_0(t, x) = \left(\int_0^x (x - s)^{n-k-1} g(t, s) \, ds \right)^{1/(n-k+1)},$$

and $h \in L_{\mathrm{loc}}([a, +\infty[)$ is subject to (13.23). Then the equation (13.1) has at least one Kneser solution vanishing at infinity and satisfying the inequalities (13.24) for sufficiently large t.

THEOREM 13.4. Let

$$(-1)^{m+n} f(t, x_1, \ldots, x_n) \geq g(t)|x_k|^\lambda$$

$$\text{for } t \geq a,\ 0 \leq (-1)^{m+i-1} x_i \leq r t^{1-i} \quad (i = 1, \ldots, n), \tag{13.30}$$

where $k \in \{1, \ldots, n\}$, $m \in \{0, 1\}$, $\lambda \in \ ]0, 1[$, $r \in \ ]0, +\infty[$, and $g \in L_{\mathrm{loc}}([a, +\infty[)$ is a nonnegative function which does not vanish on a set of positive measure in any neighborhood of $+\infty$. Suppose, in addition, that the equation (13.1) possesses a Kneser solution u vanishing at infinity and satisfying the inequalities (13.24). Then for any $\alpha \in [2/((\lambda + 1)(n - k) + 2), 1/(\lambda(n - k) + 1)[$,

$$\lim_{t \to +\infty} t^\beta \int_t^{+\infty} \tau^{(n-k\lambda+\lambda)\alpha-\beta-1}[g(\tau)]^\alpha \, d\tau = 0 \tag{13.31}$$

and

$$|u(t)| \geq \gamma_0 \left[\int_t^{+\infty} (\tau - t)^{n-2} \right.$$

$$\left(\int_\tau^{+\infty} [g(s)]^\alpha \, ds \right)^{(n-1)/((n-k\lambda+\lambda)\alpha-1)} \left. d\tau \right]^{((n-k\lambda+\lambda)\alpha-1)/(n-1)(1-\lambda)\alpha} \tag{13.32}$$

for t large, where $\beta = (n + k - 2)[1 - (1 + \lambda n - \lambda k)\alpha]/(n - k)$ and γ_0 is a constant depending on k, n, λ, and α only.

PROOF. Remark 13.1 and the condition (13.30) imply that if $t_0 \in \ ]a, +\infty[$ is sufficiently large, then

$$0 < (-1)^{m+i} u^{(i)}(t) < r t^{-i} \quad \text{for } t \geq t_0 \quad (i = 1, \ldots, n) \tag{13.33}$$

and

$$(-1)^{m+n} u^{(n)}(t) \geq g(t)|u(t)|^\lambda \quad \text{for } t \geq t_0. \tag{13.34}$$

By Lemma 11.1 and the inequality (13.34),

$$\prod_{i=k-1}^{n-1} |u^{(i)}(t)|^{2\beta/(n-k+1)(n+k-2)}$$

$$\geq \gamma \int_t^{+\infty} [g(\tau)]^\alpha |u^{(n-1)}(\tau)|^{(\beta+1-(n-k\lambda+\lambda)\alpha)/(n-1)} \, d\tau \quad \text{for } t \geq t_0. \tag{13.35}$$

This together with (13.21) yields the equality (13.31).

According to Lemma 5.2,

$$|u^{(i)}(t)| \leq 2^{(i-1/(n-1))(n-1-i)}[(n-1)!]^{(n-1-i)/(n-1)}|u(t)|^{(n-1-i)/(n-1)}$$

$$\text{for } t \geq t_0 \quad (i = 0, \ldots, n-1).$$

Thus, from (13.35) we derive

$$|u(t)|^{(1-(1+\lambda n-\lambda k)\alpha)/(n-1)}|u^{(n-1)}(t)|^{\beta/(n-1)}$$

$$\geq \gamma_1 |u^{(n-1)}(t)|^{(\beta+1-(n-k\lambda+\lambda)\alpha)/(n-1)} \int_t^{+\infty} [g(\tau)]^\alpha \, d\tau, \quad \text{for } t \geq t_0,$$

with

$$\gamma_1 = \gamma \left(\prod_{i=k-1}^{n-1} 2^{(i-1/(n-1))(n-1-i)} [(n-1)!]^{(n-1-i)/(n-1)} \right)^{-2\beta/(n-k+1)(n+k-2)}.$$

Consequently,

$$\gamma_2 \left(\int_t^{+\infty} [g(\tau)]^\alpha \, d\tau \right)^{(n-1)/((n-k\lambda+\lambda)\alpha-1)}$$

$$\leq |u(t)|^{(1-(1+\lambda n-\lambda k)\alpha)/((n-k\lambda+\lambda)\alpha-1)} |u^{(n-1)}(t)| \qquad \text{for } t \geq t_0,$$

where

$$\gamma_2 = \gamma_1^{(n-1)/((n-k\lambda+\lambda)\alpha-1)}.$$

Integrating the last inequality from t to $+\infty$ n times, we obtain

$$\frac{\gamma_2}{(n-2)!} \int_t^{+\infty} (\tau-t)^{n-2} \left(\int_\tau^{+\infty} [g(s)]^\alpha \, ds \right)^{(n-1)((n-k\lambda+\lambda)\alpha-1)} d\tau$$

$$\leq \frac{1}{(n-2)!} \int_t^{+\infty} (\tau-t)^{n-2} |u(\tau)|^{(1-(1+\lambda n-\lambda k)\alpha)/((n-k\lambda+\lambda)\alpha-1)} |u^{(n-1)}(\tau)| \, d\tau$$

$$\leq \frac{1}{(n-2)!} |u(t)|^{(1-(1+\lambda n-\lambda k)\alpha)/((n-k\lambda+\lambda)\alpha-1)} \int_t^{+\infty} (\tau-t)^{n-2} |u^{(n-1)}(\tau)| \, d\tau$$

$$= |u(t)|^{(n-1)(1-\lambda)\alpha/((n-k\lambda+\lambda)\alpha-1)} \qquad \text{for } t \geq t_0.$$

This immediately implies the estimate (13.32), with

$$\gamma_0 = \left[\frac{\gamma_2}{(n-2)!} \right]^{((n-k\lambda+\lambda)\alpha-1)/(n-1)(1-\lambda)\alpha}. \qquad \square$$

THEOREM 13.5. Let

$$|f(t,x_1,\ldots,x_n)| \leq h(t)|x_k|^\lambda$$

$$\text{for } t \geq a, \; 0 \leq (-1)^{m+i-1}x_i \leq rt^{1-i} \quad (i=1,\ldots,n), \qquad (13.36)$$

where $k \in \{1,\ldots,n\}$, $m \in \{0,1\}$, $\lambda \in \,]0,1[$, $r \in \,]0,+\infty[$, $h \in L_{\text{loc}}([a,+\infty[)$, and

$$\int_a^{+\infty} t^{n-1} h(t) \, dt < +\infty.$$

Then every Kneser solution u of the equation (13.1) vanishing at infinity and satisfying the inequalities (13.24) is subject to the estimate

$$|u^{(k-1)}(t)| \leq \left[\frac{1}{(n-k)!} \int_t^{+\infty} (\tau-t)^{n-k} h(\tau) \, d\tau \right]^{1/(1-\lambda)} \qquad (13.37)$$

for t large.

PROOF. By Remark 13.1, for a sufficiently large t_0 the inequalities (13.33) hold. So from (13.36) we can derive

$$|u^{(k-1)}(t)| = \frac{1}{(n-k)!} \left| \int_t^{+\infty} (t-\tau)^{n-k} f(\tau, u(\tau), \ldots, u^{(n-1)}(\tau)) \, d\tau \right|$$

$$\leq \frac{1}{(n-k)!} \int_t^{+\infty} (\tau - t)^{n-k} h(\tau) |u^{(k-1)}(\tau)|^\lambda \, d\tau$$

$$\leq \frac{1}{(n-k)!} \left(\int_t^{+\infty} (\tau - t)^{n-k} h(\tau) \, d\tau \right) |u^{(k-1)}(t)|^\lambda \qquad \text{for } t \geq t_0.$$

This yields the estimate (13.37). $\square$

PROBLEM 13.4*. Let $m \in \{0, 1\}$ and

$$(-1)^{m+n} f(t, x_1, \dots, x_n) \geq (-1)^{m+n} f(t, y_1, \dots, y_n)$$

$$\text{for } t \geq a, \ (-1)^{m+i-1} x_i \geq (-1)^{m+i-1} y_i \geq 0 \quad (i = 1, \dots, n). \tag{13.38}$$

Is it true that for any $c \in \,]0, +\infty[$ the equation (13.1) has at most one Kneser solution satisfying both the inequalities (13.4) and the condition $u(a) = (-1)^m c$?

PROBLEM 13.5*. Is the condition (13.38) with $m \in \{0, 1\}$ sufficient for the equation (13.1) to have at most one Kneser solution vanishing at infinity and satisfying the inequalities (13.24) for large t?

Notes. The problem of the existence and uniqueness of the Kneser solution of the differential equation $u'' = f(t, u)$ taking a given initial value was stated and solved by A. Kneser [209]. A similar problem for the equation $u'' = t^{-1/2} u^{3/2}$ appeared in the works of L.H. Thomas [347] and E. Fermi [112], which deal with the distribution of electrons in heavy atoms. Later the Kneser problem for the equation (13.1) in case $n = 2$ was considered in a series of publications of Italian mathematicians (see [203, 317] and the references quoted therein). Besides, here we will mention the subsequent studies [130, 184], which are concerned with the same case.

Theorems 13.1, 13.2, 13.2', and 13.4 are due to I.T. Kiguradze [182, 183], and Theorems 13.3 and 13.5 to G.G. Kvinikadze [245, 248, 249]. For Problems 13.1–13.3 see [200, 183, 248]. Problems 13.4* and 13.5* were posed in [197]. For third order linear differential equations a problem similar to 13.4* was solved in [170].

The question of the existence of Kneser solutions was investigated in [54, 63, 90, 131, 201, 308–310] for differential systems and in [288] for differential equations with deviating arguments.

§14. Proper Oscillatory Solutions

Consider the differential equation

$$u^{(n)} = f(t, u, \dots, u^{(n-1)}), \tag{14.1}$$

where $f \in K_{\text{loc}}([a, +\infty[\times \mathbf{R}^n)$ and $a \in \,]0, +\infty[$. In this section we give existence criteria for proper oscillatory solutions, including solutions which vanish at infinity. Note that in previous sections we have established such criteria only for the linear case.

Below n_0 denotes the integral part of the constant $n/2$.

14.1. Existence theorems.

THEOREM 14.1. Suppose on the set $[a, +\infty[\times \mathbf{R}^n$ the inequality

$$0 \le (-1)^{n-n_0-1} f(t, x_1, \dots, x_n) \operatorname{sign} x_1 \le h(t, x_1, \dots, x_{n_0}), \tag{14.2}$$

with $h \in K_{\mathrm{loc}}([a, +\infty[\times \mathbf{R}^{n_0})$, holds, and the equation (14.1) does not possess an oscillatory first kind singular solution. If, in addition, $n = 2n_0 + 1$ $(n = 2n_0)$, n_0 is odd and the equation (14.1) has either property A_1 or property $A^*_{n_0+1}$ (either property B_1 or property $B^*_{n_0}$), then there exists an $(n_0 - 1)$-parameter family of proper oscillatory solutions of (14.1). Furthermore, if $n = 2n_0 + 1$ $(n = 2n_0)$, n_0 is even and (14.1) has either property B or property $B^*_{n_0+1}$ (either property A or property $A^*_{n_0}$), then there exists an n_0-parameter family of such solutions.

THEOREM 14.2. Suppose the equation (14.1), where $n = 2n_0 + 1$, does not possess an oscillatory first kind singular solution, and on the set $[a, +\infty[\times \mathbf{R}^n$ the inequality

$$0 \le (-1)^{n_0-1} f(t, x_1, \dots, x_n) \operatorname{sign} x_1 \le h(t, x_1), \tag{14.3}$$

with $h \in K_{\mathrm{loc}}([a, +\infty[\times \mathbf{R})$, holds. If, in addition, n_0 is even and the equation (14.1) has property A_1, then there exists an n_0-parameter family of proper oscillatory solutions of (14.1). Furthermore, if n_0 is odd,

$$\int_a^{+\infty} t^{n-1} f_1(t; c)\, dt = +\infty \tag{14.4}$$

for a certain c, where

$$f_1(t; c) = \inf\{|f(t, x_1, \dots, x_n)| : |x_i| \ge ct^{n-i}\ (i = 1, \dots, n)\},$$

and (14.1) has property B, then there exists an $(n_0 + 1)$-parameter family of such solutions.

THEOREM 14.2′. Suppose the equation (14.1), where $n = 2n_0 + 1$, does not possess an oscillatory first kind singular solution, and on the set $[a, +\infty[\times \mathbf{R}^n$ the inequality (14.3), with $h \in K_{\mathrm{loc}}([a, +\infty[\times \mathbf{R})$, holds. If, in addition, n_0 is even (odd),

$$\int_a^{+\infty} t^{n-k} f_k(t; c)\, dt = +\infty$$

for a certain $k \in \{1, \dots, n - 1\}$ and any $c > 0$, where

$$f_k(t; c) = \inf\{|f(t, x_1, \dots, x_n)| :$$

$$|x_i| \ge ct^{n-k+1-i}\ (i = 1, \dots, n - k + 1),\ x_i \in \mathbf{R}\ (i = n - k + 2, \dots, n)\},$$

and the equation (14.1) has property A^*_k (property B^*_k), then there exists an n_0-parameter $((n_0 + 1)$-parameter$)$ family of proper oscillatory solutions of this equation.

First of all we will establish two lemmas, which we need in order to prove these theorems.

LEMMA 14.1. Let (14.2), with $h \in K_{\mathrm{loc}}([a, +\infty[\times \mathbf{R}^{n_0})$, hold, and let there exist $a_0 \in \,]a, a + 1]$ and $h_0 \in K([a, a_0] \times \mathbf{R})$ such that

$$|f(t, x_1, \dots, x_n)| \le h_0(t, x_1) \qquad \text{for } a \le t \le a_0,\ (x_1, \dots, x_n) \in \mathbf{R}^n. \tag{14.5}$$

Then for any $c_i \in \mathbf{R}$ $(i = 1, \ldots, n_0)$ and any positive integer m the equation (14.1) has a solution u_m satisfying the boundary conditions

$$u_m^{(i-1)}(a) = c_i \quad (i = 1, \ldots, n_0), \qquad u_m^{(i-1)}(a+m) = 0 \quad (i = 1, \ldots, n-n_0) \quad (14.6)$$

as well as the inequalities

$$\int_a^{a+m} \tau |u_m^{(n_0)}(\tau)|^2 \, d\tau < r_0 \tag{14.7}$$

and

$$\sum_{i=1}^n |u_m^{(i-1)}(t)| < r(t) \qquad \text{for } a \le t \le a+m, \tag{14.8}$$

where the constant $r_0 \in \,]0, +\infty[$ and the continuous function $r : [a, +\infty[\, \to \,]0, +\infty[$ do not depend on m.

PROOF. Let the numbers μ_{ij}^n $(i = 0, 1, \ldots; j = i, i+1, \ldots; n = i+j+1, \ldots)$ be defined by the equalities (4.6),

$$\alpha = 2(a+1)^2 \left(1 + \sum_{i=1}^{n_0} |c_i| \right)^2 \max\{\mu_{ij}^n : i = 0, \ldots, n_0;\ 2n_0 - 2 \le i+j \le n-1\},$$

$$h_0^*(t, y) = \max\{h_0(t, x) : |x| \le y\}$$

and

$$h^*(t, y_1, \ldots, y_{n_0}) = \max\{h(t, x_1, \ldots, x_{n_0}) : |x_i| \le y_i \ (i = 1, \ldots, n_0)\}.$$

Set $\beta = 2n\alpha a^{2n_0 - n - 1}$,

$$r_0 = \alpha \left[2 + 2n! \, (2n)^{n+1}(a_0 - a)^{-n} \left(\frac{\alpha}{a} \right)^{1/2} + 2n \int_a^{a_0} h_0^*(t, \beta) \, dt \right]^2,$$

$$r_i(t) = \sum_{j=i}^{n_0} \frac{|c_j|}{(j-i)!}(t-a)^{j-i} + \left(\frac{r_0}{a} \right)^{1/2} \frac{(t-a)^{n_0 - i + 1/2}}{(n-i)!} \quad (i = 1, \ldots, n_0),$$

$$r_i(t) = \frac{r_0}{\alpha} \sum_{j=i}^n \frac{(t-a)^{j-i}}{(j-i)!}$$

$$+ \frac{1}{(n-i)!} \int_a^t (t-\tau)^{n-i} h^*(\tau, r_1(\tau), \ldots, r_{n_0}(\tau)) \, d\tau \quad (i = n_0 + 1, \ldots, n),$$

$$\chi_i(t, x) = \begin{cases} x & \text{for } |x| \le r_i(t) \\ r_i(t)\,\text{sign}\,x & \text{for } |x| > r_i(t), \end{cases} \tag{14.9}$$

and

$$\tilde{f}(t, x_1, \ldots, x_n) = f(t, \chi_1(t, x_1), \ldots, \chi_n(t, x_n)). \tag{14.10}$$

By Lemma 10.1, for any positive integer m the differential equation

$$u^{(n)} = \tilde{f}(t, u, \ldots, u^{(n-1)}) \tag{14.11}$$

has a solution u_m satisfying the boundary conditions (14.6).

In view of (14.2) and (14.10),

$$p(t)|u_m^{(n)}(t) u_m(t)| + (-1)^{n-n_0} p(t) u_m^{(n)}(t) u_m(t) = 0,$$

where $p(t) = t^{n-2n_0+1}$. Integrating both sides of this equality over $[a, a+m]$ and applying Lemma 4.1, we obtain

$$\int_a^{a+m} p(\tau)|u_m^{(n)}(\tau)u_m(\tau)|\,d\tau + \int_a^{a+m} \tau|u_m^{(n_0)}(\tau)|^2\,d\tau$$

$$\leq \sum_{i=0}^{n_0-1} \sum_{j=2n_0-2-i}^{n-1-i} (-1)^{n_0-1-j}\mu_{ij}^n p^{(n-1-i-j)}(a)u_m^{(i)}(a)u_m^{(j)}(a)$$

$$< \alpha\left(1 + \sum_{j=n_0}^{n-1} |u_m^{(j)}(a)|\right). \tag{14.12}$$

Put $I_m = \{t \in [a, a_0]: |u_m(t)| < \beta\}$. By (14.5) and (14.12),

$$\int_a^{a_0} |u_m^{(n)}(t)|\,dt \leq \int_{I_m} h_0^*(\tau, \beta)\,d\tau + \frac{1}{\beta p(a)}\int_{[a,a_0]\setminus I_m} p(\tau)|u_m^{(n)}(\tau)u_m(\tau)|\,d\tau$$

$$\leq \int_a^{a_0} h_0^*(\tau, \beta)\,d\tau + \frac{1}{2n\alpha}\int_a^{a_0} p(\tau)|u_m^{(n)}(\tau)u_m(\tau)|\,d\tau$$

$$\leq \int_a^{a_0} h_0^*(\tau, \beta)\,d\tau + \frac{1}{2n}\left(1 + \sum_{j=n_0}^{n-1} |u_m^{(j)}(a)|\right). \tag{14.13}$$

Lemma 5.1 yields

$$\rho_{im} = \min\{|u_m^{(i)}(t)| : a \leq t \leq a_0\}$$

$$\leq n!\,(2n)^n(a_0-a)^{-n}a^{-1/2}\left[\int_a^{a_0} t|u_m^{(n_0)}(t)|^2\,dt\right]^{1/2} \qquad (i = n_0, \ldots, n-1).$$

Combining these estimates with the inequalities (14.12) and (14.13), we get

$$\sum_{i=n_0}^{n-1} |u_m^{(i)}(a)| \leq \sum_{i=n_0}^{n-1} \rho_{im} + n\int_a^{a_0} |u_m^{(n)}(\tau)|\,d\tau$$

$$\leq \frac{1}{2}n!\,(2n)^{n+1}(a_0-a)^{-n}\left(\frac{\alpha}{a}\right)^{1/2}\left(1 + \sum_{i=n_0}^{n-1} |u_m^{(i)}(a)|\right)^{1/2}$$

$$+ n\int_a^{a_0} h_0^*(\tau, \beta)\,d\tau + \frac{1}{2} + \frac{1}{2}\sum_{i=n_0}^{n-1} |u_m^{(i)}(a)|.$$

Hence

$$\sum_{i=n_0}^{n-1} |u_m^{(i)}(a)| < \frac{r_0}{\alpha} - 1, \tag{14.14}$$

which, together with (14.12), implies (14.7).

Applying (14.6) and (14.14) to (14.2) and (14.7), we obtain

$$|u_m^{(i-1)}(t)| < r_i(t) \qquad \text{for } a \leq t \leq a+m \qquad (i = 1, \ldots, n). \tag{14.15}$$

Therefore, the inequality (14.8), where $r(t) = \sum_{i=1}^n r_i(t)$, holds. Moreover, it follows from (14.9), (14.10) and (14.15) that u_m is a solution of the equation (14.1). $\square$

LEMMA 14.2. Let $n = 2n_0 + 1$ and let the condition (14.3), with $h \in K_{loc}([a, +\infty[\times \mathbf{R})$, hold. Furthermore, suppose that there exists $a_0 \in]a, a + 1]$ such that

$$|f(t, x_1, \dots, x_n)| \geq |x_1| \qquad \text{for } a \leq t \leq a_0, \ (x_1, \dots, x_n) \in \mathbf{R}^n. \qquad (14.16)$$

Then for any $c_i \in \mathbf{R}$ $(i = 1, \dots, n_0 + 1)$ and any positive integer m the equation (14.1) has a solution u_m satisfying the boundary conditions

$$u_m^{(i-1)}(a) = c_i \qquad (i = 1, \dots, n_0 + 1), \qquad u_m^{(i-1)}(a + m) = 0 \qquad (i = 1, \dots, n_0) \qquad (14.17)$$

as well as the inequalities (14.8) and

$$\int_a^{a+m} |u_m^{(n)}(t) u_m(t)| \, dt < r_0, \qquad (14.18)$$

where the constant $r_0 \in]0, +\infty[$ and the continuous function $r : [a, +\infty[\rightarrow]0, +\infty[$ do not depend on m.

PROOF. Set

$$\alpha = \sum_{i=1}^{n_0} |c_i| + c_{n_0+1}^2 + 1, \qquad h^*(t, y) = \max\{h(t, x) : |x| \leq y\},$$

$$r_0 = \alpha \left[2 + 2n! \, (2n)^{n+1} (a_0 - a)^{-n} + 2n \int_a^{a_0} h^*(t; 2n\alpha) \, dt \right]^2,$$

and

$$r_i(t) = r_0 \sum_{j=i}^{n} \frac{(t-a)^{j-i}}{(j-i)!} + \frac{(t-a)^{n-i}}{(n-i)!} \left[\frac{r_0}{2n\alpha} + \int_a^t h^*(\tau, 2n\alpha) \, d\tau \right] \qquad (i = 1, \dots, n).$$

Let the function $\tilde{f}$ be defined by the equalities (14.9) and (14.10). Consider the differential equation

$$u^{(n)} = p(t)u + \tilde{f}(t, u, \dots, u^{(n-1)}) - p(t)\chi_1(t, u), \qquad (14.19)$$

where

$$p(t) = \begin{cases} (-1)^{n_0-1} & \text{for } a \leq t \leq a_0 \\ 0 & \text{for } a_0 < t \leq a + m. \end{cases}$$

Since the linear homogeneous problem

$$v^{(n)} = p(t)v,$$

$$v^{(i-1)}(a) = 0 \qquad (i = 1, \dots, n_0), \qquad v^{(i-1)}(b) = 0 \qquad (i = 1, \dots, n_0 + 1)$$

has only zero solution, the equation (14.19) possesses a solution u_m satisfying the boundary conditions (14.17).[4]

According to (14.3), (14.10) and (14.16),

$$|u_m^{(n)}(t)| \leq h^*(t, |u_m(t)|) \qquad \text{for } a \leq t \leq a + m, \qquad (14.20)$$

$$|u_m^{(n)}(t) u_m(t)| \geq |u_m(t)|^2 \qquad \text{for } a \leq t \leq a_0, \qquad (14.21)$$

and

$$|u_m^{(n)}(t) u_m(t)| = (-1)^{n_0-1} u_m^{(n)}(t) u_m(t) \qquad \text{for } a \leq t \leq a + m.$$

---

[4] Using the Schauder principle, this fact can be proved similarly to Lemma 10.1.

Integration of the last equality over $[a, a+m]$ gives

$$\int_a^{a+m} |u_m^{(n)}(t)u_m(t)|\, dt = \sum_{i=1}^{n_0} (-1)^{n_0-1+i} c_i u_m^{(n-i)}(a)$$

$$+ \frac{1}{2}\left(c_{n_0+1}^2 - |u_m^{(n_0)}(a+m)|^2\right) < \alpha \left(1 + \sum_{i=n_0+1}^{n-1} |u_m^{(i)}(a)|\right). \tag{14.22}$$

By (14.20) and (14.21) this implies

$$\int_a^t |u_m^{(n)}(\tau)|\, d\tau < \int_a^t h^*(\tau, 2n\alpha)\, d\tau + \frac{1}{2n}\left(1 + \sum_{i=n_0+1}^{n-1} |u_m^{(i)}(a)|\right)$$

$$\text{for } a \leq t \leq a+m, \tag{14.23}$$

and

$$\int_a^{a_0} |u_m(t)|^2\, dt < \alpha\left(1 + \sum_{i=n_0+1}^{n-1} |u_m^{(i)}(a)|\right). \tag{14.24}$$

It follows from Lemma 5.1 that

$$\rho_{im} = \min\{|u_m^{(i)}(t)| : a \leq t \leq a_0\}$$

$$\leq n!\,(2n)^n (a_0 - a)^{-n} \left(\int_a^{a_0} |u_m(\tau)|^2\, d\tau\right)^{1/2} \qquad (i = n_0+1, \dots, n-1).$$

So, taking into account (14.23) and (14.24), we obtain

$$\sum_{i=n_0+1}^{n-1} |u_m^{(i)}(a)| \leq \sum_{i=n_0+1}^{n-1} \rho_{im} + n\int_a^{a_0} |u_m^{(n)}(\tau)|\, d\tau$$

$$\leq \frac{1}{2}n!\,(2n)^{n+1}(a_0-a)^{-n}\alpha^{1/2}\left(1 + \sum_{i=n_0+1}^{n-1} |u_m^{(i)}(a)|\right)^{1/2}$$

$$+ n\int_a^{a_0} h_0^*(\tau, 2n\alpha)\, d\tau + \frac{1}{2} + \frac{1}{2}\sum_{i=n_0+1}^{n-1} |u_m^{(i)}(a)|.$$

This yields

$$\sum_{i=n_0+1}^{n-1} |u_m^{(i)}(a)| < \frac{r_0}{\alpha} - 1.$$

Hence (14.22) and (14.23) imply the inequality (14.18) and the estimate

$$\int_a^t |u_m^{(n)}(\tau)|\, d\tau < \frac{r_0}{2n\alpha} + \int_a^t h^*(\tau, 2n\alpha)\, d\tau \qquad \text{for } a \leq t \leq a+m. \tag{14.25}$$

Since $|u_m^{(i)}(a)| < r_0$ $(i = 1, \dots, n-1)$, from (14.25) we derive (14.15). Consequently, (14.8), where $r(t) = \sum_{i=1}^n r_i(t)$, holds. It remains to note that by (14.9), (14.10) and (14.15), u is a solution of the equation (14.1). $\square$

PROOF OF THEOREM 14.1. Let c_i $(i = 1, \ldots, n_0)$ be real numbers and

$$f_0(t, x_1, \ldots, x_n) = \begin{cases} 0 & \text{for } a \leq t < a + 1 \\ f(t, x_1, \ldots, x_n) & \text{for } t \geq a + 1. \end{cases} \tag{14.26}$$

In view of Lemma 14.1, for any positive integer m the equation

$$u^{(n)} = f_0(t, u, \ldots, u^{(n-1)}) \tag{14.27}$$

has a solution u_m satisfying the boundary conditions (14.6) and the inequalities (14.7) and (14.8), where the constant $r_0 \in]0, +\infty[$ and the continuous function $r: [a, +\infty[\to]0, +\infty[$ do not depend on m.

According to Lemma 10.2, $(u_m)_{m=1}^{+\infty}$ contains a subsequence $(u_{m_j})_{j=1}^{+\infty}$ which, together with $(u_{m_j}^{(i)})_{j=1}^{+\infty}$ $(i = 1, \ldots, n-1)$, converges uniformly on each finite subinterval of $[a, +\infty[$ and, moreover, $u(t) = \lim_{j \to +\infty} u_{m_j}(t)$ for $t \geq a$ is a solution of the equation (14.27). On the other hand, (14.6) and (14.7) yield

$$u^{(i-1)}(a) = c_i \quad (i = 1, \ldots, n_0), \quad \int_a^{+\infty} \tau |u^{(n_0)}(\tau)|^2 d\tau < +\infty. \tag{14.28}$$

Since the equation (14.27) does not have an oscillatory first kind singular solution, u is a proper non-Kneser solution if either $c_1 = 0$, $\sum_{i=2}^{n_0} |c_i| > 0$ and n_0 is odd, or $\sum_{i=1}^{n_0} |c_i| > 0$ and n_0 is even. But by (14.26) and (14.28) the restriction of u to $[a + 1, +\infty[$ is a solution of the equation (14.1) and, in addition,

$$\lim_{t \to +\infty} \inf |u^{(n_0)}(t)| = 0.$$

So, taking into account Definitions 10.3–10.6 and Remark 10.1 we conclude that the assertion of the theorem is true. $\square$

Theorems 14.2 and 14.2' can be proved in the same way. The only difference is that Lemma 14.2 should be used instead of Lemma 14.1, and (14.26) should be replaced by the equality

$$f_0(t, x_1, \ldots, x_n) = \begin{cases} (-1)^{n_0-1} x_1 & \text{for } a \leq t < a + 1 \\ f(t, x_1, \ldots, x_n) & \text{for } a \leq t \geq a + 1. \end{cases}$$

THEOREM 14.3. Let either $n = 2n_0$ with n_0 even or $n = 2n_0 + 1$, and let on the set $[a, +\infty[\times \mathbf{R}^n$ the inequality

$$-h(t)\omega\left(\sum_{k=1}^n |x_k|\right) \leq f(t, x_1, \ldots, x_n) \operatorname{sign} x_1 \leq 0 \tag{14.29}$$

hold, where $h \in L_{\mathrm{loc}}([a, +\infty[)$, the continuous function $\omega: \mathbf{R}_+ \to \mathbf{R}_+$ does not decrease, and

$$\omega(x) > 0, \quad \int_x^{+\infty} \frac{ds}{\omega(s)} = +\infty \quad \text{for } x > 0. \tag{14.30}$$

If, in addition, the equation (14.1) has property A, then it possesses a $(2n_0)$-parameter family of proper oscillatory solutions.

PROOF. Let $c_i \in \mathbf{R}$ $(i = 1, \dots, n-1)$ be arbitrary constants such that

$$\sum_{i=1}^{n_0} (-1)^i c_{n-i+1} c_i + \alpha_n c_{n_0+1}^2 > 0, \tag{14.31}$$

where $\alpha_n = 0$ for $n = 2n_0$ and $\alpha_n = (-1)^{n_0-1}/2$ for $n = 2n_0 + 1$. By Theorem 11.5, any solution u of the equation (14.1) under the initial conditions $u^{(i-1)}(a) = c_i$ $(i = 1, \dots, n)$ is defined on the whole of $[a, +\infty[$. From (14.29) and (14.31), setting

$$v(t) = \sum_{i=1}^{n_0} (-1)^i u^{(n-i)}(t) u^{(i-1)}(t) + \alpha_n [u^{(n_0)}(t)]^2,$$

we get

$$v(a) > 0, \qquad v'(t) = -u^{(n)}(t) u(t) + \beta_n [u^{(n_0)}(t)]^2 \ge 0 \qquad \text{for } t \ge a,$$

where $\beta_n = 0$ for $n = 2n_0 + 1$ and $\beta_n = 1$ for $n = 2n_0$. If (14.1) has property A, this implies that u is a proper oscillatory solution. $\square$

THEOREM 14.4. Let the equation (14.1), with $n > 2$, have proprty B_1^* and let on the set $[a, +\infty[\times \mathbf{R}^n$ the inequality

$$0 \le f(t, x_1, \dots, x_n) \operatorname{sign} x_1 \le h(t) \omega \left(\sum_{i=1}^{n} |x_i| \right) \tag{14.32}$$

hold, where $h \in L_{\mathrm{loc}}([a, +\infty[)$ and the function $\omega \colon \mathbf{R}_+ \to \mathbf{R}_+$ is continuous, does not decrease and satisfies (14.30). Suppose, in addition, that either $n \ge 4$ and among the numbers n and n_0 at least one is odd or

$$\int_0^1 \frac{dx}{\omega(x)} = +\infty. \tag{14.33}$$

Then the equation (14.1) possesses an $(n-2)$-parameter family of proper oscillatory solutions.

PROOF. Let $c_i \in \mathbf{R}$ $(i = 1, \dots, n-1)$ be constants which are not all zero. Define a function f_0 by (14.26). We will show that the equation (14.27) has a solution u satisfying the conditions

$$u^{(i-1)}(a) = c_i \qquad (i = 1, \dots, n-1) \tag{14.34$_1$}$$

and

$$\lim_{t \to +\infty} \inf |u^{(n-1)}(t)| = 0. \tag{14.34$_2$}$$

Set $r_0 = \sum_{i=1}^{n-1} |c_i|$. According to (14.30) there exists a continuous function $r \colon [a, +\infty[\to]0, +\infty[$ such that

$$\int_{2nr_0(1+t)^{n-1}}^{r(t)} \frac{ds}{\omega(s)} = n(1+t)^{n-1} \int_a^t h(\tau)\, d\tau \qquad \text{for } t \ge a.$$

Put

$$\chi(t, s) = \begin{cases} 1 & \text{for } 0 \le x \le r(t) \\ 1 - \frac{x}{2r(t)} & \text{for } r(t) < x < 2r(t) \\ 0 & \text{for } x \ge 2r(t) \end{cases}$$

and
$$\tilde{f}(t, x_1, \ldots, x_n) = \chi\left(t, \sum_{i=1}^{n} |x_i|\right) f_0(t, x_1, \ldots, x_n). \tag{14.35}$$

By Lemma 10.1, for any positive integer m the equation (14.11) has a solution u_m satisfying the boundary conditions

$$u_m^{(i-1)}(a) = c_i \quad (i = 1, \ldots, n-1), \qquad u_m^{(n-1)}(a+m) = 0. \tag{14.36}$$

Assuming that $|u_m^{(n-1)}(a)| > (n-1)! \, r_0$, from (14.26) we obtain

$$u_m^{(i-1)}(a+1) u_m^{(n-1)}(a+1) = \left[\sum_{j=i}^{n-1} \frac{u_m^{(j-1)}(a)}{(j-i)!}\right] u_m^{(n-1)}(a)$$

$$\geq \left[\frac{|u_m^{(n-1)}(a)|}{(n-1)!} - r_0\right] |u_m^{(n-1)}(a)| > 0 \qquad (i = 1, \ldots, n-1).$$

Hence, (14.32) yields

$$u_m^{(i-1)}(t) u_m^{(n-1)}(t) > 0 \qquad \text{for } a+1 \leq t \leq a+m \qquad (i = 1, \ldots, n).$$

But this contradicts the last condition in (14.36). Therefore,

$$|u_m^{(n-1)}(a)| \leq (n-1)! \, r_0.$$

So, taking into account (14.32), we get

$$\sum_{i=1}^{n} |u_m^{(i-1)}(s)|$$

$$= \sum_{i=1}^{n} \left| \sum_{j=i}^{n-1} \frac{u_m^{(j-1)}(a)}{(j-i)!} (s-a)^{j-i} + \frac{1}{(n-i)!} \int_a^s (s-\tau)^{n-i} u_m^{(n)}(\tau) \, d\tau \right|$$

$$< 2n! \, r_0 (1+t)^{n-1} + n(1+t) \int_a^s h(\tau) \, d\tau \qquad \text{for } a \leq s \leq t \leq a+m.$$

By Lemma 9.2 this implies

$$\sum_{i=1}^{n} |u_m^{(i-1)}(t)| < r(t) \qquad \text{for } a \leq t \leq a+m. \tag{14.37}$$

It follows from (14.35) and (14.37) that u_m is a solution of the equation (14.27). On the other hand, $(u_m)_{m=1}^{+\infty}$ contains a subsequence $(u_{m_j})_{j=1}^{+\infty}$ such that $(u_{m_j}^{(i)})_{j=1}^{+\infty}$ $(i = 0, \ldots, n-1)$ converge uniformly on each finite interval and $u(t) = \lim_{j \to +\infty} u_{m_j}(t)$ is a solution of the equation (14.27) under the conditions (14.34_1).

We claim that (14.34_2) also holds. Assume the contrary. Then for a sufficiently large $t_0 > a$,

$$u^{(i-1)}(t_0) u^{(n-1)}(t_0) > 0 \qquad (i = 1, \ldots, n).$$

Thus there is a positive integer j_0 such that $t_0 < a + m_{j_0}$ and

$$u_{m_{j_0}}^{(i-1)}(t_0) u_{m_{j_0}}^{(n-1)}(t_0) > 0 \qquad (i = 1, \ldots, n).$$

According to (14.32) this yields

$$u_{m_{j_0}}^{(i-1)}(t)u_{m_{j_0}}^{(n-1)}(t) > 0 \qquad \text{for } t_0 \le t \le a + m_{j_0} \qquad (i = 1, \ldots, n),$$

which is impossible in view of (14.36). Hence we have proved that u is a solution of the problem (14.27), (14.34$_1$), (14.34$_2$), and the restriction of u to $[a+1, +\infty[$ is a solution of the equation (14.1).

Now suppose $n \ge 4$ and either n or n_0 is odd. Let $c_1 = 0$ and let c_i ($i = 2, \ldots, n-1$) be constants satisfying the inequality

$$\sum_{i=2}^{n_0} (-1)^{i-1} c_{n-i+1} c_i + \alpha_n c_{n_0+1}^2 > 0, \tag{14.38}$$

where $\alpha_n = 0$ for $n = 2n_0$ and $\alpha_n = (-1)^{n_0}/2$ for $n = 2n_0 + 1$. Denote by u a solution of the problem (14.27), (14.34$_1$), (14.34$_2$), and set

$$v(t) = \sum_{i=1}^{n_0} (-1)^{i-1} u^{(n-i)}(t) u^{(i-1)}(t) + \alpha_n [u^{(n_0)}(t)]^2.$$

Then (14.32) and (14.38) imply $v(a) > 0$ and $v'(t) \ge 0$ for $t \ge a$. Therefore u is a proper non-Kneser solution. Since (14.1) has property B_1^*, u is oscillatory.

It remains to consider the case when (14.33) holds. According to Theorem 11.5 the equation (14.1) does not possess a singular solution. So, if $c_1 = 0$ and $\sum_{i=2}^{n-1} |c_i| > 0$, then any solution of the problem (14.27), (14.34$_1$), (14.34$_2$) is proper and non-Kneser. Since the equation (14.1) has property B_1^*, such a solution is oscillatory. $\square$

14.2. Proper oscillatory solutions vanishing at infinity.

THEOREM 14.5. Suppose the equation (14.1) does not have an oscillatory first kind singular solution, and on the set $[a, +\infty[\times \mathbf{R}^n$ the inequality

$$\delta t^\nu |x_1|^\lambda \le (-1)^{n-n_0-1} f(t, x_1, \ldots, x_n) \operatorname{sign} x_1 \le h(t, x_1, \ldots, x_{n_0}) \tag{14.39}$$

holds, where $h \in K_{\text{loc}}([a, +\infty[\times \mathbf{R}^{n_0})$, $\lambda \ge 1$, $\delta > 0$, and $\nu > (\lambda - 1)/2 - n$. Then for n_0 even (odd) the equation (14.1) possesses an n_0-parameter (($n_0 - 1$)-parameter) family of proper oscillatory solutions satisfying the conditions

$$\lim_{t \to +\infty} t^{i+1/2} u^{(i)}(t) = 0 \qquad (i = 0, \ldots, n_0 - 1). \tag{14.40}$$

THEOREM 14.6. Suppose the equation (14.1), with $n = 2n_0 + 1$, does not have an oscillatory first kind singular solution, and on the set $[a, +\infty[\times \mathbf{R}^n$ the inequality

$$\delta t^\nu |x_1|^\lambda \le (-1)^{n_0-1} f(t, x_1, \ldots, x_n) \operatorname{sign} x_1 \le h(t, x_1) \tag{14.41}$$

holds, where $h \in K_{\text{loc}}([a, +\infty[\times \mathbf{R})$, $\lambda \ge 1$, $\delta > 0$, and $\nu > ((\lambda+3)(n-1)/2) - n$. Then for n_0 even (odd) the equation (14.1) possesses an n_0-parameter (($n_0 + 1$)-parameter) family of proper oscillatory solutions vanishing at infinity.

To justify these theorems we need two lemmas, which we prove first.

LEMMA 14.3. Let the condition (14.39) hold, where $h \in K_{\text{loc}}([a, +\infty[\times \mathbf{R}^{n_0})$, $\lambda \ge 1$, $\delta > 0$, and $\nu > (\lambda - 1)/2 - n$, and let there exist $a_0 \in]a, a+1]$ and $h_0 \in K([a, a_0] \times \mathbf{R})$

such that the inequality (14.5) is fulfilled. Then for any $c_i \in \mathbf{R}$ $(i = 1, \ldots, n_0)$ the equation (14.1) has a solution u satisfying the conditions

$$u^{(i-1)}(a) = c_i \qquad (i = 1, \ldots, n_0),$$

$$\int_a^{+\infty} t^{2k} |u^{(k)}(t)|^2 \, dt < +\infty \qquad (k = 0, \ldots, n_0). \tag{14.42}$$

PROOF. Let μ_i^k $(i = 0, 1, \ldots; k = 2i, 2i+1, \ldots)$ be the constants defined in §4,

$$\alpha_0(\gamma) = 1, \qquad \alpha_1(\gamma) = (n_0 - 1)\left(1 + \frac{\gamma}{4}\right),$$

$$\alpha_i(\gamma) = (n_0 - i)\left(1 + \frac{\gamma}{4}\right) \prod_{k=1}^{i-1} \left(\gamma - \frac{(k-1)(4k^2 + 7k + 6)}{3}\right)$$

$$(i = 2, \ldots, n_0 - 1),$$

$$\beta_0(\gamma) = 0, \qquad \beta_i(\gamma) = \prod_{k=i}^{n_0 - 1} \left(\gamma - \frac{(k-1)(4k^2 + 7k + 6)}{3}\right)^{-1}$$

$$(i = 1, \ldots, n_0 - 1).$$

Choose $\gamma > n^3$, $\epsilon > 0$ and a positive integer m_1 so that

$$n! \sum_{i=0}^{n_0 - 1} \mu_i^n \beta_i(\gamma) < \frac{1}{2} \tag{14.43}$$

and

$$\delta t^{n+\nu} \geq 2\eta t^{(\lambda-1)(1+\epsilon)/2} \qquad \text{for } t \geq m_1, \tag{14.44}$$

where

$$\eta = n! \sum_{i=0}^{n_0 - 1} \alpha_i(\gamma) \mu_i^n. \tag{14.45}$$

According to Lemmas 10.2 and 14.1, for any positive integer m the problem (14.1), (14.6) has a solution u_m, the sequence $(u_m)_{m=1}^{+\infty}$ contains a subsequence $(u_{m_j})_{j=1}^{+\infty}$ such that $(u_{m_j}^{(i)})_{j=1}^{+\infty}$ $(i = 0, \ldots, n-1)$ converge uniformly on each finite interval, and $u(t) = \lim_{j \to +\infty} u_{m_j}(t)$ is a solution of the equation (14.1). Set $t_j = a + m_j$,

$$v_j(t) = u_{m_j}(t), \qquad r_{0j} = \left[\int_{t_1}^{t_j} |v_j(t)|^2 \, dt\right]^{1/2},$$

and

$$r_j = \left[\int_{t_1}^{t_j} t^{2n_0} |v_j^{(n_0)}(t)|^2 \, dt\right]^{1/2}.$$

Lemma 4.2 implies

$$\int_{t_1}^{t_j} t^{2i} |v_j^{(i)}(t)|^2 \, dt < \gamma_0 + \alpha_i(\gamma) r_{0j}^2 + \beta_i(\gamma) r_j^2 \qquad (i = 0, \ldots, n_0 - 1), \tag{14.46}$$

with γ_0 a positive constant independent of j.

By (14.39) and (14.44),

$$(-1)^{n-n_0} t^n v_j^{(n)}(t) v_j(t) + 2\eta t^{(\lambda-1)(1+\epsilon)/2} |v_j(t)|^{\lambda+1} \leq 0 \qquad \text{for } t_1 \leq t \leq t_j.$$

Integrating both sides of this inequality over $[t_1, t_j]$ and applying Lemma 4.1, we obtain

$$\left[1 + \left(\frac{n}{2} - n_0\right)\frac{n^2}{2}\right]r_j^2 + \sum_{i=0}^{n_0-1}(-1)^{n_0+i}\frac{n!}{(2i)!}\int_{t_1}^{t_j}t^{2i}|v_j^{(i)}(t)|^2\,dt$$

$$+2\eta\int_{t_1}^{t_j}t^{(\lambda-1)(1+\epsilon)/2}|v_j(t)|^{\lambda+1}\,dt \le \gamma_1 \qquad (j=1,2,\dots),$$

where $\gamma_1 > 0$ does not depend on j. Hence, (14.43), (14.45) and (14.46) yield

$$r_j^2 + 4\eta\int_{t_1}^{t_j}t^{(\lambda-1)(1+\epsilon)/2}|v_j(t)|^{\lambda+1}\,dt \le 2\eta r_{0j}^2 + \gamma_2, \tag{14.47}$$

with

$$\gamma_2 = 2\gamma_1 + 2\gamma_0 n! \sum_{i=0}^{n_0-1}\mu_i^n.$$

According to the Young inequality,

$$|v_j(t)|^2 = t^{-(1+\epsilon)(\lambda-1)/(\lambda+1)}t^{(1+\epsilon)(\lambda-1)/(\lambda+1)}|v_j(t)|^2$$

$$\le t^{-1-\epsilon} + t^{(\lambda-1)(1+\epsilon)/2}|v_j(t)|^{\lambda+1} \qquad \text{for } t_1 \le t \le t_j,$$

and so

$$r_{0j}^2 \le \frac{1}{\epsilon} + \int_{t_1}^{t_j}t^{(\lambda-1)(1+\epsilon)/2}|v_j(t)|^{\lambda+1}\,dt.$$

Therefore (14.46) implies

$$r_j^2 + 2\eta r_{0j}^2 \le \gamma_3, \tag{14.48}$$

with $\gamma_3 = \frac{4\eta}{\epsilon} + \gamma_2$.

In view of (14.46) and (14.48),

$$\int_{t_1}^{t_j}t^{2i}|v_j^{(i)}(t)|^2\,dt \le \gamma^* \qquad (i=0,\dots,n_0; \ j=1,2,\dots),$$

where γ^* is a constant independent of j. From these inequalities, passing j to $+\infty$, we get

$$\int_{t_1}^{+\infty}t^{2i}|u^{(i)}(t)|^2\,dt \le \gamma^* \qquad (i=0,\dots,n_0).$$

Thus u is a solution of the problem (14.1), (14.42). □

LEMMA 14.4. Let $n = 2n_0 + 1$ and let the condition (14.41) hold, where $h \in K_{\text{loc}}([a,+\infty[\times \mathbf{R})$, $\lambda \ge 1$, $\delta > 0$, and $\nu > (\lambda-1)/2$. Suppose, in addition, that the inequality (14.16) with a certain $a_0 \in \,]a, a+1]$ is fulfilled. Then for any $c_i \in \mathbf{R}$ $(i=1,\dots,n_0)$ and $\sigma \in \,]n, n+\nu-(\lambda-1)/2[$ the equation (14.1) has a solution u satisfying the conditions

$$u^{(i-1)}(a) = c_i \qquad (i=1,\dots,n_0),$$

$$\tag{14.49}$$

$$\int_a^{+\infty}t^{2k-\sigma}|u^{(k)}(t)|^2\,dt < +\infty \qquad (k=0,\dots,n_0)$$

and
$$\int_a^{+\infty} t^{n-\sigma+\nu} |u(t)|^{\lambda+1}\, dt < +\infty. \tag{14.50}$$

We omit the proof of this lemma, because it is similar to the proof of Lemma 14.3. The main difference is that Lemma 14.2 should be applied instead of Lemma 14.1.

PROOF OF THEOREM 14.5. Let

$$f_0(t, x_1, \dots, x_n) = \begin{cases} (-1)^{n-n_0-1}\delta t^\nu |x_1|^\lambda \operatorname{sign} x_1 & \text{for } a \le t < a+1 \\ f(t, x_1, \dots, x_n) & \text{for } t \ge a+1, \end{cases}$$

$c_i \in \mathbf{R}$ $(i = 1, \dots, n_0)$ and $\sum_{i=1}^{n_0} |c_i| > 0$. By Lemmas 4.5 and 14.3, the problem (14.27), (14.42) possesses a solution u, and, moreover, u satisfies (14.40). Since (14.27) does not have an oscillatory first kind singular solution, Corollary 10.1 implies that u is a proper oscillatory solution if either n_0 is odd and $c_1 = 0$ or n_0 is even. It remains to note that the restriction of u to $[a+1, +\infty[$ is a solution of the equation (14.1). $\square$

PROOF OF THEOREM 14.6. Let

$$f_0(t, x_1, \dots, x_n) = \begin{cases} (-1)^{n_0-1}[\delta t^\nu |x_1|^\lambda + |x_1|] \operatorname{sign} x_1 & \text{for } a \le t < a+1 \\ f(t, x_1, \dots, x_n) & \text{for } t \ge a+1, \end{cases}$$

$c_i \in \mathbf{R}$ $(i = 1, \dots, n_0 + 1)$ and $\sum_{i=1}^{n_0+1} |c_i| > 0$. Choose $\sigma > n$ so that

$$\epsilon = 2\nu - (\lambda + 3)(\sigma - 1) + 2n > 0. \tag{14.51}$$

By Lemma 14.4 the equation (14.27) has a solution u satisfying the conditions (14.49) and (14.50).

If $\lambda = 1$, then (14.50) and (14.51) yield

$$\int_a^{+\infty} t^{\sigma-2} |u(t)|^2\, dt < +\infty. \tag{14.52}$$

For $\lambda > 1$ we again have

$$\int_a^{+\infty} t^{\sigma-2} |u(t)|^2\, dt$$
$$< \left(\int_a^{+\infty} t^{-1-\epsilon/(\lambda-1)}\, dt \right)^{(\lambda-1)/(\lambda+1)} \left(\int_a^{+\infty} t^{n-\sigma+\nu} |u(t)|^{\lambda+1}\, dt \right)^{2/(\lambda+1)} < +\infty.$$

The conditions (14.49) and (14.52) imply that $u(t) \to 0$ as $t \to +\infty$. On the other hand, according to Corollary 10.1, if either n_0 is even and $c_1 = 0$ or n_0 is odd, then u is oscillatory. Since (14.27) does not have an oscillatory first kind singular solution, it is clear that u is proper. $\square$

In conclusion of this section we consider the case when the equation (14.1) has the form

$$u^{(n)} = f(t, u), \tag{14.53}$$

where the function $f: [a, +\infty[\times \mathbf{R} \to \mathbf{R}$ is continuous, $f(\cdot, u): [a, +\infty[\to \mathbf{R}$ is locally absolutely continuous for any $u \in \mathbf{R}$, and the partial derivative of f with respect to the first variable belongs to $K_{\mathrm{loc}}([a, +\infty[\times \mathbf{R})$.

Applying the methods by which Theorems 14.3 and 14.5 were proved, we can easily justify the following assertion.

THEOREM 14.7. Suppose $n = 2n_0 \geq 4$ and the equation (14.53) does not have an oscillatory first kind singular solution. Furthermore, let

$$(-1)^{n_0} f(t, x) \operatorname{sign} x \geq 0, \qquad (-1)^{n_0} \frac{\partial(t^{n+1} f(t, x))}{\partial t} \geq \delta t^\nu |x|^\lambda$$

on $[a, +\infty[\times \mathbf{R}$, where $\lambda \geq 1$, $\delta > 0$, and $\nu > (\lambda - 1)/2$. Then for n_0 even (odd) the equation (14.53) possesses an $(n_0 - 2)$-parameter $((n_0 - 1)$-parameter) family of proper oscillatory solutions satisfying the conditions (14.40).

PROBLEM 14.1. Let $n = 4$,

$$-f(t, -x) = f(t, x) > 0, \qquad \frac{\partial f(t, x)}{\partial t} \geq 0 \qquad \text{for } t \geq a, \, x > 0, \qquad (14.54)$$

and

$$\int_0^{+\infty} f(a, x) \, dx = +\infty. \qquad (14.55)$$

Then the equation (14.53) has a continuum of proper oscillatory solutions, and every such solution is bounded.

PROBLEM 14.2. Let $n = 4$ and

$$f(t, x) \operatorname{sign} x \geq \omega(|x|), \qquad \left| \frac{\partial f(t, x)}{\partial t} \right| \leq h(t) \omega(|x|)$$

on $[a, +\infty[\times \mathbf{R}$, where $\omega \colon \mathbf{R}_+ \to \mathbf{R}_+$ is continuous, $\omega(0) = 0$, $\omega(x) > 0$ for $x > 0$,

$$\int_0^{+\infty} \omega(x) \, dx = +\infty, \qquad \int_a^{+\infty} h(t) \, dt < +\infty.$$

Then the equation (14.53) has a continuum of proper oscillatory solutions, and every such solution satisfies the conditions

$$\lim_{t \to +\infty} \sup |u^{(i)}(t)| < +\infty \qquad (i = 0, \dots, 4).$$

PROBLEM 14.3. Let $n = 3$ and

$$f(t, x_1, x_2, x_3) \operatorname{sign} x_1 \geq \omega(x_1)$$

on $[a, +\infty[\times \mathbf{R}^3$, where $\omega \colon \mathbf{R} \to \mathbf{R}_+$ is continuous, $\omega(0) = 0$, and $\omega(x) > 0$ for $x \neq 0$. Then every proper oscillatory solution of the equation (14.1) satisfies the conditions

$$\lim_{t \to +\infty} u(t) = 0, \qquad \lim_{t \to +\infty} \sup |u'(t)| < +\infty.$$

PROBLEM 14.4. Let $n = 2$,

$$-f(t, -x) = f(t, x) < 0, \qquad \frac{\partial f(t, x)}{\partial t} \leq 0 \qquad \text{for } t \geq a, \, x > 0,$$

and
$$\int_0^{+\infty} f(a, x)\, dx = -\infty.$$

Then every nonzero solution of the equation (14.1) is proper, oscillatory and bounded.

Notes. Theorems 14.1–14.7 are due to I.T. Kiguradze [190–192, 196]. For Problems 14.1–14.4 see [178, 190].

Some characteristics of proper oscillatory solutions of the equation (14.1) were studied by M. Bartušek [13–17, 19–23]. Criteria for oscillation, nonoscillation and boundedness of solutions of the equation (14.53) with $n = 2$ are contained in [138, 145–148, 150, 151, 153, 154, 161, 162, 166, 207, 290, 291, 294, 295].

CHAPTER IV

HIGHER ORDER DIFFERENTIAL EQUATIONS OF EMDEN–FOWLER TYPE

§15. Oscillatory Solutions

This chapter deals with the differential equation

$$u^{(n)} = p(t)|u|^\lambda \operatorname{sign} u,$$ (15.1)

where $p \in L_{\text{loc}}([a, +\infty[)$, $a > 0$, $\lambda > 0$, and $\lambda \neq 1$.

Recall that n_0 stands for the integral part of $n/2$.

15.1. Classification of equations with respect to their oscillation properties. Theorems 10.3, 10.4, 10.9, and 10.11 imply the following assertions.

THEOREM 15.1. Let $\lambda > 1$ and

$$p(t) \leq 0 \qquad (p(t) \geq 0) \qquad \text{for } t \geq a.$$ (15.2)

Then the equation (15.1) has property A (property B) if and only if

$$\int_a^{+\infty} t^{n-1}|p(t)|\, dt = +\infty.$$ (15.3)

THEOREM 15.2. Let $\lambda > 1$, $k \in \{1, \dots, n-1\}$, and let (15.2) hold. Then the equation (15.1) has property A_k (property B_k) if and only if

$$\int_a^{+\infty} t^{n-1+k(\lambda-1)}|p(t)|\, dt = +\infty.$$

THEOREM 15.3. Let $0 < \lambda < 1$ and let (15.2) hold. Then the equation (15.1) has property A (property B) if and only if

$$\int_a^{+\infty} t^{(n-1)\lambda}|p(t)|\, dt = +\infty.$$ (15.4)

THEOREM 15.4. Suppose $0 < \lambda < 1$, (15.2) holds, and $k \in \{1, \ldots, n-1\}$ is even (odd). Then the equation (15.1) has property A_k^* (property B_k^*) if and only if

$$\int_a^{+\infty} t^{k+(n-1-k)\lambda} |p(t)|\, dt = +\infty.$$

PROBLEM 15.1. Let $n = 3$, $\lambda > 1$, and let $p \colon [a, +\infty[\to]0, +\infty[$ be locally absolutely continuous. Moreover, suppose there exists $\sigma > 3 + (1 - 1/\sqrt{3})(\lambda - 1)$ such that $[t^\sigma p(t)]' \le 0$ for $t \ge a$. Then every proper solution of the equation (15.1) is nonoscillatory.

PROBLEM 15.2. Let $n = 3$, $0 < \lambda < 1$, $p(t) \le 0$ for $t \ge a$, and

$$\int_a^{+\infty} t^{1+\lambda} |p(t)|\, dt < +\infty.$$

Then every proper solution of the equation (15.1) is nonoscillatory.

PROBLEM 15.3. Let $n = 4$, $\lambda > 1$, and let the locally absolutely continuous function $p \colon [a, +\infty[\to]0, +\infty[$ be nonincreasing and satisfy the condition

$$\int_a^{+\infty} t^{1+2\lambda} |p(t)|\, dt < +\infty.$$

Then every proper solution of the equation (15.1) is nonoscillatory.

PROBLEM 15.4*. Let $n > 4$ and $p(t) \ne 0$ for $t \ge a$. What additional restriction should be imposed on p to guarantee nonoscillation of all proper solutions of the equation (15.1)?

15.2. Existence of proper oscillatory solutions.

THEOREM 15.5. Let $\lambda > 1$ and $(-1)^{n-n_0-1} p(t) \ge 0$ for $t \ge a$. If, in addition, n_0 is odd and

$$\int_a^{+\infty} t^{n+\lambda-2} |p(t)|\, dt = +\infty, \tag{15.5}$$

then the equation (15.1) has an $(n_0 - 1)$-parameter family of proper oscillatory solutions. Furthermore, if n_0 is even and (15.3) holds, then (15.1) has an n_0-parameter family of such solutions.

PROBLEM 15.5*. Does the equation (15.1) have proper oscillatory solutions if $n = 3$, $p(t) \le 0$ for $t \ge a$, and

$$\int_a^{+\infty} t^{1+\lambda} |p(t)|\, dt = +\infty?$$

THEOREM 15.6. Let $n = 2n_0 + 1$, $\lambda > 1$, and $(-1)^{n_0-1} p(t) \ge 0$ for $t \ge a$. If, in addition, n_0 is even and (15.5) holds, then the equation (15.1) has an n_0-parameter family of proper oscillatory solutions. Furthermore, if n_0 is odd and (15.3) holds, then (15.1) has an $(n_0 + 1)$-parameter family of such solutions.

PROBLEM 15.6*. Let $n = 2n_0$, $\lambda > 1$, and $(-1)^{n_0}p(t) \geq 0$ for $t \geq a$. Does the condition (15.3) guarantee the existence of at least one proper oscillatory solution of the equation (15.1)?

THEOREM 15.7. Let $0 < \lambda < 1$, $p(t) \leq 0$ for $t \geq a$, and let (15.4) hold. Suppose, in addition, that either n_0 is even and $n = 2n_0$ or $n = 2n_0 + 1$. Then the equation (15.1) has $a(2n_0)$-parameter family of proper oscillatory solutions.

PROBLEM 15.7*. Let $n = 2n_0$, where n_0 is odd, $0 < \lambda < 1$, and $p(t) \leq 0$ for $t \geq a$. Does the condition (15.4) guarantee the existence of at least one proper oscillatory solution of the equation (15.1)?

THEOREM 15.8. Let $n > 2$, $0 < \lambda < 1$, $p(t) \geq 0$ for $t \geq a$,

$$\int_a^{+\infty} t^{1+(n-2)\lambda}|p(t)|\,dt = +\infty, \tag{15.6}$$

and let either n or n_0 be odd. Then the equation (15.1) has an $(n-2)$-parameter family of proper oscillatory solutions.

PROBLEM 15.8*. Let $n > 2$, $0 < \lambda < 1$, $p(t) \geq 0$ for $t \geq a$, and let n be divisible by 4. Does the condition (15.6) guarantee the existence of at least one proper oscillatory solution of the equation (15.1)?

Theorems 15.5–15.8 are consequences of Theorems 14.1–14.4 and 15.1–15.4.

15.3. Proper oscillatory solutions vanishing at infinity. By restating Theorems 14.5–14.7 for the equation (15.1), we obtain the following assertions.

THEOREM 15.9. Let $\lambda > 1$ and $(-1)^{n-n_0-1}p(t) \geq \delta t^\nu$ for $t \geq a$, where $\delta > 0$ and $\nu > (\lambda - 1)/2 - n$. Then for n_0 even (odd) the equation (15.1) has an n_0-parameter $((n_0 - 1)$-parameter) family of proper oscillatory solutions satisfying the conditions

$$\lim_{t \to +\infty} t^{i+1/2}u^{(i)}(t) = 0 \qquad (i = 0,\ldots,n_0 - 1). \tag{15.7}$$

THEOREM 15.10. Let $n = 2n_0 + 1$, $\lambda > 1$, and $(-1)^{n_0-1}p(t) \geq \delta t^\nu$ for $t \geq a$, where $\delta > 0$ and $\nu > (\lambda + 3)(n - 1)/2 - n$. Then for n_0 even (odd) the equation (15.1) has an n_0-parameter $((n_0 + 1)$-parameter) family of proper oscillatory solutions vanishing at infinity.

THEOREM 15.11. Let $n = 2n_0 \geq 4$, $\lambda > 1$, and let p be locally absolutely continuous and satisfy the inequalities

$$(-1)^{n_0}p(t) > 0, \qquad (-1)^{n_0}[t^{n+1}p(t)]' \geq \delta t^\nu \qquad \text{for } t \geq a,$$

where $\delta > 0$ and $\nu > (\lambda - 1)/2$. Then for n_0 even (odd) the equation (15.1) has an (n_0-2)-parameter $((n_0-1)$-parameter) family of proper oscillatory solutions satisfying (15.7).

15.4. Oscillatory first kind singular solutions.

THEOREM 15.12. Let $0 < \lambda < 1$ and

$$0 < \delta \le (-1)^{n-n_0-1} p(t) \le \eta \qquad \text{for } t \ge a. \tag{15.8}$$

Then for n_0 even (odd) the equation (15.1) has an n_0-parameter (($n_0 - 1$)-parameter) family of oscillatory first kind singular solutions.

Below, $L^\alpha([a, +\infty[)$, where $\alpha \ge 1$, denotes the space of functions $u: [a, +\infty[\to \mathbf{R}$ for which u^α is integrable. If $u \in L^\alpha([a, +\infty[)$, we set

$$\|u\|_{\alpha,t} = \left(\int_t^{+\infty} |u(\tau)|^\alpha \, d\tau \right)^{1/\alpha}.$$

The proof of Theorem 15.12 is based on the following well-known proposition related to Kolmogorov–Gorny type inequalities.[1]

LEMMA 15.1. Let $\alpha > 1$, $\beta > 1$, $\gamma \ge \max\{\alpha, \beta\}$, and let $n \ge 2$ be a positive integer. Then there exists a positive constant $\rho = \rho(\alpha, \beta, \gamma)$ such that if a function $u \in \tilde{C}_{\text{loc}}^{n-1}([a, +\infty[)$ satisfies the conditions

$$u \in L^\alpha([a, +\infty[), \qquad u^{(n)} \in L^\beta([a, +\infty[), \tag{15.9}$$

then $u^{(j)} \in L^\gamma([a, +\infty[)$ and

$$\|u^{(j)}\|_{\gamma,t}^{n+1/\alpha-1/\beta} \le \rho \|u\|_{\alpha,t}^{n-j+1/\gamma-1/\beta} \|u^{(n)}\|_{\beta,t}^{j-1/\gamma+1/\alpha}$$

$$\text{for } t \ge a \qquad (j = 0, \dots, n-1). \tag{15.10}$$

PROOF. First of all we note that (15.9) yields

$$\lim_{t \to +\infty} u^{(j)}(t) = 0 \qquad (j = 0, \dots, n-1). \tag{15.11}$$

Without loss of generality we may assume

$$\text{meas}\{s \ge t : u^{(n)}(s) \ne 0\} > 0 \qquad \text{for } t \ge a.$$

Choose $a_i \in \mathbf{R}$ $(i = 1, \dots, n)$ such that the equation

$$\mu^n + \sum_{i=1}^n (-1)^i a_i \mu^{n-i} = 0$$

has only negative roots, and set

$$A = \begin{pmatrix} a_1 & -1 & 0 & \dots & 0 \\ a_2 & 0 & -1 & \dots & 0 \\ \cdot & \cdot & \cdot & \dots & \cdot \\ a_{n-1} & 0 & 0 & \dots & -1 \\ a_n & 0 & 0 & \dots & 0 \end{pmatrix},$$

---

[1] See [24, 125].

$$x(t) = \begin{pmatrix} u(t) \\ u'(t) \\ \cdots \\ u^{(n-1)}(t) \end{pmatrix}, \qquad y(t) = \begin{pmatrix} a_1 u(t) \\ \cdots \\ a_{n-1} u(t) \\ a_n u(t) + u^{(n)}(t) \end{pmatrix}.$$

Then $x'(t) = -Ax(t) + y(t)$. Hence

$$e^{A(\tau-t)} x(\tau) = x(t) + \int_t^\tau e^{A(s-t)} y(s)\, ds \qquad \text{for } \tau \geq t \geq a. \tag{15.12}$$

On the other hand,

$$\|e^{At}\| \leq c_0 e^{-rt} \qquad \text{for } t \geq a,$$

where c_0 and r are positive constants.[2] So (15.12), taking into account the conditions (15.9) and (15.11), implies

$$x(t) = -\int_t^{+\infty} e^{A(s-t)} y(s)\, ds$$

and

$$|u^{(j)}(t)| \leq c_1 \int_t^{+\infty} e^{-r(s-t)} (|u(s)| + |u^{(n)}(s)|)\, ds$$
$$\text{for } t \geq a \qquad (j = 0, \ldots, n-1), \tag{15.13}$$

with $c_1 = c_0 \max\{|a_1|, \ldots, |a_n|, 1\}$.

According to the Hölder inequality,

$$\int_t^{+\infty} e^{r\gamma\tau} \left(\int_\tau^{+\infty} e^{-rs} |u(s)|\, ds \right)^\gamma d\tau$$

$$\leq \int_t^{+\infty} e^{r\gamma\tau} \left(\int_\tau^{+\infty} e^{-ras/(2(\alpha-1))}\, ds \right)^{(\alpha-1)\gamma/\alpha} \left(\int_\tau^{+\infty} e^{-ras/2} |u(s)|^\alpha\, ds \right)^{\gamma/\alpha} d\tau$$

$$= \left(\frac{2(\alpha-1)}{r\alpha} \right)^{(\alpha-1)\gamma/\alpha} \int_t^{+\infty} e^{r\gamma\tau/2} \left(\int_\tau^{+\infty} e^{-ras/2} |u(s)|^\alpha\, ds \right)^{\gamma/\alpha} d\tau$$

$$\leq \frac{\gamma}{\alpha} c(\alpha) \int_t^{+\infty} e^{r(\gamma-\alpha)\tau/2} |u(\tau)|^\alpha \left(\int \tau^{+\infty} e^{-ras/2} |u(s)|^\alpha\, ds \right)^{-1+\gamma/\alpha} d\tau$$

$$\leq \frac{\gamma}{\alpha} c(\alpha) \int_t^{+\infty} |u(\tau)|^\alpha \left(\int_\tau^{+\infty} |u(s)|^\alpha\, ds \right)^{(\gamma/\alpha)-1} d\tau$$

$$= c(\alpha) \left(\int_t^{+\infty} |u(\tau)|^\alpha\, d\tau \right)^{\gamma/\alpha} \qquad \text{for } t \geq a,$$

where

$$c(\alpha) = \frac{2}{r\gamma} \left(\frac{2(\alpha-1)}{r\alpha} \right)^{(\alpha-1)\gamma/\alpha}.$$

Similarly,

$$\int_t^{+\infty} e^{r\gamma\tau} \left(\int_\tau^{+\infty} e^{-rs} |u^{(n)}(s)|\, ds \right)^\gamma d\tau \leq c(\beta) \left(\int_t^{+\infty} |u^{(n)}(\tau)|^\beta\, d\tau \right)^{\gamma/\beta} \qquad \text{for } t \geq a.$$

---

[2] By $\|e^{At}\|$ we mean the sum of the absolute values of the entries of the matrix e^{At}.

Therefore (15.13) yields

$$\|u^{(j)}\|_{\gamma,t} \le c_2(\|u\|_{\alpha,t} + \|u^{(n)}\|_{\beta,t}) \quad \text{for } t \ge a \quad (j = 0, \dots, n-1), \quad (15.14)$$

where the constant c_2 depends on n, α, β, and γ only.

Set

$$v_x(t) = u(xt) \quad \text{for } t \ge \frac{a}{x}, x > 0.$$

Then, by (15.14),

$$\|v_x^{(j)}\|_{\gamma,t/x} \le c_2(\|v_x\|_{\alpha,t/x} + \|v_x^{(n)}\|_{\beta,t/x}) \quad \text{for } t \ge a \quad (j = 0, \dots, n-1).$$

But $\|v_x^{(j)}\|_{\gamma,t/x} = x^{j-1/\gamma}\|u^{(j)}\|_{\gamma,t}$. Thus

$$x^{j-1/\gamma}\|u^{(j)}\|_{\gamma,t} \le c_2(x^{-1/\alpha}\|u\|_{\alpha,t} + x^{n-1/\beta}\|u^{(n)}\|_{\beta,t})$$

$$\text{for } t \ge a, x > 0 \quad (j = 0, \dots, n-1).$$

Putting

$$x = (\|u\|_{\alpha,t}\|u^{(n)}\|_{\beta,t}^{-1})^\nu, \quad \nu = \left(n - \frac{1}{\beta} + \frac{1}{\alpha}\right)^{-1},$$

from these inequalities we derive the estimates (15.10) with $\rho = (2c_2)^\nu$. $\square$

PROOF OF THEOREM 15.12. Let

$$\chi(x) = \begin{cases} |x|^\lambda \operatorname{sign} x & \text{for } |x| < 1 \\ \operatorname{sign} x & \text{for } |x| \ge 1. \end{cases} \quad (15.15)$$

Suppose the constants $c_i \in \mathbf{R}$ $(i = 1, \dots, n_0)$ are not all zero and, in addition, $c_1 = 0$ if n_0 is odd. By Lemma 14.3 the equation

$$u^{(n)} = p(t)\chi(u)$$

has a solution u satisfying the conditions

$$u^{(i-1)}(a) = c_i \quad (i = 1, \dots, n_0), \quad \lim_{t \to +\infty} u(t) = 0. \quad (15.16)$$

Clearly, u cannot be a nonoscillatory singular solution.

Choose $a_0 \in \,]a, +\infty[$ so that

$$\sum_{i=1}^{n_0} |u^{(i-1)}(a_0)| > 0, \quad |u(t)| < 1 \quad \text{for } t \ge a_0.$$

Then according to (15.5) the restriction of u to $[a_0, +\infty[$ is a solution of the equation (15.1). We have to show that u is an oscillatory first kind singular solution.

Assume, on the contrary, that u is proper. Set

$$v(t) = \sum_{j=0}^{n_0-1} (-1)^{j+n-n_0} u^{(n-1-j)}(t)u^{(j)}(t) + \epsilon_n[u^{(n_0)}(t)]^2, \quad (15.17)$$

with $\epsilon_n = 0$ for $n = 2n_0$ and $\epsilon_n = -1/2$ for $n = 2n_0 + 1$. It follows from (15.8) that

$$-v'(t) \ge (-1)^{n-n_0-1}u^{(n)}(t)u(t) \ge \delta|u(t)|^{\lambda+1} \quad \text{for } t \ge a_0. \quad (15.18)$$

On the other hand, by Lemma 5.2 the conditions (5.8) and (5.16) imply

$$|u^{(n)}(t)| \leq \eta|u(t)|^\lambda \qquad \text{for } t \geq a_0 \qquad (15.19)$$

and

$$\lim_{t\to+\infty} u^{(i)}(t) = 0 \qquad (i = 0, \dots, n-1).$$

Therefore

$$\lim_{t\to+\infty} v(t) = 0.$$

So from (15.18) and (15.19) we derive

$$\int_t^{+\infty} |u(s)|^\alpha \, ds \leq \delta^{-1}v(t) \qquad \text{for } t \geq a_0 \qquad (15.20)$$

and $u^{(n)} \in L^\beta([a, +\infty[)$, where $\alpha = \lambda + 1$ and $\beta = (\lambda + 1)/\lambda$.

Let $\gamma = \beta$. Then Lemma 15.1 yields the estimates (15.10). Hence, applying (15.19), we get

$$\|u^{(j)}\|_{\gamma,t}^{n+(1-\lambda)/(1+\lambda)} \leq \rho_1 \|u\|_{\alpha,t}^{n-(1-\lambda)j-\lambda(1-\lambda)/(1+\lambda)}$$

$$\text{for } t \geq a_0 \qquad (j = 0, \dots, n-1), \qquad (15.21)$$

with $\rho_1 = \text{const} > 0$.

In view of (15.17), there exists a positive constant ρ_2 such that

$$[v(t)]^{\gamma/2} \leq \rho_2 \sum_{j=0}^{\nu} |u^{(j)}(t)|^{\gamma/2}|u^{(n-1-j)}(t)|^{\gamma/2} \qquad \text{for } t \geq a_0,$$

where $\nu = n_0 - 1$ if $n = 2n_0$ and $\nu = n_0$ if $n = 2n_0 + 1$. So, by the Cauchy inequality,

$$\|v\|_{\gamma/2,t} \leq \rho_2^{2/\gamma} \sum_{j=0}^{\nu} \|u^{(j)}\|_{\gamma,t}\|u^{(n-1-j)}\|_{\gamma,t} \qquad \text{for } t \geq a_0.$$

According to (15.20) and (15.21) this estimate implies

$$\|v\|_{\gamma/2,t} \leq \rho_3[v(t)]^{1/\mu} \qquad \text{for } t \geq a_0,$$

where

$$\mu = [(1+\lambda)n + 1 - \lambda]\left[(1+\lambda)n + 1 - \lambda + \frac{2\lambda(1-\lambda)}{1+\lambda}\right]^{-1} < 1.$$

Consequently,

$$[v(t)]^{\gamma/2}\left(\int_t^{+\infty} [v(s)]^{\gamma/2} \, ds\right)^{-\mu} \geq \rho_3^{-\mu\gamma/2} \qquad \text{for } t \geq a_0.$$

Integration of the last inequality from a_0 to t gives

$$\left(\int_a^{+\infty} [v(s)]^{\gamma/2} \, ds\right)^{1-\mu} \geq (1-\mu)\rho_3^{-\mu\gamma/2}(t - a_0) \qquad \text{for } t \geq a_0,$$

which is a contradiction. $\square$

Notes. Theorems 15.1 and 15.3 for $n = 2$ are due to F.V. Atkinson [8] and Š. Belohorec [28]. For $n > 2$ and nonpositive p, Theorem 15.3 was proved by I. Ličko and M. Švec [262]. In the general case, Theorems 15.1 and 15.3, as well as Theorems 15.2, 15.4–15.6 and 15.8–15.11, were established by I.T. Kiguradze [172, 176, 180, 190, 191].

Theorems 15.7 and 15.12 sharpen results of J.W. Heidel [135] and F. Bernis [35], respectively. The question of the existence of proper and singular oscillatory solutions of the equation (15.1) was also studied by V.A. Kondratyev and V.S. Samovol [214].

For Problems 15.1–15.3 see [106, 240, 241].

§16. Nonoscillatory Solutions

In this section, as in the previous one, we consider the differential equation

$$u^{(n)} = p(t)|u|^\lambda \operatorname{sign} u, \tag{16.1}$$

where $p \in L_{\text{loc}}([a, +\infty[)$, $a > 0$, $\lambda > 0$, and $\lambda \neq 1$.

16.1. Kneser solutions. Theorems 11.2, 13.1, 13.2′, 13.3–13.5, and Corollary 9.3 imply the following assertions.

THEOREM 16.1. Let

$$(-1)^n p(t) \geq 0 \qquad \text{for } t \geq a. \tag{16.2}$$

Then the equation (16.1) has a continuum of Kneser solutions, and the condition

$$\int_a^{+\infty} t^{n-1}|p(t)|\,dt = +\infty$$

is necessary and sufficient for every such solution to vanish at infinity.

THEOREM 16.2. If $\lambda > 1$, the condition (16.2) holds and

$$\int_a^{+\infty} t^{n-1}|p(t)|\,dt < +\infty, \tag{16.3}$$

then every nonzero Kneser solution of the equation (16.1) tends to a nonzero limit as $t \to +\infty$.

THEOREM 16.3. Let $0 < \lambda < 1$,

$$\operatorname{meas}\{s \geq t : \ p(s) \neq 0\} > 0 \qquad \text{for } t > a, \tag{16.4}$$

and let the conditions (16.2) and (16.3) hold. Then for any $m \in \{0, 1\}$ the equation (16.1) has at least one Kneser solution vanishing at infinity and satisfying the inequalities

$$(-1)^{m+i} u^{(i)}(t) > 0 \qquad \text{for } t \geq a \qquad (i = 0, \dots, n-1); \tag{16.5}$$

moreover, every such solution is subject to the estimate

$$|u(t)| \leq \left[\frac{1}{(n-1)!} \int_t^{+\infty} (\tau - t)^{n-1}|p(\tau)|\,d\tau\right]^{1/(1-\lambda)} \qquad \text{for } t \geq a. \tag{16.6}$$

THEOREM 16.4. Let $0 < \lambda < 1$ and let (16.2) hold. Suppose, in addition, that for a certain $\alpha \in [2/(\lambda(n-1)+n+1), 1/(\lambda(n-1)+1)[$,

$$\lim_{t \to +\infty} \sup t^\beta \int_t^{+\infty} \tau^{n\alpha-\beta-1}|p(\tau)|^\alpha \, d\tau > 0, \tag{16.7}$$

where $\beta = 1 - (1 + \lambda n - \lambda)\alpha$. Then the equation (16.1) does not have a proper Kneser solution.

PROBLEM 16.1. For any constant $\alpha \geq 1/(\lambda(n-1)+1)$ and any piecewise continuous function $\varphi \colon [a, +\infty[\to \,]0, +\infty[$ there exists a piecewise continuous function $p \colon [a, +\infty[\to \mathbf{R}$ such that (16.2) holds,

$$\int_a^{+\infty} \varphi(\tau)|p(\tau)|^\alpha \, d\tau = +\infty$$

and the equation (16.1) has a proper Kneser solution vanishing at infinity.

THEOREM 16.5. Let $0 < \lambda < 1$ and let (16.2) hold. Suppose, in addition, that the equation (16.1) has a proper Kneser solution u vanishing at infinity. Then, whatever $\alpha \in [2/(\lambda(n-1)+n+1), 1/(\lambda(n-1)+1)[$,

$$|u(t)| \geq \gamma_0 \left[\int_t^{+\infty} (\tau-t)^{n-2} \left(\int_\tau^{+\infty} |p(s)|^\alpha \, ds \right)^{(n-1)/(n\alpha-1)} d\tau \right]^{(n\alpha-1)/(n-1)(1-\lambda)\alpha}$$

$$\text{for } t \geq a, \tag{16.8}$$

where $\beta = 1 - (1 + \lambda n - \lambda)\alpha$ and γ_0 is a positive constant depending on n, λ and α only.

THEOREM 16.6. Let $0 < \lambda < 1$ and

$$r(1 - \delta(t))t^\sigma \leq (-1)^n p(t) \leq r(1 + \delta(t))t^\sigma \qquad \text{for } t \geq a, \tag{16.9}$$

where $r \in \,]0, +\infty[$, $\sigma \in \mathbf{R}$ and the function $\delta \colon [a, +\infty[\to [0, 1[$ tends monotonically to zero as $t \to +\infty$. Then the equation (16.1) has a proper Kneser solution if and only if $\sigma < -n$. Besides, if $\sigma < -n$, then for any $m \in \{0, 1\}$ the equation (16.1) possesses at least one proper Kneser solution vanishing at infinity and satisfying the inequalities (16.5); moreover, every such solution is subject to the representation

$$u(t) = (-1)^m \rho_0 t^{(\sigma+n)/(1-\lambda)}(1 + \epsilon(t))^{\lambda/(1-\lambda)} \qquad \text{for } t \geq a, \tag{16.10}$$

where $|\epsilon(t)| \leq \delta(t)$ for $t \geq a$ and

$$\rho_0 = r^{1/(1-\lambda)} \prod_{k=0}^{n-1} \left(k - \frac{\sigma+n}{1-\lambda} \right)^{1/(\lambda-1)}. \tag{16.11}$$

PROOF. Assume that the equation (16.1) has a proper Kneser solution u. Then, according to Theorem 16.5, the condition (16.7) with $\alpha = 2/(\lambda(n-1)+n+1)$ is violated. Hence, as follows from (16.9), $\sigma < -n$.

Now let $\sigma < -n$. Then (16.9) implies (16.3) and (16.4). So, by Theorems 16.3 and 16.5, for any $m \in \{0, 1\}$ the equation (16.1) has at least one Kneser solution vanishing at infinity and satisfying the inequalities (16.5); moreover, every such solution is subject to the estimates (16.6) and (16.8). In view of (16.9) these estimates yield

$$v_1(t) = \inf\left\{s^{-(\sigma+n)/(1-\lambda)}|u(s)| : s \geq t\right\} > 0 \qquad \text{for } t \geq a,$$

$$v_2(t) = \sup\left\{s^{-(\sigma+n)/(1-\lambda)}|u(s)| : s \geq t\right\} < +\infty \qquad \text{for } t \geq a.$$

Therefore, applying the equality

$$|u(t)| = \frac{(-1)^n}{(n-1)!}\int_t^{+\infty}(\tau - t)^{n-1}p(\tau)|u(\tau)|^\lambda\,d\tau,$$

we get

$$\rho_0^{1-\lambda}(1 - \delta(t))^\lambda[v_1(t)]^\lambda \leq u(t)t^{-(\sigma+n)/(1-\lambda)} \leq \rho_0^{1-\lambda}(1 + \delta(t))^\lambda[v_2(t)]^\lambda \qquad \text{for } t \geq a,$$

and so

$$\rho_0(1 - \delta(t))^{\lambda/(1-\lambda)} \leq v_1(t) \leq v_2(t) \leq \rho_0(1 + \delta(t))^{\lambda/(1-\lambda)} \qquad \text{for } t \geq a.$$

By (16.5) this implies the representation (16.10), where $|\epsilon(t)| \leq \delta(t)$ for $t \geq a$. $\square$

COROLLARY 16.1. Let $0 < \lambda < 1$ and $r > 0$. Then the equation

$$u^{(n)} = (-1)^n r t^\sigma |u|^\lambda \operatorname{sign} u$$

has a proper Kneser solution if and only if $\sigma < -n$. Moreover, for $\sigma < -n$ this equation has exactly two proper Kneser solutions vanishing at infinity:

$$u_m(t) = (-1)^m \rho_0 t^{(\sigma+n)/(1-\lambda)} \qquad (m = 0, 1),$$

where the constant ρ_0 is defined by (16.11).

THEOREM 16.7. Suppose $0 < \lambda < 1$, (16.2) holds, and the point $t_0 \in\,]a, +\infty[$ is such that

$$\operatorname{meas}\{\tau \in [t, t_0] : p(\tau) \neq 0\} > 0 \qquad \text{for } a < t \leq t_0.$$

Then for any $m \in \{0, 1\}$ the equation (16.1) has a first kind singular solution whose carrier is $[a, t_0[$ and which satisfies the inequalities

$$(-1)^{m+i}u^{(i)}(t) > 0 \qquad \text{for } a \leq t < t_0 \qquad (i = 0, \dots, n-1); \tag{16.12}$$

moreover, every such solution is subject to the estimates

$$|u(t)| \leq \left[\frac{1}{(n-1)!}\int_t^{t_0}(\tau - t)^{n-1}|p(\tau)|\,d\tau\right]^{1/(1-\lambda)} \qquad \text{for } a \leq t \leq t_0 \tag{16.13}$$

and

$$|u(t)| \geq \gamma_0(t_0 - t)^{n-1}\left[(t_0 - t)^{2-n}\int_t^{t_0}\frac{(s-t)^{n-2}}{(s-t_0)^n}\right.$$

$$\times\left.\left(\int_s^{t_0}(t_0 - \tau)^{[n+1+(n-1)\lambda]\alpha-2}|p(\tau)|^\alpha\,d\tau\right)^{(n-1)/(n\alpha-1)}ds\right]^{(n\alpha-1)/(n-1)(1-\lambda)}$$

$$\text{for } a \leq t \leq t_0, \tag{16.14}$$

where α is an arbitrary number in the interval $[2/(\lambda(n-1)+n+1), 1/(\lambda(n-1)+1)[$ and γ_0 is a positive constant depending on n, λ and α only.

PROOF. The existence of the singular solution in question is a consequence of Theorem 11.1. To verify (16.13) and (16.14) we use the transformation

$$v(s) = s^{n-1} u \left(t_0 - \frac{1}{s} \right),$$ (16.15)

which reduces (16.1) to the equation

$$v^{(n)} = \tilde{p}(s) |v|^\lambda \operatorname{sign} v$$ (16.16)

with $\tilde{p}(s) = s^{-1-n-(n-1)\lambda} p(t_0 - 1/s)$, and then estimate v by Theorems 16.3 and 16.5. □

In a similar way we can prove

THEOREM 16.8. Let $0 < \lambda < 1$ and let (16.2) hold. Suppose, in addition, that the function p is continuous and differs from zero at a point $t_0 \in]a, +\infty[$. Then for any $m \in \{0, 1\}$ the equation (16.1) has a first kind singular solution whose carrier is $[a, t_0[$ and which satisfies the inequalities (16.12); moreover, every such solution is subject to the representation

$$u(t) = (-1)^m \rho_0(t)(t_0 - t)^{n/(1-\lambda)} \qquad \text{for } a \le t < t_0,$$

where

$$\lim_{t \to t_0} \rho_0(t) = |p(t_0)|^{1/(1-\lambda)} \prod_{k=0}^{n-1} \left(\frac{n}{1-\lambda} - k \right)^{1/(\lambda-1)}.$$

16.2. Solutions with power asymptotics. Corollaries 8.2, 8.6, 8.12, 9.3, and 9.7 imply the following assertions concerning the equation (16.1).

THEOREM 16.9. Let $k \in \{1, \dots, n\}$. Then the condition

$$\int_a^{+\infty} t^{n-k+\lambda(k-1)} |p(t)| \, dt < +\infty$$

is sufficient and, if p does not change sign, also necessary, for the equation (16.1) to possess, for any $\alpha \ne 0$, a solution having the asymptotic representation

$$u(t) = \alpha t^{k-1}(1 + o(1)).$$

THEOREM 16.10. Let $\lambda > 1$ and

$$\int_a^{+\infty} t^{\lambda(n-1)} |p(t)| \, dt < +\infty.$$ (16.17)

Then:

(a) there exists $\delta > 0$ such that every solution of the equation (16.1) satisfying the condition

$$0 < \sum_{j=0}^{n-1} |u^{(j)}(a)| < \delta$$

has the asymptotic representation

$$u(t) = \sum_{k=1}^{n} \alpha_k t^{k-1}(1 + o(1)),$$ (16.18)

where $\sum_{k=1}^{n} |\alpha_k| > 0$;

(b) the family of solutions of the type (16.18) of the equation (16.1) is stable.[3]

THEOREM 16.11. Let $0 < \lambda < 1$ and

$$\int_{a}^{+\infty} t^{n-1} |p(t)| \, dt < +\infty.$$

Then there exists $\delta > 0$ such that every solution of the equation (16.1) satisfying the condition

$$\sum_{j=0}^{n-1} |u^{(j)}(a)| > \delta$$

has the asymptotic representation (16.18).

16.3. Fast growing solutions.

THEOREM 16.12. Let $\lambda > 1$,

$$p(t) \geq 0, \operatorname{meas}\{s \geq t : p(s) > 0\} > 0 \qquad \text{for } t \geq a, \tag{16.19}$$

and let (16.17) hold. Then the equation (16.1) has an $(n-1)$-parameter family of fast growing solutions, and every such solution is subject to the estimate

$$|u^{(n-1)}| \geq \left((\lambda - 1)[(n-1)!]^{-\lambda} \int_{t}^{+\infty} \tau^{\lambda(n-1)} |p(\tau)| \, d\tau \right)^{-1/(\lambda-1)}$$

in a certain neighborhood of $+\infty$.

THEOREM 16.13. Let $\lambda > 1$ and let (16.19) hold. Furthermore, suppose that the equation (16.1) has a fast growing solution u. Then for any $\alpha \in [2/(\lambda(n-1) + n + 1), 1/n[$,

$$|u(t)| \leq \gamma_0 \left(t^{n\alpha-1} \int_{t}^{+\infty} |p(\tau)|^{\alpha} \, d\tau \right)^{-1/(\lambda-1)\alpha}$$

in a certain neighborhood of $+\infty$, where γ_0 is a positive constant depending on α, λ and n only.

Theorem 16.12 follows from Theorems 12.1 and 12.3, and Theorem 16.13 is a consequence of Theorem 12.2.

Theorems 16.12 and 16.13 immediately imply

COROLLARY 16.2. Let $\lambda > 1$ and $r > 0$. Then the equation

$$u^{(n)} = rt^{\sigma} |u|^{\lambda} \operatorname{sign} u$$

has a fast growing solution if and only if $\sigma < -(n-1)\lambda - 1$. Moreover, for $\sigma < -(n-1)\lambda - 1$ this equation has an $(n-1)$-parameter family of fast growing solutions, and every such solution is subject to the estimate

$$\gamma_1 r^{1/(1-\lambda)} t^{(\sigma+n)/(1-\lambda)} \leq |u(t)| \leq \gamma_2 r^{1/(1-\lambda)} t^{(\sigma+n)/(1-\lambda)}$$

---

[3] See Definition 7.7.

in a certain neighborhood of $+\infty$, where γ_1 and γ_2 are positive constants depending on σ, λ and n only.

PROBLEM 16.2. For any $\lambda > 1$ there exists a piecewise continuous function $p : [a, +\infty[\to]0, +\infty[$ such that

$$\int_a^{+\infty} t^{\lambda(n-1)} p(t)\, dt = +\infty$$

and the equation (16.1) has an n-parameter family of fast growing solutions.

PROBLEM 16.3. For any $\lambda > 1$ and any piecewise continuous function $\epsilon : [a, +\infty[\to]0, +\infty[$ there exists a piecewise continuous $p : [a, +\infty[\to]0, +\infty[$ such that

$$\int_a^{+\infty} \epsilon(t) t^{\lambda(n-1)} p(t)\, dt < +\infty$$

and the equation (16.1) does not have a fast growing solution.

16.4. Second kind singular solutions.

THEOREM 16.14. Let $\lambda > 1$, $t_0 \in]a, +\infty[$, and

$$p(t) \geq 0, \qquad \operatorname{meas}\{s \geq t : p(s) > 0\} > 0 \qquad \text{for } a < t < t_0.$$

Then the equation (16.1) has an $(n-1)$-parameter family of nonoscillatory second kind singular solutions satisfying the condition

$$\lim_{t \to t_0} |u(t)| = +\infty; \tag{16.20}$$

moreover, every such solution is subject to the estimates

$$|u(t)| \geq \gamma_1 (t_0 - t) \int_a^t \frac{(t-\tau)^{n-2}}{(t_0 - \tau)^n} \left(\int_\tau^{t_0} (t_0 - s)^{n-1} |p(s)|\, ds \right)^{-1/(\lambda-1)} d\tau$$

and

$$|u(t)| \leq \gamma_2 \left[(t_0 - t)^{1 - (\lambda n - \lambda + 1)\alpha} \right.$$

$$\times \left. \int_t^{t_0} (t_0 - \tau)^{(n+\lambda n - \lambda + 1)\alpha - 2} |p(\tau)|^\alpha\, d\tau \right]^{-1/(\lambda-1)\alpha}$$

in a certain left-hand neighborhood of t_0, where

$$0 < \gamma_1 < \frac{1}{(n-2)!} \left[\frac{\lambda - 1}{[(n-1)!]^\lambda} \right]^{1/(\lambda-1)}, \qquad \alpha \in \left[\frac{2}{\lambda(n-1) + n + 1}, \frac{1}{n} \right[,$$

and γ_2 is a positive constant depending on α, λ and n only.

To justify this assertion we reduce the equation (16.1) to the form (16.16) by means of the transformation (16.15) and then apply Theorems 16.12 and 16.13.

COROLLARY 16.3. Let $\lambda > 1$ and let the function p be continuous and positive at a point t_0. Then the equation (16.1) has an $(n-1)$-parameter family of second kind singular solutions satisfying the condition (16.20), and every such solution is subject to the estimate

$$\gamma_1 r^{1/(1-\lambda)}(t_0 - t)^{n/(1-\lambda)} \leq |u(t)| \leq \gamma_2 r^{1/(1-\lambda)}(t_0 - t)^{n/(1-\lambda)}$$

in a certain left-hand neighborhood of t_0, where $r = p(t_0)$ and γ_1 and γ_2 are positive constants depending on λ and n only.

PROBLEM 16.4*. Let $\lambda > 1$ and let the function p be continuous and positive at a point t_0. Is it true that every second kind singular solution of the equation (16.1) satisfying (16.20) has the representation

$$|u(t)| = \rho(t)(t_0 - t)^{n/(1-\lambda)}$$

in a certain left-hand neighborhood of t_0, where

$$\lim_{t \to t_0} \rho(t) = |p(t_0)|^{1/(1-\lambda)} \prod_{k=0}^{n-1} \left(\frac{n}{1-\lambda} - k\right)?$$

From Theorem 11.4 we can derive

THEOREM 16.15. Let $\lambda > 1$, $p(t) \geq 0$ for $t \geq a$, and let there exist $\alpha \in [2/(\lambda(n-1) + n + 1), 1/n[$ such that

$$\lim_{t \to +\infty} \sup t^{1-n} \int_t^{+\infty} \tau^{[\lambda(n-1)+n+1]\alpha - 2}|p(\tau)|^\alpha \, d\tau > 0.$$

Then every solution u of the equation (16.1) satisfying the conditions

$$u^{(i)}(t_*)u^{(n-1)}(t_*) \geq 0 \qquad (i = 1, \ldots, n-1), \qquad u^{(n-1)}(t_*) \neq 0,$$

for a certain $t_* \in [a, +\infty[$, is nonoscillatory second kind singular.

Notes. Theorems 16.1, 16.2, 16.5, 16.6, and 16.8–16.10 are due to I.T. Kiguradze [176, 179]. Theorem 16.3 was proved by G.G. Kvinikadze [245], and Theorem 16.11 by T.A. Chanturia [55]. Theorems 16.4 and 16.15 somewhat modify theorems of N.A. Izobov [141, 143], and sharpen some results from [176, 180]. Theorems 16.7 and 16.12–16.14 were established by I.T. Kiguradze and G.G. Kvinikadze [199, 250].

For Problems 16.1–16.3 see the papers of N.A. Izobov and V.A. Rabtsevich [140–144, 307].

Problem 16.4 with $n \in \{3, 4\}$ is solved by I.V. Astashova [7] (the case $n = 2$ is considered in §20 below).

CHAPTER V

SECOND ORDER DIFFERENTIAL EQUATIONS OF EMDEN–FOWLER TYPE

§17. Existence Theorems for Proper and Singular Solutions

Consider the equation

$$u'' = p(t)|u|^\lambda \operatorname{sign} u, \tag{17.1}$$

where $\lambda > 0$ but $\lambda \neq 1$, and $p \in L_{\text{loc}}(\mathbf{R}_+)$.

17.1. Existence of proper solutions.

THEOREM 17.1. Let

$$p(t) < 0 \qquad \text{for } t \geq 0, \tag{17.2}$$

and $p(t) = -p_0(t) + p_1(t)$, where $p_0, p_1 \in C(\mathbf{R}_+)$ are nonnegative and nondecreasing. Then any maximally continued solution u of the equation (17.1) is defined on $[0, +\infty[$, and for $0 \leq s < t < +\infty$ the inequality

$$\rho(u)(s) \exp\left(-\int_s^t \frac{dp_0(\tau)}{|p(\tau)|}\right) \leq \rho(u)(t) \leq \rho(u)(s) \exp\left(\int_s^t \frac{dp_1(\tau)}{|p(\tau)|}\right) \tag{17.3}$$

holds, with

$$\rho(u)(t) = \frac{|u'(t)|^2}{|p(t)|} + \frac{2}{\lambda+1}|u(t)|^{\lambda+1}. \tag{17.4}$$

PROOF. Let $u: \,]t_0, t_1[\,\to \mathbf{R}$ be a solution of the equation (17.1). To prove the theorem it suffices to show that for $t_0 < s < t < t_1$ (17.3) is fulfilled.

By (17.4), whatever $\epsilon > 0$,

$$\rho(u)(t) = (\rho(u)(s) + \epsilon) \exp\left(-\int_s^t \frac{|u'(\tau)|^2}{|p(\tau)|(\rho(u)(\tau) + \epsilon)} \frac{dp_0(\tau)}{p(\tau)}\right.$$
$$\left. + \int_s^t \frac{|u'(\tau)|^2}{|p(\tau)|(\rho(u)(\tau) + \epsilon)} \frac{dp_1(\tau)}{p(\tau)}\right) - \epsilon \qquad \text{for } t_0 < s < t < t_1.$$

Since p_0 and p_1 are nondecreasing and ϵ is arbitrary, this yields (17.3). $\square$

COROLLARY 17.1. Under the hypotheses of Theorem 17.1 the number of zeros of a nonzero solution of the equation (17.1) is at most finite on any finite interval.

COROLLARY 17.2. Under the hypotheses of Theorem 17.1,

$$\rho_1(u)(s)\exp\left(-\int_s^t \frac{dp_1(\tau)}{|p(\tau)|}\right) \le \rho_1(u)(t) \le \rho_1(u)(s)\exp\left(\int_s^t \frac{dp_0(\tau)}{|p(\tau)|}\right)$$

for $0 \le s < t < +\infty$, where

$$\rho_1(u)(t) = |u'(t)|^2 + \frac{2}{\lambda+1}|p(t)|\,|u(t)|^{\lambda+1}.$$

THEOREM 17.2. Let $p(t) = -p_0(t) + p_1(t) + p_2(t)$, where $p_0, p_1 \in C(\mathbf{R}_+)$ are non-negative and nondecreasing, $p_2 \in L_{loc}(\mathbf{R}_+)$,

$$p_0(t) - p_1(t) > 0 \qquad \text{for } t \ge 0,$$

and let, in addition,

$$\int_0^{+\infty} |p_2(t)|(p_0(t) - p_1(t))^{-1/2}\exp\left(\frac{1-\lambda}{2(\lambda+1)}\int_0^t \frac{dp_0(\tau)}{p_0(\tau) - p_1(\tau)}\right) dt < +\infty$$

if $0 < \lambda < 1$ and

$$\int_0^{+\infty} |p_2(t)|(p_0(t) - p_1(t))^{-1/2}\exp\left(\frac{\lambda-1}{2(\lambda+1)}\int_0^t \frac{dp_1(\tau)}{p_0(\tau) - p_1(\tau)}\right) dt < +\infty$$

if $\lambda > 1$. Then the equation (17.1) has proper solutions.

PROOF. We claim that any maximally continued solution of the equation (17.1) satisfying the condition

$$[\rho(u)(t_0)]^{(1-\lambda)/2(\lambda+1)} > c_\lambda(t_0)$$

for a certain $t_0 \ge 0$, where

$$\rho(u)(t) = (p_0(t) - p_1(t))^{-1}|u'(t)|^2 + \frac{2}{\lambda+1}|p(t)|\,|u(t)|^{\lambda+1},$$

$$c_\lambda(t_0) = c_0 \int_{t_0}^{+\infty} \frac{|p_2(t)|}{\sqrt{p_0(t) - p_1(t)}}\exp\left(\frac{1-\lambda}{2(\lambda+1)}\int_{t_0}^t \frac{dp_0(\tau)}{p_0(\tau) - p_1(\tau)}\right) dt$$

if $0 < \lambda < 1$,

$$c_\lambda(t_0) = c_0 \int_{t_0}^{+\infty} \frac{|p_2(t)|}{\sqrt{p_0(t) - p_1(t)}}\exp\left(\frac{\lambda-1}{2(\lambda+1)}\int_{t_0}^t \frac{dp_1(\tau)}{p_0(\tau) - p_1(\tau)}\right) dt$$

if $\lambda > 1$,

$$c_0 = 2^{-\lambda/(\lambda+1)}(\lambda+1)^{-1/(\lambda+1)}|1 - \lambda|,$$

is proper. Indeed, assume that $u\colon [t_0, t_1[\to \mathbf{R}$ is a maximally continued solution of the equation (17.1) satisfying the above-mentioned condition. Then from the equality

$$(\rho(u)(t) + \epsilon)^{(1-\lambda)/2(\lambda+1)}\exp\left(\frac{1-\lambda}{2(\lambda+1)}\int_{t_0}^t \frac{|u'(\tau)|^2\,d(p_0(t) - p_1(t))}{(\rho(u)(t) + \epsilon)(p_0(\tau) - p_1(\tau))^2}\right)$$

$$= (\rho(u)(t_0) + \epsilon)^{(1-\lambda)/2(\lambda+1)} + \frac{1-\lambda}{1+\lambda}\int_{t_0}^t \frac{u'(\tau)|u(\tau)|^\lambda \operatorname{sign} u(\tau)}{[\rho(u)(\tau)]^{(1+3\lambda)/2(\lambda+1)}}$$

$$\times \exp\left(\frac{1-\lambda}{2(\lambda+1)} \int_{t_0}^{\tau} \frac{|u'(s)|^2 \, d(p_0(s)-p_1(s))}{(\rho(u)(s)+\epsilon)(p_0(s)-p_1(s))^2}\right) d\tau,$$

where $\epsilon > 0$ is sufficiently small, taking into account the estimates

$$|u'(t)| \le (p_0(t)-p_1(t))^{1/2}(\rho(u)(t))^{1/2}$$

and

$$|u(t)| \le \left(\frac{\lambda+1}{2}\right)^{1/(\lambda+1)} (\rho(u)(t))^{1/(\lambda+1)},$$

we obtain

$$(\rho(u)(t))^{(1-\lambda)/2(\lambda+1)} \exp\left(\frac{|1-\lambda|}{2(\lambda+1)} \int_{t_0}^{t} \frac{d(p_0(\tau)+p_1(\tau))}{p_0(\tau)-p_1(\tau)}\right)$$

$$\ge (\rho(u)(t_0))^{(1-\lambda)/2(\lambda+1)} - c_\lambda(t_0) > 0$$

$$\text{for } t_0 \le t < t_1.$$

This immediately implies $t_1 = +\infty$ and $\rho(u)(t) > 0$ for $t \ge t_0$. Hence u is proper. $\square$

REMARK 17.1. It follows from Theorem 17.3, stated below, that under the hypotheses of Theorem 17.2 the equation (17.1) can have nonoscillatory first (if $0 < \lambda < 1$) or second (if $\lambda > 1$) kind singular solutions as well.

17.2. Existence of nonoscillatory singular solutions. The following proposition is a consequence of Theorems 16.7 and 16.14.

THEOREM 17.3. Let $0 < \lambda < 1$ ($\lambda > 1$) and let p be nonnegative and distinct from zero on a set of positive measure. Then the equation (17.1) has a nonoscillatory first (second) kind solution.

THEOREM 17.4. Let $0 < \lambda < 1$,

$$p(t) \ge p_0(t) > 0 \qquad \text{for } t \ge 0 \tag{17.5}$$

and

$$\int_0^{+\infty} t p_0(t) \, dt = +\infty, \tag{17.6}$$

where $p_0 \in L_{loc}(\mathbf{R}_+)$ is nonincreasing. Then any nonzero solution of the equation (17.1) satisfying the inequality

$$u(t)u'(t) \le 0 \qquad \text{for } t \ge 0 \tag{17.7}$$

is nonoscillatory first kind singular.

PROOF. First we will show that any solution of the equation

$$u'' = p_0(t)|u|^\lambda \operatorname{sign} u \tag{17.8}$$

satisfying the inequality (17.7) is singular. Assume the contrary. Then there exists a solution u of the equation (17.8) such that

$$u(t) > 0, \qquad u'(t) < 0 \qquad \text{for } t \ge 0.$$

By (17.6), $u(t) \to 0$ and $tu'(t) \to 0$ as $t \to +\infty$. So (17.1) yields

$$[u'(t)]^2 = -2 \int_t^{+\infty} p(\tau)[u(\tau)]^\lambda u'(\tau)\, d\tau. \qquad (17.9)$$

Let $s > 0$ be arbitrary. Then

$$[u'(t)]^2 \geq \frac{2}{\lambda+1} p_0(s)([u(t)]^{\lambda+1} - [u(s)]^{\lambda+1})$$

$$\geq \frac{2}{\lambda+1} p_0(s)(u(t) - u(s))^{\lambda+1} \qquad \text{for } 0 \leq t \leq s.$$

Consequently,

$$-u'(t)(u(t) - u(s))^{(\lambda+1)/2} \geq \frac{2}{\lambda+1} [p_0(s)]^{1/2} \qquad \text{for } 0 \leq t \leq s.$$

Integration of the last inequality over $[s/2, s]$ gives

$$\frac{2}{1-\lambda}\left(u\left(\frac{s}{2}\right) - u(s)\right)^{(1-\lambda)/2} \geq \frac{s}{\lambda+1}[p_0(s)]^{1/2} \qquad \text{for } s > 0.$$

Hence

$$\lim_{t \to +\infty} t^2 p_0(t) = 0,$$

and, applying (17.9), we obtain

$$t^2[u'(t)]^2 \leq [u(t)]^{\lambda+1} \qquad \text{for } t \geq t_0, \qquad (17.10)$$

with t_0 sufficiently large. Furthermore, since

$$u(t) - tu'(t) = \int_t^{+\infty} \tau p_0(\tau)[u(\tau)]^\lambda\, d\tau \qquad \text{for } t \geq 0,$$

(17.10) implies

$$x(t) \geq c \int_t^{+\infty} \tau p_0(\tau)[x(\tau)]^{2\lambda/(\lambda+1)}\, d\tau \qquad \text{for } t \geq t_0,$$

where $c > 0$ is sufficiently small and $x(t) = [u(t)]^{(\lambda+1)/2}$. But this is impossible, by Lemma 10.4. Therefore we have proved that every solution of the equation (17.8) satisfying (17.7) is singular. Using the inequality (17.5), we can easily show that the same is true for the equation (17.1). $\square$

COROLLARY 17.3. If $0 < \lambda < 1$ and

$$\lim_{t \to +\infty} \inf t^2 p(t) > 0,$$

then every nonzero solution of the equation (17.1) satisfying the inequality (17.7) is nonoscillatory first kind singular.

THEOREM 17.5. Let $\lambda > 1$ and let (17.5) hold, where $p_0 \in L_{\text{loc}}(\mathbf{R}_+)$ is nonincreasing and

$$\lim_{t \to +\infty} \sup t^{\lambda+1} p_0(t) > 0. \qquad (17.11)$$

Then any maximally continued solution u of the equation (17.1) satisfying the condition

$$u(t_0)u'(t_0) > 0, \tag{17.12}$$

for a certain $t_0 \in \mathbf{R}_+$, is nonoscillatory second kind singular.

PROOF. Assume, on the contrary, that there exists a proper solution of the equation (17.1) satisfying (17.12). For the sake of being specific suppose that $u(t_0) > 0$ and $u'(t_0) > 0$. Clearly, $u(t) > 0, u'(t) > 0$ for $t \geq t_0$ and $u(t), u'(t) \to +\infty$ as $t \to +\infty$.

Let $s > t_1 \geq t_0$ be arbitrary numbers. Then

$$[u'(t)]^2 \geq [u'(t_1)]^2 + \frac{2}{\lambda+1}p_0(s)([u(t)]^{\lambda+1} - [u(t_1)]^{\lambda+1})$$

$$\geq 2^{-\lambda-1}\left([u'(t_1)]^{2/(\lambda+1)} + \left(\frac{2}{\lambda+1}p_0(s)\right)^{1/(\lambda+1)}(u(t) - u(t_1))\right)^{\lambda+1}$$

$$\text{for } t_1 \leq t \leq s,$$

i.e.

$$-u'(t)\left[[u'(t_1)]^{2/(\lambda+1)} + \left(\frac{2}{\lambda+1}p_0(s)\right)^{1/(\lambda+1)}(u(t) - u(t_1))\right]^{-(\lambda+1)/2} \geq 2^{-(\lambda+1)/2}$$

$$\text{for } t_1 \leq t \leq s.$$

Integrating this inequality, we obtain

$$[u'(t_1)]^{(1-\lambda)/(1+\lambda)} \geq \frac{\lambda-1}{2}\left(\frac{2}{\lambda+1}p_0(s)\right)^{1/(\lambda+1)}2^{-(\lambda+1)/2}(s - t_1)$$

$$\text{for } t_0 \leq t_1 < s.$$

Since t_1 is arbitrary, the last inequality yields

$$\lim_{s\to+\infty} s^{\lambda+1}p_0(s) = 0,$$

which contradicts (17.11). $\square$

COROLLARY 17.4. If $\lambda > 1$ and

$$\lim_{t\to+\infty} \inf t^{\lambda+1}p(t) > 0,$$

then every maximally continued solution of the equation (17.1) satisfying the condition (17.12) for a certain $t_0 \in \mathbf{R}_+$ is nonoscillatory second kind singular.

For $n = 2$ Theorems 16.4 and 16.15 imply

THEOREM 17.6. Let $0 < \lambda < 1$ ($\lambda > 1$), and let p be nonnegative and satisfy the condition

$$\int_1^{+\infty} t^{(\lambda+3)\alpha-2}[p(t)]^\alpha \, dt = +\infty$$

for a certain $\alpha \in \,]0, 1/(1+\lambda)[$ ($\alpha \in \,]0, 1/2[$). Then every nonzero solution of the equation (17.1) satisfying the condition (17.7) (the condition (17.12)) is nonoscillatory first (second) kind singular.

COROLLARY 17.5. If $0 < \lambda < 1$, p is nonnegative, and for a certain $\alpha \in]1/2,, 2/(\lambda + 3)]$ $(\alpha \in]1/(\lambda + 1), 2/(\lambda + 3)])$,

$$\int_0^{+\infty} [p(t)]^\alpha \, dt = +\infty,$$

then every nonzero solution of the equation (17.1) satisfying the condition (17.7) (the condition (17.12)) is nonoscillatory first (second) kind singular.

17.3. Existence of oscillatory singular solutions. It follows from Theorem 17.1 that if $p \in C(\mathbf{R}_+) \cap V_{\mathrm{loc}}(\mathbf{R}_+)$ is negative, then the equation (17.1) does not have an oscillatory singular solution. In this subsection we will show that the restriction $p \in V_{\mathrm{loc}}(\mathbf{R}_+)$ is essential.

THEOREM 17.7. Let $0 < \lambda < 1$ $(\lambda > 1)$. There exists a negative function $p \in C(\mathbf{R}_+)$ such that the equation (17.1) has an oscillatory first (second) kind singular solution.

To prove this theorem we need

LEMMA 17.1. Let

$$c = 8(\lambda + 1) \left(\int_0^1 \frac{dx}{\sqrt{1 - x^{\lambda+1}}} \right)^2.$$

Then for any positive integer k there exists $q_k \in C([0, 1])$ such that

$$q_k(0) = q_k(1) = 0, \tag{17.13}$$

$$\lim_{k \to +\infty} q_k(t) = 0 \qquad \text{uniformly on } [0, 1], \tag{17.14}$$

and the equation

$$u'' = -(c + q_k(t))|u|^\lambda \operatorname{sign} u \tag{17.15}$$

under the conditions

$$u(0) = 1, \qquad u(1) = \left(\frac{k+1}{k} \right)^{4/(\lambda-1)}, \qquad u'(0) = u'(1) = 0 \tag{17.16}$$

has a solution with exactly two zeros in the interval $]0, 1[$.

PROOF. Let w be the solution of the problem

$$w'' = -c|w|^\lambda \operatorname{sign} w, \qquad w(0) = 1, \qquad w'(0) = 0.$$

It is easily verified that w is a periodic function of period 1 and has exactly two zeros in $]0, 1[$.

Choose $t_0 \in]0, 1[$ so that $w(t) > 0$ for $t_0 \leq t \leq 1$, and set

$$u_k(t) = \begin{cases} w(t) & \text{for } t \in [0, t_0] \\ \left(\frac{k+1}{k} \right)^{4/(\lambda-1)} - 1 + w(t) - \int_t^1 (t - s) f_k(s) \, ds & \text{for } t \in]t_0, 1], \end{cases} \tag{17.17}$$

where

$$f_k(t) = \alpha_k(t - t_0)^3 + \beta_k(t - t_0)^2 + \gamma_k(t - t_0) \tag{17.18}$$

and the constants α_k, β_k and γ_k are defined by the system

$$\frac{\alpha_k}{4}(1 - t_0)^4 + \frac{\beta_k}{3}(1 - t_0)^3 + \frac{\gamma_k}{2}(1 - t_0)^2 = 0, \tag{17.19}$$

$$\frac{\alpha_k}{5}(1-t_0)^5 + \frac{\beta_k}{4}(1-t_0)^4 + \frac{\gamma_k}{3}(1-t_0)^3 = 1 - \left(\frac{k+1}{k}\right)^{4/(\lambda-1)}, \qquad (17.20)$$

$$\alpha_k(1-t_0)^3 + \beta_k(1-t_0)^2 + \gamma_k(1-t_0) = c\left[1 - \left(\frac{k+1}{k}\right)^{4\lambda/(\lambda-1)}\right]. \qquad (17.21)$$

Clearly,

$$|\alpha_k| + |\beta_k| + |\gamma_k| = O\left(\frac{1}{k}\right) \qquad \text{as } k \to +\infty. \qquad (17.22)$$

According to (17.17)–(17.20), $u_k \in C^2([0,1])$. Furthermore, in view of (17.22) we may assume without loss of generality that

$$u_k(t) > 0, \qquad u_k'(t) < 0, \qquad u_k''(t) < 0 \qquad \text{for } t \in [t_0,1] \qquad (k=1,2,\dots).$$

We now define the function q_k as

$$q_k(t) = \begin{cases} 0 & \text{for } t \in [0,t_0] \\ -c - [u_k(t)]^{-\lambda} - u_k''(t) & \text{for } t \in \,]t_0,1]. \end{cases} \qquad (17.23)$$

It is obvious that $q_k \in C([0,1])$ and u_k is a solution of the equation (17.15). Moreover, from (17.17), (17.18) and (17.21) we obtain (17.16). Finally, since

$$\lim_{k \to +\infty} \frac{u_k(t)}{w(t)} = 1$$

uniformly on $[t_0,1]$ and $u_k''(t) = w''(t) + f_k(t)$, (17.22) and (17.23) imply (17.14). $\square$

PROOF OF THEOREM 17.7. Consider the sequence $(t_k)_{k=1}^{+\infty}$, where

$$t_1 = 0, \qquad t_k = \sum_{i=1}^{k-1} \frac{1}{i^2} \qquad (k=2,3,\dots).$$

We have $\lim_{k \to +\infty} t_k = \pi/6$.

Let q_k be the function constructed in Lemma 17.1 and let u_k be a solution of the problem (17.15), (17.16) having exactly two zeros in $]0,1[$. Define in the interval $[0,\pi/6[$ the functions p and u by the equalities

$$p(t) = -(c + q_k(k^2(t-t_k))), \qquad u(t) = k^{4/(\lambda-1)} u_k(k^2(t-t_k))$$

$$\text{for } t_k \leq t < t_{k+1} \qquad (k=1,2,\dots).$$

According to (17.13) and (17.14), $p \in C([0,\pi/6[)$ and $\lim_{t \to \pi/6} p(t) = -c$, where c is the constant appearing in Lemma 17.1. Furthermore, u is a solution of the equation (17.1) on each interval $[t_k,t_{k+1}]$, and $u \in C^2([0,\pi/6[)$. So u is a solution of (17.1) on $[0,\pi/6[$.

It follows from (17.20) that

$$\lim_{t \to \pi/6} u(t) = \lim_{t \to \pi/6} u'(t) = 0 \qquad \text{for } 0 < \lambda < 1$$

and

$$\lim_{t \to \pi/6} \sup |u(t)| = +\infty \qquad \text{for } \lambda > 1.$$

Hence, if p is extended to the whole of $\mathbf{R}_+$ so that $p(t) = -c$ for $t \geq \pi/6$, then in case $0 < \lambda < 1$

$$u(t) = \begin{cases} k^{4/(\lambda-1)}u_k(k^2(t - t_k)) & \text{for } t_k \leq t < t_{k+1} \\ 0 & \text{for } t_k \geq \pi/6 \end{cases} \qquad (k = 1, 2, \dots)$$

is an oscillatory first kind singular solution of the equation (17.1), and in case $\lambda > 1$

$$u(t) = k^{4/(\lambda-1)}u_k(k^2(t - t_k)) \qquad \text{for } t_k \leq t < t_{k+1} \qquad (k = 1, 2, \dots)$$

is an oscillatory second kind singular solution of this equation defined on $[0, \pi/6[$. $\square$

Notes. For Theorems 17.1 and 17.2 see [154, 181]. Theorems 17.4 and 17.5 are due to T.A. Chanturia and Theorem 17.6 is due to N.A. Izobov [140, 143]. Theorem 17.7 was proved by C.V. Coffman and D.F. Ullrich [91] for $\lambda > 1$, and by J.W. Heidel [136] for $0 < \lambda < 1$. S.P. Hastings [133] and A.D. Myshkis [284] have also constructed examples of this type.

§18. Oscillation and Nonoscillation Criteria for Proper Solutions

In this section we will establish for the differential equation

$$u'' = p(t)|u|^\lambda \operatorname{sign} u, \tag{18.1}$$

with $\lambda > 0$, $\lambda \neq 1$, and $p \in L_{\text{loc}}(\mathbf{R}_+)$, sufficient conditions under which all proper solutions are oscillatory, there exists at least one oscillatory proper solution, all maximally continued or all proper solutions are nonoscillatory. The conditions for oscillation of all proper solutions given here differ from those in §15, and cover the case when the function p changes sign.

18.1. Oscillation of all proper solutions.

THEOREM 18.1. Let $0 < \lambda < 1$ and

$$\int_0^{+\infty} t^\mu p(t)\, dt = -\infty \tag{18.2}$$

for a certain $\mu \in [0, \lambda]$. Then all proper solutions of the equation (18.1) are oscillatory. PROOF. Assume, on the contrary, that the equation (18.1) has a nonoscillatory proper solution $u: [t_0, +\infty[\to \mathbf{R}$. For the sake of being specific suppose

$$u(t) > 0 \qquad \text{for } t \geq t_0. \tag{18.3}$$

If

$$w(t) = (t^{-1}u(t))^{1-\lambda},$$

then, according to (18.1),

$$t^{\mu-1-\lambda}(t^2 w'(t))' = (1 - \lambda)t^\mu u''(t)u^{-\lambda}(t)$$

$$-\lambda(1 - \lambda)t^{\mu-2}u^{-\lambda-1}(t)(tu'(t) - u(t))^2 \leq (1 - \lambda)t^\mu p(t) \qquad \text{for } t \geq t_0.$$

Hence

$$t^{\mu-\lambda+1}w'(t) + (\lambda - \mu + 1)t^{\mu-\lambda}w(t) + (1 + \lambda - \mu)(\lambda - \mu)\int_{t_0}^t s^{\mu-\lambda-1}w(s)\, ds$$

$$\leq c_0 + (1-\lambda)\int_{t_0}^t s^\alpha p(s)\,ds \qquad \text{for } t \geq t_0,$$

where

$$c_0 = t_0^{\mu-\lambda+1}w'(t_0) + (\lambda-\mu+1)t_0^{\mu-\lambda}w(t_0).$$

By (18.2) and (18.3), the last inequality implies the existence of $t_1 > t_0$ such that $w'(t) \leq -t^{-1}$ for $t \geq t_1$. Therefore, $w(t) \to -\infty$ as $t \to +\infty$. This contradiction shows that the equation (18.1) does not have a nonoscillatory proper solution. $\square$

THEOREM 18.2. Let $\lambda > 1$ and let for a certain $\mu \in [0,1]$ the condition (18.2) hold. Then all proper solutions of the equation (18.1) are oscillatory.

PROOF. Assume that the equation (18.1) has a solution satisfying (18.3). Multiplying both sides of (18.1) by $t^\mu u^{-\lambda}(t)$ and integrating over $[t_0,t]$, we get

$$t^\mu u^{-\lambda}(t)u'(t) - \mu\int_{t_0}^t s^{\mu-1}u^{-\lambda}(s)u'(s)\,ds + \lambda\int_{t_0}^t s^\mu u^{-\lambda-1}(s)u'^2(s)\,ds$$

$$\leq t_0^\mu u^{-\lambda}(t_0)u'(t_0) + \int_{t_0}^t s^\mu p(s)\,ds \qquad \text{for } t \geq t_0.$$

So, the second mean value theorem implies

$$t^\mu u^{-\lambda}(t)u'(t) + \frac{\mu}{\lambda-1}t_0^{\mu-1}u^{1-\lambda}(\tau) + \lambda\int_{t_0}^t s^\mu u^{-\lambda-1}(s)u'^2(s)\,ds$$

$$\leq c_0 + \int_{t_0}^t s^\mu p(s)\,ds,$$

where $\tau \in [t_0,t]$ and

$$c_0 = t_0^\mu u^{-\lambda}(t_0)u'(t_0) + \frac{\mu}{\lambda-1}t_0^{\mu-1}u^{1-\lambda}(t_0).$$

In view of (18.2), there exists $t_1 > t_0$ such that

$$c_0 + \int_{t_0}^t s^\mu p(s)\,ds \leq -1 \qquad \text{for } t \geq t_1.$$

Hence

$$-t^\mu u^{-\lambda}(t)u'(t) \geq 1 + \lambda\int_{t_1}^t s^\mu u^{-\lambda-1}(s)u'^2(s)\,ds \qquad \text{for } t \geq t_1 \tag{18.4}$$

and

$$\lambda t^\mu \frac{u'^2(t)}{u^{\lambda+1}(t)}\left[1 + \lambda\int_{t_1}^t s^\mu \frac{u'^2(s)}{u^{\lambda+1}(s)}\,ds\right]^{-1} \geq -\lambda\frac{u'(t)}{u(t)} \qquad \text{for } t \geq t_1.$$

Integration of the last inequality gives

$$\ln\left(1 + \lambda\int_{t_1}^t s^\mu \frac{u'^2(s)}{u^{\lambda+1}(s)}\,ds\right) \geq \ln\frac{u^\lambda(t_1)}{u^\lambda(t)} \qquad \text{for } t \geq t_1.$$

By (18.4) this yields

$$u'(t) \leq -t^{-\mu}u^\lambda(t_1) \leq -t^{-1}u^\lambda(t_1) \qquad \text{for } t \geq 1 + t_1.$$

Consequently,

$$u(t) \le u(1 + t_1) - u^\lambda(t_1) \ln \frac{t}{1 + t_1} \to -\infty \qquad \text{as } t \to +\infty,$$

which is a contradiction. $\square$

PROBLEM 18.1*. Does the equation (18.1) have at least one oscillatory proper solution under the hypotheses of Theorem 18.1 or 18.2?

PROBLEM 18.2. Let $0 < \lambda < 1$, $p(t) = p_0(t) + p_1(t)$, $p_i \in L_{\text{loc}}(\mathbf{R}_+)$ $(i = 0, 1)$, $p_0(t) \le 0$ for $t \ge 0$, and

$$\int_0^{+\infty} t^\lambda |p_1(t)| \, dt < +\infty.$$

Then all proper solutions of the equation (18.1) are oscillatory if and only if

$$\int_0^{+\infty} t^\lambda p_0(t) \, dt = -\infty.$$

PROBLEM 18.3. Let $\lambda > 1$, $p(t) = p_0(t) + p_1(t)$, $p_i \in L_{\text{loc}}(\mathbf{R}_+)$ $(i = 0, 1)$, $p_0(t) \le 0$ for $t \ge 0$, and

$$\int_0^{+\infty} t |p_1(t)| \, dt < +\infty.$$

Then all proper solutions of the equation (18.1) are oscillatory if and only if

$$\int_0^{+\infty} t p_0(t) \, dt = -\infty.$$

PROBLEM 18.4. Let $0 < \lambda < 1$ and

$$\lim_{t \to +\infty} \inf \frac{1}{t} \int_0^t \int_0^\tau s^\mu p(s) \, ds \, d\tau = -\infty$$

for a certain $\mu \in [0, \lambda]$. Then all proper solutions of the equation (18.1) are oscillatory.

PROBLEM 18.5. Let $\lambda > 1$ and

$$\lim_{t \to +\infty} \sup \int_0^t s^\mu p(s) \, ds < +\infty,$$

$$\lim_{t \to +\infty} \inf \frac{1}{t} \int_0^t \int_0^\tau s^\mu p(s) \, ds \, d\tau = -\infty$$

for a certain $\mu \in [0, 1]$. Then all proper solutions of the equation (18.1) are oscillatory.

THEOREM 18.3. Let there exist $c \in \mathbf{R}$ and $q \in L_{\text{loc}}(\mathbf{R}_+)$ such that

$$-p(t) \ge q(t) \qquad \text{for } t \ge 0, \tag{18.5}$$

$$\lim_{t \to +\infty} \inf \int_0^t q(s) \, ds > -\infty, \tag{18.6}$$

$$\int_1^{+\infty} \frac{1}{t} \left[\int_0^t q(s)\, ds - c \right]_+^2 dt = +\infty, \tag{18.7}$$

and

$$\int_1^{+\infty} \frac{1}{t} \left[\int_0^t q(s)\, ds - c \right]_-^2 dt = +\infty. \tag{18.8}$$

Then every proper solution of the equation (18.1) is oscillatory.

PROOF. Assume, on the contrary, that the equation (18.1) has a proper solution u which is positive on some interval $[t_0, +\infty[$. Then

$$(-u'(t)[u(t)]^{-\lambda})' = -p(t) + \lambda [u'(t)]^2 [u(t)]^{-\lambda-1} \qquad \text{for } t \geq t_0.$$

If

$$\int_{t_0}^{+\infty} [u'(t)]^2 [u(t)]^{-\lambda-1}\, dt = +\infty,$$

then by (18.5) and (18.6) we obtain

$$-u'(t)[u(t)]^{-\lambda} \geq 1 + \lambda \int_{t_1}^t [u'(s)]^2 [u(s)]^{-\lambda-1}\, ds \qquad \text{for } t \geq t_1,$$

with a certain $t_1 > t_0$. By the Gronwall–Bellman lemma this yields

$$-u'(t)[u(t)]^{-\lambda} \geq \exp\left(-\lambda \int_{t_1}^t \frac{u'(s)}{u(s)}\, ds \right) = \left(\frac{u(t_1)}{u(t)} \right)^\lambda \qquad \text{for } t \geq t_1.$$

Hence

$$u'(t) \leq -[u(t_1)]^\lambda \qquad \text{for } t \geq t_1$$

and

$$u(t) \leq u(t_1) - [u(t_1)]^\lambda (t - t_1) \to -\infty \qquad \text{as } t \to +\infty.$$

This contradiction shows that

$$\int_{t_0}^{+\infty} [u'(t)]^2 [u(t)]^{-\lambda-1}\, dt < +\infty.$$

Thus, the Cauchy inequality implies

$$[u(t)]^{(1-\lambda)/2} - [u(\tau)]^{(1-\lambda)/2} = \frac{1-\lambda}{2} \int_\tau^t u'(s)[u(s)]^{-(\lambda+1)/2}\, ds$$

$$\leq \frac{|\lambda - 1|}{2} \left(\int_\tau^{+\infty} [u'(s)]^2 [u(s)]^{-\lambda-1}\, ds \right)^{1/2} (t - \tau)^{1/2} \qquad \text{for } t \geq \tau \geq t_0.$$

Consequently,

$$\lim_{t \to +\infty} t^{-1} [u(t)]^{1-\lambda} = 0,$$

i.e. if $t_1 \geq t_0$ is sufficiently large,

$$t^{-1} [u(t)]^{1-\lambda} \leq \lambda \qquad \text{for } t \geq t_1. \tag{18.9}$$

Since $v = u^\lambda$ is a positive solution of the linear equation

$$([u(t)]^{1-\lambda} v')' = \lambda p(t) v$$

on $[t_0, +\infty[$, using (18.5) and (18.9) we can easily show that the equation

$$(tv')' + q(t)v = 0$$

also has a solution v which is positive on $[t_1, +\infty[$.

Let $r(t) = -tv'(t)[v(t)]^{-1}$. Then

$$r'(t) = q(t) + t^{-1}[r(t)]^2. \tag{18.10}$$

As above we conclude that

$$\int_{t_0}^{+\infty} \frac{1}{t}[r(t)]^2\, dt < +\infty. \tag{18.11}$$

Therefore, by (18.10),

$$\int_0^t q(s)\, ds - r(t) - c_0 = \int_t^{+\infty} \frac{1}{s}[r(s)]^2\, ds \qquad \text{for } t \geq t_0,$$

where

$$c_0 = \lim_{t \to +\infty} \left(\int_0^t q(s)\, ds - r(t) \right).$$

Consequently,

$$\int_0^t q(s)\, ds - c_0 \geq r(t) \qquad \text{for } t \geq t_0$$

and, moreover, for any $\epsilon > 0$ there exists $t_1 \geq t_0$ such that

$$\int_0^t q(s)\, ds - c_0 - \epsilon \leq r(t) \qquad \text{for } t \geq t_1.$$

The last two inequalities imply

$$\left[\int_0^t q(s)\, ds - c_0 \right]_- \leq |r(t)| \qquad \text{for } t \geq t_0$$

and

$$\left[\int_0^t q(s)\, ds - c_0 - \epsilon \right]_+ \leq |r(t)| \qquad \text{for } t \geq t_1.$$

Hence, in view of (18.11),

$$\int_{t_0}^{+\infty} \frac{1}{t} \left[\int_0^t q(s)\, ds - c_0 \right]_-^2 dt < +\infty$$

and

$$\int_{t_0}^{+\infty} \frac{1}{t} \left[\int_0^t q(s)\, ds - c_0 - \epsilon \right]_+^2 dt < +\infty.$$

So (18.7) and (18.8) yield $c_0 < c$ and $c_0 + \epsilon > c$. But since ϵ is arbitrary, these inequalities contradict each other. $\square$

COROLLARY 18.1. Let the inequality (18.5) hold, where q is a periodic function of a period $\omega > 0$. Suppose, in addition, that q differs from zero on a set of positive measure and satisfies the condition

$$\int_0^\omega q(t)\, dt \geq 0.$$

Then every proper solution of the equation (18.1) is oscillatory.

PROOF. Without loss of generality we may assume

$$\int_0^\omega q(t)\,dt = 0.$$

Let c be such that

$$\min\left\{\int_0^t q(s)\,ds : t \in [0, +\infty[\right\} < c < \max\left\{\int_0^t q(s)\,ds : t \in [0, +\infty[\right\}.$$

Then

$$\int_0^\omega \left[\int_0^t q(s)\,ds - c\right]_\pm^2 dt = \alpha_\pm > 0.$$

So

$$\int_1^{+\infty} \frac{1}{t}\left[\int_0^t q(s)\,ds - c\right]_\pm^2 dt = \sum_{k=1}^{+\infty} \int_{1+(k-1)\omega}^{1+k\omega} \frac{1}{t}\left[\int_0^t q(s)\,ds - c\right]_\pm^2 dt$$

$$\geq \sum_{k=1}^{+\infty} \frac{\alpha_\pm}{1+k\omega} = +\infty,$$

and all the hypotheses of Theorem 18.3 are satisfied. $\square$

18.2. Existence of at least one oscillatory proper solution.

THEOREM 18.4. Let $p \in C(\mathbf{R}_+)$ be negative and let the function $t^{(\lambda+3)/2}|p(t)|$ be nondecreasing. Then the equation (18.1) has at least one oscillatory proper solution.
PROOF. By Theorem 17.1 every maximally continued solution of the equation (18.1) is defined on $\mathbf{R}_+$ and proper. Let $\epsilon > 0$, $t_0 > 0$ and $c_0 > 0$ be chosen so that

$$4c_0^{\lambda-1}|p(t)|t^{(\lambda+3)/2} \geq 1 + \epsilon \qquad \text{for } t \geq t_0. \tag{18.12}$$

First suppose $0 < \lambda < 1$ and consider a solution u of the equation (18.1) satisfying the conditions

$$u(t_0) = 0, \qquad 0 < 4t_0 u'^2(t_0) < c_0^2 \epsilon.$$

We claim

$$|u(t)| < c_0 t^{1/2} \qquad \text{for } t \geq t_0. \tag{18.13}$$

Indeed, otherwise there would exist $t_1 > t_0$ such that

$$|u(t_1)| = c_0 t_1^{1/2}, \qquad |u(t)| < c_0 t^{1/2} \qquad \text{for } t \in [t_0, t_1[.$$

Then setting $t = t_1$ in the equality

$$\int_{t_0}^t d\left(\tau u'^2(\tau) - u(\tau)u'(\tau) - \frac{2}{\lambda+1}\tau p(\tau)|u(\tau)|^{\lambda+1}\right)$$

$$= \frac{2}{\lambda+1}\int_{t_0}^t \tau^{-(\lambda+1)/2}|u(\tau)|^{\lambda+1}\,d(\tau^{(\lambda+3)/2}|p(\tau)|) \qquad \text{for } t \geq t_0, \tag{18.14}$$

we would have

$$\left(t_1^{1/2}u'(t_1) - \frac{1}{2}t_1^{-1/2}u(t_1)\right)^2 - \frac{1}{4}t_1^{-1}u^2(t_1) - \frac{2}{\lambda+1}t_1 p(t_1)|u(t_1)|^{\lambda+1} - t_0 u'^2(t_0)$$

$$\leq \frac{2}{\lambda+1}c_0^{\lambda+1}t_1^{(\lambda+3)/2}|p(t_1)| - \frac{2}{\lambda+1}c_0^{\lambda+1}t_0^{(\lambda+3)/2}|p(t_0)|,$$

i.e.

$$\frac{2}{\lambda+1}c_0^{\lambda+1}t_0^{(\lambda+3)/2}|p(t_0)| < \frac{c_0^2\epsilon}{4} + \frac{c_0^2}{4},$$

which contradicts (18.12). So (18.13) is true.

If $u(t) \neq 0$ for $t \geq t_2 \geq t_0$, then u is a solution of the linear equation

$$v'' = p^*(t)v \tag{18.15}$$

on $[t_2, +\infty[$, where $p^*(t) = p(t)|u(t)|^{\lambda-1}$. Since

$$-t^2 p^*(t) \geq -t^2 p(t)c_0^{\lambda-1}t^{(\lambda-1)/2} \geq \frac{1+\epsilon}{4} \qquad \text{for } t \geq t_2,$$

by the well-known Kneser theorem every solution of the equation (18.15) is oscillatory. This contradiction completes the proof in case $0 < \lambda < 1$.

Now let $\lambda > 1$. We will show that every solution of the equation (18.1) satisfying the condition

$$t_0 u'^2(t_0) - u(t_0)u'(t_0) - \frac{2}{\lambda+1}t_0 p(t_0)|u(t_0)|^{\lambda+1} \geq c_0^2 \tag{18.16}$$

is oscillatory.

Assume the contrary. Let u be a nonoscillatory solution of the equation (18.1) under the condition (18.16). Suppose

$$u(t) > 0, \qquad u'(t) > 0, \qquad u''(t) < 0 \qquad \text{for } t \geq t_1,$$

where $t_1 \geq t_0$ is sufficiently large. Then

$$u'(t) \geq -\int_t^{+\infty} p(s)[u(s)]^\lambda \, ds \geq -[u(t)]^\lambda t^{(\lambda+3)/2}p(t)\int_t^{+\infty} s^{-(\lambda+3)/2} \, ds$$

$$= -\frac{2}{\lambda+1}tp(t)[u(t)]^\lambda \qquad \text{for } t \geq t_1.$$

On the other hand, from (18.14) we can derive

$$tu'^2(t) - u(t)u'(t) - \frac{2}{\lambda+1}tp(t)|u(t)|^{\lambda+1} \geq c_0^2 \qquad \text{for } t \geq t_0,$$

and so

$$tu'^2(t) \geq c_0^2 \qquad \text{for } t \geq t_1.$$

Hence

$$u(t) \geq c_0 t^{1/2} \qquad \text{for } t \geq t_2,$$

where t_2 is sufficiently large. Arguing now as in the previous case, we obtain a contradiction. $\square$

REMARK 18.1. Let $p(t) = -t^{-\alpha}$ for sufficiently large t, where $\alpha \in]\lambda+1, (\lambda+3)/2]$ if $0 < \lambda < 1$ and $\alpha \in]2, (\lambda+3)/2]$ if $\lambda > 1$. Then Corollary 8.2 and Theorem 18.4 imply that the equation (18.1) has both nonoscillatory solutions and oscillatory solutions.

18.3. Nonoscillation of all proper solutions.

THEOREM 18.5. Let $\lambda > 1$. Suppose $p \in C(\mathbf{R}_+)$ is negative and for some $\epsilon > 0$ the function $t^{(\lambda+3)/2+\epsilon}|p(t)|$ is nonincreasing on a certain interval $[t_0, +\infty[$. Then the equation (18.1) does not have an oscillatory proper solution.

PROOF. Assume the contrary. Let $u: [t_0, +\infty[\to \mathbf{R}$ be an oscillatory proper solution of the equation (18.1) and let $(t_k)_{k=0}^{+\infty}$ be the unboundedly increasing sequence of its zeros. Then

$$\int_{t_k}^{t_{k+1}} |p(t)|\, |u(t)|^{\lambda+1}\, dt = \int_{t_k}^{t_{k+1}} |u'(t)|^2\, dt \qquad (k = 0, 1, \dots). \tag{18.17}$$

Taking into account the inequality

$$|u(t)| = \left| \int_{t_k}^{t} u'(s)\, ds \right| \le t^{1/2} \left(\int_{t_k}^{t} |u'(s)|^2\, ds \right)^{1/2} \qquad \text{for } t \ge t_k,$$

from (18.17) we get

$$\int_{t_k}^{t_{k+1}} |u'(t)|^2\, dt \le \int_{t_k}^{t_{k+1}} t^{(\lambda+1)/2}|p(t)|\, dt \left(\int_{t_k}^{t_{k+1}} |u'(t)|^2\, dt \right)^{(\lambda+1)/2},$$

i.e.

$$1 \le \int_{t_0}^{+\infty} t^{(\lambda+1)/2}|p(t)|\, dt \left(\int_{t_k}^{t_{k+1}} |u'(t)|^2\, dt \right)^{(\lambda-1)/2} \qquad (k = 0, 1, \dots).$$

This implies

$$\int_{t_0}^{+\infty} |u'(t)|^2\, dt = +\infty,$$

and so, by (18.17),

$$\int_{t_0}^{+\infty} |p(t)|\, |u(t)|^{\lambda+1}\, dt = +\infty.$$

On the other hand, the equality (18.14) with $t = t_k$ yields

$$t_k|u'(t_k)|^2 - t_0|u'(t_0)|^2$$

$$= \frac{2}{\lambda+1} \int_{t_0}^{t_k} \tau^{-\epsilon-(\lambda+1)/2}|u(\tau)|^{\lambda+1}\, d\left(\tau^{(\lambda+3)/2+\epsilon}|p(\tau)|\right)$$

$$- \frac{2\epsilon}{\lambda+1} \int_{t_0}^{t_k} |p(\tau)|\, |u(\tau)|^{\lambda+1}\, d\tau \qquad (k = 1, 2, \dots).$$

Therefore,

$$\frac{2\epsilon}{\lambda+1} \int_{t_0}^{+\infty} |p(\tau)|\, |u(\tau)|^{\lambda+1}\, d\tau \le t_0|u'(t_0)|^2,$$

and we arrive at a contradiction. $\square$

PROBLEM 18.6*. Does Theorem 18.5 remain true if $0 < \lambda < 1$?

THEOREM 18.6. Let $0 < \lambda < 1$ and $p(t) = -p_0(t) + p_1(t)$, where the functions $p_i \in C(\mathbf{R}_+)$ $(i = 0, 1)$ are nondecreasing. Suppose, in addition, that p is negative,

$$\lim_{t \to +\infty} t^2 p(t) = 0, \tag{18.18}$$

and

$$\int_0^{+\infty} \frac{dp_0(t)}{|p(t)|} < +\infty. \tag{18.19}$$

Then every nonzero maximally continued solution of the equation (18.1) is nonoscillatory.

PROOF. By Theorem 17.1, every maximally continued solution of the equation (18.1) is proper. Assume that $u: \mathbf{R}_+ \to \mathbf{R}$ is an oscillatory solution of the equation (18.1) and that the sequences $(t_k)_{k=1}^{+\infty}$ and $(s_k)_{k=1}^{+\infty}$ satisfy the conditions

$$0 < t_k < s_k < t_{k+1}, \qquad \lim_{k\to+\infty} t_k = +\infty,$$

$$u(t) > 0, \qquad u'(t) > 0 \qquad \text{for } t \in [t_k, s_k],$$

$$u(t_k) = 0, \qquad u'(s_k) = 0 \qquad (k = 1, 2, \dots).$$

Then

$$u(t) \le u'(t_k)t \qquad \text{for } t \in [t_k, s_k]$$

and

$$u'(t_k) = -\int_{t_k}^{s_k} p(t)[u(t)]^\lambda \, dt \le [u'(t_k)]^\lambda \int_{t_k}^{+\infty} t^\lambda |p(t)| \, dt. \tag{18.20}$$

Moreover, according to (17.3) and (18.19),

$$\lim_{k\to+\infty} \inf |p(t_k)|^{-1}[u'(t_k)]^2 > 0. \tag{18.21}$$

It follows from (18.20) and (18.21) that

$$\lim_{k\to+\infty} \inf |p(t_k)|^{(\lambda-1)/2} \int_{t_k}^{+\infty} t^\lambda |p(t)| \, dt > 0. \tag{18.22}$$

On the other hand, by (18.18), for any $\epsilon > 0$ there exists $t_0 \in \mathbf{R}_+$ such that

$$\tau(t) = \epsilon^{1/2} |p(t)|^{-1/2} > t \qquad \text{for } t \ge t_0.$$

Hence, taking into account the equality

$$|p(t)| = |p(0)| \exp\left(\int_0^t \frac{1}{p(s)} \, dp_1(s)\right) \exp\left(-\int_0^t \frac{1}{p(s)} \, dp_0(s)\right),$$

we obtain

$$\int_t^{\tau(t)} s^\lambda |p(s)| \, ds \le \exp\left(\int_0^{+\infty} \frac{1}{p(s)} \, dp_0(s)\right) \frac{\epsilon^{(\lambda+1)/2}}{\lambda+1} |p(t)|^{(1-\lambda)/2} \qquad \text{for } t \ge t_0.$$

Furthermore,

$$\int_{\tau(t)}^{+\infty} s^\lambda |p(s)| \, ds \le \epsilon \int_{\tau(t)}^{+\infty} s^{\lambda-2} \, ds = \frac{\epsilon^{(\lambda+1)/2}}{\lambda-1} |p(t)|^{(1-\lambda)/2} \qquad \text{for } t \ge t_0.$$

The last two inequalities yield

$$\lim_{t\to+\infty} |p(t)|^{(\lambda-1)/2} \int_t^{+\infty} s^\lambda |p(s)| \, ds = 0,$$

which contradicts (18.22). $\square$

THEOREM 18.7. Let $\lambda > 1$ and $p(t) = -p_0(t) + p_1(t)$, where the functions $p_i \in C(\mathbf{R}_+)$ $(i = 0, 1)$ are nondecreasing. Suppose, in addition, that p is negative, the condition (18.19) holds, and

$$\lim_{t \to +\infty} t^{\lambda+1} p(t) = 0. \tag{18.23}$$

Then every nonzero maximally continued solution of the equation (18.1) is nonoscillatory.

PROOF. By Theorem 17.1, every maximally continued solution u of the equation (18.1) is proper. Moreover, (18.19) and Corollary 17.2 yield

$$|u'(t)| \le c, \qquad |u(t)| \le ct \qquad \text{for } t \ge t_0 > 0.$$

Clearly, u is a solution of the linear equation (18.15), where

$$0 < -p^*(t) = -p(t)|u(t)|^{\lambda-1} \le c^{\lambda-1} t^{\lambda-1} |p(t)| \qquad \text{for } t \ge t_0.$$

In view of (18.23),

$$\lim_{t \to +\infty} t^2 p^*(t) = 0$$

and so the Kneser theorem implies that u is nonoscillatory. □

PROBLEM 18.7. The conditions (18.18) and (18.23) in Theorems 18.6 and 18.7 are equivalent to the conditions

$$\lim_{t \to +\infty} t \int_t^{+\infty} p(s)\, ds = 0 \qquad \text{and} \qquad \lim_{t \to +\infty} t \int_t^{+\infty} s^{\lambda-1} p(s)\, ds = 0,$$

respectively.

Notes. Theorem 18.1 is due to Š. Belohorec [28, 33]. Theorem 18.2 was proved by F.V. Atkinson [8] in case p does not change sign, by P. Waltman [365] for $\mu = 0$, and by I.T. Kiguradze [181] in the general case. For Problems 18.2–18.5 see [28, 164, 181, 239, 302, 373], and for Theorem 18.3 see [251, 252]. Corollary 18.1 is due to G.J. Butler [46]. Theorem 18.4 was established by J. Kurzweil and M. Jasný [159, 243] for $\lambda > 1$ (the proof given here follows [173]), and by Chiou Kuo-Liang [88] for $0 < \lambda < 1$. Theorem 18.5 is due to I.T. Kiguradze [173], while Theorems 18.6 and 18.7 are due to J.S.W. Wong [375].

Results similar to Theorems 18.1–18.5 were obtained by D.V. Izyumova [145], J.W. Heidel and I.T. Kiguradze [138], D.V. Izyumova and I.A. Toroshelidze [157] for the equation $u'' = f(t, u)$, and by J.D. Mirzov [272–281], D.V. Izyumova and J.D. Mirzov [156] and Ch.A. Skhalyakho [329, 331–334, 337, 338] for two-dimensional differential systems.

§19. Unbounded and Bounded Solutions. Solutions Vanishing at Infinity

In this section we study the problems concerning the existence of bounded solutions, the boundedness of all solutions, and the vanishing at infinity or unboundedness of all oscillatory solutions of the equation

$$u'' = p(t)|u|^\lambda \operatorname{sign} u, \tag{19.1}$$

where $\lambda > 0$, $\lambda \ne 1$, and $p \in L_{\text{loc}}(\mathbf{R}_+)$.

19.1. Bounded solutions. Theorem 17.1 and the proof of Theorem 17.2 imply the following assertions.

THEOREM 19.1. Let

$$p(t) = -p_0(t) + p_1(t), \qquad p_0, p_1 \in C(\mathbf{R}_+) \quad \text{being nondecreasing,}$$
$$p_i(t) \geq 0 \quad (i = 0, 1), \qquad p(t) < 0 \quad \text{for } t \geq 0, \tag{19.2}$$
$$\int_0^{+\infty} \frac{dp_1(t)}{|p(t)|} < +\infty.$$

Then all solutions of the equation (19.1) are bounded.

THEOREM 19.2. If the hypotheses of Theorem 17.2 are satisfied and if, in addition,

$$\int_0^{+\infty} \frac{dp_1(t)}{p_0(t) - p_1(t)} < +\infty,$$

then the equation (19.1) has bounded solutions.

19.2. Solutions vanishing at infinity.

THEOREM 19.3. Let (19.2) hold,

$$\lim_{t \to +\infty} p(t) = -\infty, \tag{19.3}$$

and let there exist a positive constant ϵ such that for any sequence $(\tau_k)_{k=1}^{+\infty}$ satisfying the conditions

$$0 < \tau_k < \tau_{k+1} \quad (k = 1, 2, \dots), \qquad \lim_{k \to +\infty} \tau_k = +\infty, \tag{19.4}$$

$$\lim_{k \to +\infty} \int_{\tau_{2k}}^{\tau_{2k+1}} \sqrt{p_0(t)} \, dt \in \,]0, +\infty[, \tag{19.5}$$

$$\lim_{k \to +\infty} \inf \sqrt{p_0(\tau_{2k})}(\tau_{2k} - \tau_{2k-1}) > 0, \tag{19.6}$$

$$\lim_{k \to +\infty} \sup \sqrt{p_0(\tau_{2k-1})}(\tau_{2k} - \tau_{2k-1}) < \epsilon, \tag{19.7}$$

we have

$$\sum_{k=1}^{+\infty} (\ln p_0(\tau_{2k+1}) - \ln p_0(\tau_{2k})) = +\infty. \tag{19.8}$$

Then every maximally continued solution u of the equation (19.1) is subject to the equalities

$$\lim_{t \to +\infty} |p(t)|^{-1/2} u'(t) = 0, \qquad \lim_{t \to +\infty} u(t) = 0. \tag{19.9}$$

PROOF. First of all we note that[1]

$$\lim_{t \to +\infty} \frac{p_1(t)}{p_0(t)} = 0. \tag{19.10}$$

So

$$\frac{1}{2} p_0(t) \leq -p(t) \leq p_0(t) \qquad \text{for } t \geq t_0, \tag{19.11}$$

---

[1] See the proof of Theorem 4.9.

whenever t_0 is sufficiently large.

Let u be a maximally continued solution of the equation (19.1). By Theorem 17.1, u is defined on $\mathbf{R}_+$ and $\rho(u)(t) > 0$ for $t \geq 0$, where

$$\rho(u)(t) = \frac{|u'(t)|^2}{|p(t)|} + \frac{2}{\lambda + 1}|u(t)|^{\lambda+1}.$$

Hence, according to (19.2), the equality

$$\rho(u)(t) = \rho(u)(0) \exp\left(-\int_0^t \frac{|u'(s)|^2}{|p(s)|^2 \rho(u)(s)} \, d(p_0(s) - p_1(s))\right) \tag{19.12}$$

implies the existence of the finite limit

$$\lim_{t \to +\infty} \rho(u)(t) = c_0 \geq 0.$$

Our purpose is to show that $c_0 = 0$. Assume the contrary: $c_0 > 0$. Then (19.12), taking into account (19.11), yields

$$\int_0^{+\infty} \frac{|u'(s)|^2}{|p(s)|} \, d\ln p_0(s) < +\infty. \tag{19.13}$$

Moreover, it is clear that u is oscillatory. Let $(t_{2k-1})_{k=1}^{+\infty}$ and $(t_{2k})_{k=1}^{+\infty}$, respectively, be the sequences of zeros of the functions u and u' in $[t_0, +\infty[$ and let $t_k < t_{k+1}$ $(k = 1, 2, \ldots)$. Choose $\nu \in \,]0, 1[$ so that

$$2(\lambda + 1)^{1/(\lambda+1)} \left(\frac{c_0}{2}\right)^{(1-\lambda)/2(\lambda+1)} \int_\nu^1 \frac{dx}{\sqrt{1 - x^{\lambda+1}}} < \epsilon. \tag{19.14}$$

Each interval $]t_{2k-1}, t_{2k+1}[$ contains a pair of points τ_{2k-1}, τ_{2k} for which

$$\tau_{2k-1} < t_{2k} < \tau_{2k}, \qquad \frac{2}{\lambda + 1}|u(\tau_k)|^{\lambda+1} = \nu^{\lambda+1}\rho(u)(\tau_k) \tag{19.15}$$

and

$$\frac{2}{\lambda + 1}|u(t)|^{\lambda+1} < \nu^{\lambda+1}\rho(u)(t) \qquad \text{for } t \in \,]\tau_{2k-1}, \tau_{2k+1}[. \tag{19.16}$$

In view of (19.15),

$$\frac{|u'(t)|^2}{|p(t)|\rho(u)(t)} > 1 - \nu^{\lambda+1} > 0 \qquad \text{for } t \in \,]\tau_{2k}, \tau_{2k+1}[.$$

Hence (19.13) implies

$$\sum_{k=1}^{+\infty}(\ln p_0(\tau_{2k+1}) - \ln p_0(\tau_{2k})) < +\infty. \tag{19.17}$$

It is easy to verify that

$$\int_{\tau_{2k}}^{\tau_{2k+1}} \sqrt{|p(t)|} \, dt$$

$$= \int_{\tau_{2k}}^{\tau_{2k+1}} \frac{[\rho(u)(t)]^{(1-\lambda)/2(1+\lambda)} \operatorname{sign}(u(t)u'(t))}{\sqrt{1 - \frac{2|u(t)|^{\lambda+1}}{(\lambda+1)\rho(u)(t)}}} \, d\frac{|u(t)|}{\sqrt[\lambda+1]{\rho(u)(t)}}$$

$$-\frac{1}{\lambda+1}\int_{\tau_{2k}}^{\tau_{2k+1}}\frac{u(t)u'(t)}{\sqrt[\lambda+1]{\rho(u)(t)}\sqrt{|p(t)|\rho(u)(t)}}[\rho(u)(t)]^{(1-\lambda)/2(\lambda+1)}\frac{dp(t)}{p(t)}.$$

By (19.15) and (19.16), the first integral at the right-hand side of this equality tends to

$$2\left(\frac{\lambda+1}{2}\right)^{1/(\lambda+1)}c_0^{(1-\lambda)/2(\lambda+1)}\int_0^\nu\frac{dx}{\sqrt{1-x^{\lambda+1}}},$$

while by (19.2), (19.10) and (19.17), the second one tends to zero as $k\to+\infty$. Consequently, (19.5) is fulfilled.

From (19.1), using (19.11), we obtain

$$|u'(t)|^2=-2\int_t^{t_{2k}}p(s)|u(s)|^{\lambda+1}u'(s)\,\mathrm{sign}\,u(s)\,ds$$

$$\geq\frac{1}{\lambda+1}p_0(\tau_{2k-1})[|u(t_{2k})|^{\lambda+1}-|u(t)|^{\lambda+1}]\qquad\text{for }t\in\,]\tau_{2k-1},t_{2k}[.$$

This yields

$$\sqrt{p_0(\tau_{2k-1})}(t_{2k}-\tau_{2k-1})$$

$$\leq\sqrt{\lambda+1}\int_{\tau_{2k-1}}^{t_{2k}}\frac{|u'(t)|\,dt}{\sqrt{|u(t_{2k})|^{\lambda+1}-|u(t)|^{\lambda+1}}}.\tag{19.18}$$

Since

$$\lim_{k\to+\infty}|u(t_{2k})|=\lim_{k\to+\infty}\frac{1}{\nu}|u(\tau_{2k-1})|=\left(\frac{\lambda+1}{2}c_0\right)^{1/(\lambda+1)},$$

we have

$$\lim_{k\to+\infty}\sqrt{\lambda+1}\int_{\tau_{2k-1}}^{t_{2k}}\frac{|u'(t)|\,dt}{\sqrt{|u(t_{2k})|^{\lambda+1}-|u(t)|^{\lambda+1}}}$$

$$=(\lambda+1)^{1/(\lambda+1)}\left(\frac{c_0}{2}\right)^{(1-\lambda)/2(\lambda+1)}\int_\nu^1\frac{dx}{\sqrt{1-x^{\lambda+1}}}.$$

So, in view of (19.14) and (19.18),

$$\lim_{k\to+\infty}\sup\sqrt{p_0(\tau_{2k-1})}(t_{2k}-\tau_{2k-1})<\frac{\epsilon}{2}.$$

Similarly,

$$\lim_{k\to+\infty}\sup\sqrt{p_0(\tau_{2k-1})}(\tau_{2k}-t_{2k})<\frac{\epsilon}{2}.$$

Thus, (19.7) holds.

In the same way the validity of the inequality (19.6) can be proved.

Therefore the sequence $(\tau_k)_{k=1}^{+\infty}$ satisfies the conditions (19.4)–(19.7) and (19.17), which contradicts (19.8). $\square$

COROLLARY 19.1. Let (19.2) and (19.3) hold, and let there exist a positive constant ϵ such that for any sequence $(\tau_k)_{k=1}^{+\infty}$ satisfying the conditions (19.4) and

$$\lim_{k\to+\infty}\sup\frac{\tau_{2k+2}-\tau_{2k+1}}{\tau_{2k+1}-\tau_{2k}}<\epsilon,\qquad\lim_{k\to+\infty}\sup\sqrt{p_0(\tau_{2k+1})}(\tau_{2k+1}-\tau_{2k})<+\infty,$$

the equality (19.8) holds. Then every solution of the equation (19.1) is subject to (19.9).

COROLLARY 19.2. Let (19.2) and (19.3) hold, $p_0 \in \tilde{C}_{loc}(\mathbf{R}_+)$, and let there exist a positive nonincreasing function $\varphi \in C(\mathbf{R}_+)$ such that

$$p_0'(t) \geq \varphi(t)p_0(t) \quad \text{for } t \geq 0, \qquad \int_0^{+\infty} \varphi(t)\,dt = +\infty.$$

Then every solution of the equation (19.1) satisfies (19.9).

PROBLEM 19.1*. Is it true that if (19.2) and (19.3) hold, then the equation (19.1) has a solution satisfying the conditions (19.9)?

PROBLEM 19.2. If $p \in \tilde{C}^1_{loc}(\mathbf{R}_+)$ is negative and nonincreasing,

$$\lim_{t \to +\infty} p'(t)|p(t)|^{-3/2} = 0, \qquad \int_0^{+\infty} \left| dp'(t)|p(t)|^{-3/2} \right| < +\infty,$$

and (19.3) holds, then every maximally continued solution of the equation (19.1) satisfies the conditions (19.9).

PROBLEM 19.3. Let $p \in \tilde{C}^2_{loc}(\mathbf{R}_+)$ be negative and nonincreasing,

$$\lim_{t \to +\infty} \frac{1}{\sqrt{|p(t)|}} \int_0^t \left| d(|p(s)|^{-3/2})'' \right| = 0,$$

and let (19.3) hold. Then every maximally continued solution of the equation (19.1) satisfies the conditions (19.9).

19.3. Unbounded oscillatory solutions. Arguing as in the proof of Theorem 19.3, we can justify

THEOREM 19.4. Let

$$p(t) = -p_0(t) + p_1(t), \qquad p_0, p_1 \in C(\mathbf{R}_+) \quad \text{being nonincreasing,}$$

$$(19.19)$$

$$p_i(t) \geq 0 \quad (i = 0, 1), \qquad p(t) < 0 \quad \text{for } t \geq 0,$$

$$\lim_{t \to +\infty} p(t) = 0, \qquad \int_0^{+\infty} \frac{dp_1(t)}{|p(t)|} < +\infty, \qquad (19.20)$$

and let there exist a positive constant ϵ such that for any sequence $(\tau_k)_{k=1}^{+\infty}$ satisfying the conditions (19.4), (19.5),

$$\lim_{k \to +\infty} \inf \sqrt{p_0(\tau_{2k-1})}(\tau_{2k} - \tau_{2k-1}) > 0$$

and

$$\lim_{k \to +\infty} \sup \sqrt{p_0(\tau_{2k})}(\tau_{2k} - \tau_{2k-1}) < \epsilon,$$

we have

$$\sum_{k=1}^{+\infty} (\ln p_0(\tau_{2k}) - \ln p_0(\tau_{2k+1})) = +\infty. \qquad (19.21)$$

Then every oscillatory solution u of the equation (19.1) is subject to the equality

$$\lim_{t\to+\infty}\left(\frac{|u'(t)|^2}{|p(t)|}+\frac{2}{\lambda+1}|u(t)|^{\lambda+1}\right)=+\infty. \tag{19.22}$$

COROLLARY 19.3. Let (19.19) and (19.20) hold, and let there exist a positive constant ϵ such that for any sequence $(\tau_k)_{k=1}^{+\infty}$ satisfying the conditions (19.4) and

$$\lim_{k\to+\infty}\sup\frac{\tau_{2k}-\tau_{2k-1}}{\tau_{2k+1}-\tau_{2k}}<\epsilon, \qquad \lim_{k\to+\infty}\inf\sqrt{p_0(\tau_{2k-1})}(\tau_{2k}-\tau_{2k-1})>0$$

the equality (19.21) holds. Then every nonzero solution of the equation (19.1) is subject to (19.22).

COROLLARY 19.4. Let (19.19) and (19.20) hold, $p_0\in\tilde{C}_{loc}(\mathbf{R}_+)$, and let there exist a positive nonincreasing function $\varphi\in C(\mathbf{R}_+)$ such that

$$-p_0'(t)\ge\varphi(t)p_0(t) \qquad\text{for }t\ge0, \qquad \int_0^{+\infty}\varphi(t)\,dt=+\infty.$$

Then every oscillatory solution of the equation (19.1) satisfies (19.22).

PROBLEM 19.4*. Do the conditions (19.19) together with (19.20) guarantee the existence of a solution of the equation (19.1) satisfying (19.22)?

Notes. For Theorems 19.3 and 19.4 see [56, 171, 154], and for Problems 19.2 and 19.3 see [174, 372].

§20. Asymptotic Formulas

Below we will derive asymptotic formulas for oscillatory and nonoscillatory solutions of the equation

$$u''=p(t)|u|^\lambda\operatorname{sign}u, \tag{20.1}$$

where $\lambda>0$, $\lambda\ne1$ and, unless otherwise stipulated, $p\in L_{loc}(\mathbf{R}_+)$.

20.1 Asymptotic representations for oscillatory solutions.

THEOREM 20.1. Let $p\in\tilde{C}_{loc}(\mathbf{R}_+)$ be negative, and

$$g(t)=p'(t)|p(t)|^{-(\lambda+5)/(\lambda+3)}, \qquad g\in V(\mathbf{R}_+), \qquad \lim_{t\to\infty}g(t)=0. \tag{20.2}$$

Then every solution of the equation (20.1) either is subject to the condition

$$\lim_{t\to+\infty}r(u)(t)=c_0\in\,]0,+\infty[, \tag{20.3}$$

or satisfies, for $t\ge0$, the inequality

$$r(u)(t)\ge M\left(\int_t^{+\infty}|dg(s)|\right)^{2(\lambda+1)/(\lambda-1)} \qquad\text{if }0<\lambda<1, \tag{20.4}$$

$$r(u)(t)\le M\left(\int_t^{+\infty}|dg(s)|\right)^{2(\lambda+1)/(\lambda-1)} \qquad\text{if }\lambda>1, \tag{20.5}$$

where
$$r(u)(t) = |p(t)|^{-2/(\lambda+3)}|u'(t)|^2 + \frac{2}{\lambda+1}|p(t)|^{(\lambda+1)/(\lambda+3)}|u(t)|^{\lambda+1}, \qquad (20.6)$$

$$M = \left(\frac{8}{\lambda+3}\right)^{2(\lambda+1)/(\lambda-1)}(\lambda+1)^{2/(\lambda-1)}.$$

PROOF. By Theorem 17.1, every maximally continued solution u of the equation (20.1) is defined on $\mathbf{R}_+$ and, moreover, $r(u)(t) > 0$ for $t \geq 0$. From (20.1) we obtain

$$\int_s^t d\left(r(u)(\tau) - \frac{2}{\lambda+3}g(\tau)u(\tau)u'(\tau)\right) = -\frac{2}{\lambda+3}\int_s^t u(\tau)u'(\tau)\,dg(\tau). \qquad (20.7)$$

Thus
$$r(u)(t) = r(u)(0) - \frac{2}{\lambda+3}g(0)u(0)u'(0) + \frac{2}{\lambda+3}g(t)u(t)u'(t)$$
$$-\frac{2}{\lambda+3}\int_0^t u(\tau)u'(\tau)\,dg(\tau) \qquad \text{for } t \geq 0. \qquad (20.8)$$

Since
$$|u(t)| \leq \left(\frac{\lambda+1}{2}\right)^{1/(\lambda+1)}|p(t)|^{-1/(\lambda+1)}[r(u)(t)]^{1/(\lambda+1)},$$

$$|u'(t)| \leq |p(t)|^{1/(\lambda+3)}[r(u)(t)]^{1/2} \qquad (20.9)$$

for $t \geq 0$, (20.2) and (20.8) imply that if $r(u)$ is bounded, the finite limit $\lim_{t\to+\infty} r(u)(t) = c_0 \geq 0$ exists.

Assume $c_0 = 0$. Then for any $t_1 \in \mathbf{R}_+$ we can find $t_3 > t_2 \geq t_1$ so that
$$2r(u)(t_3) = r(u)(t_2) = r(u)(t_1),$$
$$r(u)(t_3) \leq r(u)(t) \leq r(u)(t_2) \qquad \text{for } t \in [t_2, t_3]. \qquad (20.10)$$

Setting $s = t_2$ and $t = t_3$ in (20.7), by (20.9) and (20.10) we get
$$\frac{1}{2}r(u)(t_2) \leq \frac{2}{\lambda+3}\left(\frac{\lambda+1}{2}\right)^{1/(\lambda+1)}\left(\int_{t_2}^{t_3}|dg(\tau)| + |g(t_3)| + |g(t_2)|\right)[r(u)(t_2)]^{(\lambda+3)/2(\lambda+1)}.$$

So
$$[r(u)(t_1)]^{(\lambda-1)/2(\lambda+1)} \leq \frac{8}{\lambda+3}(\lambda+1)^{1/(\lambda+1)}\int_{t_1}^{+\infty}|dg(\tau)|.$$

Since t_1 is arbitrary, this inequality is contradictory if $0 < \lambda < 1$ and yields (20.5) if $\lambda > 1$.

Let now $r(u)(t)$ be unbounded. Then for any $t_1 \in \mathbf{R}_+$ there exist $t_3 > t_2 \geq t_1$ such that
$$\frac{1}{2}r(u)(t_3) = r(u)(t_2) = r(u)(t_1)$$

and
$$r(u)(t_2) \leq r(u)(t) \leq r(u)(t_3) \qquad \text{for } t \in [t_2, t_3].$$

As above we obtain the inequality
$$[r(u)(t_1)]^{(\lambda-1)/2(\lambda+1)} \leq \frac{8}{\lambda+3}(\lambda+1)^{1/(\lambda+1)}\int_{t_1}^{+\infty}|dg(\tau)|,$$

which is contradictory when $\lambda > 1$ and yields (20.4) when $0 < \lambda < 1$. $\square$

Theorem 20.1 implies that if the conditions (20.2) hold, then the equation (20.1) has a solution satisfying (20.3). Moreover, the following theorem is true.

THEOREM 20.2. Under the hypotheses of Theorem 20.1, for any $c_0 \in]0, +\infty[$ there exists a solution of the equation (20.1) satisfying (20.3).

PROOF. Let $c_0 \in]0, +\infty[$, $c > c_0 + 1$, and let $t_0 \in \mathbf{R}_+$ be chosen so that

$$c_0 + c^{(\lambda+3)/2(\lambda+1)} \frac{4}{\lambda+3} \left(\frac{\lambda+1}{2}\right)^{1/(\lambda+1)} \int_{t_0}^{+\infty} |dg(t)| < c. \tag{20.11}$$

Consider a sequence $(t_k)_{k=1}^{+\infty}$, $t_{k+1} > t_k > t_0$, $\lim_{k \to +\infty} t_k = +\infty$, and denote by u_k a solution of the equation (20.1) subject to the condition $r(u_k)(t_k) = c_0$. First we will show that

$$r(u_k)(t) < c \qquad \text{for } t \geq t_0. \tag{20.12}$$

Assume the contrary. Then there exist $s_2 > s_1 \geq t_0$ such that

$$|r(u_k)(s_2) - r(u_k)(s_1)| = c - c_0$$

and

$$c_0 < r(u)(t) < c \qquad \text{for } t \in]s_1, s_2[.$$

Taking into account (20.7) with $u = u_k$, $s = s_1$, $t = s_2$ and using (20.9), we get

$$c - c_0 = \frac{4}{\lambda+3} \left(\frac{\lambda+1}{2}\right)^{1/(\lambda+1)} \int_{s_1}^{+\infty} |dg(\tau)| c^{(\lambda+3)/2(\lambda+1)}.$$

But this contradicts the inequality (20.11). Hence (20.12) is proved.

According to (20.7) with $u = u_k$, $s = t_k$, the inequality (20.12) implies

$$|r(u_k)(t) - c_0| < \frac{4}{\lambda+3} \left(\frac{\lambda+1}{2}\right)^{1/(\lambda+1)} c^{(\lambda+3)/2(\lambda+1)} \int_t^{+\infty} |dg(\tau)|$$

$$\text{for } t \in [t_0, t_k]. \tag{20.13}$$

Therefore, the sequence of solutions $(u_k)_{k=1}^{+\infty}$ is uniformly bounded and, thus, equicontinuous on each subinterval in $[t_0, +\infty[$. By the Arzelá–Ascoli lemma we may assume that this sequence converges locally uniformly together with $(u_k')_{k=1}^{+\infty}$. Let $u = \lim_{k \to +\infty} u_k$. Then, clearly, u is a solution of the equation (20.1). Furthermore, (20.13) yields

$$|r(u)(t) - c_0| \leq \frac{4}{\lambda+3} \left(\frac{\lambda+1}{2}\right)^{1/(\lambda+1)} c^{(\lambda+3)/2(\lambda+1)} \int_t^{+\infty} |dg(\tau)| \qquad \text{for } t \geq t_0.$$

Consequently, u satisfies (20.3). □

THEOREM 20.3. Under the hypotheses of Theorem 20.1 every solution u of the equation (20.1) satisfying the condition (20.3) can be represented in the form

$$u(t) = |p(t)|^{-1/(\lambda+3)} \rho(t) w(\alpha(t)),$$

$$\tag{20.14}$$

$$u'(t) = |p(t)|^{1/(\lambda+3)} [\rho(t)]^{(\lambda+1)/2} w'(\alpha(t)),$$

where w is the solution of the problem

$$w'' = -|w|^\lambda \operatorname{sign} w, \qquad w(0) = 0, \qquad w'(0) = 1, \qquad (20.15)$$

and ρ and α are such that

$$\lim_{t \to +\infty} \rho(t) = \rho_0 \in \,]0, +\infty[, \qquad \lim_{t \to +\infty} \frac{\alpha(t)}{\int_0^t |p(\tau)|^{2/(\lambda+3)} \, d\tau} = \rho_0^{(\lambda-1)/2}. \qquad (20.16)$$

PROOF. Let u be a solution of the equation (20.1) under the condition (20.3). Define functions ρ and α by the equalities (20.14). Since $w'^2(t) + (2/(\lambda+1))|w(t)|^{\lambda+1} \equiv 1$, (20.14) implies that $[\rho(t)]^{\lambda+1} = r(u)(t)$ and that α satisfies the differential equation

$$\alpha'(t) = |p(t)|^{2/(\lambda+3)}[r(u)(t)]^{(\lambda-1)/2(\lambda+1)} + \frac{1}{\lambda+1}\frac{p'(t)}{p(t)}w(\alpha(t))w'(\alpha(t)).$$

Clearly,

$$\lim_{t \to +\infty} \rho(t) = \rho_0 = c_0^{1/(\lambda+1)}, \qquad \lim_{t \to +\infty} \alpha'(t)|p(t)|^{-2/(\lambda+3)} = \rho_0^{(\lambda-1)/2}. \qquad (20.17)$$

Also, from (20.2) we readily obtain

$$\int_0^{+\infty} |p(t)|^{2/(\lambda+3)} \, dt = +\infty.$$

So (20.17) yields the second equality in (20.16). $\square$

COROLLARY 20.1. Let the hypotheses of Theorem 20.1 be satisfied and let u be a solution of the equation (20.1) for which (20.3) holds. If $(t_k)_{k=1}^{+\infty}$ and $(\tau_k)_{k=1}^{+\infty}$ are the sequences of zeros and extreme points of u, then

$$\lim_{k \to +\infty} |p(\tau_k)|^{1/(\lambda+3)}|u(\tau_k)| = \left(\frac{\lambda+1}{2}\right)^{1/(\lambda+1)} \rho_0,$$

$$(20.18)$$

$$\lim_{k \to +\infty} |p(t_k)|^{-1/(\lambda+3)}|u'(t_k)| = \rho_0^{(\lambda+1)/2},$$

and

$$\lim_{k \to +\infty} \int_{t_k}^{t_{k+1}} |p(t)|^{2/(\lambda+3)} \, dt$$

$$= 2\left(\frac{\lambda+1}{2}\right)^{1/(\lambda+1)} \rho_0^{(1-\lambda)/2} \int_0^1 \frac{dx}{\sqrt{1-x^{\lambda+1}}}. \qquad (20.19)$$

PROOF. The solution w of the problem (20.15) is an oscillatory periodic function of period

$$4\left(\frac{\lambda+1}{2}\right)^{1/(\lambda+1)} \int_0^1 \frac{dx}{\sqrt{1-x^{\lambda+1}}}.$$

Hence, if $(x_k)_{k=1}^{+\infty}$ and $(\xi_k)_{k=1}^{+\infty}$ are the sequences of zeros and extreme points of w, then

$$|w'(x_k)| = 1, \qquad |w(\xi_k)| = \left(\frac{\lambda+1}{2}\right)^{1/(\lambda+1)},$$

$$x_{k+1} - x_k = 2 \left(\frac{\lambda+1}{2}\right)^{1/(\lambda+1)} \int_0^1 \frac{dx}{\sqrt{1 - x^{\lambda+1}}}.$$

Since $\alpha(t_k) = x_k$ and $\alpha(\tau_k) = \xi_k$, from (20.14), (20.16) we get (20.18), (20.19). $\square$

Consider the equation (20.1), where $p(t) = t^{-\sigma}$ for $t \geq 1$, $\sigma \in]-(\lambda+3)/2, -\lambda-1[$ if $0 < \lambda < 1$ and $\sigma \in]-(\lambda+3)/2, -2[$ if $\lambda > 1$. Then (20.2) holds. But since $\int_1^{+\infty} t^\lambda |p(t)|\, dt < +\infty$ when $0 < \lambda < 1$ and $\int_1^{+\infty} t|p(t)|\, dt < +\infty$ when $\lambda > 1$, the equation (20.1) has a solution u such that $u(t) = (1 + o(1))t$ in the former case and $u(t) = 1 + o(1)$ in the latter case. So, under the hypotheses of Theorem 20.1, not all solutions of the equation (20.1) satisfy the asymptotic relations (20.14), (20.16) and, thus, (20.3). $\square$

In the following theorems we establish conditions under which the asymptotic formulas (20.14), (20.16) are valid for all solutions of the equation (20.1).

THEOREM 20.4. Let $p \in \tilde{C}_{\mathrm{loc}}(\mathbf{R}_+)$ be negative and nonincreasing,

$$\int_0^{+\infty} |d(p'(t)|p(t)|^{-3/2})| < +\infty \qquad \text{if } 0 < \lambda < 1$$

and

$$\int_0^{+\infty} |d(p'(t)|p(t)|^{-(\lambda+2)/(\lambda+1)})| < +\infty \qquad \text{if } \lambda > 1.$$

Then every solution of the equation (20.1) can be represented in the form (20.14), where w is the solution of the problem (20.15) and ρ and α satisfy (20.16).

PROOF. First we consider the case $0 < \lambda < 1$. Clearly, $\lim_{t\to+\infty} p'(t)|p(t)|^{-3/2} = 2\delta \leq 0$, which yields $\lim_{t\to+\infty} t^{-1}|p(t)|^{-1/2} = \delta \geq 0$. Hence $\delta = 0$ and, moreover, $g(t) \equiv p'(t)|p(t)|^{-(\lambda+5)/(\lambda+3)} = o(1)$. So, using the equality

$$\int_s^t dg(\tau) = \int_s^t |p(\tau)|^{(\lambda-1)/2(\lambda+3)}\, d(p'(\tau)|p(\tau)|^{-3/2})$$

$$+ \frac{1-\lambda}{2(\lambda+3)} \int_s^t |p'(\tau)|^2 |p(\tau)|^{-2(\lambda+4)/(\lambda+3)}\, d\tau, \qquad (20.20)$$

we can conclude that

$$\int_0^{+\infty} |p'(t)|^2 |p(t)|^{-2(\lambda+4)/(\lambda+3)}\, dt < +\infty \qquad (20.21)$$

and $g \in V(\mathbf{R}_+)$. Therefore the hypotheses of Theorem 20.1 are satisfied.

Suppose the inequality (20.4) holds for a certain solution of the equation (20.1). Since $|p|$ is nondecreasing and $0 < \lambda < 1$, we have

$$\int_t^{+\infty} |dg(\tau)| \leq |p(t)|^{(\lambda-1)/2(\lambda+3)} \int_t^{+\infty} |d(p'(\tau)|p(\tau)|^{-3/2})|$$

$$+ \frac{1-\lambda}{2(\lambda+3)} \int_t^{+\infty} |p'(\tau)|^2 |p(\tau)|^{-2(\lambda+4)/(\lambda+3)}\, d\tau \qquad \text{for } t \geq 0.$$

By (20.2) and (20.20) this implies

$$\int_t^{+\infty} |dg(\tau)| \leq 2|p(t)|^{(\lambda-1)/2(\lambda+3)} \int_t^{+\infty} |d(p'(\tau)|p(\tau)|^{-32})| - g(t)$$

$$\leq 3|p(t)|^{(\lambda-1)/2(\lambda+3)} \int_t^{+\infty} |d(p'(\tau)|p(\tau)|^{-3/2})| \quad \text{for } t \geq 0.$$

Hence the estimate (20.4) becomes

$$r(u)(t) \geq M|p(t)|^{(\lambda+1)/(\lambda+3)} \left(\int_t^{+\infty} \left| d(p'(\tau)|p(\tau)|^{-3/2}) \right| \right)^{2(\lambda+1)/(\lambda-1)} \quad \text{for } t \geq 0.$$

Then, in view of (20.6),

$$\lim_{t \to +\infty} \left(|p(t)|^{-1}|u'(t)|^2 + \frac{2}{\lambda+1}|u(t)|^{\lambda+1} \right) = +\infty.$$

But since p is nonincreasing, this contradicts Theorem 17.1.

Therefore, for every solution of the equation (20.1) the condition (20.3) holds, and by Theorem 20.3 the asymptotic formulae (20.14), (20.16) are true.

Now let $\lambda > 1$. As above we can show that

$$p'(t)|p(t)|^{-(\lambda+2)/(\lambda+1)} = o(1),$$

the conditions (20.2) are fulfilled and

$$\int_t^{+\infty} |dg(\tau)| \leq 3|p(t)|^{(1-\lambda)/(\lambda+1)(\lambda+3)} \int_t^{+\infty} \left| d(p'(\tau)|p(\tau)|^{(\lambda+2)/(\lambda+1)}) \right|$$

$$\text{for } t \geq 0. \tag{20.22}$$

According to Theorem 20.3 it suffices to prove that every solution of the equation (20.1) satisfies (20.3).

Assume the contrary. Then by Theorem 20.1 there exists a solution of the equation (20.1) for which the inequality (20.5) holds. So, in view of (20.22),

$$\lim_{t \to +\infty} |p(t)|^{2/(\lambda+3)} r(u)(t) = +\infty.$$

But since p is nonincreasing, this contradicts Corollary 17.2. $\square$

The following assertion can be proved in a similar manner.

THEOREM 20.5. Let $p \in \tilde{C}_{\text{loc}}(\mathbf{R}_+)$ be negative and nondecreasing,

$$\int_0^{+\infty} \left| d(p'(t)|p(t)|^{-(\lambda+2)/(\lambda+1)}) \right| < +\infty \quad \text{if } 0 < \lambda < 1,$$

$$\int_0^{+\infty} \left| d(p'(t)|p(t)|^{-3/2}) \right| < +\infty \quad \text{if } \lambda > 1,$$

and let (20.2) hold. Then every solution of the equation (20.1) has the asymptotic representations (20.14), (20.16).

Note that if $p(t) = -t^\sigma$ for $t \geq 1$, then the hypotheses of Theorem 20.4 are satisfied whenever $\sigma \geq 0$, while the hypotheses of Theorem 20.5 are fulfilled when either $-(\lambda+1) \leq \sigma < 0$ and $0 < \lambda < 1$ or $-2 \leq \sigma < 0$ and $\lambda > 1$.

20.2. Auxiliary assertions. In the sequel we will need some asymptotic properties of solutions of the equation

$$v'' = a_1(s)v' + a_2(s)v + \alpha(s)|v|^\lambda \operatorname{sign} v, \qquad (20.23)$$

where $\lambda > 0$, $\lambda \neq 1$, $a_1 \in \tilde{C}_{\mathrm{loc}}(\mathbf{R}_+)$, $a_2, \alpha \in L_{\mathrm{loc}}(\mathbf{R}_+)$, and

$$a_1(s) = s^\sigma(a_{10} + o(1)), \qquad a_1'(s) = s^{\sigma-1}(\sigma a_{10} + o(1)), \qquad 0 \leq \sigma \leq 1, \qquad (20.24)$$

$$a_2(s) = a_{20} + o(1), \qquad \alpha(s) = \alpha_0 + o(1) \qquad (20.25)$$

for certain $a_{10}, a_{20}, \alpha_0 \in \mathbf{R}$.

LEMMA 20.1. Let $0 < \lambda < 1$, $a_1 \in \tilde{C}_{\mathrm{loc}}(\mathbf{R}_+)$, $a_2, \alpha \in L_{\mathrm{loc}}(\mathbf{R}_+)$, and let the conditions (20.24) and (20.25) with $a_{10} \neq 0$, $a_{20} > 0$ and $\alpha_0 = -1$ hold. Then for every nonoscillatory proper solution of the equation (20.23) either

$$v(s) = \pm a_{20}^{1/(\lambda-1)} + o(1) \qquad (20.26)$$

or

$$v(s) = v_\infty(s)(1 + o(1)), \qquad (20.27)$$

where v_∞ is a solution of the equation

$$v'' = a_1(s)v' + a_2(s)v \qquad (20.28)$$

such that

$$\lim_{s \to +\infty} |v_\infty(s)| = +\infty. \qquad (20.29)$$

PROOF. Let v be a positive solution of the equation (20.23) defined on $[s_0, +\infty[$. We will show that either $v(s) = a_{20}^{1/(\lambda-1)} + o(1)$ or $v(s) \to +\infty$ as $s \to +\infty$.

It follows from (20.25) that for any $\delta \in\,]0, a_{20}[$ there exists $s_\delta \geq s_0$ such that

$$(a_{20} - \delta)|\alpha(s)| \leq a_2(s) \leq (a_{20} + \delta)|\alpha(s)| \qquad \text{for } s \geq s_\delta.$$

Hence, by (20.23), an extremum of the function v is a minimum if it lies above the line $v = (a_{20} - \delta)^{1/(\lambda-1)}$, and a maximum if it lies below the line $v = (a_{20} + \delta)^{1/(\lambda-1)}$. So, without loss of generality, assume that for $s \geq s_\delta$ the solution v does not cross these lines, i.e. one of the inequalities

$$v(s) \geq (a_{20} - \delta)^{1/(\lambda-1)}, \qquad (20.30)$$

$$(a_{20} + \delta)^{1/(\lambda-1)} \leq v(s) \leq (a_{20} - \delta)^{1/(\lambda-1)}, \qquad (20.31)$$

or

$$0 < v(s) \leq (a_{20} + \delta)^{1/(\lambda-1)} \qquad (20.32)$$

holds on $[s_\delta, +\infty[$.

First we will prove that if for at least one $\delta \in\,]0, a_{20}[$ (20.30) is fulfilled, then $v(s) \to +\infty$ as $s \to +\infty$. Since v is monotone, the finite or infinite limit $\lim_{s \to +\infty} v(s)$ exists. Suppose $\lim_{s \to +\infty} v(s) < +\infty$. Then, in view of (20.23), (20.24) and (20.30),

$$v''(s) + c_0 s v'(s) \geq \delta_1 > 0 \qquad \text{for } s \geq s_*,$$

where s_* is a sufficiently large constant, $c_0 = 0$ if $a_{10}v$ is nondecreasing and $c_0 = -2a_{10}$ if $a_{10}v$ is nonincreasing. Multiplying both sides of the last inequality by s^{-1} and integrating over $[s_*, s]$, we obtain

$$s^{-1}v'(s) + (s^{-2} + c_0)v(s) + 2\int_{s_*}^{s} x^{-3}v(x)\, dx$$

$$\geq \delta_1 \ln\frac{s}{s_*} + s_*^{-1}v'(s_*) + (s_*^{-2} + c_0)v(s_*) \qquad \text{for } s \geq s_*,$$

which contradicts the boundedness of v. Hence, in the case of (20.30) we have $\lim_{s\to+\infty} v(s) = +\infty$.

If for any arbitrarily small $\delta > 0$ the inequality (20.31) is valid on $[s_\delta, +\infty[$, then $\lim_{s\to+\infty} v(s) = a_{20}^{1/(\lambda-1)}$.

We claim that (20.32) is impossible. Indeed, otherwise (20.23), taking into account the monotonicity of v and the conditions (20.24), (20.25) and (20.32), would imply

$$v''(s) + c_1 s v'(s) \leq -\delta_2 v^\lambda(s) \qquad \text{for } s \geq s_*,$$

where $\delta_2 > 0$, s_* is a sufficiently large constant, $c_1 = 0$ if $a_{10}v$ is nonincreasing and $c_1 = -2a_{10}$ if $a_{10}v$ is nondecreasing. Multiplying both sides of this inequality by $s^{-1}v^{-\lambda}$ and integrating over $[s_*, s]$, we get

$$\frac{v'(s)}{sv^\lambda(s)} - \frac{v'(s_*)}{s_* v^\lambda(s_*)} + \lambda\int_{s_*}^{s} \frac{|v'(x)|^2}{xv^{\lambda+1}(x)}\, dx + \frac{v^{1-\lambda}(s)}{(1-\lambda)s^2}$$

$$-\frac{v^{1-\lambda}(s_*)}{(1-\lambda)s_*^2} + \frac{2}{1-\lambda}\int_{s_*}^{s} \frac{v^{1-\lambda}(x)}{x^3}\, dx + \frac{c_1}{1-\lambda}v^{1-\lambda}(s)$$

$$-\frac{c_1}{1-\lambda}v^{1-\lambda}(s_*) \leq -\delta_2 \ln\frac{s}{s_*} \qquad \text{for } s \geq s_*.$$

Thus

$$\lim_{s\to+\infty} \frac{v'(s)}{v^\lambda(s)} = -\infty, \qquad \lim_{s\to+\infty} v^{1-\lambda}(s) = +\infty,$$

which is impossible.

Finally we will show that every solution of the equation (20.23) tending to infinity has the representation (20.27). For this it suffices to prove (see Problem 6.2) the inequality

$$\int_{s_*}^{+\infty} \exp\left(-\int_{s_0}^{s} a_1(x)\, dx\right) v_1(s)v_2(s)|v(s)|^{\lambda-1}\, ds < +\infty, \qquad (20.33)$$

where s_* is a sufficiently large constant, v_1 and v_2 are linearly independent solutions of the equation (20.28) such that

$$v_1(s) > 0 \qquad \text{for } s \geq s_*,$$

$$\int_{s_*}^{+\infty} \exp\left(\int_{s_0}^{s} a_1(x)\, dx\right) |v_1(s)|^{-2}\, ds < +\infty, \qquad (20.34)$$

$$v_2(s) = v_1(s)\int_{s}^{+\infty} \exp\left(\int_{s_*}^{x} a_1(\xi)\, d\xi\right) |v_1(x)|^{-2}\, dx, \qquad (20.35)$$

and v is an arbitrary solution of the equation (20.23) tending to infinity.

By means of the transformation

$$v(s) = \exp\left(\frac{1}{2}\int_0^s a_1(x)\,dx\right)w(s) \qquad (20.36)$$

the equation (20.28) can be reduced to

$$w'' = b(s)w, \qquad (20.37)$$

where

$$b(s) = \frac{1}{4}a_1^2(s) - \frac{1}{2}a_1'(s) + a_2(s). \qquad (20.38)$$

Clearly,

$$\lim_{s\to+\infty} s^{-2\sigma}b(s) = \begin{cases} \frac{1}{4}a_{10}^2 & \text{if } \sigma > 0 \\ \frac{1}{4}a_{10}^2 + a_{20} & \text{if } \sigma = 0. \end{cases} \qquad (20.39)$$

Consider the functions

$$\varphi_{\pm\epsilon}(s) = \exp\left(\int_{s_*}^s \sqrt{\frac{1}{4}a_1^2(x) + a_{20} \pm \epsilon}\,dx\right),$$

where $\epsilon > 0$ is a sufficiently small constant. It is clear that

$$\varphi_{\pm\epsilon}''(s) = b_{\pm\epsilon}(s)\varphi_{\pm\epsilon}(s),$$

where

$$b_{\pm\epsilon}(s) = \frac{1}{4}a_1^2(s) + a_{20} \pm \epsilon + \frac{a_1'(s)a_1(s)}{4\sqrt{\frac{1}{4}a_1^2(s) + a_{20} \pm \epsilon}}.$$

In view of (20.24) we may assume that

$$b_{-\epsilon}(s) < b(s) < b_{+\epsilon}(s) \qquad \text{for } s \geq s_*.$$

Let w_1 be a solution of the equation (20.37) under the conditions

$$w_1(s_*) = 1, \qquad \varphi_{-\epsilon}'(s_*) < w_1'(s_*) < \varphi_{+\epsilon}'(s_*).$$

It is readily verified that

$$\varphi_{-\epsilon}(s) \leq w_1(s) \leq \varphi_{+\epsilon}(s) \qquad \text{for } s \geq s_*.$$

Hence the corresponding solution v_1 of the equation (20.28) satisfies (20.34) and the estimate

$$\exp\left(\int_{s_*}^s \left(\frac{1}{2}a_1(x) + \sqrt{\frac{1}{4}a_1^2(x) + a_{20} - \epsilon}\right)dx\right) \leq v_1(s)$$

$$\leq \exp\left(\int_{s_*}^s \left(\frac{1}{2}a_1(x) + \sqrt{\frac{1}{4}a_1^2(x) + a_{20} + \epsilon}\right)dx\right) \qquad \text{for } s \geq s_*. \qquad (20.40)$$

So (20.34) yields

$$c_1 s^{-\sigma}\exp\left(-2\int_{s_*}^s \sqrt{\frac{1}{4}a_1^2(x) + a_{20} + \epsilon}\,dx\right) \leq \frac{v_2(s)}{v_1(s)}$$

$$\leq c_2 s^{-\sigma} \exp\left(-2\int_{s_*}^{s}\sqrt{\frac{1}{4}a_1^2(x) + a_{20} - \epsilon}\,dx\right) \qquad \text{for } s \geq s_*, \qquad (20.41)$$

with $c_2 > c_1 > 0$.

Let v be a solution of the linear equation

$$v'' = a_1(s)v' + (a_2(s) + \beta(s))v, \qquad (20.42)$$

where $\beta(s) = \alpha(s)|v(s)|^{\lambda-1} = o(1)$. As above we obtain

$$v(s) \geq c_3 \exp\left(\int_{s_*}^{s}\left(\frac{1}{2}a_1(x) + \sqrt{\frac{1}{4}a_1^2(x) + a_{20} - \epsilon}\right)dx\right) \qquad \text{for } s \geq s_*, \qquad (20.43)$$

with $c_3 > 0$. By (20.24), $\epsilon > 0$ and $\delta > 0$ can be chosen so small and s_* so large that

$$2\left(\sqrt{\frac{1}{4}a_1^2(s) + a_{20} + \epsilon} - \sqrt{\frac{1}{4}a_1^2(s) + a_{20} - \epsilon}\right)$$

$$+ (\lambda - 1)\left(\frac{1}{2}a_1(s) + \sqrt{\frac{1}{4}a_1^2(s) + a_{20} - \epsilon}\right) < \delta s^{-\sigma} \qquad \text{for } s \geq s_*.$$

Therefore (20.40), (20.41) and (20.43) imply (20.33). $\square$

REMARK 20.1. If $0 < \lambda \leq 1$, $a_1(s) \equiv a_{10} > 0$, $a_2(s) \equiv a_{20} > 0$, and $\alpha(s) \equiv -1$, then every proper solution of the equation (20.23) satisfies either (20.26) or (20.27).

Indeed, let $v\colon [s_0, +\infty[\to \mathbf{R}$ be a bounded oscillatory proper solution of (20.23) and let $(s_k)_{k=1}^{+\infty}$ and $(s_k')_{k=1}^{+\infty}$ be the sequences of its zeros and extreme points. Multiply both sides of (20.23) by v' and integrate over $[s_0, s_k']$. Taking into account the boundedness of v, we obtain $v' \in L^2([s_0, +\infty[)$. On the other hand, by integrating over $[s_{k-1}, s_k]$ we may conclude that the sequence $(|v'(s_k)|)_{k=1}^{+\infty}$ is bounded and nondecreasing. So v' and, therefore, v'' are bounded. Hence, the equality

$$[v'(s)]^3 = -3\int_s^{s_k'} [v'(x)]^2 v''(x)\,dx \qquad \text{for } s_0 \leq s \leq s_k'$$

yields $v'(s) = o(1)$. We have obtained a contradiction, which indicates that the equation (20.23) does not have a bounded oscillatory proper solution. Thus, Lemma 20.1 implies the validity of Remark 20.1.

PROBLEM 20.1*. Is the assertion of Lemma 20.1 true for any proper solution of the equation (20.23) if we additionally require that $a_{10} > 0$?

LEMMA 20.2. Let $\lambda > 1$, $a_1 \in \tilde{C}_{\mathrm{loc}}(\mathbf{R}_+)$, $a_2, \alpha \in V(\mathbf{R}_+)$, and let the conditions (20.24) and (20.25) with $a_{10} < 0$, $a_{20} > 0$ and $\alpha_0 = -1$ hold. Then for every proper solution of the equation (20.23) we have either (20.26) or

$$v(s) = v_0(s)(1 + o(1)), \qquad (20.44)$$

where v_0 is a solution of the equation (20.28) such that

$$v_0(s) = o(1). \qquad (20.45)$$

PROOF. Suppose $v \colon [s_0, +\infty[\to \mathbf{R}$ is a solution of the equation (20.23). First we will show that v is bounded. Assume the contrary. Then there exists an increasing sequence $(s_k)_{k=1}^{+\infty}$ satisfying the conditions

$$\lim_{k \to +\infty} |v(s_k)| = +\infty, \qquad |v(s)| \le |v(s_k)| \quad \text{for } s \in [s_1, s_k] \qquad (k = 1, 2, \dots).$$

The equality

$$\int_\tau^s d \left(|v'(x)|^2 - a_2(x)|v(x)|^2 - \frac{2}{\lambda+1} \alpha(x)|v(x)|^{\lambda+1} \right)$$

$$= \int_\tau^s 2a_1(x)|v'(x)|^2 \, dx - \int_\tau^s |v(x)|^2 \, da_2(x) - \frac{2}{\lambda+1} \int_\tau^s |v(x)|^{\lambda+1} \, d\alpha(x) \qquad (20.46)$$

with $\tau = s_m$, $s = s_k$ and $k > m$ yields

$$-a_2(s_k)|v(s_k)|^2 - \frac{2}{\lambda+1} \alpha(s_k)|v(s_k)|^{\lambda+1}$$

$$\le |v'(s_m)|^2 - \frac{2}{\lambda+1} \alpha(s_m)|v(s_m)|^{\lambda+1}$$

$$+ |v(s_k)|^2 \int_{s_m}^{s_k} |da_2(x)| + \frac{2}{\lambda+1} |v(s_k)|^{\lambda+1} \int_{s_m}^{s_k} |d\alpha(x)|.$$

Dividing both sides of this inequality by $|v(s_k)|^{\lambda+1}$ and passing k to $+\infty$, we get

$$1 \le \int_{s_m}^{+\infty} |d\alpha(x)| \qquad (m = 1, 2, \dots).$$

This contradiction proves that v is bounded.

Therefore, according to (20.23) and (20.46) with $\tau = s_0$,

$$|v'(s)| \le c_1, \qquad |v''(s)| \le c_1 s^\sigma \qquad \text{for } s \ge s_0,$$

$$\int_{s_1}^{+\infty} x^\sigma |v'(x)|^2 \, dx < +\infty, \qquad (20.47)$$

where $c_1 > 0$ is a constant. Hence there exists a sequence $(s'_k)_{k=1}^{+\infty}$ such that $s'_k \to +\infty$ and $v'(s'_k) \to 0$ as $k \to +\infty$. But since

$$[v'(s)]^3 = [v'(s'_k)]^3 - 3 \int_s^{s'_k} |v'(x)|^2 v''(x) \, dx \qquad \text{for } s_0 \le s \le s'_k,$$

(20.47) implies

$$|v'(s)|^3 \le 3c_1 \int_s^{+\infty} x^\sigma |v'(x)|^2 \, dx \qquad \text{for } s \ge s_0.$$

Consequently, $v'(s) = o(1)$. So, by (20.46) with $\tau = s_0$, the function v tends to a finite limit c as $s \to +\infty$. We will prove that $|c|^\lambda \operatorname{sign} c + a_{20}c = 0$.

Assume the contrary. Then we may suppose that v is monotone on $[s_0, +\infty[$ and either

$$v''(s) - a_1(s)v'(s) \ge \delta_1 > 0 \qquad \text{for } s \ge s_0$$

or

$$v''(s) - a_1(s)v'(s) \le -\delta_2 < 0 \qquad \text{for } s \ge s_0.$$

In the first case we have

$$v''(s) + c_0 s v'(s) \geq \delta_1 > 0 \qquad \text{for } s \geq s_0,$$

where $c_0 = 0$ if v is nonincreasing and $c_0 = -2a_{10}$ if v is nondecreasing. Multiplying both sides of the last inequality by s^{-1} and integrating over $[s_0, s]$, we get

$$s^{-1} v'(s) + (s^{-2} + c_0) v(s) + 2 \int_{s_0}^{s} x^{-3} v(x) \, dx$$

$$\geq \delta_1 \ln \frac{s}{s_0} + s_0^{-1} v'(s_0) + (s_0^{-2} + c_0) v(s_0) \qquad \text{for } s \geq s_0,$$

which is impossible.

In the second case we can obtain a contradiction by a similar argument. So either v is subject to (20.26) or $v(s) = o(1)$.

As in the proof of Lemma 20.1, we can show that if $v(s) = o(1)$, then (20.44) holds. Notice, however, that in this case v will satisfy an estimate which is analogous to the estimate satisfied by v_2. □

LEMMA 20.3. Let $\lambda > 1$, $a_1 \in \tilde{C}_{\text{loc}}(\mathbf{R}_+)$, $a_2, \alpha \in V(\mathbf{R}_+)$, and let the conditions (20.24) and (20.25) with $a_{10} > 0$, $a_{20} > 0$ and $\alpha_0 = -1$ hold. Then for every nonoscillatory proper solution of the equation (20.23) we have either (20.26) or (20.44), where v_0 is a solution of the equation (20.28) satisfying (20.45).

PROOF. First we will prove that every nonoscillatory proper solution of the equation (20.23) tends to a finite or infinite limit as $s \to +\infty$. Suppose, on the contrary, that the equation (20.23) has a solution $v: [s_0, +\infty[\to]0, +\infty[$ which does not tend to any limit. Denote by $(\xi_k)_{k=1}^{+\infty}$ a sequence of extreme points of v converging to $+\infty$. If the function v is unbounded, then without loss of generality we may assume that

$$v(s) \leq v(\xi_{2k}) \qquad \text{for } s \in [\xi_{2k}, \xi_{2k+1}],$$

$$\lim_{k \to +\infty} v(\xi_{2k+1}) = c_0 < +\infty, \qquad \lim_{k \to +\infty} v(\xi_{2k}) = +\infty.$$

On the other hand, in view of (20.46) with $\tau = \xi_{2k}$, $s = \xi_{2k+1}$,

$$-\frac{2}{\lambda+1} \alpha(\xi_{2k+1}) |v(\xi_{2k+1})|^{\lambda+1} \geq -\frac{2}{\lambda+1} \left(\alpha(\xi_{2k}) + \int_{\xi_{2k}}^{+\infty} |d\alpha(s)| \right) |v(\xi_{2k})|^{\lambda+1}$$

$$- \left(a_2(\xi_{2k}) + \int_{\xi_{2k}}^{+\infty} |da_2(s)| \right) |v(\xi_{2k})|^2,$$

which becomes impossible as $k \to +\infty$. Hence v is bounded.

From (20.46) with $s = \xi_k$, $\tau = s_0$, taking into account the boundedness of v, we obtain

$$\int_{s_0}^{+\infty} a_1(s) |v'(s)|^2 \, ds < +\infty.$$

So (20.46) with $\tau = s_0$ implies that v' is bounded as well. Therefore, v satisfies the conditions (20.47). As has been shown in the proof of Lemma 20.2, this yields $v'(s) = o(1)$. Now applying the identity (20.46) with $\tau = s_0$ once again, we conclude that $\lim_{s \to +\infty} v(s)$ exists and is finite, which contradicts our assumption.

Thus, every nonoscillatory proper solution of the equation (20.23) tends to a finite or infinite limit as $s \to +\infty$.

Suppose

$$\lim_{s \to +\infty} v(s) = +\infty. \tag{20.48}$$

Then

$$\alpha(s)|v(s)|^\lambda + a_2(s)v(s) < 0 \qquad \text{for } s \geq s_1,$$

where s_1 is a sufficiently large constant, and from the equality

$$v'(s) = \exp\left(\int_{s_1}^{s} a_1(x)\, dx\right)$$

$$\times \left[v'(s_1) + \int_{s_1}^{s} (\alpha(x)|v(x)|^\lambda + a_2(x)v(x)) \exp\left(-\int_{s_1}^{x} a_1(\xi)d\xi\right) dx\right] \tag{20.49}$$

we get

$$v'(s_1) \geq -\int_{s_1}^{+\infty} (\alpha(x)|v(x)|^\lambda + a_2(x)v(x)) \exp\left(-\int_{s_1}^{x} a_1(\xi)d\xi\right) dx.$$

If

$$v'(s_1) = -\int_{s_1}^{+\infty} (\alpha(x)|v(x)|^\lambda + a_2(x)v(x)) \exp\left(-\int_{s_1}^{x} a_1(\xi)\,d\xi\right) dx,$$

then

$$v'(s) = -\int_{s}^{+\infty} (\alpha(x)|v(x)|^\lambda + a_2(x)v(x)) \exp\left(-\int_{s}^{x} a_1(\xi)\,d\xi\right) dx \tag{20.50}$$

for $s \geq s_1$. Since $v'(s) = o(1)$, we have $s^{-1}v(s) = o(1)$. Hence, according to (20.24), (20.25) and (20.50) $v'(s) = O(\exp(-(1/2)a_{10}s))$. This yields $v(s) = O(1)$, which contradicts (20.48). Consequently,

$$v'(s_1) > -\int_{s_1}^{+\infty} (\alpha(x)|v(x)|^\lambda + a_2(x)v(x)) \exp\left(-\int_{s_1}^{x} a_1(\xi)\,d\xi\right) dx. \tag{20.51}$$

Therefore, in view of (20.24) and (20.25),

$$v'(s) \geq c_1 \exp\left(\int_{s_1}^{s} a_1(x)\, dx\right), \qquad v(s) \geq c_2 s^{-\sigma} \exp\left(\int_{s_1}^{s} a_1(x)\, dx\right)$$

for $s \geq s_1$. But then (20.51) implies the contradictory inequality

$$\int_{s_1}^{+\infty} s^{-\lambda\sigma} \exp\left((\lambda - 1)\int_{s_1}^{s} a_1(x)\, dx\right) ds < +\infty.$$

So (20.48) is false.

Let $v(s) \to c$ as $s \to +\infty$. In this case from (20.49) we readily get (20.50). If $a_{20}c - c^\lambda \neq 0$, then by (20.24),

$$|v'(s)| > c_1 \int_{s}^{+\infty} \exp\left(-\int_{s}^{x} a_1(\xi)\,d\xi\right) dx \geq \frac{c_2}{s}$$

for s sufficiently large. Hence $v(s) \to +\infty$ as $s \to +\infty$. We have arrived at a contradiction, which indicates that either $|v(s)|^{\lambda-1} \to a_{20}$ or $v(s) \to 0$ as $s \to +\infty$. As in the proof of Lemma 20.2 we can show that in the second case v satisfies (20.44). $\square$

LEMMA 20.4. Let $0 < \lambda < 1$, $a_1 \in \tilde{C}_{\mathrm{loc}}(\mathbf{R}_+)$, $a_2, \alpha \in V(\mathbf{R}_+)$, and let the conditions (20.24) and (20.25) with $a_{10} > 0$, $a_{20} < 0$ and $\alpha_0 = 1$ hold. If, in addition, the equation (20.28) does not have an oscillatory solution, then for every nonoscillatory proper solution of the equation (20.23), we have either

$$\lim_{s \to +\infty} |v(s)| = |a_{20}|^{1/(\lambda-1)} \tag{20.52}$$

or (20.27), where v_∞ is a solution of the equation (20.28) satisfying (20.29).

PROOF. Let $v: [s_0, +\infty[\to \mathbf{R}$ be a nonoscillatory solution of the equation (20.23). Without loss of generality we may assume that $v(s) > 0$ for $s \geq s_0$. Arguing as in the proof of Lemma 20.3, we can show that v tends to a finite or infinite limit as $s \to +\infty$. Moreover, if $v(s) \to c \in \mathbf{R}$ as $s \to +\infty$, then either $c = 0$ or $c = |a_{20}|^{1/(\lambda-1)}$.

If $c = 0$, then

$$0 < v(s) < (4|a_{20}|)^{1/(\lambda-1)}, \qquad v'(s) < 0 \qquad \text{for } s \geq s_1, \tag{20.53}$$

whenever $s_1 > s_0$ is sufficiently large. Besides, we may assume that

$$0 < a_1(s) \leq 2a_{10}s^\sigma, \qquad 2a_{20} < a_2(s) < 0, \qquad \frac{3}{4} < \alpha(s) < \frac{5}{4} \tag{20.54}$$

for $s \geq s_1$. Hence, (20.23) yields

$$v''(s) \leq \frac{5}{4}|v(s)|^\lambda, \qquad v''(s) \geq 2a_{10}sv'(s) + \frac{1}{4}|v(s)|^\lambda \qquad \text{for } s \geq s_1.$$

Multiplying both sides of the first inequality by $v'(s)$ and of the last inequality by $s^{-1}|v(s)|^{-\lambda}$ and integrating over $[s, +\infty[$ and $[s_1, s]$, respectively, we obtain

$$|v'(s)| \leq \left(\frac{5}{2(\lambda+1)}\right)^{1/2} |v(s)|^{(\lambda+1)/2} \qquad \text{for } s \geq s_1$$

and

$$\frac{v'(s)}{s|v(s)|^\lambda} - \frac{v'(s_1)}{s_1|v(s_1)|^\lambda} + \int_{s_1}^s \frac{v'(x)}{x^2|v(x)|^2}\,dx + \lambda \int_{s_1}^s \frac{|v'(x)|^2}{x|v(x)|^{\lambda+1}}\,dx$$

$$- \frac{2a_{10}}{1-\lambda}|v(s)|^{1-\lambda} + \frac{2a_{10}}{1-\lambda}|v(s_0)|^{1-\lambda} \geq \frac{1}{4}\ln\frac{s}{s_1} \qquad \text{for } s \geq s_1.$$

In view of the first inequality, the second one implies

$$\int_{s_1}^{+\infty} |v(x)|^{-(\lambda+1)/2}|v'(x)|\,dx = +\infty,$$

which is a contradiction.

So either $v(s) \to |a_{20}|^{1/(\lambda-1)}$ or $v(s) \to +\infty$ as $s \to +\infty$, and it remains to show that in the second case (20.27) holds. For this we will prove that (20.33) holds, where s_* is a sufficiently large constant, v_1 and v_2 are positive solutions of the equation (20.28) satisfying the conditions (20.34) and (20.35), and v is an arbitrary solution of the equation (20.23) tending to infinity.

The transformation (20.36) reduces the equation (20.28) to (20.37), where b is defined by (20.38). Since the solutions of (20.28) are nonoscillatory, (20.39) implies that the finite or infinite limit $\lim_{s \to +\infty} b(s) \geq 0$ exists.

Suppose $b(s) = o(1)$. This occurs only if $\sigma = 0$. Let

$$0 < \epsilon < \left(\frac{(1-\lambda)a_{10}}{3-\lambda} \right)^2.$$

Assume that

$$b(s) \le \epsilon, \qquad 2(3-\lambda)\sqrt{\epsilon} - (1-\lambda)a_1(s) \le \delta < 0 \qquad \text{for } s \ge s_*. \tag{20.55}$$

Then, according to Corollary 11.6.5 in [127], the solutions w_1 and w_2 of the equation (20.37) corresponding to v_1 and v_2 satisfy the estimates

$$c_1 \exp(-\sqrt{\epsilon}s) \le w_i(s) \le c_2 \exp(\sqrt{\epsilon}s) \qquad \text{for } s \ge s_*,$$

where c_1 and c_2 are positive constants. Therefore,

$$c_1 \exp \left(\frac{1}{2} \int_0^s a_1(x)\, dx - \sqrt{\epsilon}s \right) \le v_i(s) \le c_2 \exp \left(\frac{1}{2} \int_0^s a_1(x)\, dx + \sqrt{\epsilon}s \right)$$

$$\text{for } s \ge s_*. \tag{20.56}$$

Similarly, since v can be represented as a solution of the linear equation (20.42) with $\beta(s) = o(1)$,

$$v(s) \ge c_1 \exp \left(\frac{1}{2} \int_0^s a_1(x)\, dx - \sqrt{\epsilon}s \right) \qquad \text{for } s \ge s_*. \tag{20.57}$$

Applying (20.55)–(20.57), we conclude that (20.33) holds.

Let now $\lim_{s \to +\infty} b(s) > 0$. Then, as in the proof of Lemma 20.1, we obtain the estimates (20.40) and (20.41). Furthermore, note that the function

$$\psi(s) = \exp \left(-\int_{s_*}^s \sqrt{\frac{1}{4}a_1^2(x) + a_{20} + \epsilon}\, dx \right)$$

satisfies the equation

$$\psi'' = b_\epsilon(s)\psi,$$

where

$$b_\epsilon(s) = \frac{1}{4}a_1^2(s) + a_{20} + \epsilon - \frac{a_1'(s)a_1(s)}{4\sqrt{\frac{1}{4}a_1^2(s) + a_{20} + \epsilon}}.$$

By (20.24) we may assume that

$$b(s) < b_\epsilon(s) \qquad \text{for } s \ge s_*,$$

and so for any solution of the equation (20.37) Corollary 11.6.5 in [127] yields

$$w(s) \ge c \exp \left(-\int_{s_*}^s \sqrt{\frac{1}{4}a_1^2(x) + a_{20} + \epsilon}\, dx \right)$$

on $[s_*, +\infty[$, where c is a positive constant. Consequently, for any solution of (20.28) we have

$$v(s) \ge c \exp \left(\int_{s_*}^s \left(\frac{1}{2}a_1(x) - \sqrt{\frac{1}{4}a_1^2(x) + a_{20} + \epsilon} \right) dx \right) \tag{20.58}$$

on $[s_*, +\infty[$. Since every solution of the equation (20.23) tending to infinity can be represented as a solution of the linear equation (20.42) with $\beta(s) = o(1)$, such a solution also satisfies (20.58).

Applying (20.40), (20.41) and (20.58), we conlude that (20.33) is true. $\square$

LEMMA 20.5. Let $0 < \lambda < 1$, $a_1 \in L_{loc}(\mathbf{R}_+)$, $a_2, \alpha \in V(\mathbf{R}_+)$,

$$a_1(s) = s^\sigma(1 + o(1)), \qquad 0 \leq \sigma \leq 1, \tag{20.59}$$

and let the conditions (20.25) with $a_{20} < 0$ and $\alpha_0 = 1$ hold. Then every nonoscillatory proper solution of the equation (20.23) satisfies (20.52).

PROOF. Let $v\colon [s_0, +\infty[\to \mathbf{R}$ be a positive solution of the equation (20.23). As in the proof of Lemma 20.2, we can show that $v(s)$ tends to a finite limit c as $s \to +\infty$. According to (20.23), c is a root of the equation $a_{20}c + c^\lambda = 0$.

It remains to prove that $c \neq 0$. If $c = 0$, then $v'(s) < 0$ for $s \geq s_1$, where $s_1 \geq s_0$ is a constant, and we may assume that

$$v''(s) \geq \frac{1}{2}|v(s)|^\lambda \qquad \text{for } s \geq s_1.$$

But by Theorem 17.4 this is impossible. Hence $c \neq 0$. $\square$

LEMMA 20.6. Let $\lambda > 1$, $a_1 \in \tilde{C}_{loc}(\mathbf{R}_+)$, $a_2, \alpha \in L_{loc}(\mathbf{R}_+)$, and let the conditions (20.24) and (20.25) with $a_{10} < 0$, $a_{20} < 0$ and $\alpha_0 = 1$ hold. If, in addition, the equation (20.28) does not have an oscillatory solution, then for every nonoscillatory proper solution of the equation (20.23) we have either (20.52) or (20.44), where v_0 is a nonzero solution of (20.28) satisfying (20.45).

PROOF. Suppose $v\colon [s_0, +\infty[\to \,]0, +\infty[$ is a solution of the equation (20.23). By (20.25), whatever $\delta \in \,]0, |a_{20}|[$, there exists $s_\delta \geq s_0$ such that

$$(|a_{20}| - \delta)\alpha(s) \leq |a_2(s)| \leq (|a_{20}| + \delta)\alpha(s) \qquad \text{for } s \geq s_\delta.$$

Thus, according to (20.23) each extremum of the function v lying above the line $v = (|a_{20}| + \delta)^{1/(\lambda-1)}$ should be a minimum, while each extremum of v lying below the line $v = (|a_{20}| - \delta)^{1/(\lambda-1)}$ should be a maximum. We may assume that for $s \geq s_\delta$ the solution v does not cross these lines. Hence, on $[s_\delta, +\infty[$ one of the following inequalities holds: either

$$v(s) \geq (|a_{20}| + \delta)^{1/(\lambda-1)}, \tag{20.60}$$

$$(|a_{20}| - \delta)^{1/(\lambda-1)} \leq v(s) \leq (|a_{20}| + \delta)^{1/(\lambda-1)}, \tag{20.61}$$

or

$$0 < v(s) \leq (|a_{20}| - \delta)^{1/(\lambda-1)}. \tag{20.62}$$

First we will show that (20.60) is impossible. Indeed, otherwise, by the monotonicity of v, there would exist the finite or infinite limit $\lim_{s \to +\infty} v(s)$. Suppose $v(s) \to +\infty$ as $s \to +\infty$. We may assume that

$$2a_{10}s < a_1(s) < 0, \qquad \frac{1}{2} < \alpha(s) < \frac{3}{2} \qquad \text{for } s \geq s_\delta.$$

Consequently, (20.23) yields

$$v''(s) \le \frac{3}{2}v^\lambda(s), \qquad v''(s) \ge 2a_{10}sv'(s) + \frac{1}{2}v^\lambda(s) \qquad \text{for } s \ge s_\delta.$$

Multiplying both sides of the first inequality by $v'(s)$ and of the last inequality by $s^{-1}v^{-\lambda}(s)$ and integrating over $[s_\delta, s]$, we obtain

$$|v'(s)|^2 \le |v'(s_\delta)|^2 + \frac{2}{\lambda+1}|v(s)|^{\lambda+1}.$$

and

$$\frac{v'(s)}{s|v(s)|^\lambda} + \lambda \int_{s_\delta}^s \frac{|v'(x)|^2}{x|v(x)|^{\lambda+1}}\,dx \ge \frac{1}{2}\ln\frac{s}{s_\delta} + \frac{2a_{10} - s_\delta^{-2}}{\lambda-1}|v(s_\delta)|^{1-\lambda}.$$

The first of these relations implies

$$v'(s) \ge c_1|v(s)|^{(\lambda+1)/2} \qquad \text{for } s \ge s_\delta,$$

with $c_1^2 = 3/(\lambda+1) + |v'(s_\delta)|^2|v(s_\delta)|^{-\lambda-1}$. Thus, the second relation gives the contradiction

$$\int_{s_\delta}^{+\infty} v'(x)|v(x)|^{-(\lambda+1)/2}\,dx = +\infty.$$

Furthermore, let $v(s) \to c \in \,](|a_{20}| + \delta)^{1/(\lambda-1)}, +\infty[$ as $s \to +\infty$. Then, in view of (20.23),

$$v''(s) + c_0 sv'(s) \ge \delta_1 \qquad \text{for } s \ge s_*,$$

where $\delta_1 > 0, s_*$ is a sufficiently large constant, $c_0 = 0$ if v is nonincreasing and $c_0 = -2a_{10}$ if v is nondecreasing. Multiplying both sides of this inequality by s^{-1} and integrating over $[s_*, s]$, we get

$$s^{-1}v'(s) + (s^{-2} + c_0)v(s) + 2\int_{s_*}^s x^{-3}v(x)\,dx \ge \delta_1\ln\frac{s}{s_*} + s_*^{-1}v'(s_*) \qquad \text{for } s \ge s_*,$$

which contradicts the boundedness of v.

Therefore, (20.60) is impossible.

If for any sufficiently small $\delta > 0$, (20.61) holds on $[s_\delta, +\infty[$, then $v(s) = |a_{20}|^{1/(\lambda-1)} + o(1)$.

We claim that if for at least one $\delta \in \,]0, |a_{20}|[$ the inequality (20.62) holds, then $v(s) = o(1)$.

Suppose $v(s) \to c \in \,]0, |a_{20}|^{1/(\lambda-1)}[$ as $s \to +\infty$. According to (20.23),

$$v''(s) + c_0 s^{-1}v'(s) \le -\delta_2 < 0 \qquad \text{for } s \ge s_*,$$

where s_* is a sufficiently large constant, $c_0 = 0$ if v is nondecreasing and $c_0 = -2a_{10}$ if v is nonincreasing. This yields

$$s^{-1}v'(s) + (s^{-2} + c_0)v(s) \le -\delta_2\ln\frac{s}{s_*} + s_*^{-1}v'(s_*) + (s_*^{-2} + c_0)v(s_*) \qquad \text{for } s \ge s_*.$$

Hence $v'(s) \to -\infty$ as $s \to +\infty$, which contradicts (20.62). So $v(s) = o(1)$.

It remains to show that every solution of the equation (20.23) tending to zero is of the type (20.44). This can be done by the argument we have used to prove that every solution of (20.23) tending to infinity satisfies (20.27) (see the proof of Lemma 20.4). $\square$

LEMMA 20.7. Let $\lambda > 1$, $a_1, a_2, \alpha \in L_{\text{loc}}(\mathbf{R}_+)$, and let the conditions (20.25) and (20.59) with $a_{20} < 0$ and $\alpha_0 = 1$ hold. Then every nonoscillatory proper solution of the equation (20.23) satisfies (20.52).

PROOF. Let $v: [s_0, +\infty[\to]0, +\infty[$ be a solution of the equation (20.23). As above, we consider three cases: (20.60), (20.61) and (20.62).

In the first case, if $v(s) \to +\infty$ as $s \to +\infty$, then

$$v''(s) \geq \delta_1 |v(s)|^\lambda \qquad \text{for } s \geq s_*,$$

where s_* is a sufficiently large constant and $\delta_1 > 0$. But by Theorem 17.5 this is impossible. Moreover, if $v(s) \to c \in](|a_{20}| + \delta)^{1/(\lambda-1)}, +\infty[$, we also obtain a contradiction (see the proof of Lemma 20.6). So (20.60) cannot occur.

Assuming that the inequality (20.62) is true for a sufficiently small $\delta > 0$, as in the proof of Lemma 20.6 we get $v(s) = o(1)$. Then we can find s_* so that $v'(s) < 0$, $v''(s) < 0$ for $s \geq s_*$. Hence $v'(s) \leq v'(s_*)$ and $v(s) \leq v'(s_*)(s - s_*) + v(s_*)$ for $s \geq s_*$, which contradicts (20.62).

Therefore, whatever $\delta \in]0, |a_{20}|[$, the inequality (20.61) holds for $s \geq s_\delta$, where $s_\delta \geq s_0$ is a certain constant. This yields $v(s) = |a_{20}|^{1/(\lambda-1)} + o(1)$. $\square$

20.3. Asymptotic representations for nonoscillatory solutions (the case of a negative coefficient). Corollary 8.2 implies that if

$$\int_0^{+\infty} t^\lambda |p(t)|\, dt < +\infty, \tag{20.63}$$

then for any $c \neq 0$ the equation (20.1) has a solution u such that $u(t) = ct(1 + o(1))$; moreover, if

$$\int_0^{+\infty} t|p(t)|\, dt < +\infty, \tag{20.64}$$

then for any $c \neq 0$ this equation possesses a solution of the type $u(t) = c + o(1)$.

THEOREM 20.6. Let $0 < \lambda < 1$, and let p be negative and satisfy (20.64). Suppose, in addition, that $p(t) = -p_0(t) + p_1(t)$, where the functions $p_0, p_1: \mathbf{R}_+ \to \mathbf{R}_+$ are nondecreasing and

$$\int_0^{+\infty} \frac{dp_0(t)}{|p(t)|} < +\infty. \tag{20.65}$$

Then every proper solution of the equation (20.1) is of the form

$$u(t) = c_1(1 + o(1)) + c_2 t(1 + o(1)), \qquad |c_1| + |c_2| \neq 0. \tag{20.66}$$

PROOF. By Theorem 17.1, every maximally continued solution u of the equation (20.1) is proper. Besides, according to (20.65),

$$\frac{|u'(t)|^2}{|p(t)|} + \frac{2}{\lambda+1}|u(t)|^{\lambda+1} \geq c_0 > 0 \qquad \text{for } t \geq 0.$$

So if the equation (20.1) has an oscillatory solution u_1, we may assume that

$$\lim_{t \to +\infty} \sup u_1(t) = \alpha_1 > 0.$$

Suppose $c \in]0, \alpha_1[$. In view of (20.64), the equation (20.1) possesses a solution u_2 for which $u_2(t) = c + o(1)$. Clearly, there exist t_1 and t_2, $0 < t_1 < t_2 < +\infty$, such that

$$u_1(t_1) = u_2(t_1), \qquad u_1(t_2) = u_2(t_2),$$

$$u_1(t) > u_2(t) > 0 \qquad \text{for } t \in]t_1, t_2[.$$

However, applying the equality

$$(u_1(t)u_2'(t) - u_1'(t)u_2(t))'$$

$$= p(t)u_1^\lambda(t)u_2^\lambda(t)(u_1^{1-\lambda}(t) - u_2^{1-\lambda}(t)) < 0 \qquad \text{for } t \in]t_1, t_2[,$$

we obtain

$$0 \le u_1(t_2)(u_2'(t_2) - u_1'(t_2)) = u_1(t_2)u_2'(t_2) - u_2(t_2)u_1'(t_2)$$

$$< u_1(t_1)u_2'(t_1) - u_2(t_1)u_1'(t_1) = u_1(t_1)(u_2'(t_1) - u_1'(t_1)) \le 0.$$

This contradiction shows that (20.1) does not have an oscillatory solution.

Let u be a proper solution of the equation (20.1). As already proved, we can choose $t_0 \ge 0$ so that $u(t) \ne 0$ for $t \ge t_0$ and, since p is negative,

$$u'(t)u(t) > 0, \qquad u''(t)u(t) < 0 \qquad \text{for } t \ge t_0.$$

This yields $|u(t)| \ge c > 0$ for $t \ge t_0$. Hence, on $[t_0, +\infty[$ we can represent u as a solution of the linear equation (18.15), where

$$0 < -p^*(t) = -p(t)|u(t)|^{\lambda-1} \le c^{\lambda-1}|p(t)| \qquad \text{for } t \ge t_0.$$

By (20.64),

$$\int_{t_0}^{+\infty} t|p^*(t)| \, dt < +\infty. \tag{20.67}$$

Therefore Corollary 6.4 implies that u satisfies (20.66). $\square$

THEOREM 20.7. Let $\lambda > 1$. If p is negative, $p(t) = -p_0(t) + p_1(t)$, where the functions $p_0, p_1 \colon \mathbf{R}_+ \to \mathbf{R}_+$ are nondecreasing, and the conditions (20.63) and (20.65) hold, then every proper solution of the equation (20.1) is of the form (20.66).

PROOF. According to Theorem 17.1, every maximally continued solution u of the equation (20.1) is proper. Besides, (20.65) and Corollary 17.2 yield

$$|u'(t)|^2 + \frac{2}{\lambda+1}|p(t)||u(t)|^{\lambda+1} \le c_0 < +\infty \qquad \text{for } t \ge 0.$$

Thus $|u'(t)| \le c$, $|u(t)| \le ct$ for $t \ge t_0 > 0$. Clearly, u is a solution of the linear equation (18.15), where

$$0 < -p^*(t) = -p(t)|u(t)|^{\lambda-1} \le c^{\lambda-1}t^{\lambda-1}|p(t)| \qquad \text{for } t \ge t_0.$$

From (20.63) we easily get (20.67), and by Corollary 6.4 u has the representation (20.66). $\square$

REMARK 20.2. In Theorems 20.6 and 20.7 the condition (20.65) is essential. Indeed, it follows from the proof of Theorem 17.7 that there exists a function $q \in C([0,1[) \cap V_{loc}([0,1[)$ such that $\lim_{s \to 1} q(s) = -1$ and the equation

$$v'' = q(s)|v|^\lambda \operatorname{sign} v$$

has an oscillatory solution $v: [0,1[\to \mathbf{R}$. Set

$$t = \frac{1}{1-s}, \qquad v(s) = t^{-1}u(t).$$

Then $u: [1, +\infty[\to \mathbf{R}$ is an oscillatory proper solution of the equation (20.1), where $p(t) = t^{-\lambda-3}q(1 - (1/t))$. It is obvious that

$$\int_1^{+\infty} t|p(t)|\,dt < +\infty, \qquad \int_1^{+\infty} t^\lambda |p(t)|\,dt < +\infty,$$

but (20.1) has both an oscillatory proper solution and nonoscillatory solutions.

PROBLEM 20.2. If $\lambda > 1$, p is negative, $p(t) = -p_0(t) + p_1(t)$, where the functions $p_0, p_1: \mathbf{R}_+ \to \mathbf{R}_+$ are nondecreasing, and

$$\int_0^{+\infty} t^\lambda |p(t)| \exp\left(\frac{\lambda-1}{2}\int_0^{+\infty} \frac{dp_0(\tau)}{|p(\tau)|}\right) dt < +\infty,$$

then every proper solution of the equation (20.1) is of the form (20.66).

PROBLEM 20.3*. Is a similar assertion valid in case $0 < \lambda < 1$?

Below we assume that

$$p(t) = -p_0(t) + p_1(t), \qquad p_0 \in \tilde{C}^1_{loc}(\mathbf{R}_+), \qquad p_1 \in L_{loc}(\mathbf{R}_+),$$

$$(20.68)$$

$$p_0(t) > 0 \quad \text{for } t \geq 0, \qquad \lim_{t \to +\infty} \frac{p_1(t)}{p_0(t)} = 0.$$

Moreover, in Theorems 20.8–20.15 and 20.21–20.28 we use the following notation:

$$q(t) = [p_0(t)]^{-1/(\lambda+3)},$$

$$q_1(t) = \frac{2(\lambda+1)}{(\lambda-1)^2} + q''(t)[q(t)]^3 \left(\int_t^{+\infty} [q(\tau)]^{-2}\,d\tau\right)^2,$$

$$q_2(t) = q_1(t)\left|\ln \int_t^{+\infty} [q(\tau)]^{-2}\,d\tau\right| - \frac{\lambda+3}{(\lambda-1)^2},$$

$$q_3(t) = \frac{2(\lambda+1)}{(\lambda-1)^2} + q''(t)[q(t)]^3 \left(\int_0^t [q(\tau)]^{-2}\,d\tau\right)^2,$$

$$q_4(t) = q_3(t)\ln \int_0^t [q(\tau)]^{-2}\,d\tau + \frac{\lambda+3}{(\lambda-1)^2}.$$

THEOREM 20.8. If $0 < \lambda < 1$, the conditions (20.68) hold and

$$\int_0^{+\infty} [q(t)]^{-2}\, dt < +\infty, \qquad \lim_{t\to+\infty} q_1(t) = q_{10} < 0,$$

then every nonoscillatory proper solution of the equation (20.1) is either of the form

$$u(t) = (c_0 + o(1))t, \qquad c_0 \neq 0, \tag{20.69}$$

or of the form

$$u(t) = \pm(|q_{10}|^{1/(\lambda-1)} + o(1))q(t) \left(\int_t^{+\infty} [q(\tau)]^{-2}\, d\tau \right)^{2/(1-\lambda)}.$$

PROOF. Applying the transformation

$$s = \left| \ln \int_t^{+\infty} [q(\tau)]^{-2}\, d\tau \right|,$$

$$u(t) = q(t) \left(\int_t^{+\infty} [q(\tau)]^{-2}\, d\tau \right)^{2/(1-\lambda)} v(s), \tag{20.70}$$

we reduce the equation (20.1) to (20.23) with

$$a_1(s) = \frac{\lambda+3}{1-\lambda}, \qquad a_2(s) = -q_1(t), \qquad \alpha(s) = -1 + \frac{p_1(t)}{p_0(t)}.$$

Thus, the hypotheses of Lemma 20.1 are satisfied. It is readily verified that the functions

$$v_1(s) = tv_2(s), \qquad v_2(s) = [q(t)]^{-1} \left(\int_t^{+\infty} [q(\tau)]^{-2}\, d\tau \right)^{2/(\lambda-1)} \tag{20.71}$$

are linearly independent solutions of the equation (20.28). Since $a_{20} = -q_{10} > 0$, we have $v_1(s) \to +\infty$ as $s \to +\infty$, i.e. $v_\infty = c_0 v_1$, $c_0 \neq 0$. Therefore, Theorem 20.8 follows from Lemma 20.1. □

Let $q_1(t) = o(1)$ and $q_2(t) = q_{20} + o(1)$, $q_{20} < 0$. Setting

$$s = 2 \left| \ln \int_t^{+\infty} [q(\tau)]^{-2}\, d\tau \right|^{1/2},$$

$$u(t) = \left(\frac{s}{2} \right)^{2/(1-\lambda)} q(t) \left(\int_t^{+\infty} [q(\tau)]^{-2}\, d\tau \right)^{2/(1-\lambda)} v(s),$$

for v we obtain the equation (20.23) with

$$a_1(s) = \frac{\lambda+3}{2(1-\lambda)} \left(s - \frac{2}{s} \right), \qquad a_2(s) = -q_2(t) - \frac{4\lambda}{(\lambda-1)^2} s^{-2},$$

$$\alpha(s) = -1 + \frac{p_1(t)}{p_0(t)}.$$

Note that the functions

$$v_1(s) = tv_2(s), \qquad v_2(s) = \left(\frac{s}{2} \right)^{2/(\lambda-1)} [q(t)]^{-1} \left(\int_t^{+\infty} [q(\tau)]^{-2}\, d\tau \right)^{2/(\lambda-1)}$$

are linearly independent solutions of the equation (20.28). So Lemma 20.1 implies

THEOREM 20.9. If $0 < \lambda < 1$, the conditions (20.68) hold and

$$\int_0^{+\infty} [q(t)]^{-2}\,dt < +\infty, \qquad \lim_{t\to+\infty} q_2(t) = q_{20} < 0,$$

then every nonoscillatory proper solution of the equation (20.1) is either of the form (20.69) or of the form

$$u(t) = \pm(|q_{20}|^{1/(\lambda-1)} + o(1)) \left| \ln \int_t^{+\infty} [q(\tau)]^{-2}\,d\tau \right|^{1/(1-\lambda)}$$

$$\times q(t) \left(\int_t^{+\infty} [q(\tau)]^{-2}\,d\tau \right)^{2/(1-\lambda)}.$$

Let $q_1(t) = q_{10} + o(1)$, $q_{10} > 0$ and $0 < \lambda < 1$. We claim that

$$\int_0^{+\infty} tp_0(t)\,dt < +\infty. \qquad (20.72)$$

Clearly, $b_0 = -q_{10} + (\lambda+3)^2/4(\lambda-1)^2 \geq 0$. First we consider the case $b_0 > 0$. Since $2(1-\lambda)\sqrt{b_0} > \lambda+3$, we can choose $\epsilon \in {]}0, b_0[$ so small that $(\lambda+3)\mu_\epsilon + 2 < 0$ where $4(\sqrt{b_0}-\epsilon-2\sqrt{b_0+\epsilon})\mu_\epsilon = 1 - 2\sqrt{b_0+\epsilon} + 4\sqrt{b_0}-\epsilon$. By (20.41) and (20.70),

$$[q(t)]^{-1} \left(\int_t^{+\infty} [q(\tau)]^{-2}\,d\tau \right)^{-(1/2)+\sqrt{b_0-\epsilon}-2\sqrt{b_0+\epsilon}} \geq c_1,$$

$$[q(t)]^{-1} \leq c_2 \left(\int_t^{+\infty} [q(\tau)]^{-2}\,d\tau \right)^{(1/2)-\sqrt{b_0+\epsilon}+2\sqrt{b_0-\epsilon}}$$

for $t \geq t_*$, where c_1 and c_2 are positive constants and t_* is a sufficiently large number. Integrating the first inequality we get

$$\int_t^{+\infty} [q(\tau)]^{-2}\,d\tau \leq c_3 t^{1/(2\sqrt{b_0-\epsilon}-4\sqrt{b_0+\epsilon})} \qquad \text{for } t \geq t_*.$$

Hence the second inequality yields $[q(t)]^{-1} \leq c_4 t^{\mu_\epsilon}$ for $t \geq t_*$, i.e. $p_0(t) \leq c_4^{\lambda+3} t^{(\lambda+3)\mu_\epsilon}$ for $t \geq t_*$. Consequently, (20.72) is true.

Furthermore, let $b_0 = 0$. Then for t sufficiently large we have $q''(t) > 0$ and $q'(t) > 0$. So $q(t) \geq ct$, $c > 0$, and $p_0(t) \leq c^{-\lambda-3} t^{-\lambda-3}$. Thus, here (20.72) is valid as well.

Therefore, if $0 < \lambda < 1$ and $q_{10} > 0$, then (20.64) is fulfilled. But this case has already been considered in Theorem 20.6.

THEOREM 20.10. If $0 < \lambda < 1$, the conditions (20.68) hold and

$$\int_0^{+\infty} [q(t)]^{-2}\,dt = +\infty, \qquad \lim_{t\to+\infty} q_3(t) = q_{30} < 0,$$

then for every nonoscillatory proper solution u of the equation (20.1) we have either (20.69) or

$$u(t) = \pm(|q_{30}|^{1/(\lambda-1)} + o(1))q(t) \left(\int_0^t [q(\tau)]^{-2}\,d\tau \right)^{2/(1-\lambda)}.$$

PROOF. Set

$$s = \ln \int_0^t [q(\tau)]^{-2}\,d\tau, \qquad u(t) = q(t) \left(\int_0^t [q(\tau)]^{-2}\,d\tau \right)^{2/(1-\lambda)} v(s).$$

Then for v we obtain the equation (20.23) with

$$a_1(s) = \frac{\lambda + 3}{\lambda - 1}, \qquad a_2(s) = -q_3(t), \qquad \alpha(s) = -1 + \frac{p_1(t)}{p_0(t)}.$$

So the functions

$$v_1(s) = tv_2(s), \qquad v_2(s) = [q(t)]^{-1} \left(\int_0^t [q(\tau)]^{-2}\, d\tau \right)^{2/(\lambda-1)}$$

are linearly independent solutions of the equation (20.28). If v is a nonoscillatory proper solution of the equation (20.23), then by Lemma 20.1 either

$$|v(s)| = |q_{30}|^{1/(\lambda-1)} + o(1)$$

or

$$v(s) = (c_0 + o(1))[q(t)]^{-1} \left(\int_0^t [q(\tau)]^{-2}\, d\tau \right)^{2/(\lambda-1)},$$

where $c_0 \neq 0$. This yields the assertion of the theorem. $\square$

Let $q_3(t) = o(1)$ and $\lim_{t\to+\infty} q_4(t) = q_{40} < 0$. By means of the transformation

$$s = 2 \left(\ln \int_0^t [q(\tau)]^{-2}\, d\tau \right)^{1/2},$$

$$u(t) = \left(\frac{s}{2} \right)^{2/(1-\lambda)} q(t) \left(\int_0^t [q(\tau)]^{-2}\, d\tau \right)^{2/(1-\lambda)} v(s),$$

the equation (20.1) can be reduced to (20.23), where

$$a_1(s) = \frac{\lambda + 3}{2(1 - \lambda)} \left(s + \frac{2}{s} \right), \qquad a_2(s) = -q_4(t) - \frac{4\lambda}{(\lambda-1)^2} s^{-2},$$

$$\alpha(s) = -1 + \frac{p_1(t)}{p_0(t)}.$$

So Lemma 20.1 immediately implies

THEOREM 20.11. If $0 < \lambda < 1$, the conditions (20.68) hold and

$$\int_0^{+\infty} [q(t)]^{-2}\, dt = +\infty, \qquad \lim_{t\to+\infty} q_4(t) = q_{40} < 0,$$

then for every nonoscillatory proper solution u of the equation (20.1) we have either (20.69) or

$$u(t) = \pm(|q_{40}|^{(1/(\lambda-1)} + o(1)) \left(\ln \int_0^t [q(\tau)]^{-2}\, d\tau \right)^{1/(1-\lambda)}$$

$$\times q(t) \left(\int_0^t [q(\tau)]^{-2}\, d\tau \right)^{2/(1-\lambda)}.$$

Furthermore, if $q_3(t) = q_{30} + o(1)$, $q_{30} > 0$, and $0 < \lambda < 1$, then, as above, we can prove that $t^{\lambda+1-\epsilon} p(t) \leq c < 0$ for sufficiently large t. Hence, by Theorem 18.1 the equation (20.1) does not have nonoscillatory proper solutions in this case.

It follows from Corollary 20.2, stated below, that under the hypotheses of Theorems 20.10 and 20.11 the equation (20.1) may also have oscillatory solutions.

PROBLEM 20.4*. Is every proper solution of the equation (20.1) nonoscillatory under the hypotheses of Theorem 20.8 or 20.9?

The proofs of the following propositions are similar to those of Theorems 20.8–20.11, but instead of Lemma 20.1 one should apply Lemmas 20.2 and 20.3.

THEOREM 20.12. Let $\lambda > 1$,

$$p(t) = -p_0(t) + p_1(t), \qquad p_0 \in \tilde{C}^2_{loc}(\mathbf{R}_+), \qquad p_0(t) > 0 \qquad \text{for } t \geq 0,$$
$$(20.73)$$

$$\frac{p_1}{p_0} \in V(\mathbf{R}_+), \qquad \lim_{t \to +\infty} \frac{p_1(t)}{p_0(t)} = 0,$$

and

$$\int_0^{+\infty} [q(t)]^{-2} dt < +\infty, \qquad q_1 \in V(\mathbf{R}_+), \qquad \lim_{t \to +\infty} q_1(t) = q_{10} < 0.$$

Then every proper solution of the equation (20.1) is either of the form

$$u(t) = c_0 + o(1), \qquad c_0 \neq 0, \tag{20.74}$$

or of the form

$$u(t) = \pm(|q_{10}|^{1/(\lambda-1)} + o(1))q(t) \left(\int_t^{+\infty} [q(\tau)]^{-2} d\tau \right)^{2/(1-\lambda)}.$$

In case $q_1(t) = o(1)$ we obtain

THEOREM 20.13. If $\lambda > 1$, the conditions (20.73) hold, and

$$\int_0^{+\infty} [q(t)]^{-2} dt < +\infty, \qquad q_2 \in V(\mathbf{R}_+), \qquad \lim_{t \to +\infty} q_2(t) = q_{20} < 0,$$

then every proper solution of the equation (20.1) is either of the form (20.74) or of the form

$$u(t) = \pm(|q_{20}|^{1/(\lambda-1)} + o(1)) \left| \ln \int_t^{+\infty} [q(\tau)]^{-2} d\tau \right|^{1/(1-\lambda)}$$
$$\times q(t) \left(\int_t^{+\infty} [q(\tau)]^{-2} d\tau \right)^{2/(1-\lambda)}.$$

If $q_1(t) = q_{10} + o(1)$, $q_{10} > 0$, and $\lambda > 1$, then we can easily show that (20.63) is fulfilled. So this case is already covered by Theorem 20.7.

THEOREM 20.14. If $\lambda > 1$, the conditions (20.73) hold, and

$$\int_0^{+\infty} [q(t)]^{-2} dt = +\infty, \qquad q_3 \in V(\mathbf{R}_+), \qquad \lim_{t \to +\infty} q_3(t) = q_{30} < 0,$$

then for every nonoscillatory proper solution of the equation (20.1) we have either (20.74) or

$$u(t) = \pm(|q_{30}|^{1/(\lambda-1)} + o(1))q(t) \left(\int_0^t [q(\tau)]^{-2} d\tau \right)^{2/(1-\lambda)}.$$

For $q_3(t) = o(1)$ the following assertion is true.

THEOREM 20.15. If $\lambda > 1$, the conditions (20.73) hold, and

$$\int_0^{+\infty} [q(t)]^{-2}\, dt = +\infty, \qquad q_4 \in V(\mathbf{R}_+), \qquad \lim_{t \to +\infty} q_4(t) = q_{40} < 0,$$

then for every nonoscillatory proper solution of the equation (20.1) we have either (20.74) or

$$u(t) = \pm(|q_{40}|^{1/(\lambda-1)} + o(1)) \left(\ln \int_0^t [q(\tau)]^{-2}\, d\tau \right)^{1/(1-\lambda)}$$

$$\times q(t) \left(\int_0^t [q(\tau)]^{-2}\, d\tau \right)^{2/(1-\lambda)}.$$

If $q_3(t) = q_{30} + o(1)$, $q_{30} > 0$, and $\lambda > 1$, then by the argument used above we obtain $t^2 p(t) \geq c > 0$ for sufficiently large t. Hence, according to Theorem 18.2 the equation (20.1) does not have a nonoscillatory proper solution.

COROLLARY 20.2. Let $0 < \lambda < 1$ ($\lambda > 1$), $p(t) = -t^\sigma$ for $t \geq 1$, and let u be a proper solution of the equation (20.1). Then:

(a) if $\sigma < -2$ ($\sigma < -\lambda - 1$), we have (20.66);
(b) if $-2 < \sigma < -(\lambda+3)/2$ ($-\lambda - 1 < \sigma < -(\lambda+3)/2$), we have either (20.66) with $c_1 = 0$ ($c_2 = 0$) or

$$u(t) = \pm \left(\left[\frac{|(\sigma+2)(\sigma+\lambda+1)|}{(\lambda-1)^2} \right]^{1/(\lambda-1)} + o(1) \right) t^{(\sigma+2)/(1-\lambda)}; \qquad (20.75)$$

(c) if $-(\lambda+3)/2 < \sigma < -\lambda - 1$ ($-(\lambda+3)/2 < \sigma < -2$), we have either (20.66) with $c_1 = 0$ ($c_2 = 0$), the representation (20.75), or

$$u(t) = t^{-\sigma/(\lambda+3)}(c_0 + o(1)) w \left([|c_0|^{(\lambda-1)/2} + o(1)] \frac{\lambda+3}{2\sigma + \lambda + 3} t^{1+2\sigma/(\lambda+3)} \right),$$

where $c_0 \neq 0$ and w is the solution of the problem (20.15).

PROOF. The assertion (a) follows from Theorem 20.6 (from Theorem 20.7), the assertion (b) from Theorem 20.8 and Remark 20.1 (from Theorem 20.12), and the assertion (c) from Theorems 20.4 and 20.10 (from Theorems 20.4 and 20.14). □
Theorem 20.9 (Theorem 20.13) yields

COROLLARY 20.3. If $0 < \lambda < 1$ ($\lambda > 1$), $p(t) = -t^\sigma$ for $t \geq 1$, and $\sigma = -2$ ($\sigma = -\lambda - 1$), then for every nonoscillatory (every proper) solution u of the equation (20.1) we have either (20.66) with $c_1 = 0$ ($c_2 = 0$) or

$$u(t) = \pm(|1 - \lambda|^{1/(1-\lambda)} + o(1))(\ln t)^{1/(1-\lambda)}.$$

PROBLEM 20.5. If $0 < \lambda < 1$ and $p(t) = -t^{-(\lambda+3)/2}$ for $t \geq 1$, then for every nonoscillatory solution u of the equation (20.1) we have either

$$u(t) = t(c_0 + o(1)), \qquad c_0 \neq 0,$$

or

$$u(t) = \pm t^{1/2}(4^{1/(1-\lambda)} + o(1)).$$

PROBLEM 20.6*. Derive asymptotic formulas for nonoscillatory solutions of the equation (20.1), where $\lambda > 1$ and $p(t) = -t^{-(\lambda+3)/2}$ on $[1, +\infty[$.

Theorem 18.4 implies that if $p(t) = -t^{-(\lambda+3)/2}$ for $t \geq 1$, then the equation (20.1) has oscillatory solutions.

PROBLEM 20.7*. Derive asymptotic formulas for oscillatory solutions of the equation

$$u'' = -t^{-(\lambda+3)/2}|u|^\lambda \operatorname{sign} u \qquad (t \geq 1).$$

20.4. Asymptotic representations for nonoscillatory solutions (the case of a positive coefficient). Throughout this subsection we assume that the function p is nonnegative and distinct from zero on a set of positive measure in any neighborhood of $+\infty$. Without loss of generality we may study the solutions of the equation (20.1) starting from points in the first quadrant of the (t, u)-plane only, because any other solution of this equation can be transformed into such by the change of sign.

Fix $t_0 \geq 0$ and $\beta_0 \geq 0$. Let $\tilde{u}$ be the solution of the equation (20.1) under the conditions $u(t_0) = \beta_0$, $u(t) \geq 0$, $u'(t) \leq 0$ for $t \geq t_0$, and let $\gamma \in \mathbf{R}$ be arbitrary. Consider the Cauchy problem $u(t_0) = \beta_0$, $u'(t_0) = \gamma$ for the equation (20.1). This problem is uniquely solvable unless $0 < \lambda < 1$ and $\gamma = \tilde{u}'(t_0)$. In case of uniqueness we denote by u_γ the solution of this problem that is maximally continued to the right. For $0 < \lambda < 1$ and $\gamma = \tilde{u}'(t_0)$ we set $u_\gamma(t) = \tilde{u}(t)$.

THEOREM 20.16. If $0 < \lambda < 1$, then for every $t_0 \geq 0$ and $\beta_0 \geq 0$ there exists $\gamma_0 \in \mathbf{R}$ such that u_{γ_0} is nonnegative and nonincreasing. Moreover,

$$\lim_{t \to +\infty} u_\gamma(t) = \begin{cases} -\infty & \text{for } \gamma < \gamma_0 \\ +\infty & \text{for } \gamma > \gamma_0, \end{cases} \qquad (20.76)$$

and for sufficiently small positive β_0 the solution u_{γ_0} is nonoscillatory first kind singular.

To prove this theorem we need

LEMMA 20.8. If u and v are solutions of the equation (20.1) defined on $[t_0, t_1[$ and satisfying the conditions

$$u(t_0) \geq v(t_0), \qquad u'(t_0) > v'(t_0),$$

then for $t \in [t_0, t_1[$ we have

$$u(t) - v(t) \geq u(t_0) - v(t_0) + (u'(t_0) - v'(t_0))(t - t_0),$$

$$\qquad (20.77)$$

$$u'(t) - v'(t) \geq u'(t_0) - v'(t_0).$$

PROOF. Let t_2 be the least upper bound of the set of those $t \in]t_0, t_1[$ for which

$$u(t) > v(t), \qquad u'(t) > v'(t). \qquad (20.78)$$

Note that $t_2 = t_1$, because otherwise we would obtain $u'(t_2) = v'(t_2)$, which is impossible since

$$u'(t_2) - v'(t_2) = u'(t_0) - v'(t_0)$$
$$+ \int_{t_0}^{t_2} p(t)(|u(t)|^\lambda \operatorname{sign} u(t) - |v(t)|^\lambda \operatorname{sign} v(t))\, dt > 0.$$

Hence, the inequalities (20.78) hold in the interval $]t_0, t_1[$. So $u''(t) > v''(t)$ for $t \in$ $]t_0, t_1[$. This immediately implies the validity of (20.77). $\square$

PROOF OF THEOREM 20.16. The existence of γ_0 follows from Theorem 13.1 and the relation (20.76) can be readily derived from Lemma 20.8.

Let $t_1 > t_0$. By Theorem 17.3 there exists a solution u such that

$$u(t) > 0, \quad u'(t) < 0 \quad \text{for } t \in [t_0, t_1[, \quad u(t) \equiv 0 \quad \text{on } [t_1, +\infty[.$$

We will show that if $\beta_0 \in \,]0, u(t_0)[$, then u_{γ_0} is a nonoscillatory first kind singular solution. Assume the contrary. Then we can find a point $t_2 \in \,]t_0, t_1[$ for which $u_{\gamma_0}(t_2) = u(t_2)$, $u'_{\gamma_0}(t_2) > u'(t_2)$. According to Lemma 20.8,

$$u_{\gamma_0}(t) \geq u_{\gamma_0}(t_2) - u(t_2) + (u'_{\gamma_0}(t_2) - u'(t_2))(t - t_2) \quad \text{for } t \geq t_2.$$

Thus $u_{\gamma_0}(t) \to +\infty$ as $t \to +\infty$. This contradiction shows that for $\beta_0 > 0$ sufficiently small, the solution u_{γ_0} is first kind singular. $\square$

THEOREM 20.17. If $0 < \lambda < 1$ and

$$\lim_{t \to +\infty} \inf t^2 p(t) > 0,$$

then u_{γ_0} is a nonoscillatory first kind singular solution; if, in addition,

$$\int_0^{+\infty} t^\lambda p(t)\, dt < +\infty, \tag{20.79}$$

then for $\gamma \neq \gamma_0$ we have

$$u_\gamma(t) = (c + o(1))t, \quad c \neq 0.$$

PROOF. The first part of the theorem is a consequence of Corollary 17.3 and the last part follows from Corollary 8.2 and (20.76). $\square$

THEOREM 20.18. If $\lambda > 1$, then for every $t_0 \geq 0$ and $\beta_0 \geq 0$ there exist constants $\underline{\gamma} \leq \gamma_0 \leq \bar{\gamma}$ such that for $\gamma > \bar{\gamma}$ the solution u_γ is positive and second kind singular, for $\gamma < \underline{\gamma}$ this solution is second kind singular, but becomes negative from a certain value of t onwards, and for all other γ it is proper. Furthermore, if $\beta_0 = 0$, then $u_{\gamma_0}(t) \equiv 0$, and if $\beta_0 > 0$, then u_{γ_0} is positive and monotonically decreases. If, in addition, $\underline{\gamma} < \gamma_0 < \bar{\gamma}$, then

$$\lim_{t \to +\infty} u_\gamma(t) = \begin{cases} -\infty & \text{for } \underline{\gamma} \leq \gamma < \gamma_0 \\ +\infty & \text{for } \gamma_0 < \gamma \leq \bar{\gamma}. \end{cases}$$

PROOF. Theorem 13.1 implies the existence of γ_0. Let $t_1 \in \,]t_0, t_0 + 1]$. By Theorem 17.3 we can find $\gamma_1 > 0$ so that the solution of the equation (20.1) under the conditions $u(t_1) = 0$, $u'(t_1) = \gamma_1$ $(u'(t_1) = -\gamma_1)$ is positive (negative) and second kind singular.

It follows from Lemma 20.8 that u_γ with $\gamma > \gamma_1$ is a positive second kind singular solution.

Let

$$\gamma < -\gamma_1 - c_0 - c_0^\lambda \int_{t_0}^{t_1} p(t)\, dt.$$

Assuming that $u(t) > 0$ for $t \in [t_0, t_1[$, we get

$$0 < u_\gamma(t_1) = c_0 + \gamma + \int_{t_0}^{t_1} (t_1 - t)p(t)|u_\gamma(t)|^\lambda\, dt \leq c_0 + \gamma + c_0^\lambda \int_{t_0}^{t_1} (t_1 - t)p(t)\, dt < 0.$$

Thus, u_γ vanishes for a certain $t_2 \in\,]t_0, t_1[$. Besides,

$$u_\gamma'(t_2) = \gamma + \int_{t_2}^{t_1} p(t)|u_\gamma(t)|^\lambda\, dt \leq \gamma + c_0^\lambda \int_{t_0}^{t_1} p(t)\, dt < -\gamma_1.$$

So, by Lemma 20.8 the solution u_γ is negative for $t > t_2$ and second kind singular.

Therefore we have proved that if $|\gamma|$ is sufficiently large, then u_γ is a nonoscillatory second kind singular solution. Denote by $\bar\gamma$ the greatest lower bound of the set of those $\gamma > \gamma_0$, and by $\underline\gamma$ the least upper bound of the set of those $\gamma < \gamma_0$, for which u_γ is singular.

We claim that $u_{\bar\gamma}$ is a proper solution. Assume the contrary. Then there exists $t_1 > t_0$ such that

$$\lim_{t \to t_1} u_{\bar\gamma}(t) = \lim_{t \to t_1} u_{\bar\gamma}'(t_1) = +\infty. \tag{20.80}$$

According to the definition of $\bar\gamma$ and Lemma 20.8, the solution u_γ with $\gamma \in\,]\gamma_0, \bar\gamma[$ is proper. Let u be a positive singular solution which starts from the point $(t_1, 0)$. Applying (20.80), we can easily show that if $\gamma \in\,]\gamma_0, \bar\gamma[$ and $\bar\gamma - \gamma$ is sufficiently small, then $u_\gamma(t_1) > 0$ and $u_\gamma'(t_1) > u'(t_1)$. Hence, Lemma 20.8 yields $u_\gamma(t) > u(t)$ for $t \geq t_1$, which is impossible, because u is a singular solution. Consequently, $u_{\bar\gamma}$ is proper.

We can similarly prove that $u_{\underline\gamma}$ is also proper.

Lemma 20.8 implies that if either $\gamma > \bar\gamma$ or $\gamma < \underline\gamma$, then u_γ is singular, while if $\underline\gamma \leq \gamma \leq \bar\gamma$, then u_γ lies between $u_{\bar\gamma}$ and $u_{\underline\gamma}$ and so is proper.

Let $\underline\gamma < \gamma_0 < \bar\gamma$. By Lemma 20.8 $u_\gamma(t) \geq (\gamma - \gamma_0)(t - t_0) \to +\infty$ as $t \to +\infty$ for $\gamma_0 < \gamma \leq \bar\gamma$ and $u_\gamma(t) \leq u_{\gamma_0}(t) + (\gamma - \gamma_0)(t - t_0) \to -\infty$ as $t \to +\infty$ for $\underline\gamma \leq \gamma < \gamma_0$. $\square$

THEOREM 20.19. If $\lambda > 1$ and (20.79) holds, then $\underline\gamma < \gamma_0 < \bar\gamma$.

PROOF. We have to show that the solution u_γ is proper whenever $|\gamma - \gamma_0|$ is sufficiently small. Let $t_1 > t_0$ be so large that

$$\frac{|u_{\gamma_0}(t_1)|}{t_1} + |u_{\gamma_0}'(t_1)| < \epsilon < 1$$

and

$$c = \epsilon^{1-\lambda} - (\lambda - 1) \int_{t_1}^{+\infty} t^\lambda p(t)\, dt > 0. \tag{20.81}$$

Choose γ close to γ_0 such that the solution u_γ exists on an interval $[t_0, t_2[$ with $t_2 > t_1$ and satisfies the inequality

$$\frac{|u_\gamma(t_1)|}{t_1} + |u_\gamma'(t_1)| \leq \epsilon. \tag{20.82}$$

By (20.82), the equality

$$u_\gamma(t) = u_\gamma(t_1) + u'_\gamma(t_1)(t - t_1) + \int_{t_1}^t (t - \tau)p(\tau)|u_\gamma(\tau)|^\lambda \operatorname{sign} u_\gamma(\tau)\, d\tau$$

implies

$$\frac{|u_\gamma(t)|}{t} \leq \epsilon + \int_{t_1}^t p(\tau)|u_\gamma(\tau)|^\lambda\, d\tau \qquad \text{for } t \in [t_1, t_2[. \tag{20.83}$$

This yields

$$p(t)|u_\gamma(t)|^\lambda \left[\epsilon + \int_{t_1}^t p(\tau)|u_\gamma(\tau)|^\lambda\, d\tau \right]^{-\lambda} \leq t^\lambda p(t) \qquad \text{for } t \in [t_1, t_2[.$$

Integrating this inequality from t_1 to t, we obtain

$$\epsilon^{1-\lambda} \leq \left[\epsilon + \int_{t_1}^t p(\tau)|u_\gamma(\tau)|^\lambda\, d\tau \right]^{1-\lambda} + (\lambda - 1)\int_{t_1}^t \tau^\lambda p(\tau)\, d\tau \qquad \text{for } t \in [t_1, t_2[.$$

Then, according to (20.81) and (20.83),

$$|u_\gamma(t)| \leq c^{1/(1-\lambda)} t \qquad \text{for } t \in [t_1, t_2[.$$

So $t_2 = +\infty$ and u_γ is a proper solution. $\square$

THEOREM 20.20. Let $\lambda > 1$ and

$$\lim_{t \to +\infty} \inf t^{\lambda+1} p(t) > 0. \tag{20.84}$$

Then $\gamma = \gamma_0 = \bar\gamma$. If, in addition,

$$\int_0^{+\infty} tp(t)\, dt < +\infty,$$

then

$$u_{\gamma_0}(t) = c_0 + o(1), \qquad c_0 > 0.$$

PROOF. If (20.84) holds, then by Corollary 17.4 every proper solution of the equation (20.1) is bounded. Hence $\gamma = \gamma_0 = \bar\gamma$.

The second part of the theorem immediately follows from Corollary 8.2. $\square$

THEOREM 20.21. If $0 < \lambda < 1$,

$$p(t) = p_0(t) + p_1(t), \qquad p_0 \in \tilde C^2_{\text{loc}}(\mathbf{R}_+), \qquad p_0(t) > 0 \qquad \text{for } t \geq 0, \tag{20.85}$$

$$\frac{p_1}{p_0} \in V(\mathbf{R}_+), \qquad \lim_{t \to +\infty} \frac{p_1(t)}{p_0(t)} = 0,$$

and

$$\int_0^{+\infty} [q(t)]^{-2}\, dt < +\infty, \qquad q_1 \in V(\mathbf{R}_+), \qquad \lim_{t \to +\infty} q_1(t) = q_{10} > 0,$$

then either u_{γ_0} is a nonoscillatory first kind singular solution, the representation

$$u_{\gamma_0}(t) = c_0 + o(1), \qquad c_0 > 0, \tag{20.86}$$

is valid, or

$$u_{\gamma_0}(t) = (q_{10}^{1/(\lambda-1)} + o(1))q(t)\left(\int_t^{+\infty} [q(\tau)]^{-2} d\tau\right)^{2/(1-\lambda)};$$

moreover,

$$u_\gamma(t) = \begin{cases} (c_1 + o(1))t, & c_1 > 0, \quad \text{for } \gamma > \gamma_0 \\ (-c_2 + o(1))t, & c_2 > 0, \quad \text{for } \gamma < \gamma_0. \end{cases} \tag{20.87}$$

PROOF. By means of the transformation (20.70) we can reduce the equation (20.1) to the equation (20.23), where

$$a_1(s) = \frac{\lambda+3}{1-\lambda}, \qquad a_2(s) = -q_1(t), \qquad \alpha(s) = 1 + \frac{p_1(t)}{p_0(t)}.$$

Note that the functions (20.71) are linearly independent solutions of the equation (20.28). So it remains to apply Lemma 20.4. $\square$

If $q_1(t) = o(1)$, but $q_2(t) = q_{20} + o(1)$ and $q_{20} > 0$, then the following assertion can be proved similarly to Theorem 20.9.

THEOREM 20.22. Let $0 < \lambda < 1$ and let the conditions (20.85) be fulfilled. Suppose, in addition, that

$$\int_0^{+\infty} [q(t)]^{-2} dt < +\infty, \qquad q_2 \in V(\mathbf{R}_+), \qquad \lim_{t \to +\infty} q_2(t) = q_{20} > 0.$$

Then either u_{γ_0} is a nonoscillatory first kind singular solution, (20.86) holds, or

$$u_{\gamma_0}(t) = (q_{20}^{1/(\lambda-1)} + o(1))\left|\ln \int_t^{+\infty} [q(\tau)]^{-2} d\tau\right|^{1/(1-\lambda)}$$

$$\times q(t)\left(\int_t^{+\infty} [q(\tau)]^{-2} d\tau\right)^{2/(1-\lambda)};$$

moreover, for the other solutions we have (20.87).

If $q_1(t) = q_{10} + o(1)$, $q_{10} < 0$, and $0 < \lambda < 1$, then, as above, we can show that for a certain $\epsilon > 0$ and sufficiently large t the inequalities $t^{\lambda+1+\epsilon}p(t) \le c_1$ and $t^{2-\epsilon}p(t) \ge c_2 > 0$ are true. But this case has been already considered in Theorem 20.17.

The following two theorems can be proved similarly to Theorems 20.10 and 20.11, provided one uses Lemma 20.5 instead of Lemma 20.1.

THEOREM 20.23. If $0 < \lambda < 1$, the conditions (20.85) are fulfilled, and

$$\int_0^{+\infty} [q(t)]^{-2} dt = +\infty, \qquad q_3 \in V(\mathbf{R}_+), \qquad \lim_{t \to +\infty} q_3(t) = q_{30} > 0,$$

then u_{γ_0} is a nonoscillatory first kind singular solution and

$$u_\gamma(t) = \pm(q_{30}^{1/(\lambda-1)} + o(1))q(t)\left(\int_0^t [q(\tau)]^{-2} d\tau\right)^{2/(1-\lambda)} \qquad \text{for } \gamma \ne \gamma_0.$$

THEOREM 20.24. If $0 < \lambda < 1$, the conditions (20.85) are fulfilled, and

$$\int_0^{+\infty} [q(t)]^{-2}\, dt = +\infty, \qquad q_4 \in V(\mathbf{R}_+), \qquad \lim_{t \to +\infty} q_4(t) = q_{40} > 0,$$

then u_{γ_0} is a nonoscillatory first kind singular solution and

$$u_\gamma(t) = \pm(q_{40}^{1/(\lambda-1)} + o(1))\left(\ln \int_0^t [q(\tau)]^{-2}\, d\tau\right)^{1/(1-\lambda)}$$

$$\times q(t)\left(\int_0^t [q(\tau)]^{-2}\, d\tau\right)^{2/(1-\lambda)} \qquad \text{for } \gamma \neq \gamma_0.$$

Clearly, under the hypotheses of Theorem 20.24, $q_3(t) = o(1)$. If $q_3(t) = q_{30} + o(1)$, $q_{30} < 0$ and $0 < \lambda < 1$, then for sufficiently large t we have $t^{\lambda+1+\epsilon}p(t) \leq c_1$ and $t^{2-\epsilon}p(t) \geq c_2 > 0$, where $\epsilon > 0$ is a sufficiently small constant. But this case is covered by Theorem 20.17.

THEOREM 20.25. Let $\lambda > 1$,

$$p(t) = p_0(t) + p_1(t), \qquad p_0 \in \tilde{C}_{\text{loc}}^1(\mathbf{R}_+), \qquad p_1 \in L_{\text{loc}}(\mathbf{R}_+),$$

$$p_0(t) > 0 \quad \text{for } t \geq 0, \qquad \lim_{t \to +\infty} \frac{p_1(t)}{p_0(t)} = 0, \tag{20.88}$$

and

$$\int_0^{+\infty} [q(t)]^{-2}\, dt < +\infty, \qquad \lim_{t \to +\infty} q_1(t) = q_{10} > 0. \tag{20.89}$$

Then $\underline{\gamma} < \gamma_0 < \bar{\gamma}$ and

$$u_{\bar{\gamma}}(t) = (q_{10}^{1/(\lambda-1)} + o(1))q(t)\left(\int_t^{+\infty} [q(\tau)]^{-2}\, d\tau\right)^{2/(1-\lambda)},$$

$$u_\gamma(t) = (c_1 + o(1))t, \qquad c_1 = c_1(\gamma) > 0, \qquad \text{for } \gamma_0 < \gamma < \bar{\gamma},$$

$$u_{\gamma_0}(t) = c_0 + o(1), \qquad c_0 > 0,$$

$$u_\gamma(t) = (-c_2 + o(1))t, \qquad c_2 = c_2(\gamma) > 0, \qquad \text{for } \underline{\gamma} < \gamma < \gamma_0,$$

$$u_{\underline{\gamma}}(t) = (-q_{10}^{1/(\lambda-1)} + o(1))q(t)\left(\int_t^{+\infty} [q(\tau)]^{-2}\, d\tau\right)^{2/(1-\lambda)}.$$

PROOF. Recall that (20.89) yields (20.79). So, according to Theorem 20.19, $\underline{\gamma} < \gamma_0 < \bar{\gamma}$.

If u is a solution of the equation (20.1), then by applying the transformation (20.70), we find that v satisfies the equation (20.23) with

$$a_1(s) = \frac{\lambda+3}{1-\lambda}, \qquad a_2(s) = -q_1(t), \qquad \alpha(s) = 1 + \frac{p_1(t)}{p_0(t)}.$$

Without loss of generality we may assume that

$$\int_{t_0}^{+\infty} [q(t)]^{-2}\, dt < 1.$$

Let

$$s_0 = \left| \ln \int_{t_0}^{+\infty} [q(t)]^{-2} \, dt \right|, \qquad \beta_1 = [q(t_0)]^{-1} \left(\int_{t_0}^{+\infty} [q(t)]^{-2} \, dt \right)^{2/(\lambda-1)} \beta_0.$$

Denote by v_μ the solution of the equation (20.23) under the conditions $v(s_0) = \beta_1$, $v'(s_0) = \mu$, and by $v_{\underline{\mu}}$, v_{μ_0} and $v_{\bar{\mu}}$ the solutions of this equation obtained from u_γ, u_{γ_0} and $u_{\bar{\gamma}}$ via the transformation (20.70). It is obvious that for $\mu \notin [\underline{\mu}, \bar{\mu}]$ the solutions v_μ are singular. Besides, if $\mu \in [\mu_0, \bar{\mu}]$, then $v_\mu(s) > 0$ for $s \geq s_0$, while if $\mu \in [\underline{\mu}, \mu_0[$, then $v_\mu(s) < 0$ for s large.

Clearly, in the case under consideration the hypotheses of Lemma 20.6 are satisfied. Since the functions v_1 and v_2 defined by (20.71) are linearly independent solutions of the equation (20.28), for any $\mu \in [\underline{\mu}, \bar{\mu}]$ this implies that either $v_\mu(s) = \pm q_{10}^{1/(\lambda-1)} + o(1)$ or $v_\mu(s) = o(1)$ and

$$v_\mu(s) = [q(t)]^{-1} \left(\int_t^{+\infty} [q(\tau)]^{-2} \, d\tau \right)^{2/(\lambda-1)} (c_1 + o(1) + c_2(1 + o(1))t), \qquad (20.90)$$

where $|c_1| + |c_2| \neq 0$.

We will show that

$$v_{\bar{\mu}}(s) = q_{10}^{1/(\lambda-1)} + o(1), \qquad v_{\underline{\mu}}(s) = -q_{10}^{1/(\lambda-1)} + o(1), \qquad (20.91)$$

and that for $\mu \in]\underline{\mu}, \bar{\mu}[$ (20.90) holds.

Suppose $v_{\bar{\mu}}(s) = o(1)$. Choose $s_1 > s_0$ and $c > 0$ so that

$$v_{\bar{\mu}}(s_1) < c, \qquad v'_{\bar{\mu}}(s_1) < 0, \qquad |a_2(s)| \geq c^{\lambda-1}|\alpha(s)| \qquad \text{for } s \geq s_1.$$

Then $v_\mu(s_1) < c$, $v'_\mu(s_1) < 0$, provided $\mu > \bar{\mu}$ and $\mu - \bar{\mu}$ is sufficiently small. It follows from (20.23) that for $s \geq s_1$ every extremum of a solution of the equation (20.23) lying between the lines $v = 0$ and $v = c$ is a maximum. Therefore $v'_\mu(s) < 0$ and $0 < v_\mu(s) < c$ for $s \geq s_1$, which is impossible, because the solution v_μ is singular whenever $\mu > \bar{\mu}$. This proves the first relation in (20.91). The proof of the second relation is similar.

Let $\mu_0 \leq \mu < \bar{\mu}$. We claim that $v_\mu(s) = o(1)$. Assume the contrary. Then $v_\mu(s) = q_{10}^{1/(\lambda-1)} + o(1)$. Suppose that for $s \geq s_1 = s(t_1)$ the solutions $v_{\bar{\mu}}$ and v_μ satisfy (20.61), where $a_{20} = -q_{10}$ and δ is chosen so that

$$(q_{10} - \delta) \left[1 - \left(\frac{q_{10} - \delta}{q_{10} + \delta} \right)^{\lambda/(\lambda-1)} \right] > (q_{10} + \delta) \left[1 - \left(\frac{q_{10} - \delta}{q_{10} + \delta} \right)^{1/(\lambda-1)} \right]. \qquad (20.92)$$

Assume that u_γ and $u_{\bar{\gamma}}$ are the solutions of the equation (20.1) corresponding to v_μ and $v_{\bar{\mu}}$, $\bar{u}$ is a solution of the same equation under the conditions

$$\bar{u}(t_1) = u_{\bar{\gamma}}(t_1), \qquad u'_\gamma(t_1) < \bar{u}'(t_1) < u'_{\bar{\gamma}}(t_1), \qquad (20.93)$$

and $\bar{v}$ is the solution of (20.23) corresponding to $\bar{u}$. By Lemma 20.8, $u_\gamma(t) < \bar{u}(t) < u_{\bar{\gamma}}(t)$ for $t \geq t_1$. This implies $v_\mu(s) < \bar{v}(s) < v_{\bar{\mu}}(s)$ for $s \geq s_1$. So the function $\bar{v}$

satisfies, together with v_μ and $v_{\bar\mu}$, the condition (20.61) for $s \geq s_1$. Hence (20.92) yields

$$\frac{[v_{\bar\mu}(s)]^\lambda - [\bar v(s)]^\lambda}{v_{\bar\mu}(s) - \bar v(s)} = [v_{\bar\mu}(s)]^{\lambda-1}\left(1 - \left[\frac{\bar v(s)}{v_{\bar\mu}(s)}\right]^\lambda\right)\left(1 - \frac{\bar v(s)}{v_{\bar\mu}(s)}\right)^{-1}$$

$$\geq (q_{10} - \delta)\left(1 - \left[\frac{q_{10} - \delta}{q_{10} + \delta}\right]^{\lambda/(\lambda-1)}\right)\left(1 - \left[\frac{q_{10} - \delta}{q_{10} + \delta}\right]^{1/(\lambda-1)}\right)^{-1} > q_{10} + \delta$$

for $s \geq s_1$. Then assuming that $|a_2(s)| \leq (q_{10} + \delta)\alpha(s)$ for $s \geq s_1$, from (20.23) we derive

$$(v_{\bar\mu}(s) - \bar v(s))'' - a_{10}(v_{\bar\mu}(s) - \bar v(s))'$$

$$\geq \alpha(s)(v_{\bar\mu}(s) - \bar v(s))\left(q_{10} + \delta + \frac{a_2(s)}{\alpha(s)}\right) \geq 0 \quad \text{for } s \geq s_1.$$

Since by (20.93), $v'_{\bar\mu}(s_1) > \bar v'(s_1)$, the last inequality implies

$$(v_{\bar\mu}(s) - \bar v(s))' \geq \exp(a_{10}(s - s_1)(v'_{\bar\mu}(s_1) - \bar v'(s_1))) > 0,$$

which is impossible, because $0 < v_{\bar\mu}(s) - \bar v(s) = o(1)$. This contradiction shows that $v_\mu(s) = o(1)$, and in view of Lemma 20.6 we obtain (20.90). For $\mu \in]\underline\mu, \mu_0[$ (20.90) can be proved in a similar manner.

It now remains to apply (20.70) to (20.90) and (20.91). $\square$

If $q_1(t) = o(1)$, but $q_2(t) = q_{20} + o(1)$, $q_{20} > 0$, then, as in the proofs of Theorems 20.9 and 20.25, we can justify the following proposition.

THEOREM 20.26. Let $\lambda > 1$ and let the conditions (20.88) hold. Suppose, in addition, that

$$\int_0^{+\infty} [q(t)]^{-2}\, dt < +\infty, \qquad \lim_{t\to+\infty} q_2(t) = q_{20} > 0.$$

Then $\underline\gamma < \gamma_0 < \bar\gamma$, and

$$u_{\bar\gamma}(t) = (q_{20}^{1/(\lambda-1)} + o(1))\left|\ln\int_t^{+\infty} [q(\tau)]^{-2}\, d\tau\right|^{1/(1-\lambda)}$$

$$\times q(t)\left(\int_t^{+\infty} [q(\tau)]^{-2}\, d\tau\right)^{2/(1-\lambda)},$$

$$u_\gamma(t) = (c_1 + o(1))t, \qquad c_1 = c_1(\gamma) > 0, \qquad \text{for } \gamma_0 < \gamma < \bar\gamma,$$

$$u_{\gamma_0}(t) = c_0 + o(1), \qquad c_0 > 0,$$

$$u_\gamma(t) = (-c_2 + o(1))t, \qquad c_2 = c_2(\gamma) > 0, \qquad \text{for } \underline\gamma < \gamma < \gamma_0,$$

$$u_{\underline\gamma}(t) = (-q_{20}^{1/(\lambda-1)} + o(1))\left|\ln\int_t^{+\infty} [q(\tau)]^{-2}\, d\tau\right|^{1/(1-\lambda)}$$

$$\times q(t)\left(\int_t^{+\infty} [q(\tau)]^{-2}\, d\tau\right)^{2/(1-\lambda)}.$$

If $q_1(t) = q_{10} + o(1)$, $q_{10} < 0$, and $\lambda > 1$, then $t^{\lambda+1-\epsilon}p(t) \geq c_1 > 0$ and $t^{2+\epsilon}p(t) \leq c_2$ for a certain $\epsilon > 0$ and all sufficiently large t. So this case is already covered by Theorem 20.20.

THEOREM 20.27. Let $\lambda > 1$ and let the conditions (20.88) hold. Suppose, in addition, that

$$\int_0^{+\infty} [q(t)]^{-2} \, dt = +\infty, \qquad \lim_{t \to +\infty} q_3(t) = q_{30} > 0.$$

Then $\underline{\gamma} = \gamma_0 = \bar{\gamma}$, and

$$u_{\gamma_0}(t) = (q_{30}^{1/(\lambda-1)} + o(1))q(t) \left(\int_0^t [q(\tau)]^{-2} \, d\tau \right)^{2/(1-\lambda)}.$$

PROOF. Let u be a proper solution of the equation (20.1). Setting

$$s = \ln \int_0^t [q(\tau)]^{-2} \, d\tau, \qquad u(t) = q(t) \left(\int_0^t [q(\tau)]^{-2} \, d\tau \right)^{2/(1-\lambda)} v(s),$$

we obtain for v the equation (20.23) with

$$a_1(s) = \frac{\lambda + 3}{\lambda - 1}, \qquad a_2(s) = -q_3(t), \qquad \alpha(s) = 1 + \frac{p_1(t)}{p_0(t)}.$$

Obviously, in this case the hypotheses of Lemma 20.7 are satisfied. Thus, $v(s) = q_{30}^{1/(\lambda-1)} + o(1)$. $\square$

Similarly to Theorems 20.11 and 20.27 we can derive

THEOREM 20.28. If $\lambda > 1$, the conditions (20.88) hold, and

$$\int_0^{+\infty} [q(t)]^{-2} \, dt = +\infty, \qquad \lim_{t \to +\infty} q_4(t) = q_{40} > 0,$$

then $\underline{\gamma} = \gamma_0 = \bar{\gamma}$ and

$$u_{\gamma_0}(t) = (q_{40}^{1/(\lambda-1)} + o(1)) \left(\ln \int_0^t [q(\tau)]^{-2} \, d\tau \right)^{1/(1-\lambda)} q(t) \left(\int_0^t [q(\tau)]^{-2} \, d\tau \right)^{2/(1-\lambda)}.$$

Under the hypotheses of Theorem 20.28, $q_3(t) = o(1)$. If $q_3(t) = q_{30} + o(1)$, $q_{30} < 0$ and $\lambda > 1$, then, as above, we are in the situation covered by Theorem 20.20.

20.5. Asymptotic representations for nonoscillatory singular solutions. Suppose $t_0 \geq 0$, $\beta_0 \geq 0$ and u_γ is the solution of the equation (20.1) defined in the previous subsection.

THEOREM 20.29. Let $0 < \lambda < 1$ and let $p \in V_{\text{loc}}(\mathbf{R}_+) \cap C(\mathbf{R}_+)$ be positive. Then for any $t_1 > t_0$ there exists a constant β_0 such that the solution u_{γ_0}, where $\gamma_0 = \gamma_0(\beta_0)$ is the constant appearing in Theorem 20.16, satisfies the condition

$$u_{\gamma_0}(t) = \left(\left[\frac{p(t)(1-\lambda)^2}{2(\lambda+1)} \right]^{1/(1-\lambda)} + o(1) \right) (t_1 - t)^{2/(1-\lambda)} \tag{20.94}$$

as $t \to t_1$.

PROOF. It follows from the proof of Theorem 20.16 that we can find a constant β_0 such that the solution u_{γ_0} is singular and

$$u_{\gamma_0}(t) > 0 \qquad \text{for } t \in [t_0, t_1[, \qquad u_{\gamma_0}(t) \equiv 0 \qquad \text{on } [t_1, +\infty[. \tag{20.95}$$

Put

$$s = \frac{1}{t_1 - t}, \qquad u_{\gamma_0}(t) = s^{-1}v_0(s).$$ (20.96)

Clearly, v_0 is a solution of the equation

$$v'' = (p_0(s) + p_1(s))|v|^\lambda \operatorname{sign} v,$$ (20.97)

where

$$p_0(s) = p(t_1)s^{-\lambda-3}, \qquad p_1(s) = (p(t) - p(t_1))s^{-\lambda-3}.$$ (20.98)

For the equation (20.97) the hypotheses of Theorem 20.21 are satisfied. Moreover, by (20.95) and (20.96), $v_0(s) = o(1)$. Hence, according to Theorem 20.21,

$$v_0(s) = \left(\left[\frac{p(t)(1-\lambda)^2}{2(\lambda+1)} \right]^{1/(1-\lambda)} + o(1) \right) s^{(1+\lambda)/(\lambda-1)}.$$

Therefore, (20.96) yields (20.94). □

THEOREM 20.30. If $\lambda > 1$ and $p \in C(\mathbf{R}_+)$ is positive, then for any $t_1 > t_0$ there exist constants $\gamma_*(t_1)$ and $\gamma^*(t_1)$ such that the solution u_γ is bounded on $[t_0, t_1]$ for $\gamma \in]\gamma_*(t_1), \gamma^*(t_1)[$ and has a vertical asymptote at a certain point $t \in]t_0, t_1[$ for $\gamma \notin [\gamma_*(t_1), \gamma^*(t_1)]$; moreover, the solutions $u_{\gamma_*(t_1)}$ and $u_{\gamma^*(t_1)}$ satisfy the conditions

$$u_{\gamma_*(t_1)}(t) = \left(-\left[\frac{2(\lambda+1)}{p(t_1)(\lambda-1)^2} \right]^{1/(\lambda-1)} + o(1) \right) (t_1 - t)^{2/(1-\lambda)},$$

$$u_{\gamma^*(t_1)}(t) = \left(\left[\frac{2(\lambda+1)}{p(t_1)(\lambda-1)^2} \right]^{1/(\lambda-1)} + o(1) \right) (t_1 - t)^{2/(1-\lambda)}$$

as $t \to t_1$.

PROOF. The substitution

$$s = \frac{1}{t_1 - t}, \qquad u(t) = s^{-1}v(s)$$ (20.99)

transforms the equation (20.1) into the equation (20.97), where p_0 and p_1 are defined by (20.98).

Let $s_0 = 1/(t_1 - t_0)$, $\beta_1 = s_0\beta_0$, and let v_μ, $\mu \in \mathbf{R}$, be the solution of the equation (20.97) under the conditions $v(s_0) = \beta_1$, $v'(s_0) = \mu$. It is easy to show that for this equation all the hypotheses of Theorem 20.25 are satisfied, and so there exist $\underline{\mu}$ and $\bar{\mu}$ such that

$$v_{\underline{\mu}}(s) = \left(-\left[\frac{2(\lambda+1)}{p(t_1)(\lambda-1)^2} \right]^{1/(\lambda-1)} + o(1) \right) s^{(\lambda+1)/(\lambda-1)},$$

$$v_{\bar{\mu}}(s) = \left(\left[\frac{2(\lambda+1)}{p(t_1)(\lambda-1)^2} \right]^{1/(\lambda-1)} + o(1) \right) s^{(\lambda+1)/(\lambda-1)},$$

$$v_\mu(s) = O(s) \qquad \text{if } \mu \in]\underline{\mu}, \bar{\mu}[,$$

and for the other values of μ the solutions v_μ are singular. Setting $\gamma_*(t_1) = -\beta_1 + s_0\underline{\mu}$, $\gamma^*(t_1) = -\beta_1 + s_0\bar{\mu}$ and applying (20.99) complete the proof. □

Notes. Asymptotic properties of solutions of the equation $u'' = ct^\sigma |u|^\lambda \operatorname{sign} u$ were studied by R. Emden [103] and R.H. Fowler [114, 115]. The results obtained in this field are summed up in the monographs of R. Bellman [27] and G. Sansone [317].

Theorems 20.1–20.5, 20.8–20.30 are due to I.T. Kiguradze [174, 175, 179] and T.A. Chanturia [52, 71]. Theorems 20.6 and 20.7 were proved by Š. Belohorec [29] and F.V. Atkinson [8], respectively.

Asymptotic formulas for solutions of second order differential equations with power nonlinearities of various types were derived by M.M. Aripov [2, 3], L.A. Beklemisheva [25], V.M. Evtukhov [108, 110], L.B. Klebanov [206], and A.V. Kostin [233, 234], and for solutions of a two-dimensional differential system by J.D. Mirzov [276]. See also the survey papers [189, 374].

REFERENCES

1. Ananyeva, G.V.; Balagansky, V.I. On the oscillation of solutions of some higher order differential equations (in Russian). *Uspekhi Mat. Nauk,* **14** (1959), *no.* 1, 135-140.

2. Aripov, M.M. The method of 'standard' equations (WKB-method) for nonlinear second order equations (in Russian). *Izv. Akad. Nauk UzSSR. Ser. Fiz.-Mat. Nauk,* 1970, *no.* 4, 3-8.

3. Aripov, M.M. On solution of a second order ordinary nonlinear equation (in Russian). *Izv. Akad. Nauk UzSSR. Ser. Fiz.-Mat. Nauk,* 1970, *no.* 7, 6-8.

4. Armellini, G. Sopra un'equazione differenziale della Dinamica. *Rend. R. Accad. Naz. Lincei,* **21** (1935), 111-116.

5. Ascoli, G. Sul comportamento asintotico degli integrali delle equazioni differenziali lineari di $2°$ ordine. *Rend. R. Accad. Naz. Lincei,* **22** (1935), 234-243.

6. Ascoli, G. Sulla decomposizione degli operatori differenziali lineari in fattori lineari e sopra alcune questioni geometriche ve si riconnettono. *Rev. Mat. Fis. Theor. (Tucuman),* **1** (1940), 180-215.

7. Astashova, I.V. On the asymptotic behavior of solutions of some nonlinear differential equations (in Russian). *Rep. Enlarged Sessions Sem. I.N. Vekua Inst. Appl. Math.,* **1** (1985), *no.* 3, 9-11.

8. Atkinson, F.V. On second-order non-linear oscillations. *Pacific J. Math.,* **5** (1955), *no.* 1, 643-647.

9. Atkinson, F.V. A stability problem with algebraic aspects. *Proc. Roy. Soc. Edinburgh,* **78A** (1978), 299-314.

10. Azbelev, N.V. On zeros of solutions of a second order linear differential equation with retarded argument (in Russian). *Differentsial'nye Uravneniya,* **7** (1971), *no.* 7, 1147-1157.

11. Azbelev, N.V.; Tsalyuk, Z.B. Concerning the distribution of zeros of solutions of a third order linear differential equation (in Russian). *Mat. Sb.,* **51** (1960), *no.* 4, 475-486.

12. Barrett, J.H. Oscillation theory of ordinary linear differential equations. *Adv. Math.,* **3** (1969), *no.* 3, 415-509.

13. Bartušek, M. On zeros of oscillatory solutions of the equation $(p(t)x')' +$

$f(t, x, x') = 0$ (in Russian). *Differentsial'nye Uravneniya*, **12** (1976), *no.* 4, 621-625.

14. Bartušek, M. Monotonicity theorems concerning differential equation $y'' + f(t, y, y') = 0$. *Arch. Math. (Brno)*, **12** (1976), *no.* 4, 169-178.

15. Bartušek, M. On asymptotic properties and distribution of zeros of solutions of $y'' + f(t, y, y') = 0$. *Arch. Math. (Brno)*, **14** (1978), *no.* 1, 1-12.

16. Bartušek, M. On zeros of solutions of the differential equation $y'' + f(t, y)g(y') = 0$. *Arch. Math. (Brno)*, **15** (1979), *no.* 3, 129-132.

17. Bartušek, M. On properties of oscillatory solutions of two-dimensional differential systems (in Russian). *Proc. I.N.Vekua Inst. Appl. Math.*, **8** (1980), 5-12.

18. Bartušek, M. On existence of oscillatory solutions of the system of differential equations. *Arch. Math. (Brno)*, **17** (1981), *no.* 1, 7-10.

19. Bartušek, M. On asymptotic properties of oscillatory solutions of the system of differential equations of fourth order. *Arch. Math. (Brno)*, **17** (1981), *no.* 3, 125-136.

20. Bartušek, M. On properties of oscillatory solutions of third order ordinary differential equations (in Russian). *Differentsial'nye Uravneniya*, **17** (1981), *no.* 5, 771-777.

21. Bartušek, M. The asymptotic behaviour of solutions of the differential equations of the third order. *Arch. Math. (Brno)*, **20** (1984), *no.* 3, 101-112.

22. Bartušek, M. On oscillatory solutions of the differential equation of the n-th order. *Arch. Math. (Brno)*, **22** (1986), *no.* 3, 145-156.

23. Bartušek, M. On properties of oscillatory solutions of ordinary differential inequalities and equations (in Russian). *Differentsial'nye Uravneniya*, **23** (1987), *no.* 2, 187-191.

24. Beckenbach, E.F.; Bellman, R. Inequalities. Springer, Berlin, 1961.

25. Beklemisheva, L.A. On a second order nonlinear differential equation (in Russian). *Mat. Sb.*, **56** (1962), *no.* 2, 207-236.

26. Bellman, R. The boundedness of solutions of linear differential equations. *Duke Math. J.*, **14** (1947), *no.* 1, 83-97.

27. Bellman, R. Stability theory of differential equations. McGraw-Hill, New York, 1953.

28. Belohorec, Š. Oscilatorické riešenia istej nelineárnej diferenciálnej rovnice druhého rádu. *Mat.-Fyz. Časopis*, **11** (1961), *no.* 4, 250-255.

29. Belohorec, Š. Neoscilatorické riešenia istej nelineárnej diferenciálnej rovnice druhého rádu. *Mat.-Fyz. Časopis*, **12** (1962), *no.* 4, 253-262.

30. Belohorec, Š. On some properties of the equation $y''(x) + f(x)y^{\alpha}(x) = 0, 0 < \alpha < 1$. *Mat. Časopis*, **17** (1967), *no.* 1, 10-19.

31. Belohorec, Š. Monotone and oscillatory solutions of a class of nonlinear differential equations. *Mat. Časopis*, **19** (1969), *no.* 3, 169-187.

32. Belohorec, Š. A criterion for oscillation and nonoscillation. *Acta F.R.N. Univ. Comen. Math.*, **20** (1969), 75-79.

33. Belohorec, Š. Two remarks on the properties of solutions of a nonlinear differential equation. *Acta F.R.N. Univ. Comen. Math.*, **22** (1969), 19-26.

34. Bernis, F. Compactness of the support in convex and non-convex fourth order elasticity problems. *Nonlinear Anal., Theory, Meth. and Appl.*, **6** (1982), *no.* 11, 1221-1243.

35. Bernis, F. Asymptotic rates of decay for some nonlinear ordinary differential equations and variational problems of arbitrary order. *Ann. Fac. Sci. Toulouse*, **6** (1984), 121-151.

36. Biernacki, M. Sur l'équation différentielle $x'' + A(t)x = 0$. *Prace Mat. Fiz.*, **43** (1933), 163-171.

37. Biernacki, M. Sur l'équation différentielle $y^{(4)} + A(x)y = 0$. *Ann. Univ. M.Curie-Sklodowska*, **6** (1952), 65-78.

38. Bihari, I. A generalization of a lemma of Bellman and its applications to uniqueness problems of differential equations. *Acta Math. Acad. Sci. Hungar.*, **7** (1956), *no.* 1, 71-94.

39. Bobrowski, D. On asymptotic properties of solutions of some differential equations in Banach spaces. *Bull. Acad. Pol. Sci. Ser. Sci. Math. Astr. Phys.*, **15** (1967), *no.* 6, 401-406.

40. Bobrowski, D. O pewnych wlasnośćiach rozwiazan równania rózniczkowego nieliniowego trzeciego rzedu. Politechnica Poznanska Rozprawy, Poznan, 1967.

41. Bôcher, M. On regular singular points of linear differential equations of the second order whose coefficients are not necessarily analytic. *Trans. Amer. Math. Soc.*, **1** (1900), 40-53.

42. Bulgakov, A.I. On the oscillation of solutions of a second order differential system (in Russian). *Rep. Enlarged Sessions Sem. I.N.Vekua Inst. Appl. Math.*, **1** (1985), *no.* 3, 19-22.

43. Bulgakov, A.I. On the oscillation of solutions of systems of second order differential equations (in Russian). *Differentsial'nye Uravneniya*, **23** (1987), *no.* 2, 204-217.

44. Bulgakov, A.I.; Sergeev, B.A. Oscillatory properties of solutions of a nonlinear system of second order ordinary differential equations (in Russian). *Differentsial'nye Uravneniya*, **20** (1984), *no.* 2, 207-214.

45. Butler, G.J. On the oscillatory behaviour of a second order nonlinear differential equation. *Ann. Mat. Pura Appl.*, **105** (1975), 73-92.

46. Butler, G.J. Oscillation theorems for a non-linear analogue of Hille's equation. *Quart. J. Math.*, **27** (1976), *no.* 106, 159-171.

47. Butler, G.J. Hille-Wintner type comparison theorems for second-order ordinary differential equations. *Proc. Amer. Math. Soc.*, **76** (1979), *no.* 1, 51-59.

48. Chanturia, T.A. On nonoscillatory solutions of second order nonlinear differential equations (in Russian). *Soobshch. Akad. Nauk Gruzin. SSR*, **55** (1969), *no.* 1, 17-20.

49. Chanturia, T.A. The asymptotics of solutions of some nonlinear second order differential equations (in Russian). *Soobshch. Akad. Nauk Gruzin. SSR*, **57** (1970), *no.* 2, 289-292.

50. Chanturia, T.A. On the asymptotic representation of solutions of nonlinear second order differential equations (in Russian). *Differentsial'nye Uravneniya*, **6** (1970), *no.* 6, 948-961.

51. Chanturia, T.A. On asymptotic properties of solutions of perturbed linear systems of differential equations. *Ann. Mat. Pura Appl.*, **94** (1972), 41-62

52. Chanturia, T.A. On the asymptotic representation of solutions of the equation

$u'' = a(t)|u|^n \operatorname{sign} u$ (in Russian). *Differentsial'nye Uravneniya*, **8** (1972), *no.* 7, 1195-1206.

53. Chanturia, T.A. On the asymptotic representation of oscillatory solutions of the perturbed equation of Emden–Fowler type (in Russian). *Issledovaniya nekotorykh uravn. mat. fiz.* Tbilisi Univ. Press, Tbilisi, 1972, 5-15.

54. Chanturia, T.A. On a problem of Kneser type for a system of ordinary differential equations (in Russian). *Mat. Zametki*, **15** (1974), *no.* 6, 897-906.

55. Chanturia, T.A. A remark on the asymptotic behavior of perturbed linear systems of differential equations (in Russian). *Proc. I.N. Vekua Inst. Appl. Math.*, **4** (1975), 29-34.

56. Chanturia, T.A. On the asymptotic behavior of oscillatory solutions of second order ordinary differential equations (in Russian). *Differentsial'nye Uravneniya*, **11** (1975), *no.* 7, 1232-1245.

57. Chanturia, T.A. On a comparison theorem for linear differential equations (in Russian). *Izv. Akad. Nauk SSSR. Ser. Mat.*, **40** (1976), *no.* 5, 1128-1142.

58. Chanturia, T.A. On singular solutions of nonlinear systems of ordinary differential equations. *Colloq. Math. Soc. János Bolyai*, **15** (1976), 107-119.

59. Chanturia, T.A. On singular solutions of strongly nonlinear systems of ordinary differential equations. *Funct. Theor. Meth. in Diff. Equat.* Pitman Publ., London, 1976, 196-204.

60. Chanturia, T.A. On some asymptotic properties of solutions of ordinary differential equations (in Russian). *Dokl. Akad. Nauk SSSR*, **235** (1977), *no.* 5, 1049-1052.

61. Chanturia, T.A. Some comparison theorems for higher order ordinary differential equations (in Russian). *Bull. Acad. Pol. Sci. Ser. Sci. Math. Astr. Phys.*, **25** (1977), *no.* 8, 749-756.

62. Chanturia, T.A. On some asymptotic properties of solutions of linear ordinary differential equations (in Russian). *Bull. Acad. Pol. Sci. Ser. Sci. Math. Astr. Phys.*, **25** (1977), *no.* 8, 757-762.

63. Chanturia, T.A. On monotone solutions of a system of nonlinear differential equations (in Russian). *Ann. Polon. Math.*, **37** (1980), *no.* 1, 59-69.

64. Chanturia, T.A. On monotone and oscillatory solutions of ordinary differential equations (in Russian). *Ann. Polon. Math.*, **37** (1980), *no.* 1, 93-111.

65. Chanturia, T.A. Comparison theorems of Sturm type for higher order differential equations (in Russian). *Soobshch. Akad. Nauk Gruzin. SSR*, **99** (1980), *no.* 2, 289-291.

66. Chanturia, T.A. Some remarks on oscillation of solutions of linear differential equations with retarded argument (in Russian). *Differentsial'nye Uravneniya*, **16** (1980), *no.* 2, 264-272.

67. Chanturia, T.A. Integral criteria for oscillation of solutions of higher order linear differential equations. I (in Russian). *Differentsial'nye Uravneniya*, **16** (1980), *no.* 3, 470-482.

68. Chanturia, T.A. Integral criteria for oscillation of solutions of higher order linear differential equations. II (in Russian). *Differentsial'nye Uravneniya*, **16** (1980), *no.* 4, 635-644.

69. Chanturia, T.A. On the oscillation of all solutions of odd order linear differential

equations (in Russian). *Mat. Zametki,* **28** (1980), *no.* 4, 565-569.

70. Chanturia, T.A. Comparison theorems for systems of ordinary differential equations (in Russian). *Rep. Sem. I.N. Vekua Inst. Appl. Math.,* **14** (1980), 13-17.

71. Chanturia, T.A. On the asymptotic representation of oscillatory solutions of an equation of Emden–Fowler type (in Russian). *Differentsial'nye Uravneniya,* **17** (1981), *no.* 6, 1035-1040.

72. Chanturia, T.A. Asymptotic properties of solutions of some classes of nonautonomous ordinary differential equations (in Russian). *Mat. Zametki,* **32** (1982), *no.* 4, 577-588.

73. Chanturia, T.A. On the oscillation of solutions of linear ordinary differential equations (in Russian). *Differentsial'nye Uravneniya,* **18** (1982), *no.* 6, 1095.

74. Chanturia, T.A. On the oscillation of solutions of higher order linear differential equations (in Russian). *Rep. Sem. I.N. Vekua Inst. Appl. Math.,* **16** (1982), 3-72.

75. Chanturia, T.A. On unbounded solutions of higher order linear differential equations (in Russian). *Uspekhi Mat. Nauk,* **38** (1983), *no.* 5, 131-132.

76. Chanturia, T.A. On oscillation properties of systems of nonlinear ordinary differential equations (in Russian). *Proc. I.N. Vekua Inst. Appl. Math.,* **14** (1983), 163-206.

77. Chanturia, T.A. On unbounded solutions of linear ordinary differential equations (in Russian). *Mat. Zametki,* **35** (1984), *no.* 2, 231-242.

78. Chanturia, T.A. On integral comparison theorems of Hille type for higher order differential equations (in Russian). *Soobshch. Akad. Nauk Gruzin. SSR,* **117** (1985), *no.* 2, 241-244.

79. Chanturia, T.A. On specific criteria for oscillation of solutions of differential equations of Emden–Fowler type with retarded argument (in Russian). *Rep. Enlarged Sessions Sem. I.N. Vekua Inst. Appl. Math.,* **1** (1985), *no.* 3, 150-153.

80. Chanturia, T.A. On the relationship between various definitions of oscillation of solutions of linear differential equations (in Russian). *Uspekhi Mat. Nauk,* **40** (1985), *no.* 5, 231-232.

81. Chanturia, T.A. On the oscillation of solutions of linear ordinary differential equations (in Russian). *Uspekhi Mat. Nauk,* **40** (1985), *no.* 5, 242.

82. Chanturia, T.A. A note on the oscillation of solutions of linear ordinary differential equations (in Russian). *Trudy Tbiliss. Univ.,* **259** (1986), 330-342.

83. Chanturia, T.A. On specific criteria for oscillation of solutions of linear differential equations with retarded argument (in Russian). *Ukrain. Mat. Zh.,* **38** (1986), *no.* 5, 662-665.

84. Chanturia, T.A. On the oscillation of solutions of general linear ordinary differential equations (in Russian). *Differentsial'nye Uravneniya,* **22** (1986), *no.* 11, 1905-1915.

85. Chanturia, T.A. On oscillation of solutions of linear ordinary differential equations. *Equadiff 6. Proc. Inter. Conf. Different. Equat. Appl.* Brno, 1986, 431-434.

86. Chanturia, T.A.; Khvedelidze, N.N. On the existence of oscillatory solutions of third order linear ordinary differential equations (in Russian). *Uspekhi Mat. Nauk,* **42** (1987), *no.* 4, 150.

87. Chaplygin, S.A. A new method for approximate integration of differential equations (in Russian). Gostekhizdat, Moscow, 1950.

88. Chiou Kuo-Liang. The existence of oscillatory solutions for the equation $d^2y/dt^2 + q(t)y^r = 0$, $0 < r < 1$. *Proc. Amer. Math. Soc.*, **35** (1972), no. 1, 120-122.

89. Coddington, E.A.; Levinson, N. Theory of ordinary differential equations. McGraw-Hill, New York, 1955.

90. Coffman, C.V. Non-linear differential equations on cones in Banach spaces. *Pacific J. Math.*, **14** (1964), no. 1, 9-15.

91. Coffman, C.V.; Ullrich, D.F. On the continuation of solutions of a certain non-linear differential equation. *Monatsh. Math.*, **71** (1967), no. 5, 385-392.

92. Coffman, C.V.; Wong, J.S.W. On a second order nonlinear oscillation problem. *Trans. Amer. Math. Soc.*, **147** (1970), no. 2, 357-366.

93. Coppel, W.A. Disconjugacy. Springer, Berlin, 1971.

94. De Kleine, H.A. A counterexample to conjecture in second order linear equations. *Michigan Math. J.*, **17** (1970), 29-32.

95. Dolan, J.M. On the relationship between the oscillatory behaviour of a linear third-order differential equation and its adjoint. *J. Different. Equat.*, **7** (1970), no. 2, 367-388.

96. Doležal, V. Asymptotické vzorce pro rešeni diferenciálni rovnice $y'' + f(t)y = 0$. *Časopis Pěst. Mat.*, **83** (1958), no. 4, 451-463.

97. Domshlak, Yu.I. A comparison method by Sturm for investigation of behavior of solutions of differential-operator equations (in Russian). Elm, Baku, 1986.

98. Drakhlin, M.E. On oscillatory properties of some functional-differential equations (in Russian). *Differentsial'nye Uravneniya*, **22** (1986), no. 3, 396-402.

99. Elias, U. The extremal solutions of the equation $Ly + p(x)y = 0$. *J. Math. Anal. Appl.*, **50** (1975), no. 3, 447-457.

100. Elias, U. Nonoscillation and eventual disconjugacy. *Proc. Amer. Math. Soc.*, **66** (1977), no. 2, 269-275.

101. Elias, U. A classification of the solutions of a differential equation according to their asymptotic behaviour. *Proc. Roy. Soc. Edinburgh*, **83A** (1979), no. 1-2, 25-38.

102. Elias, U. Generalizations of an inequality of Kiguradze. *J. Math. Anal. Appl.*, **97** (1983), no. 1, 277-290.

103. Emden, R. Gaskugeln, Anwendungen der mechanischen Warmentheorie auf Kosmologie und metheorologische Probleme. Leipzig, 1907.

104. Erbe, L. Comparison theorems of Hille–Wintner type for third order linear differential equations. *Bull. Austral. Math. Soc.*, **21** (1980), 175-188.

105. Erbe, L. Hille–Wintner type comparison theorem for self-adjoint fourth order linear differential equations. *Proc. Amer. Math. Soc.*, **80** (1980), no. 3, 417-422.

106. Erbe, L.H.; Rao, V.S.H. Nonoscillation results for third order nonlinear differential equations. *J. Math. Anal. Appl.*, **125** (1987), no. 2, 471-482.

107. Erugin, N.P. A book for readings in the general course of differential equations (in Russian). Nauka i tekhnika, Minsk, 1970.

108. Evtukhov, V.M. On a nonlinear second order differential equation (in Russian). *Dokl. Akad. Nauk SSSR*, **233** (1977), *no.* 4, 531-534.

109. Evtukhov, V.M. Asymptotic representations of solutions of a class of second order nonlinear differential equations (in Russian). *Soobshch. Akad. Nauk Gruzin. SSR*, **106** (1982), *no.* 3, 473-476.

110. Evtukhov, V.M. Asymptotic properties of solutions of a class of second order differential equations (in Russian). *Math. Nachr.*, **115** (1984), *no.* 2, 215-236.

111. Fedoryuk, M.V. Asymptotic methods for linear ordinary differential equations. Nauka, Moscow, 1983.

112. Fermi, E. Un metodo statistico per la determinazione di alcune proprietá dell'atomo. *Rend. R. Accad. Naz. Lincei*, **6** (1927), 602-607.

113. Fite, W.B. Concerning the zeros of the solutions of certain differential equation. *Trans. Amer. Math. Soc.*, **19** (1918), *no.* 4, 341-352.

114. Fowler, R.H. The solutions of Emden's and similar differential equations. *Monthly Notices Roy. Astronom. Soc.*, **91** (1930), 63-91.

115. Fowler, R.H. Further studies of Emden's and similar differential equations. *Quart. J. Math.*, **2** (1931), *no.* 2, 259-288.

116. Futák, J. On the asymptotic behavior of solutions of functional-differential equations (in Russian). *Proc. I.N.Vekua Inst. Appl. Math.*, **8** (1980), 68-78.

117. Galbraith, A.S.; McShane, E.J.; Parrish, G.B. On the solutions of linear second order differential equations. *Proc. Nat. Acad. Sci. USA*, **53** (1965), *no.* 2, 247-249.

118. Glazman, I.M. Direct methods of qualitative spectral analysis of singular differential operators (in Russian). Fizmatgiz, Moscow, 1963.

119. Grammatikopoulos, M.K. Oscillatory and asymptotic behaviour of differential equations with deviating arguments. *Hiroshima Math. J.*, **6** (1976), *no.* 1, 31-53.

120. Greguš, M. O niektorých nových vlastnostiach riešení diferenciálnej rovnice $y'' + Qy' + Q'y = 0$. *Spisy Přir. Fak. MU (Brno)* (1955), *no.* 5, 237-254.

121. Greguš, M. Third order linear differential equations. D.Reidel Publ. Co., Dordrect, 1987.

122. Grobman, D.M. The asymptotics of solutions of almost linear systems of differential equations (in Russian). *Dokl. Akad. Nauk SSSR*, **158** (1964), *no.* 4, 774-776.

123. Gustafson, G.B.; Sedziwy, S. Solution space decompositions for n-th order linear differential equations. *Canad. J. Math.*, **27** (1975), *no.* 3, 508-512.

124. Hallam, T.G.; Heidel, J.W. The asymptotic manifolds of a perturbed linear system of differential equations. *Trans. Amer. Math. Soc.*, **149** (1970), *no.* 5, 233-241.

125. Hardy, G.H.; Littlewood, J.E.; Pólya, G. Inequalities. Cambridge Univ. Press, Cambridge, 1934.

126. Hartman, P. The existence of large or small solutions of linear differential equations. *Duke Math. J.*, **28** (1961), *no.* 3, 421-429.

127. Hartman, P. Ordinary differential equations. John Wiley, New York, 1964.

128. Hartman, P. Principal solutions of disconjugate n-th order linear differential equations. *Amer. J. Math.*, **91** (1969), *no.* 2, 306-362.

129. Hartman, P. Corrigendum and addendum: principal solutions of disconjugate n-th order linear differential equations. *Amer. J. Math.*, **93** (1971), *no.* 2, 439-451.

130. Hartman, P.; Wintner, A. On the non-increasing solutions of $y'' = f(x,y,y')$. *Amer. J. Math.*, **73** (1951), *no.* 2, 390-404.

131. Hartman, P.; Wintner, A. On monotone solutions of systems of non-linear differential equations. *Amer. J. Math.*, **76** (1954), *no.* 4, 860-866.

132. Hartman, P.; Wintner, A. Asymptotic integrations of linear differential equations. *Amer. J. Math.*, **77** (1955), *no.* 1, 45-87.

133. Hastings, S.P. Boundary value problems in one differential equation with a discontinuity. *J. Different. Equat.*, **1** (1965), *no.* 3, 346-369.

134. Haupt, O. Über das asymptotische Verhalten der Lösungen gewisser linearer gewöhnlicher Differentialgleichungen. *Math. Z.*, **48** (1943), *no.* 2, 289-292.

135. Heidel, J.W. The existence of oscillatory solutions for a nonlinear odd order differential equation. *Czechoslovak Math. J.*, **20** (1970), *no.* 1, 93-97.

136. Heidel, J.W. Uniqueness, continuation and nonoscillation for a second order differential equation. *Pacific J. Math.*, **32** (1970), *no.* 3, 715-721.

137. Heidel, J.W.; Hinton, D.B. The existence of oscillatory solutions for a nonlinear differential equation. *SIAM J. Math Anal.*, **3** (1972), *no.* 2, 344-351.

138. Heidel, J.W.; Kiguradze, I.T. Oscillatory solutions for a generalized sublinear second order differential equation. *Proc. Amer. Math. Soc.*, **38** (1973), *no.* 1, 80-82.

139. Hille, E. Non-oscillation theorems. *Trans. Amer. Math. Soc.*, **64** (1948), 234-252.

140. Izobov, N.A. On the Emden–Fowler equations with unbounded infinitely continuable solutions (in Russian). *Mat. Zametki*, **35** (1984), *no.* 2, 189-198.

141. Izobov, N.A. On continuable and noncontinuable solutions of an arbitrary order nonlinear differential equation (in Russian). *Mat. Zametki*, **35** (1984), *no.* 6, 829-839.

142. Izobov, N.A. On continuable and noncontinuable solutions of the Emden–Fowler equation (in Russian). *Rep. Enlarged Sessions Sem. I.N.Vekua Inst. Appl. Math.*, **1** (1985), *no.* 3, 43-46.

143. Izobov, N.A. On Kneser solutions (in Russian). *Differentsial'nye Uravneniya*, **21** (1985), *no.* 4, 581-588.

144. Izobov, N.A.; Rabtsevich, V.A. On the unimprovability of the I.T.Kiguradze–G.G.Kvinikadze condition for existence of unbounded proper solutions of the Emden–Fowler equation (in Russian). *Differentsial'nye Uravneniya*, **23** (1987), *no.* 11, 1872-1881.

145. Izyumova, D.V. On oscillation and nonoscillation conditions for solutions of nonlinear second order differential equations (in Russian). *Differentsial'nye Uravneniya*, **11** (1966), *no.* 12, 1572-1586.

146. Izyumova, D.V. A note on the oscillation of solutions of nonlinear second order differential equations (in Russian). *Soobshch. Akad. Nauk Gruzin. SSR*, **17** (1967), *no.* 1, 19-24.

147. Izyumova, D.V. On the asymptotic behavior of solutions of some nonlinear second order ordinary differential equations (in Russian). *Trudy Tbiliss. Univ.*, **129** (1968), 157-178.

148. Izyumova, D.V. On oscillatory solutions of nonlinear second order differential inequalities (in Russian). *Issledovaniya nekotorykh uravn. mat. fiz.* Tbilisi Univ. Press,

Tbilisi, 1972, 17-22.

149. Izyumova, D.V. Concerning the oscillation of solutions of a linear even order differential equation whose coefficient changes sign (in Russian). *Asimptoticheskoe povedenie reshen. differents.-funkts. uravn.* Ukrain.SSR Acad. Sci. Press, Kiev, 1972, 70-77.

150. Izyumova, D.V. Concerning the oscillation of solutions of nonlinear second order differential inequalities and equations (in Russian). *Issledovaniya nekotorykh uravn. mat. fiz.* Tbilisi Univ. Press, Tbilisi, 1976, 63-70.

151. Izyumova, D.V. On the boundedness and stability of solutions of some nonlinear second order functional-differential equations (in Russian). *Proc. I.N. Vekua Inst. Appl. Math.,* **14** (1983), 52-60.

152. Izyumova, D.V. On solutions of systems of functional-differential equations vanishing at infinity (in Russian). *Proc. I.N. Vekua Inst. Appl. Math.,* **17** (1986), 94-103.

153. Izyumova, D.V. On positive and bounded solutions of nonlinear second order ordinary differential equations (in Russian). *Proc. I.N. Vekua Inst. Appl. Math.,* **22** (1987), 100-109.

154. Izyumova, D.V.; Kiguradze, I.T. Some remarks on solutions of the equation $u'' + a(t)f(u) = 0$ (in Russian). *Differentsial'nye Uravneniya,* **4** (1968), no. 4, 589-605.

155. Izyumova, D.V.; Kiguradze, I.T. On oscillation properties of a class of differential equations with deviating argument (in Russian). *Differentsial'nye Uravneniya,* **21** (1985), no. 4, 588-596.

156. Izyumova, D.V.; Mirzov, J.D. On the oscillation and nonoscillation of solutions of nonlinear differential systems (in Russian). *Differentsial'nye Uravneniya,* **12** (1976), no. 7, 1187-1193.

157. Izyumova, D.V.; Toroshelidze, I.A. Oscillation theorems for some nonlinear second order ordinary differential equations (in Russian). *Proc. I.N. Vekua Inst. Appl. Math.,* **8** (1980), 13-23.

158. Izyumova, D.V.; Toroshelidze, I.A. On the asymptotic behavior of solutions of nonlinear differential equations with deviating argument (in Russian). *Proc. I.N. Vekua Inst. Appl. Math.,* **14** (1983), 62-68.

159. Jasný, M. On the existence of oscillatory solutions of the second order nonlinear differential equation $y'' + f(x)y^{2n-1} = 0$ (in Russian). *Časopis Pěst. Mat.,* **85** (1960), no. 1, 78-83.

160. Johnson, G.W. The k-th conjugate point function for an even order linear differential equation. *Proc. Amer. Math. Soc.,* **42** (1974), no. 2, 563-568.

161. Kamenev, I.V. On the oscillation of solutions of a nonlinear equation with multiplicatively separated rigt-hand side (in Russian). *Differentsial'nye Uravneniya,* **6** (1970), no. 8, 1510-1513.

162. Kamenev, I.V. On the oscillation of solutions of a second order nonlinear equation whose coefficient changes sign (in Russian). *Differentsial'nye Uravneniya,* **6** (1970), no. 9, 1718-1721.

163. Kamenev, I.V. On a sufficient condition for oscillation of solutions of a higher order differential equation (in Russian). *Mat. Zametki,* **9** (1971), no. 4, 421-423.

164. Kamenev, I.V. On some specifically nonlinear oscillation theorems (in Russian). *Mat. Zametki,* **10** (1971), *no.* 2, 129-134.

165. Kamenev, I.V. On a specific nonlinear oscillation theorem (in Russian). *Differentsial'nye Uravneniya,* **9** (1973), *no.* 3, 574-576.

166. Kamenev, I.V. Criteria for oscillation of solutions of second order ordinary differential equations related with the averaging (in Russian). *Differentsial'nye Uravneniya,* **10** (1974), *no.* 2, 246-252.

167. Kamynin, L.I. On the boundedness of solutions of the differential equation $y'' + F(x)y = 0$ (in Russian). *Vestnik Moskov. Univ. Ser. Fiz. Mat.,* 1951, *no.* 3, 3-12.

168. Kantorovich, L.V.; Akilov, G.P. Functional analysis (in Russian). Nauka, Moscow, 1977.

169. Kartsatos, A.G. On n-th order differential inequalities. *J. Math. Anal. Appl.,* **52** (1975), *no.* 1, 1-9.

170. Khvedelidze, N.N. On the uniqueness of solutions of the Kneser problem for third order linear differential equations (in Russian). *Proc. I.N.Vekua Inst. Appl. Math.,* **17** (1986), 180-194.

171. Kiguradze, I.T. On a generalization of the Armellini–Tonelli–Sansone theorem (in Georgian). *Trudy Tbiliss. Univ.,* **84** (1961), 233-238.

172. Kiguradze, I.T. On the oscillation of solutions of some ordinary differential equations (in Russian). *Dokl. Akad. Nauk SSSR,* **144** (1962), *no.* 1, 33-36.

173. Kiguradze, I.T. On oscillation conditions for solutions of the equation $u'' + a(t)|u|^n \operatorname{sign} u = 0$ (in Russian). *Časopis Pěst. Mat.,* **87** (1962), *no.* 4, 492-495.

174. Kiguradze, I.T. On asymptotic properties of solutions of the equation $u'' + a(t)u^n = 0$ (in Russian). *Soobshch. Akad. Nauk Gruzin. SSR,* **30** (1963), *no.* 2, 129-136.

175. Kiguradze, I.T. On nonoscillatory solutions of the equation $u''+a(t)|u|^n \operatorname{sign} u = 0$ (in Russian). *Soobshch. Akad. Nauk Gruzin. SSR,* **35** (1964), *no.* 1, 15-22.

176. Kiguradze, I.T. On the oscillation of solutions of the equation $d^m u/dt^m + a(t)|u|^n \operatorname{sign} u = 0$ (in Russian). *Mat. Sb.,* **65** (1964), *no.* 2, 172-187.

177. Kiguradze, I.T. On the asymptotic representation of solutions of linear differential equations (in Russian). *Trudy Tbiliss. Univ.,* **102** (1964), 149-167.

178. Kiguradze, I.T. A note on the boundedness of solutions of differential equations (in Russian). *Trudy Tbiliss. Univ.,* **110** (1965), 103-108.

179. Kiguradze, I.T. Asymptotic properties of solutions of a nonlinear differential equation of Emden–Fowler type (in Russian). *Izv. Akad. Nauk SSSR. Ser. Mat.,* **29** (1965), *no.* 5, 965-986.

180. Kiguradze, I.T. Concerning the oscillation of solutions of nonlinear differential equations (in Russian). *Differentsial'nye Uravneniya,* **1** (1965), *no.* 8, 995-1006.

181. Kiguradze, I.T. A note on the oscillation of solutions of the equation $u'' + a(t)|u|^n \operatorname{sign} u = 0$ (in Russian). *Časopis Pěst. Mat.,* **92** (1967), *no.* 3, 343-350.

182. Kiguradze, I.T. On monotone solutions of nonlinear n-th order ordinary differential equations (in Russian). *Dokl. Akad. Nauk SSSR,* **181** (1968), *no.* 5, 1054-1057.

183. Kiguradze, I.T. On monotone solutions of nonlinear n-th order ordinary differential equations (in Russian). *Izv. Akad. Nauk SSSR. Ser. Mat.,* **33** (1969), *no.* 6,

1373-1398.

184. Kiguradze, I.T. On the non-negative non-increasing solutions of non-linear second order differential equations. *Ann. Mat. Pura Appl.*,**81** (1969), 169-192.

185. Kiguradze, I.T. On the oscillation of solutions of nonlinear ordinary differential equations (in Russian). *Proc. V Inter. Conf. Nonlinear Oscillations. V.1.* Kiev, 1970, 293-298.

186. Kiguradze, I.T. On oscillation conditions for solutions of nonlinear ordinary differential equations. I (in Russian). *Differentsial'nye Uravneniya*, **10** (1974), *no.* 8, 1387-1399.

187. Kiguradze, I.T. On oscillation conditions for solutions of nonlinear ordinary differential equations. II (in Russian). *Differentsial'nye Uravneniya*, **10** (1974), *no.* 9, 1586-1594.

188. Kiguradze, I.T. Some singular boundary value problems for ordinary differential equations (in Russian). Tbilisi Univ. Press, Tbilisi, 1975.

189. Kiguradze, I.T. On the oscillatory and monotone solutions of ordinary differential equations. *Arch. Math. (Brno)*, **14** (1978), *no.* 1, 21-44.

190. Kiguradze, I.T. On asymptotic behavior of solutions of non-linear non-autonomous ordinary differential equations. *Qual. Theory Different. Equat. V.1.* Amsterdam, 1981, 507-554.

191. Kiguradze, I.T. On some singular boundary value problems for ordinary differential equations. *Equadiff 5. Proc. 5 Czech. Conf. Different. Equat. Appl.* Leipzig, 1982, 174-178.

192. Kiguradze, I.T. On solutions of nonautonomous ordinary differential equations vanishing at infinity (in Russian). *Uspekhi Mat. Nauk*, **37** (1982), *no.* 4, 122.

193. Kiguradze, I.T. On vanishing at infinity of solutions of ordinary differential equations. *Czechoslovak Math. J.*, **33** (1983), *no.* 4, 613-646.

194. Kiguradze, I.T. On the nonlocal continuability of solutions of the Emden–Fowler equation (in Russian). *Uspekhi Mat. Nauk*, **38** (1983), *no.* 5, 171.

195. Kiguradze, I.T. On the asymptotic behavior of solutions of linear high order differential equations (in Russian). *Uspekhi Mat. Nauk*, **39** (1984), *no.* 4, 144.

196. Kiguradze, I.T. On a boundary value problem with a condition at infinity for higher order ordinary differential equations (in Russian). *Proc. All-Union Symp. Partial Different. Equat.* Tbilisi Univ. Press, Tbilisi, 1986, 91-105.

197. Kiguradze, I.T. On Kneser solutions of ordinary differential equations (in Russian). *Uspekhi Mat. Nauk*, **41** (1986), *no.* 4, 211.

198. Kiguradze, I.T.; Chanturia, T.A. A remark on the asymptotic behavior of solutions of the equation $u'' + a(t)u = 0$ (in Russian). *Differentsial'nye Uravneniya*, **6** (1970), *no.* 6, 1115-1117.

199. Kiguradze, I.T.; Kvinikadze, G.G. On strongly increasing solutions of nonlinear ordinary differential equations. *Ann. Mat. Pura Appl.*, **130** (1982), 67-87.

200. Kiguradze, I.T.; Rachůnková, I. On the solvability of a nonlinear problem of Kneser type (in Russian). *Differentsial'nye Uravneniya*, **15** (1979), *no.* 10, 1754-1765.

201. Kiguradze, I.T.; Rachůnková, I. On a certain nonlinear problem for two-dimensional differential systems. *Arch. Math. (Brno)*, **15** (1980), *no.* 1, 15-38.

202. Kiguradze, I.T.; Rozov, N.Kh. On the absolute stability of nonlinear non-stationary automatic control systems (in Russian). *Differentsial'nye Uravneniya*, **16** (1980), *no.* 4, 755-756.

203. Kiguradze, I.T.; Shekhter, B.L. Singular boundary value problems for second order ordinary differential equations (in Russian). *Sovremennye Problemy Mat. Noveishie Dostizheniya*, **30** (1987), 105-201.

204. Kim, W.J. On the fundamental system of solutions of $y^{(n)} + py = 0$. *Funkcial. Ekvac.*, **25** (1982), *no.* 1, 1-17.

205. Kim, W.J. Oscillation and nonoscillation criteria for n-th order linear differential equations. *J. Different. Equat.*, **64** (1986), *no.* 3, 317-335.

206. Klebanov, L.B. Local behavior of solutions of ordinary differential equations (in Russian). *Differentsial'nye Uravneniya*, **7** (1971), *no.* 8, 1393-1397.

207. Klokov, Yu.A. Some theorems on the boundedness of solutions of ordinary differential equations (in Russian). *Uspekhi Mat. Nauk*, **80** (1958), *no.* 2, 189-194

208. Kneser, A. Untersuchungen über die reelen Nullstellen der Integrale linearer Differentialgleichungen. *Math. Ann.*, **42** (1893), *no.* 3, 409-435.

209. Kneser, A. Untersuchung und asymptotische Darstellung der Integrale gewisser Differentialgleichungen bei grossen reelen Werten des Arguments. *J. Reine Angew. Math.*, **116** (1896), 178-212.

210. Kondratyev, V.A. On the oscillation of solutions of third and fourth order equations (in Russian). *Dokl. Akad. Nauk SSSR*, **118** (1958), *no.* 1, 23-24.

211. Kondratyev, V.A. On zeros of solutions of the equation $y^{(n)} + p(x)y = 0$ (in Russian). *Dokl. Akad. Nauk SSSR*, **120** (1958), *no.* 6, 1180-1182.

212. Kondratyev, V.A. On the oscillation of solutions of third and fourth order linear equations (in Russian). *Trudy Moskov. Mat. Obshch.*, **8** (1959), 259-282.

213. Kondratyev, V.A. On the oscillation of solutions of the equation $y^{(n)}+p(x)y = 0$ (in Russian). *Trudy Moskov. Mat. Obshch.*, **10** (1961), 419-436.

214. Kondratyev, V.A.; Samovol, V.S. On some asymptotic properties of solutions of equations of Emden–Fowler type (in Russian). *Differentsial'nye Uravneniya*, **17** (1981), *no.* 4, 749-750.

215. Koplatadze, R.G. On the asymptotic behavior of solutions of a system of two linear differential equations (in Russian). *Trudy Tbiliss. Univ.*, **129** (1968), 179-194.

216. Koplatadze, R.G. A note on the oscillation of solutions of differential equations with retarded argument (in Russian). *Mat. Časopis*, **22** (1972), *no.* 3, 253-261.

217. Koplatadze, R.G. On the oscillation of solutions of first order nonlinear differential equations (in Russian). *Soobshch. Akad. Nauk Gruzin. SSR*, **70** (1973), *no.* 1, 17-20.

218. Koplatadze, R.G. On oscillatory solutions of second-order delay differential inequalities. *J. Math. Anal. Appl.*, **42** (1973), *no.* 1, 148-157.

219. Koplatadze, R.G. On the existence of oscillatory solutions of second order nonlinear differential equations with retarded argument (in Russian). *Dokl. Akad. Nauk SSSR*, **210** (1973), *no.* 2, 260-262.

220. Koplatadze, R.G. A note on the oscillation of solutions of higher order differential inequalities and equations with retarded argument (in Russian). *Differentsial'nye*

Uravneniya, **10** (1974), *no.* 8, 1400-1405.

221. Koplatadze, R.G. On the oscillation of solutions of second order differential inequalities and differential equations with retarded argument (in Russian). *Math. Balkanica,* **5** (1975), 163-172.

222. Koplatadze, R.G. On the oscillation of solutions of an *n*-th order differential inequality with retarded argument (in Russian). *Ukrain. Mat. Zh.,* **28** (1976), *no.* 2, 233-237.

223. Koplatadze, R.G. On some properties of solutions of nonlinear differential inequalities and equations with retarded argument (in Russian). *Differentsial'nye Uravneniya,* **12** (1976), *no.* 11, 1971-1984.

224. Koplatadze, R.G. On the asymptotic behavior of solutions of second order linear differential equations with retarded argument (in Russian). *Differentsial'nye Uravneniya,* **16** (1980), *no.* 11, 1963-1966.

225. Koplatadze, R.G. On zeros of solutions of first order differential equations with retarded argument (in Russian). *Proc. I.N. Vekua Inst. Appl. Math.,* **14** (1983), 128-134.

226. Koplatadze, R.G. Oscillation criteria for solutions of second order differential inequalities and equations with retarded argument (in Russian). *Proc. I.N. Vekua Inst. Appl. Math.,* **17** (1986), 104-120.

227. Koplatadze, R.G. On oscillation properties of *n*-th order differential equations with retarded argument (in Russian). *Uspekhi Mat. Nauk,* **41** (1986), *no.* 4, 167.

228. Koplatadze, R.G. Integral criteria for oscillation of solutions of *n*-th order delay differential inequalities and equations (in Russian). *Proc. I.N. Vekua Inst. Appl. Math.,* **22** (1987), 110-135.

229. Koplatadze, R.G.; Chanturia, T.A. On oscillation properties of differential equations with deviating argument (in Russian). Tbilisi Univ. Press, Tbilisi, 1977.

230. Koplatadze, R.G.; Chanturia, T.A. On oscillatory and monotone solutions of first order differential equations with deviating argument (in Russian). *Differentsial'nye Uravneniya,* **18** (1982), *no.* 8, 1463-1465.

231. Korshikova, N.L. On zeros of solutions of higher order linear equations (in Russian). *Differentsial'nye uravneniya i ikh prilozheniya.* Moscow Univ. Press, Moscow, 1984, 143-148.

232. Korshikova, N.L. On zeros of solutions of a class of *n*-th order linear equations (in Russian). *Differentsial'nye Uravneniya,* **21** (1985), *no.* 5, 757-764.

233. Kostin, A.V. On the asymptotics of continuable solutions of an equation of Emden–Fowler type (in Russian). *Dokl. Akad. Nauk SSSR,* **200** (1971), *no.* 1, 28-31.

234. Kostin, A.V.; Evtukhov, V.M. The asymptotics of solutions of a nonlinear differential equation (in Russian). *Dokl. Akad. Nauk SSSR,* **231** (1976), *no.* 5, 1059-1062.

235. Krasnoselsky, M.A.; Rutitsky, Ya.B. Convex functions and Orlicz spaces (in Russian). Fizmatgiz, Moscow, 1958.

236. Kreith, K. Oscillation theory. Springer, Berlin, 1973.

237. Kreith, K. Oscillation properties of weakly nonlinear differential equations. *Lecture Notes Math.,* **846** (1981), 203-209.

238. Kreith, K.; Swanson, C.A. Kiguradze classes for characteristic initial value problems. *Comput. and Math.*, **11** (1985), *no.* 1-3, 239-247.

239. Kura, T. Oscillation theorems for a second order sublinear ordinary differential equation. *Proc. Amer. Math. Soc.*, **84** (1982), *no.* 4, 535-538.

240. Kura, T. Existence of oscillatory solutions for fourth order superlinear ordinary differential equations. *Hiroshima Math. J.*,**13** (1983), *no.* 3, 653-664.

241. Kura, T. Nonoscillation criteria for nonlinear ordinary differential equations of the third order. *Nonlinear Anal., Theory, Meth. and Appl.*, **8** (1984), *no.* 4, 369-379.

242. Kurzweil, J. Sur l'équation $x'' + f(t)x = 0$. *Časopis Pěst. Mat.*, **82** (1957), *no.* 2, 218-226.

243. Kurzweil, J. A note on oscillatory solutions of the equation $y'' + f(x)y^{2n-1} = 0$ (in Russian). *Časopis Pěst. Mat.*, **85** (1960), *no.* 3, 357-358.

244. Kusano, T.; Onose, H. On the oscillation of solutions of nonlinear functional differential equations. *Hiroshima Math. J.*, **6** (1976), *no.* 3, 635-645.

245. Kvinikadze, G.G. Some remarks on solutions of the Kneser problem (in Russian). *Differentsial'nye Uravneniya*, **14** (1978), *no.* 10, 1775-1783.

246. Kvinikadze, G.G. On singular solutions of nonlinear ordinary differential equations (in Russian). *Rep. Sem. I.N.Vekua Inst. Appl. Math.*, **17** (1983), 36-49.

247. Kvinikadze, G.G. On monotone proper and singular solutions of nonlinear ordinary differential equations (in Russian). *Differentsial'nye Uravneniya*, **20** (1984), *no.* 2, 360-361.

248. Kvinikadze, G.G. On solutions of the Kneser problem vanishing at infinity (in Russian). *Soobshch. Akad. Nauk Gruzin. SSR*, **118** (1985), *no.* 2, 241-244.

249. Kvinikadze, G.G. On Kneser solutions of nonlinear ordinary differential equations (in Russian). *Rep. Enlarged Sessions Sem. I.N.Vekua Inst. Appl. Math.*, **1** (1985), *no.* 3, 47-53.

250. Kvinikadze, G.G.; Kiguradze, I.T. On fast growing solutions of nonlinear ordinary differential equations (in Russian). *Soobshch. Akad. Nauk Gruzin. SSR*, **106** (1982), *no.* 3, 465-468.

251. Kwong, M.K.; Wong, J.S.W. Linearization of second-order nonlinear oscillation theorems. *Trans. Amer. Math. Soc.*, **279** (1983), *no.* 2, 705-712.

252. Kwong, M.K.; Zettl, A. Asymptotically constant functions and second order linear oscillation. *J. Math. Anal. Appl.*, **83** (1983), *no.* 2, 475-494.

253. Ladas, G.; Ladde, G.; Papadakis, J.S. Oscillations of functional-differential equations generated by delays. *J. Diff. Equat.*, **12** (1972), *no.* 2, 385-395.

254. Ladas, G.; Lakshmikantham, V. Oscillations caused by retarded actions. *Appl. Anal.*, **4** (1974), *no.* 1, 9-15.

255. Ladas, G.; Lakshmikantham, V.; Leela, S. On the perturbability of the asymptotic manifold of perturbed system of differential equations. *Proc. Amer. Math. Soc.*, **27** (1971), *no.* 1, 65-71.

256. Lazer, A.C. A stability condition for the differential equation $y'' + p(x)y = 0$. *Michigan Math. J.*, **12** (1965), 193-196.

257. Leighton, W.; Nehari, Z. On the oscillation of solutions of self-adjoint linear differential equations of the fourth order. *Trans. Amer. Math. Soc.*, **89** (1958), *no.* 2,

325-377.

258. Levin, A.Yu. Some questions of the oscillation of solutions of linear differential equations (in Russian). *Dokl. Akad. Nauk SSSR*, **148** (1963), *no.* 3, 512-515.

259. Levin, A.Yu. Nonoscillation of solutions of the equation $x^{(n)} + p_1(t)x^{(n-1)} + \ldots + p_n(t)x = 0$ (in Russian). *Uspekhi Mat. Nauk*, **24** (1969), *no.* 2, 43-96.

260. Levinson, N. The asymptotic nature of the solutions of linear systems of differential equations. *Duke Math. J.*, **15** (1948), *no.* 1, 111-126.

261. Lewis, R.T. The existence of conjugate points for selfadjoint differential equations of even order. *Proc. Amer. Math. Soc.*, **56** (1976), 162-166.

262. Ličko, I.; Švec, M. La caractère oscillatoire des solutions de l'équation $y^{(n)} + f(x)y^\alpha = 0$, $n > 1$. *Czechoslovak Math. J.*, **13** (1963), *no.* 4, 481-491.

263. Lomtatidze, A.G. On oscillation properties of solutions of second order linear differential equations (in Russian). *Rep. Sem. I.N. Vekua Inst. Appl. Math.*, **19** (1985), 39-52.

264. Mařik, J.; Ráb, M. Asymptotische Eigenschaften von Lösungen der Differentialgleichung $y'' = A(x)y$ im nichtoszillatorischen Fall. *Czechoslovak Math. J.*, **10** (1960), *no.* 7, 501-522.

265. Marušiak, P. Oscillation of solutions of delay differential equations. *Czechoslovak Math. J.*, **24** (1974), *no.* 3, 284-291.

266. Matell, M. Asymptotische Eigenschaften gewisser linearer Differentialgleichungen. Applbergs Boktryckeri Aktiebolag, Uppsala, 1924.

267. McShane, E.J. On the solutions of the differential equation $y'' + p^2 y = 0$. *Proc. Amer. Math. Soc.*, **17** (1966), *no.* 1, 55-61.

268. Meir, A.; Willet, D.; Wong, J.S.W. On the asymptotic behavior of the solutions of $x'' + a(t)x = 0$. *Michigan Math. J.*, **14** (1967), *no.* 1, 47-52.

269. Mikusiński, J.G. On Fite's oscillation theorems. *Colloq. Math.*, **2** (1951), *no.* 1, 34-39.

270. Mikusiński, J.G. Sur l'équation $x^{(n)} + A(t)x = 0$. *Ann. Polon. Math.*, **1** (1955), *no.* 2, 207-221.

271. Milloux, H. Sur l'équation différentielle $x'' + A(t)x = 0$. *Prace Mat. Fiz.*, **41** (1934), 39-54.

272. Mirzov, J.D. A note on the oscillation of solutions of a system of ordinary differential equations (in Russian). *Soobshch Akad. Nauk Gruzin. SSR*, **61** (1971), *no.* 2, 277-279.

273. Mirzov, J.D. On the oscillation of solutions of a system of nonlinear differential equations (in Russian). *Differentsial'nye Uravneniya*, **9** (1973), *no.* 3, 581-583.

274. Mirzov, J.D. Concerning the oscillation of solutions of a system of nonlinear differential equations (in Russian). *Mat. Zametki*, **16** (1974), *no.* 2, 571-576.

275. Mirzov, J.D. Some remarks on the asymptotic behavior of solutions of two-dimensional nonlinear differential systems (in Russian). *Issledovaniya nekotorykh uravn. mat. fiz.* Tbilisi Univ. Press, Tbilisi, 1976, 131-152.

276. Mirzov, J.D. On an analogue of the Kurzweil-Jasný theorem for a system of two differential equations (in Russian). *Časopis Pěst. Mat.*, **101** (1976), *no.* 1, 45-52.

277. Mirzov, J.D. On some analogs of Sturm's and Kneser's theorems for nonlinear systems. *J. Math. Anal. Appl.*, **53** (1976), *no.* 2, 418-425.

278. Mirzov, J.D. On asymptotic properties of solutions of two-dimensional differential systems (in Russian). *Differentsial'nye Uravneniya*, **13** (1977), *no.* 12, 2183-2198.

279. Mirzov, J.D. On zeros of solutions of a nonlinear differential system (in Russian). *Differentsial'nye Uravneniya*, **20** (1984), *no.* 10, 1726-1732.

280. Mirzov, J.D. Asymptotic properties of solutions of a nonlinear system of Emden–Fowler type (in Russian). *Rep. En larged Sessions Sem. I.N.Vekua Inst. Appl. Math.*, **1** (1985), *no.* 3, 101-104.

281. Mirzov, J.D. On asymptotic properties of solutions of a system of Emden–Fowler type (in Russian). *Differentsial'nye Uravneniya*, **21** (1985), *no.* 9, 1498-1504.

282. Mitropolsky, Yu.A.; Shevelo, V.N. On the development of the oscillation theory of solutions of differential equations with retarded argument (in Russian). *Ukrain. Mat. Zh.*, **29** (1977), *no.* 3, 313-323.

283. Müller-Pfeiffer, E. An oscillation theorem for self-adjoint differential equations. *Math. Nachr.*, **108** (1982), 79-92.

284. Myshkis, A.D. An example of a solution of a second order oscillatory type differential equation which is noncontinuable to the whole of the axis (in Russian). *Differentsial'nye Uravneniya*, **5** (1969), *no.* 12, 2267-2268.

285. Myshkis, A.D. Linear differential equations with retarded argument (in Russian). Nauka, Moscow, 1972.

286. Myshkis, A.D.; Bainov, D.D.; Zahariev, A.I. Oscillatory and asymptotic properties of a class of operator-differential inequalities. *Proc. Roy. Soc. Edinburgh*, **96A** (1984), *no.* 1, 5-13.

287. Nadareishvili, V.A. On oscillatory and monotone solutions of first order differential equations with retarded argument (in Russian). *Rep. Enlarged Sessions Sem. I.N.Vekua Inst. Appl. Math.*, **1** (1985), *no.* 3, 111-115.

288. Nadareishvili, V.A. On the existence of monotone solutions of higher order linear differential equations with deviating argument (in Russian). *Proc. I.N.Vekua Inst. Appl. Math.*, **22** (1987), 180-194.

289. Naito, M. Oscillations of differential inequalities with retarded arguments. *Hiroshima Math. J.*, **5** (1975), *no.* 2, 187-192.

290. Nehari, Z. On a class of nonlinear second-order differential equations. *Trans. Amer. Math. Soc.*, **95** (1960), *no.* 1, 101-123.

291. Nehari, Z. A nonlinear oscillation theorem. *Duke Math. J.*, **42** (1975), *no.* 1, 183-189.

292. Nehari, Z. Green's functions and disconjugacy. *Arch. Rational Mech. Anal.* **62** (1976), *no.* 1, 53-76.

293. Neuman, F. On two problems on oscillations of linear differential equations of the third order. *J. Different. Equat.*, **15** (1974), *no.* 3, 589-596.

294. Olekhnik, S.N. Asymptotic behavior of solutions of a second order ordinary differential equation (in Russian). *Differentsial'nye Uravneniya*, **5** (1969), *no.* 11, 2093-2095.

295. Olekhnik, S.N. On the boundedness and unboundedness of solutions of a sec-

ond order ordinary differential equation (in Russian). *Differentsial'nye Uravneniya*, **8** (1972), *no.* 9, 1701-1704.

296. Onose, H. A comparison theorem and the forced oscillation. *Bull. Austral. Math. Soc.*, **13** (1975), *no.* 1, 13-19.

297. Opial, Z. Sur l'équation différentielle $u'' + a(t)u = 0$. *Ann. Polon. Math.*, **5** (1959), *no.* 1, 77-93.

298. Opial, Z. Nouvelles remarques sur l'équation différentielle $u'' + a(t)u = 0$. *Ann. Polon. Math.*, **6** (1959), *no.* 1, 75-81.

299. Paatashvili, D.V. On unbounded solutions of second order linear differential equations with retarded argument (in Russian). *Rep. Enlarged Sessions Sem. I.N. Vekua Inst. Appl. Math.*, **1** (1985), *no.* 3, 123-126.

300. Perron, O. Über lineare Differentialgleichungen, bei denen die unabhängige Variable reell ist. *J. Reine Angew. Math.*, **142** (1913), 254-270.

301. Perron, O. Ueber ein vermeintliches Stabilitätskriterium. *Nachr. Ges. Wiss. Göttingen Math.-Phys. Kl. Fachgruppe I*, 1930, 28-29.

302. Philos, Ch.G. Integral averages and second order superlinear oscillation. *Math. Nachr.*, **120** (1985), *no.* 1, 127-138.

303. Púža, B. On a comparison theorem for systems of two differential equations (in Russian). *Differentsial'nye Uravneniya*, **13** (1977), *no.* 7, 1336-1338.

304. Ráb, M. Asymptotische Eigenschaften der Lösungen der Differentialgleichung $y'' + A(x)y = 0$. *Czechoslovak Math. J.*, **8** (1958), *no.* 4, 513-519.

305. Ráb, M. Kriterien für die Oszillation der Lösungen der Differentialgleichung $[p(x)y']' + q(x)y = 0$. *Časopis Pěst. Mat.*, **84** (1959), *no.* 3, 335-370.

306. Ráb, M. Asymptotische Formeln für die Lösungen der Differentialgleichung $y'' + q(x)y = 0$. *Czechoslovak Math. J.*, **14** (1964), *no.* 2, 203-221.

307. Rabtsevich, V.A. On unbounded proper solutions of the Emden–Fowler equation (in Russian). *Differentsial'nye Uravneniya*, **22** (1986), *no.* 5, 780-789.

308. Rachůnková, I. On a Kneser problem for systems of nonlinear ordinary differential equations (in Russian). *Soobshch. Akad. Nauk Gruzin. SSR*, **94** (1979), *no.* 3, 545-548.

309. Rachůnková, I. On a Kneser problem for a system of nonlinear ordinary differential equations. *Czechoslovak Math. J.*, **31** (1981), *no.* 1, 114-126.

310. Rachůnková, I. On a nonlinear problem for n-th order differential systems (in Russian). *Czechoslovak Math. J.*, **34** (1984), *no.* 2, 285-297.

311. Rapoport, I.M. On some asymptotic methods in the theory of differential equations (in Russian). Ukrain. SSR Acad. Sci. Press, Kiev, 1954.

312. Reid, W.T. Sturmian theory for ordinary differential equations. Springer, New York, 1980.

313. Ryder, G.H.; Wend, D.V.V. Oscillation of solutions of certain ordinary differential equations of n-th order. *Proc. Amer. Math. Soc.*, **25** (1970), *no.* 3, 463-469.

314. Sansone, G. Sopra il comportamento asintotico delle soluzioni di un'equazione differenziale della dinamica. *Scritti Matematici offerti a Luigi Berzolari*. Pavia, 1936, 385-403.

315. Sansone, G. Studi sulla equazioni differenziali lineari omogenee di terzo ordine nel campo reale. *Rev. Mat. Fis. Theor. (Tucuman)*, **6** (1948), 195-253.

316. Sansone, G. Equazioni differenziali nel campo reale. I. Zanichelli, Bologna, 1948.

317. Sansone, G. Equazioni differenziali nel campo reale. II. Zanichelli, Bologna, 1949.

318. Schrader, K. Oscillation and comparison for second order differential equations. *Proc. Amer. Math. Soc.*, **51** (1975), *no.* 1, 131-135.

319. Šeda, V. Nonoscillatory solutions of differential equations with deviating argument. *Czechoslovak Math. J.*, **36** (1986), *no.* 1, 93-107.

320. Šeda, V. On nonlinear differential systems with deviating argument. *Czechoslovak Math. J.*, **36** (1986), *no.* 4, 450-466.

321. Sficas, Y.G.; Staikos, V.A. The effect of retarded actions of nonlinear oscillations. *Proc. Amer. Math. Soc.*, **46** (1974), *no.* 2, 259-264.

322. Sharman, T.L. Properties of solutions of n-th order linear differential equations. *Pacific J. Math.*, **15** (1965), *no.* 3, 1045-1060.

323. Shevelo, V.N. Problems, methods and main results of the oscillation theory of solutions of nonlinear nonautonomous ordinary differential equations (in Russian). *Proc. II All-Union Congress Theor. Appl. Mech.*, **2** (1965), 142-152.

324. Shevelo, V.N. Oscillation of solutions of differential equations with deviating argument (in Russian). Naukova Dumka, Kiev, 1978.

325. Shevelo, V.N.; Odarich, O.N. Some questions of the oscillation (nonoscillation) theory of solutions of second order differential equations with retarded argument (in Russian). *Ukrain. Mat. Zh.*, **23** (1971), *no.*4, 508-516.

326. Shevelo, V.N.; Shtelik, V.G. Some questions of oscillation of solutions of nonlinear nonautonomous second order equations (in Russian). *Dokl. Akad. Nauk SSSR*, **149** (1963), *no.* 2, 276-279.

327. Shevelo, V.N.; Varekh, N.V. On some properties of solutions of delay differential equations (in Russian). *Ukrain. Mat. Zh.*, **24** (1972), *no.* 6, 807-813.

328. Skhalyakho, Ch.A. On some comparison theorems for systems of differential equations of the Emden–Fowler type (in Russian). *Soobshch. Akad. Nauk Gruzin. SSR*, **100** (1980), *no.* 1, 49-52.

329. Skhalyakho, Ch.A. On the oscillation and nonoscillation of solutions of a system of nonlinear differential equations (in Russian). *Differentsial'nye Uravneniya*, **16** (1980), *no.* 8, 1523-1526.

330. Skhalyakho, Ch.A. On comparison theorems for systems of ordinary differential equations (in Russian). *Proc. I.N.Vekua Inst. Appl. Math.*, **8** (1980), 29-49.

331. Skhalyakho, Ch.A. Oscillation conditions for solutions of a two-dimensional differential system (in Russian). *Kraevye zadachi*. Perm Polytechn. Inst. Press, Perm, 1980, 113-116.

332. Skhalyakho, Ch.A. On the oscillation of solutions of some two-dimensional systems of differential equations (in Russian). *Izv. Severo-Kavkaz. Nauchn. Tsentra Vyssh. Shkoly. Estestv. Nauk.*, 1981, *no.* 2, 19-22.

333. Skhalyakho, Ch.A. On the oscillation and nonoscillation of solutions of a two-

dimensional system of nonlinear differential equations (in Russian). *Differentsial'nye Uravneniya*, **17** (1981), *no.* 9, 1702-1705.

334. Skhalyakho, Ch.A. On the nonoscillation of solutions of a system of two differential equations (in Russian). *Časopis Pěst. Mat.*, **107** (1982), *no.* 2, 139-142.

335. Skhalyakho, Ch.A. On proper solutions of multi-dimensional systems of nonlinear ordinary differential equations (in Russian). *Proc. I.N. Vekua Inst. Appl. Math.*, **14** (1983), 146-162.

336. Skhalyakho, Ch.A. Necessary and sufficient conditions for oscillation of solutions of second order linear differential equations (in Russian). *Soobshch. Akad. Nauk Gruzin. SSR*, **120** (1985), *no.* 3, 469-471.

337. Skhalyakho, Ch.A. On zeros of solutions of a system of two differential equations whose coefficients change signs (in Russian). *Rep. Enlarged Sessions Sem. I.N. Vekua Inst. Appl. Math.*, **1** (1985), *no.* 3, 138-141.

338. Skhalyakho, Ch.A. On zeros of solutions of two-dimensional linear differential systems whose coefficients change signs (in Russian). *Proc. I.N. Vekua Inst. Appl. Math.*, **17** (1986), 135-152.

339. Sobol, I.M. On the asymptotic behavior of solutions of linear differential equations (in Russian). *Dokl. Akad. Nauk SSSR*, **61** (1948), *no.* 2, 219-222.

340. Stachurska, B. On the asymptotic behavior of solutions of the second-order linear differential equations. *Ann. Polon. Math.*, **24** (1971), *no.* 2, 203-208.

341. Sturm, C. Sur les équations différentielles linéaires de second ordre. *J. Math. Pures Appl.*, **1** (1836), 101-186.

342. Švec, M. Sur une propriété des intégrales de l'équation $y^{(n)} + Q(x)y = 0$, $n = 3, 4$. *Czechoslovak Math. J.*, **7** (1957), *no.* 4, 450-462.

343. Švec, M. Sur le comportement asymptotique des intégrales de l'équation différentielle $y^{(4)} + Q(x)y = 0$. *Czechoslovak Math. J.*, **8** (1958), *no.* 2, 230-245.

344. Švec, M. Monotone solutions of some differential equations. *Colloq. Math.*, **18** (1967), 7-21.

345. Švec, M. Les propriétés asymptotiques des solutions d'une équation différentielle nonlinéaire d'ordre *n*. *Czechoslovak Math. J.*, **17** (1967), *no.* 4, 550-557.

346. Swanson, C.A. Comparison and oscillation theory of linear differential equations. Acad. Press, New York, 1968.

347. Thomas, L.H. The calculation of atomic fields. *Proc. Cambridge Phil. Soc.*, **23** (1927), 542-548.

348. Tonelli, L. Estratto di lettera al prof. Giovanni Sansone. *Scritti Matematici offerti a Luigi Berzolari.* Pavia, 1936, 404-405.

349. Toroshelidze, I.A. On the asymptotic representation of solutions of some systems of nonlinear differential equations (in Russian). *Soobshch. Akad. Nauk Gruzin. SSR*, **42** (1966), *no.* 2, 286-292.

350. Toroshelidze, I.A. On the asymptotic behavior of solutions of some nonlinear differential equations (in Russian). *Differentsial'nye Uravneniya*, **3** (1967), *no.* 6, 926-940.

351. Toroshelidze, I.A. On the asymptotic behavior of solutions of some systems of nonlinear differential equations (in Russian). *Trudy Gruzin. Polytechn. Inst.*, 1969,

no. 5, 117-125.

352. Toroshelidze, I.A. On the asymptotic representation of solutions of quasi-linear systems of ordinary differential equations (in Russian). *Trudy Tbiliss. Univ.*, **1A** (1971), 89-102.

353. Toroshelidze, I.A. A remark on solutions of quasi-linear systems of ordinary differential equations (in Russian). *Trudy Tbiliss. Univ.*, **5A** (1972), 31-35.

354. Toroshelidze, I.A. On a theorem of G.J.Butler (in Russian). *Differentsial'nye Uravneniya*, **14** (1978), *no.* 10, 1903-1904.

355. Tramov, M.I. Oscillation conditions for solutions of first order differential equations with retarded argument (in Russian). *Izv. Vyssh. Uchebn. Zaved. Mat.*, 1975, *no.* 3, 92-96.

356. Tramov, M.I. Some questions of the oscillation theory of solutions of differential equations with deviating argument (in Russian). *Izv. Vyssh. Uchebn. Zaved. Mat.*, 1979, *no.* 4, 46-52.

357. Tramov, M.I. On zeros of solutions of differential equations with retarded argument (in Russian). *Vestnik Moskov. Univ. Ser. Mat. Mekh.*, 1979, *no.* 4, 3-7.

358. Tramov, M.I. Oscillation of solutions of differential equations with deviating argument (in Russian). *Differentsial'nye Uravneniya*, **18** (1982), *no.* 2, 245-253.

359. Trench, W.F. Asymptotic integration of linear differential equations subject to integral smallness conditions involving ordinary convergence. *SIAM J. Math Anal.*, **7** (1976), *no.* 2, 213-221.

360. Trench, W.F. Eventual disconjugacy of a linear differential equation. *Proc. Amer. Math. Soc.*, **89** (1983), *no.* 3, 461-466.

361. Trevisan, G. Sul l'equazione differenziale $y'' + A(x) = 0$. *Rend. Sem. Mat. Univ. Padova*, **23** (1954), 340-342.

362. Tyshkevich, V.A. Some questions of the stability theory of functional-differential equations (in Russian). Naukova Dumka, Kiev, 1981.

363. Usami, H. Global existence and asymptotic behavior of solutions of second order nonlinear differential equations. *J. Math. Anal. Appl.*, **122** (1987), *no.* 1, 152-171.

364. Villari, C. Contributi allo studio asintotico dell'equazione $x'''(t) + p(t)x(t) = 0$. *Ann. Mat. Pura Appl.*, **51** (1960), 301-328.

365. Waltman, P. An oscillation criterion for a nonlinear second order equation. *J. Math. Anal. Appl.*, **10** (1965), *no.* 2, 439-441.

366. Werbowski, J. On the oscillation of solutions of differential inequalities with deviating arguments of a mixed type (in Russian). *Proc. I.N.Vekua Inst. Appl. Math.*, **14** (1983), 5-12.

367. Werbowski, J. Oscillations of differential inequalities caused by several delay arguments. *J. Math. Anal. Appl.*, **124** (1987), *no.* 1, 200-212.

368. Wilkins, J.E. On the growth of solutions of linear differential equations. *Bull. Amer. Math. Soc.*, **50** (1944), *no.* 6, 388-394.

369. Willet, D. On an example in second order linear ordinary differential equations. *Proc. Amer. Math. Soc.*, **17** (1966), *no.* 6, 1263-1266.

370. Willet, D. Classification of second order linear differential equations with respect to oscillation. *Adv. Math.*, **3** (1969), *no.* 4, 594-623.

371. Wintner, A. On the normalization of characteristic differentials in continuous spectra. *Phys. Rev.*, **72** (1947), 516-517.

372. Wong, J.S.W. Some stability conditions for $x'' + a(t)f(x) = 0$. *SIAM J. Appl. Math.*, **15** (1967), *no.* 4, 889-892.

373. Wong, J.S.W. A second order nonlinear oscillation theorem. *Proc. Amer. Math. Soc.*, **40** (1973), *no.* 2, 487-491.

374. Wong, J.S.W. On the generalized Emden–Fowler equation. *SIAM Rev.*, **17** (1975), *no.* 2, 339-360.

375. Wong, J.S.W. Remarks on nonoscillation theorems for a second order nonlinear differential equation. *Proc. Amer. Math. Soc.*, **83** (1981), *no.* 3, 541-546.

376. Yakubovich, V.Ya. On the asymptotic behavior of systems of differential equations (in Russian). *Dokl. Akad. Nauk SSSR*, **24** (1948), *no.* 7, 363-366.

377. Zlamal, M. Über asymptotische Eigenschaften der Lösungen der linearen Differentialgleichungen zweiter Ordnung. *Czechoslovak Math. J.*, **6** (1956), *no.* 1, 75-91.

Author Index

Subject Index